MEDIA AND THE MIND

Media & the Mind

ART, SCIENCE, AND NOTEBOOKS AS
PAPER MACHINES, 1700–1830

Matthew Daniel Eddy

The University of Chicago Press
Chicago and London

The University of Chicago Press, Chicago 60637
The University of Chicago Press, Ltd., London
© 2023 by The University of Chicago
All rights reserved. No part of this book may be used or reproduced in any manner
whatsoever without written permission, except in the case of brief quotations
in critical articles and reviews. For more information, contact the University of
Chicago Press, 1427 E. 60th St., Chicago, IL 60637.
Published 2023
Printed in the United States of America

32 31 30 29 28 27 26 25 24 23   1 2 3 4 5

ISBN-13: 978-0-226-18386-2 (cloth)
ISBN-13: 978-0-226-82075-0 (e-book)
DOI: https://doi.org/10.7208/chicago/9780226820750.001.0001

Library of Congress Cataloging-in-Publication Data

Names: Eddy, Matthew, 1972– author.
Title: Media and the mind : art, science, and notebooks as paper machines,
1700–1830 / Matthew Daniel Eddy.
Description: Chicago : University of Chicago Press, 2023. | Includes bibliographical
references and index.
Identifiers: LCCN 2022009827 | ISBN 9780226183862 (cloth) | ISBN 9780226820750
(ebook)
Subjects: LCSH: Enlightenment. | Philosophy—Scotland—History. | Note-taking. |
Philosophy, Modern—18th century.
Classification: LCC B1402.E55 E33 2023 | DDC 192—dc23/eng/20220325
LC record available at https://lccn.loc.gov/2022009827

♾ This paper meets the requirements of ANSI/NISO Z39.48-1992 (Permanence
of Paper).

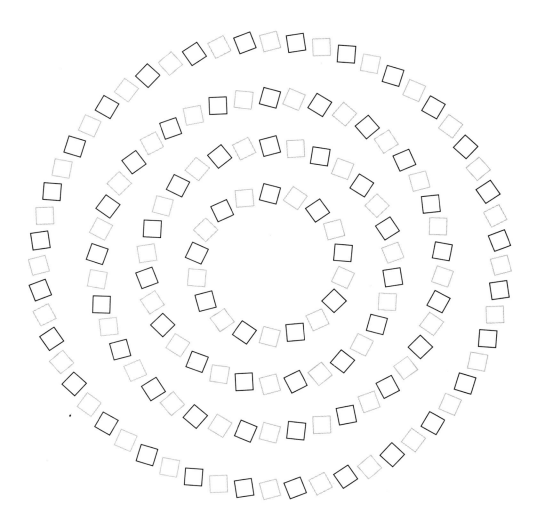

I dedicate this book to
my son, George, who loves to draw notes.

I have seldom or never observed any one to get the skill of reasoning
well, or speaking handsomely, by studying those rules which pretend
to teach it.

JOHN LOCKE

Often did my dear and amiable instructress listen with mingled
solicitude and delight to my senseless, though accurate, recitation
of passages, which excited in her mind a train of ideas very different
from those they raised in mine.

ELIZABETH HAMILTON

The art of memory is a clear case of a marginal subject, not recognized as belonging to any of the normal disciplines, having been
omitted because it was no one's business. And yet it has turned out
to be, in a sense, everyone's business.

FRANCES A. YATES

Much of our human intelligence results from our ability to construct
artifacts . . . Our technologies also make us smart: They educate and
entertain us.

DONALD A. NORMAN

Contents

Bibliographic Note | xiii

Prologue | xv

Introduction

1. **RECRAFTING NOTEBOOKS** | 3

   The *Tabula Rasa* and Media Interface | 3

   Notebooks as Artifacts | 8

   Notekeeping as Artificing | 14

   Notekeepers as Artificers | 22

   Thought as a Realtime Activity | 29

   Science as a System | 36

   Book Outline | 39

Part I: Inside the *Tabula Rasa*

2. **WRITING** | 45

   Writing as a Knowledge-Creating Tool | 45

   The Place of Writing within Literacy | 49

   Script and Observational Learning | 59

   Grids and Verbal Pictures | 68

   Copies and the Exercise of Memory | 78

3. **CODEXING** | 85

   Paper Machines as Material Artifacts | 85

Paper as an Informatic Medium | 89
Quires and Knowledge Management | 96
Books and Customized Packaging | 106

4. ANNOTATING | 114
Revisibilia Made through Annotation | 114
Marginalia as Scribal Interface | 117
Paratexts and Editorial Training | 124
Ciphers and the Acquisition of Numeracy | 136

Part II: Around the *Tabula Rasa*

5. CATEGORIZING | 151
Headings as Realtime Categories | 151
Headings as Mnemonic Labels | 155
Headings as Visual Cues | 162
Headings as Coordinates for Scanpaths and Sightlines | 170

6. DRAWING | 181
Description and Movement across a Page | 181
Learning to Draw a Picture | 184
Figures as Developmental Tools | 195
Scenes as Observational Training | 210
Observation and the Utility of Perception | 218

7. MAPPING | 224
Mapkeepers and Knowledge Systems on Paper | 224
Map-Mindedness and Embodied Experience | 227
Desk Maps as Crafted Constructions | 233
Field-Mindedness in the Classroom | 238
Field Maps and Visualized Data | 250
Maps as Mnemonic Devices | 256

Part III: Beyond the *Tabula Rasa*

8. SYSTEMIZING | 269
The Syllabus as a System and a Machine | 269
Lecture Notebooks and Knowledge Formation | 274
The Syllabus and Its Organizational Technologies | 282

Scroll Books and the Strategies of Realtime Learning | 292

Transcripts and the Extension of Memory | 305

Lines and the Media of the Mind | 316

9. DIAGRAMMING | 324

Paths and Diagrammatic Knowledge | 324

Schemata as Useful Mnemonic Aids | 327

Shapes as Repurposed Perceptual Devices | 335

Pictograms and Visual Judgment | 351

Tables as Kinesthetic Diagrams | 366

Traces and Realtime Observation | 374

10. CIRCULATING | 382

Local and Global Networks | 382

Personal and Institutional Libraries | 384

Commodities within Knowledge Economies | 392

Courts of Law and Public Opinion | 399

Conclusion

11. RETHINKING MANUSCRIPTS | 415

The *Tabula Rasa* and Manuscripts | 415

Manuscripts as Dynamic Artifacts | 416

Manuscript Skills as Artifice | 418

Manuscript Keepers as Artificers | 420

Acknowledgments | 425

Bibliography | 429

Abbreviations | 429

Primary Sources | 430

Manuscripts and Ephemera | 430

Printed Primary Sources | 437

Secondary Sources | 452

Index | 495

Bibliographic Note

Whenever I cite a manuscript in the footnotes, I give the author, name, date, the signification "Bound MS" or "Unbound MS," and folio number ("f." singular, "ff." plural). For example, Fowler Bound MS (1780, ff. 28–30) refers to folios 28–30 of the following source given in the manuscript section of the bibliography: Fowler, James. 1780. *Schoolbook of James Fowler, Strathpeffer*, Bound MS, NLS MS 1428. Where necessary, I also state whether I am referring to the right ("r," recto) or left ("v," verso) side of the page. For university notebooks cited in the footnotes, I have included all of the foregoing information as well as the name of the professor at the start and the student notekeeper at the end. In cases when bound copies of student lectures have no folio numbers, I give the number of the section ("section 1") or lecture ("lecture 1").

Prologue

What was your favorite book as a child? Mine was *The Silver Chair* by C. S. Lewis. My mother read it to me on snowy evenings. Its beautiful illustrations by Pauline Baynes captivated my imagination. But the thing I liked most about the book was its words. They moved. If I looked long enough, they even formed vibrating shapes.

Later I learned that I have specific learning disabilities that affect the ways I read, write, and calculate. I have never seen the words, numbers, and lines of texts as fixed objects. For me, they are a form of media that dance, sparkle, ripple, and realign like the squares in the intertwining illusion created by the Italian psychologist Biangio Pinna.[1] Figure 1 shows an example of an image that Pinna uses to create the illusion.

Pinna's image consists of tilted squares arranged in concentric rings. Staring at the center of the image creates a sense of movement around the squares and realigns the concentric circles into overlapping swirls. Though the image is called an "illusion," the experience that it evokes in many viewers is very real and sometimes disorienting. The movement and realignment of the squares within the lines of the circle in Pinna's image are very similar to how I experience words arranged as lines in sentences grouped into paragraphs on the screen or page.

Like everyone with specific learning conditions, I was born with them and they affect me every day. I will have them for the rest of my life. They impact the way I process knowledge, including the rate at which I interface with

---

1. Pinna and Brelstaff (2000). The historical emergence of visual illusions and textured counterchange patterns is summarized in Gombrich (1979, 115–148).

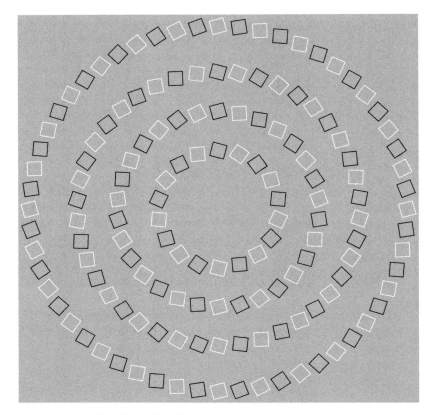

FIGURE 1. Biangio Pinna's "intertwining illusion."

printed, written, and digital media. They slow my reading abilities; but they also increase my capability to spot visual patterns and changes in meaning that remain unseen, or perhaps unnoticed, by others. They allow me to remember and rotate 3D structures in my mind and to read texts as pictures of ordered space that can flip meaning.

I was not diagnosed until I was an adult. During my school years I was always the last to finish a test, the one whose handwriting was feared most by teachers, and the one whose interpretation of test questions was technically correct, but not in the way that the examiner had meant. Reading would have most likely been a monumental challenge were it not for a discovery that I made when I was in primary school. One evening while I was enjoying a candy cane, I realized that I could use its tip to guide my eye across the words of the sentences featured in a picture book that I was perusing. I started to use the practice on a regular basis and my parents captured it in a picture taken of me and my sister (figure 2).

FIGURE 2. Childhood photograph of the author (right) and his sister. Note how he is using the candy cane in his right hand as a manual reading tool.

The candy cane discovery was a turning point in the development of my ability to interface with word-based media. It allowed me to see that I could repurpose everyday objects in a way that enabled me to interface with texts in a manner that partially improved my reading abilities.

Mitigating my struggles with writing took longer. I ended up reteaching myself to write in high school. My own cursive handwriting was difficult for me to read. I switched to deep blue ink and began to print my words, designing each letter in a way that made it easier for me to see when I read it. I used blue plastic rulers to guide my eye across the page and I used colored highlighters to code terms in my textbooks and notes. In so doing, I became aware of the mnemonic power of communications media.

In university, I noticed that I could use notekeeping as a kinesthetic mode of learning to create accessible patterns of words on the page that helped me to remember information. I experimented with different modes of writing and developed material and visual notekeeping skills that matched my own abilities. Most of us experience this kind of discovery in one way or another somewhere along our educational journeys. We seek to develop skills that help us learn better and in a way not foreseen by our teachers. It was this realization that led me to write this book.

After my undergraduate studies, I eventually earned a PhD in history and was hired by a university. During my early years of teaching, I began to think about how I might use my training as a historian to examine the history of notekeeping as a collection of media skills. I realized quickly that there was much work to be done on the history of how notekeepers acquired the writing, drawing, and codexing abilities required to make notebooks.

To be sure, there were many fine studies that treated notes as important forms of historical evidence that spoke to what students were supposed to learn. But I had noticed that there was a delightful variety in the notes kept by predigital notekeepers. What was the relationship that they and their teachers drew between media and the adaptability of the mind? I wanted to explore this topic in more detail so that I could better understand the historical development of the skills that allowed learners to make and use notebooks as artifacts in realtime. After fifteen years of research, the final result is this book.

I used to see my disability as a liability, as something that I needed to hide from colleagues and research collaborators. But I now see it as a strength. I have embraced it as a unique tool, one that gives me a different perspective—one that allows me to see my field in a new, fascinating light.

Introduction

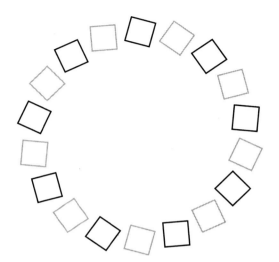

CHAPTER 1

RECRAFTING NOTEBOOKS

The *Tabula Rasa* and Media Interface

Think back to your childhood. What did you take with you when you went to school? Many of us would have set off with a backpack brimming with notebooks, erasers, a ruler, pens, pencils, and perhaps a handful of sweets to enjoy when the teacher wasn't looking. Even in today's digital classrooms, students use these tools to learn how to write, draw, and calculate on sheets of paper that, when clipped, stapled, or glued together, become a notebook.

Scientists who research the cognitive impact of inscription emphasize that writing or drawing notes facilitates hand and eye movements that significantly aid how we remember information.[1] Despite this fact, the time and resources that schools devote to teaching handwriting skills have declined in recent years. Students are now dazzled with new digital technologies that are marketed as being *like* a notebook and which operate on laptop computers called *notebooks*.

The paper notebook is of course an ancient learning tool. But is it in any way similar to the digital notebooks that we use in today's classrooms? Is it appropriate to compare a computer screen to the page of a student notebook rendered centuries ago? The truth is that these questions are difficult to answer because scholars have devoted relatively little attention to how students learned to make and use notebooks in the past. Though it is clear that notebooks were important, the day-to-day skills required to write, rewrite, draw, and redraw them have only just begun to be explored.

As a historian, I have spent much of my career holding hundreds of note-

---

1. Mueller and Oppenheimer (2014) and Mangen, Anda, Oxborough, and Brønnick (2015). For the cognitive impact of writing and drawing, see the many works by Barbara Tversky. See also Taylor and Zacks (2017, esp. chap. 10).

books created by students living in Europe and the Americas from the mid-seventeenth century to the early nineteenth century, a time in which literacy and scholarly learning became more accessible to middle-class families who sent their children to schools, academies, and universities.[2] These notebooks have taught me that the spread of knowledge was closely linked to how a person could organize facts and observations on paper in a useful manner. They have shown me that notekeeping was an art, a performance, that took place on, around, and across paper.

As explained in the 1771 edition of the *Encyclopaedia Britannica*, an "art" was understood to be "a system of rules serving to facilitate the performance of certain actions."[3] In this book I wish to explore this enactive notion of an art by taking a closer look at notebooks kept by students. More specifically, I want to treat them as material and visual artifacts, as paper machines, that can be used to recover the skills through which students learned to order and reorder their understanding of the world. The interpretation that I offer is one in which notebooks were media technologies that helped their makers and users interface with information in a meaningful and purposeful way. I focus on notebooks kept by students attending educational institutions in Scotland circa 1700 to 1830, drawing comparisons to other locations in Britain and, more broadly, in Europe and its former colonies.

Framing the notebooks designed by students as artifacts that operated as paper machines also connects to the way in which many Europeans reconceived the mind as an interactive entity during the long eighteenth century. Key to this reconceptualization was the *tabula rasa*, a metaphor that compared the mind of a learner to a blank piece of paper. The metaphor built on longstanding comparisons that had been drawn between the mind and other forms of material culture such as wooden boxes, clay tablets, vellum maps and stone buildings. But the material facets of paper from the sixteenth century forward increasingly captured the attention of those seeking to move away from medieval models of the human mind. The *tabula rasa* metaphor emerged as part of this effort. Its usage accelerated during the late seventeenth century when it was famously articulated by the English physician and philosopher John Locke. From that point forward, it became one of the most

---

2. The relationship between the rise of literacy and the print explosion of the mid-eighteenth century has been documented in many classic studies including Febvre and Martin (1976), Darnton (1982a), Martin (1995), Cavallo and Chartier (1999), Houston (2002). For Britain, see Houston (1985), Vincent (1989), and St. Clair (2004).
3. Smellie (1771, I: 427).

recognizable cognitive metaphors used over the next three centuries to describe the relationship between thinking and learning.

Though scholars acknowledge the metaphor's impact on the conceptualization of communications media in the ensuing centuries, they seldom consider why it was so popular for those who used it in predigital times. Throughout this book, I show that treating student notebooks as artifacts that functioned as personalized technologies allows us to gain deeper insight into the material and visual conditions that gave rise to the metaphor's usage. During the eighteenth century, the Latin term "tabula" (and its vernacular form of "table") was used to describe any visualization made on a square or rectangular surface.[4] In Britain, schoolbooks regularly used this wider notion of a table until the early nineteenth century. For instance, *The Royal Standard Dictionary*, published in 1775 by the Edinburgh Academy master William Perry for the use of schoolchildren, defined a "table" as "any flat surface" and "tabular" as an object "formed in squares or plates."[5] At the most basic level, then, every page was a *tabula*.

When used to describe the minds of young learners, the *tabula rasa* metaphor implied the possibility of achieving a state of filledness, one in which the writtenness of script served both as a form of order and a bearer of meaning. Consequently, in addition to comparing two objects—the mind and a piece of paper—the *tabula rasa* also compared two modes of interface—thinking and writing. Thinking transformed an empty mind into a filled mind. Writing transformed a blank page into a written page.[6] For the past two centuries scholars have conceptualized the *tabula rasa* primarily as an object, drawing comparisons between a sheet of paper and other object-based metaphors that liken the mind to a cabinet, a theater, a room, or a house.[7] But what if we moved the focus from objects to modes of media interface? That is to say, what if we flipped the metaphor so that the skills of writing as thinking served as the starting point for understanding the mind as a piece of paper? What if, instead of focusing solely on textbooks that stated what schoolchildren *should* write, we went even further and concentrated on the handwritten artifacts they actually made and used?

Taking these questions as a starting point, this book reveals that the success of Locke's metaphor was founded on its appeal to everyday notekeep-

---

4. Ferguson (1987).
5. Perry (1775, 373).
6. Pasanek (2015, 227–248).
7. Pasanek (2015, 205–226).

ing activities performed on and around the medium of paper and other associated forms of material culture. These activities not only included writing, drawing, annotating, and mapping, but also what I call codexing and categorizing, that is to say, processes that organized components of a notebook into a paper machine that made it a user-friendly media device, one that makes it possible to see the historical role played by education in the social reproduction of knowledge. Focusing on such processes allows me to show that the design and construction of notebooks can help us to better understand the developmental preconditions that sustained the *tabula rasa* metaphor.

Throughout the book, I treat notebooks as paper machines that operated as realtime technologies when they were made and then when they were used as reference tools. According to the media historian Markus Krajewski, when viewed as a material artifact, a "paper machine" is a technology that consists of different paper components—slips, sheets, scraps—that are both crafted and set in motion by the human hand.[8] I will have more to say about this topic in chapter 3, where I expand Krajewski's thesis and explain that there were many such machines in the predigital era. In later chapters I go on to reveal that many notekeepers living at the dawn of the modern world viewed the embodied forms of "motion" associated with the act of notekeeping as an ongoing event that simultaneously combined manual and mental skills. As such an activity, creating and using a notebook constituted a mode of realtime interface.

The story I tell about notebooks is based solidly on the lives of notekeepers whom time has forgotten and upon fascinating forms of evidence that they bequeathed to posterity. Like archaeologists who study the affordances of prehistoric artifacts, I use the material culture of notebooks to gain insight into how they functioned as thinking and learning platforms. Using this approach, I suggest that, rather than being a marginal object made by learners, the student notebook was a central artifact in the cultural and intellectual history of Europe, and in the history of modern human development more generally.[9] Making one extended the mind across each page in its capacity as an adaptable form of communications media.[10] Once mastered, the processes of

---

8. Krajewski (2011; 2015).
9. For the cognitive importance of material culture in human history, see Boivin (2010), Malafouris (2013), and Knappett and Malafouris (1998).
10. The idea of a notebook, both metaphorical and literal, has long played a significant role in the work of those who write about the history of the extended mind, external memory, and distributed cognition. One of the most widely cited examples is Clark and Chalmers's (1998) discussion of a notebook kept by an Alzheimer's patient named Otto. Their use of the notebook is also addressed in Preston (2010). For the larger historical context of the relationship between storage media, memory and the mind, see Anderson, Rousseau, and Wheeler (2019).

keeping a notebook enabled students to creatively mitigate the effects of media overload.[11]

For reasons that I will explain more fully in the following sections, there are three principles that guide my investigation of notebooks as media technologies. My first principle is that notebooks were artifacts assembled from different components over time. Second, notekeeping was a form of artificing, that is, it involved learning and applying a collection of interrelated skills that need to be more clearly identified by historians. Third, notekeepers were artificers in the sense that they were makers and users who lived within a specific educational community. These principles might be visualized in the following manner:

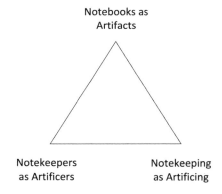

The diagram depicts the principles as being separate but linked directly to each other. Asking questions about the notebook as an artifact, for instance, leads to questions about its notekeeper as an artificer and to questions concerning the modes of artifice used to make it. Likewise, asking questions about notekeepers as artificers or notekeeping as artificing leads to questions about the topics depicted on the other points of the triangle. Throughout this book, I use these principles to explore how notebooks functioned as powerful paper machines.

But here it is important to note that the interconnected nature of the principles reveals that I am not only interested in the notebook as a medium of knowledge transmission. I also seek to identify and understand how notekeeping skills were acquired and to explore what it meant to be a notekeeper. As artifacts, student notebooks were valued as material objects. As artificers, student notekeepers were agents involved in their own education. As artificing,

---

[11]. Predigital media overload is addressed throughout Rosenberg (2003), Blair (2003), Ogilvie (2003), Blair and Stallybrass (2010), and Müller-Wille and Charmantier (2012a).

student notekeeping was an embodied and enabling act through which important capability-building skills were learned.

The foregoing principles allow me to take a decidedly interactive approach to the notebook as a paper machine that was crafted and used over time. Instead of treating notebooks as repositories of disembodied ideas that drifted effortlessly across time and space, I use their material components to champion the everyday skills that gave rise to ideas. Put more plainly, this book advances the position that notebooks were not solely evidence of reason as it was understood in the predigital era. Rather, I suggest that they were part of reason itself. This meant that making a notebook enabled students to experience reason in motion as they wrote, drew, folded, and sewed the components of pages, leading them to see a reciprocal relationship between reasoning and notekeeping.

Throughout the book I underscore the notion that rethinking the nature of predigital student notebooks offers deep insight into the historical relationship between learning and notekeeping. It presents the opportunity for historians to reflect directly on the formative, and oftentimes kinesthetic, role played by the material and visual facets of manuscript culture as it existed at the dawn of the modern world.[12] In an age preoccupied with the future, it is wise to intelligently judge the relationship between past and present forms of knowledge. Perhaps most importantly, it helps us to conceptualize how reason was literally rewritten, giving us the ability to understand the value of future media technologies as well.

Notebooks as Artifacts

I was originally trained in England and Germany as a historian of science.[13] I have since spent many years reflecting on the role played by science, medicine, and technology in the emergence of the modern understanding of objectivity, probability, testimony, and observation.[14] My original interest in

---

12. Here I am extending the larger view that manuscript culture played a central role in organizing modern knowledge. Chartier (1994). See also Vine (2018). For an exploration of manuscripts as material artifacts, see Lake (2020, 109–136).

13. Technically, I was trained as historian of *technoscience*. The meanings of this term are reviewed in Hottois (2006, 21–38). The term was popularized by Latour (1987), but my views are equally influenced by Klein (2020; 2016; 2005a; 2005b) and Bensaude-Vincent (2013; 2009).

14. The literature on these concepts is large. For objectivity, see Gillispie (1960), Cartwright (1983), Harding (1991), Porter (1996), and Daston and Galison (2007). For probability, see Daston (1988), Hacking (1984; 1990), Porter (1986; 2010), and Cartwright (1989). For testimony (including the creation and communication of scientific "facts"), see Fleck (1979), Shapin and Schaffer (1989), Latour and Woolgar (1979), and Poovey (1998). For observation, see Crary (1992), Canales (2010), and Daston and Lunbeck (2011).

notebooks stemmed from the pivotal role that they played in organizing, representing, and disseminating scientific knowledge within European universities from the seventeenth century to the early twentieth century. My early understanding of them was influenced by the ways in which environmental and medical knowledge was packaged into systems that were intelligible to those operating in commercial, industrial, agricultural and educational contexts.[15]

Put another way, my interest in notebooks arose from research that led me to see the various branches of science as systems that unfolded into further systems over time. In this reading, sciences were not solely reducible to causes such as laws of motion or categories such as evolutionary species. Rather, they were the sum of *all* the concepts, objects, and actions that gave rise to them. Without everyday things such as knots, keyboards, and paper, science as we know it would not have evolved.[16] When seen from this perspective, notebooks played a crucial role in the emergence of scientific knowledge. In the words of Hans-Jörg Rheinberger, "Notebooks do not simply serve as passive carriers of data, they are not merely the locus where events are recorded. The notebook can bring the experiments, both planned and executed, in short, the whole work of the laboratory, into a kind of condensed and compressed disposability on paper. . . . They make the enterprise available at a glance, ubiquitously present and easily transportable in pocket format."[17]

My first book addressed the relationship between medicine and the emergence of environmental science. While writing it, I had noticed that manuscripts were indispensable media technologies that facilitated the flow of information during the modern period. When it came time to write my second book, I was looking for a way to study notebooks as transformative tools that facilitated the creation of knowledge systems. Inspired by studies on the role played by paperwork within modern laboratory culture, a key moment for me occurred when I realized that notebooks were artifacts that helped to create new insights from old facts and observations. They featured affordances calibrated to operate within vast manuscript systems. Crucially, many of the skills that scientists used to make notebooks were learned *before* they became scientists.[18]

---

15. Eddy (2007; 2008; 2010c).

16. The importance of the material culture of knots, keyboards, archives, and notebooks to modern science is explored in Turkle (2011).

17. Rheinberger (2001, 57).

18. The classic studies of laboratory paperwork are Latour and Woolgar (1979) and Latour (1987). The emergence of new facts from old observations is underscored in Damerow (1996, 203). The larger generative relationship between training, inscription and the formation of scientific knowledge is addressed in more recent works, including Rheinberger (1997; 2003), Holmes, Renn, and Rheinberger (2003), Warwick (2003), Eddy (2010a), Krauthausen and Nasim (2010), Hoffmann (2013), and Nasim (2013). For a synthesis of the foregoing works, see Renn (2020).

Further investigation revealed that for much of modern history the paper notebook was an essential media technology, an important mode of communication that continued to thrive alongside the growing influence of print culture.[19] Student notebooks in particular focused the attention and assisted the memory. They operated in a world in which "information" was, in the words of Paul Duguid, "input (stuff delivered), process (the action resulting from that delivery) or outcome (the content of the ensuing mental state, having been informed)."[20] In a context where the line between information and knowledge was understood to be more fluid, notebooks helped students package facts, rules, and observations into accessible formats that were easier for them to understand and manipulate. In other words, student notebooks were realtime learning tools. They were sources about the past *and* artifacts that were part of the past.[21]

Though notebooks are fascinatingly informative artifacts, they are sometimes not the easiest forms of historical evidence to find, particularly those that were made by students. As pointed out in the work of historians of gender and childhood such as Anita Kurimay, Durba Mitra, and Marah Gubar, the documents, books, and artifacts that reside in today's museums and archives are often there because of medicalized, criminalized, or nationalized interpretations of history that existed at the time they were collected or donated.[22] The result is a striking paucity of evidence that speaks to the everyday lives and experiences of the many different kinds of people who lived in the past. In the cases where evidence of such historical actors exists, it was often written or collected by others with different motivations or perspectives. In doing the research for this book, I encountered a similar challenge. For young students, particularly schoolchildren, the traditional evidence housed by libraries is primers and conduct books written by adults, most of whom were men. Such books usually address what was supposed to be learned and not necessarily what was actually being learned. For older students, particularly those learning at home or in universities, the most common evidence held by libraries is the printed educational treatises or lecture outlines written by their teachers.

Early in the research that I conducted for this book, I realized that school and university notebooks were artifacts that could be treated as a form of evidence that was definitively used or made by students living in modern Britain

---

19. Heesen (2005) and Yeo (2014).
20. Duguid (2015, 351).
21. Knoles, Kennedy, and Knoles (2003), Nelles (2007), Blair (2010b), Grafton (2012), Grafton and Weinberg (2016), Schotte (2013), and Goeing (2016).
22. Kurimay (2020), Mitra (2020), and Gubar (2010). The destruction of sources distorts the historical record as well. See Ovenden (2020).

and its former colonies. Though finding notebooks was a challenge at first, it became easier as my research progressed. I learned that libraries often house lecture notebooks on account of their relevance to the professor in whose course they were kept. Finding notebooks and marginalia crafted by young children or adolescents proved to be more difficult, mainly because such documents were not valued as collectable objects by Victorian or Edwardian librarians. In the end, it was family collections that turned the tide. The process of finding school notebooks involved sifting through box after box of family archives that had remained relatively untouched since their donation. It was through reviewing the thousands of papers and objects contained in neglected collections that I was able to assemble the rich corpus of manuscript ephemera that helped me understand the notekeeping skills learned by schoolchildren.

One of my surprising discoveries was that a school or university notebook was one of the largest artifacts that many students made before they got married or entered a profession. I realized that holding one in many respects placed me in contact with the material and visual world that shaped their learning experiences. Consider the beautifully crafted school notebook depicted in figure 1.1. It was designed and assembled by an anonymous teenage student who attended Scotland's renowned Perth Academy during the 1790s. At first glance, it might appear to be a singular object, a manuscript book presented in a neat leather binding. But when we begin to thumb through its pages, we quickly learn that it was anything but a static object. The tell-tale clue occurs on page 103, which, due to overuse, is now detached from the binding.[23]

The page is actually the front side of what manuscript historians call a bifolium (plural: bifolia), that is, a sheet folded in half to make a booklet of four pages. Like many manuscripts at the time, scores of bifolia were used to make the Perth notebook. Each bifolium began as a blank sheet, a *tabula rasa* onto which students wrote their notes. If mistakes were made, the problematic bifolium was simply discarded and a new, rewritten version was made to take its place. Once a collection of bifolia was finished, students stacked them into a desirable order to be bound. They then numbered the pages of the notebook from start to finish.

The detached bifolium offers insight into a world of flexible information management skills that transformed student notebooks into adaptable and rearrangeable platforms suited to carry useful knowledge for young learners. It reveals how students learned to see paper as a knowledge-creating mate-

---

23. Anonymous Bound MS (1790, 103–104).

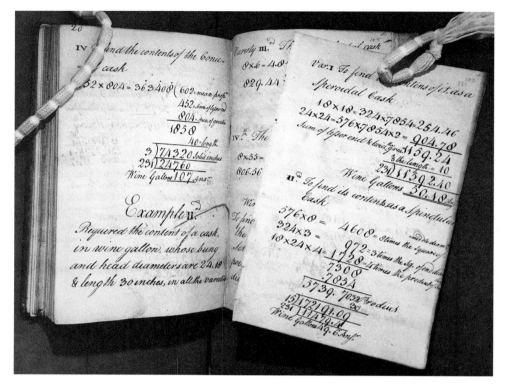

FIGURE 1.1. Anonymous notekeeper, *Perth Academy Notebook* (1790), Bound MS, NLS MS 14291, ff. 103–104. Reproduced with permission of the National Library of Scotland.

rial that could be used to store and retrieve facts, observations, and rules. Its organization tells us much about the hand and the mind that organized it. It speaks to a world in which students were allowed to make mistakes, to try out the different kinds of writing, folding, and drawing skills that they had practiced on other sheets of paper, thereby making each page a noteworthy object of historical enquiry with its own backstory.[24]

It is, however, very difficult to excavate the backstories of manuscripts without realizing the importance of spending much time with them. In the words of the medievalist and librarian Christopher de Hamel: "There will always be details which no one has seen before. You will make discoveries every time. Unnoticed evidence may be wrested from signs of manufacture, erasures, scratches, overpainting, offsets, patches, sewing-holes, bindings, and nu-

---

24. The historical epistemology of paper as a media technology is examined in Krajewski (2011; 2015), Hunter (2013), Gitelman (2014), Heesen (2014), Friedman (2018), and Bittel, Leong, and Oertzen (2019).

ances of color and texture, all entirely invisible in any reproduction."[25] Like de Hamel, I have learned to see manuscripts as collocations that were assembled over time in different ways with a variety of materials.

De Hamel's observation reveals that the different parts of a notebook, its components, are in fact individual objects of enquiry that can be employed to investigate the skills used to make them. Throughout this book, my approach to unpacking the materiality of notebook components is inspired by the anthropology of artifacts developed in the work of Tim Ingold and André Leroi-Gourhan. Both emphasize that artifacts are often made out of discrete components, each of which speaks to a cluster of skills that took time and effort to learn and implement.[26] This viewpoint transforms each component of a notebook into evidence of a kinesthetic notekeeping process that took place along the way. Ingold uses the term "wayfaring" to capture this kind of thinking through making. In this reading, wayfaring represents a dynamic process of learning or enacting the skills required to make the various components that add up to a singular artifact. When viewed as this kind of wayfaring, notekeeping becomes something that was not strictly pursued inside a notebook, but through, around, to, and from its many components.[27]

In addition to using the wayfaring metaphor to understand the materiality of artifacts as they exist in the present, Ingold develops it further to explore the ways in which historically minded scholars might explore the processes through which artificers made artifacts in the past. He explores this temporal facet of wayfaring by using footpaths to retrace how artifacts are created and experienced. Though such paths initially seem to be singular lines to an observer, they are, in fact, a compilation of many lines created by many wayfarers over a period of time. The metaphor treats the path as a material representation of the process through which lines were laid across the surface of a field over days, years, or even centuries. In this reading, the path itself functions as a material artifact that offers evidence of those who made and used it.

Ingold intimates that the kinesthetic logic of a path's formation emerges more fully to observers when they actively seek one out, wander along its contours, and experience the environment surrounding the journey. Throughout this book I use this embodied understanding of wayfaring to help reclaim the material and visual processes of modern notekeeping that are instantiated

---

25. De Hamel (2018, 3).
26. Leroi-Gourhan (1993) and Ingold (2000; 2007; 2015; 2017).
27. Tim Ingold discusses wayfaring along paths and lines throughout Ingold (2011; 2009). Cognitive scientists seeking to understand how artifacts are made and used sometimes use the term "wayfinding" as well. See Golledge (1999).

in the material components of a notebook. Like the lines of a path running across a field of grass, the strokes, cuts, and folds of a sheet of paper offer rich insight into how notekeepers learned to experience, remember, and value knowledge about themselves and the world around them.

Notebook components were adapted to fit the needs and abilities of individual notekeepers. Components and their associated skills also could be repeated or adapted based on a desired outcome. Students used some components more than others, making what might be seen as a shorter journey toward coming to understand how knowledge might be represented. In this book I show that, whether the journey be long or short, the components of a notebook reveal that notekeeping was a materially grounded form of wayfaring that enabled students to become self-organized learners.

The Perth Notebook depicted in figure 1.1 offers many examples that illustrate how a notebook comprised different components that added up to a sophisticated codex. Its pages were composed of paper, ink, and bindings. Like many school students, its creator had to design the notebook from the bottom up. Its sheets of paper were shaped into bifolia. These were then stacked into gatherings that ultimately became a codex. Ink was written as words, numbers and figures that were in turn ordered as paragraphs, formulae, and tableaux. Bindings came in the form of folds, string, and thread, all of which were used at different stages in the artificing process. All of these components added up to the final version of a bound notebook.

One of the aims of this book is to show how the many components of a notebook can be used to characterize it as an artifact, a paper machine, in motion that was assembled and used through time. Rather than existing as a fixed object, the components of a student's notebook thrived in an adaptable world of representation, one in which misdrawn lines and misapplied watercolors could be re-inscribed and repainted on a fresh piece of paper. This adaptability allowed students to understand the styles of notekeeping that underpinned the acquisition of literacy and numeracy in Europe at the time.

Notekeeping as Artificing

Scholars working on the history of the material culture of writing, drawing, and painting in Europe and Asia have long emphasized the important role played by notetaking and annotating in the transmission of knowledge.[28] Likewise,

---

28. For Europe, see Saenger (1997), Jackson (2002), Daston (2004), Sherman (2009), Blair (2010a; 2010b), Soll (2010), Grafton (2012), Yeo (2014). For Asia, see Cho (2020), LaMarre (2000), Richter (2013), Williams (2018), and Lurie (2011). See also Bausi, Friedrich, and Maniaci (2019).

they have pointed out that different kinds of paper, vellum, ink, chalk, and paint affected how words and figures were created, experienced, and valued.[29] Historians of print culture have made similar observations about the layout and media of predigital and postdigital texts.[30] Add to this the fact that historians of the mnemotechnic tradition, the art of memory, have repeatedly underscored the importance of observing how the shapes and patterns presented on a printed or inscribed page functioned as gestalt images, that is, as visualizations that made memorable impressions in the mind.[31] When considered in tandem, the foregoing studies reveal that the material and visual culture of script and print was also a mnemonic culture, one in which the acts of making and remembering functioned as mutually beneficial cognitive partners.

Knowing how to craft a blank notebook page so that it functioned as an accessible knowledge platform took much practice. It required students to interface with tools and materials in a way that allowed them to learn a portfolio of artificing skills. At present, we know very little about what these skills were or how they were acquired via the everyday communications media available to school students living prior to the early twentieth century, especially those whose learning experiences took place outside elite settings. This is not only the case for Britain, but for other European and colonial contexts as well.

As shown in the work of Richard Sennett, skills are forms of "learned practice."[32] They provide a helpful entry point into the history of daily life and thought, but they are difficult to historicize. Within the world of predigital manuscript culture, bringing a page into existence, a process we might call foliation, required the skills of folding, ripping, pasting, gridding, ranging, and tabling. Bringing script into existence through the process of inscription was accomplished through enacting the skills of writing, drawing, ciphering, sketching, pictogramming, figuring, and diagramming. Of equal importance, there was the process of selecting, assembling, and fixing pages into a codex, which, for reasons that will become more apparent in chapter 3, I call *codection*. Though often overlooked as an object of enquiry outside manuscript studies, the process of creating a codex from start to finish involved a host of

---

29. Baxandall (1972; 1985; 1997), Alpers (1983), Alpers and Baxandall (1996), Haskell (1993), Camille (1992; 1998), and Freedberg (1989; 2003).
30. A classic book on the fonts and layouts of the hand press is Moxon (1963). The meanings and varieties of modern graphic design and layout are addressed in Cramsie (2010), Barchas (2003), and Gaskell (1995). For late modern layouts, see Hartley (1988), Ambrose and Harris (2007), and Johnson (2010). Chang, Grafton, and Most (2021) address the historical overlap between manuscript and printed layouts in European and Asian contexts.
31. Ong (2004), Yates (1966), Spence (1985), Carruthers (1998; 2008), Blair (1997), and Rossi (2000).
32. Sennett (2008, 52).

skills including stacking, quiring, punching, sewing, collating, binding, covering, and more.

The skills of artifice which underpinned foliation, inscription, and codection sometimes overlapped or worked in conversation to enhance a student's ability to craft a notebook into a customized paper machine. Throughout the book, I suggest that the skills involved in these processes enabled notekeepers to negotiate the materials of manuscript media on a case-by-case basis in a way that familiarized them with the power of keeping an organized notebook. Though students living during the long eighteenth century learned in diverse settings such as homes, schools, academies, universities, shopfloors, and even the fields surrounding cities, acquiring specialized notekeeping skills took practice and had to be executed in a way that created a userfriendly artifact.

The pages of the Perth Notebook in figure 1.1, for instance, are structured according to an easily accessible layout. Its script is cleanly written. Its bifolia are thematically ordered and elegantly bound. The exquisite and skillfully rendered script further reveals that its maker saw in advance how a rough sheet of paper needed to be laid out. Such skills speak to the long training process through which students learned how to visualize the placement of words, numbers, and symbols on a grid in advance. The notebook's customized surveying illustrations of local architecture and landscapes, several of which were rendered in fresh pastel pink and blue watercolors, speak to the knowledge of a geometric sense of space (figure 1.2).[33]

If we step back for a moment and consider the broad range of processes involved in foliation, inscription, and codection, it can be seen that notekeeping was a mode of artificing that wove together a rich variety of skills. This being the case, if we are to fully understand how students created the components of notebooks, then we must diversify, or perhaps reclaim, the wider notion of "notekeeping" that existed in the predigital era. In other words, the technical complexity of eighteenth-century student notebooks as artifacts requires us to recalibrate the verbs traditionally used to discuss the skills through which they were created.

Writing, for instance, does not describe adequately all the skills that students learned to master while making a notebook, especially those such as folding, tabling, and sketching. These specialized, and fundamentally material, skills traditionally have remained largely on the periphery of studies that

---

33. Eddy (2013).

FIGURE 1.2. Anonymous notekeeper, *Perth Academy Notebook* (1790), Bound MS, NLS MS 14291, f. 20. Reproduced with permission of the National Library of Scotland.

address notekeeping as a kind of *writing*. Similarly, *notetaking* is an awkward metonym for the combined skills of bifoliating, writing, and drawing. Looking beyond notetaking and writing to the tools and skills of *notekeeping*, that is, instead of focusing solely on how notes were *taken* in a manner that fixed them as content on paper, it is possible—and important—to reconstruct how they were *kept* across multiple sheets and through different modes of representation in a way that created an interactive system of human interface.[34]

Though the difference between *taking* notes and *keeping* notes might at first seem subtle, it is a helpful distinction because it turns our attention to the nature of notebooks as dynamic artifacts that speak to the knowledge management techniques possessed by their creators.[35] It transforms notebooks into objects that are eminently suited to tantalizing questions about the capabilities or intentions surrounding the materiality and visuality of the manuscript cultures that emerged in modern Europe. It also draws our attention to a notebook's status as a self-organized memory device that was made with a variety of skills for a number of purposes. In sum, treating notebooks as kept artifacts makes it easier to see the role that they played as mediators of knowledge.

Throughout the book I emphasize that students used the space in and around a notebook's components to acquire skills that allowed them to transform movement into a thinking and learning tool. Put more clearly, my interpretation of notekeeping treats *keeping* as a kind of movement-oriented artifice that facilitated spatial thinking, a process that involved, in the words of Barbara Tversky, "the perception of space and action in it."[36] When approached from this perspective, notekeeping becomes a multisensory and kinesthetic activity that contributes significantly to a notekeeper's ability to use the components of notebooks as tools through which embodied skills of learning and thinking can be acquired and practiced.[37]

In recent years, the relationships between historical artifacts and the skills of spatial thinking have received increased attention, particularly in studies that seek to understand the cognitive capabilities of prehistoric humans. Neil

---

34. Here my conception of human interface as a user-focused, dynamic process that operates across systems is influenced by Raskin (2000).

35. Some scholars prefer the term "notemaking" as well. Corey (1935), Tomlinson (1997).

36. Tversky (2019, 3). Tversky has published many illuminating articles on spatial thinking, two of which are Tversky (2011; 2001).

37. Kinaesthesia, also called proprioception, is the awareness, the sense, of bodily movement acquired through experience. My understanding of kinaesthetic of learning is influenced by Barbara Tversky's notion that there is a very close relationship between memory, abstraction, and bodily action in space. She calls this collective process of interface "spraction." For her discussion of the term, see Tversky (2011; 2019, 277–288).

MacGregor, the director of the British Museum, for example, offered a striking presentation of the relationship in his BBC Radio 4 series *A History of the World in 100 Objects*. In his commentary about a prehistoric hand axe found at the Olduvai Gorge in Tanzania, he exclaimed: "What makes this stone axe so interesting is how much it tells us, not just about the hand, but about the mind that made it."[38] In this reading, every chip presents a clue of how the maker's body moved around the space surrounding the axe while it was being created. To use Tversky's terminology, the axe is an example of an interactive mind in motion.

Like the skills used to create a prehistoric hand axe, the skills of notekeeping occurred in a series of events that added up to a process. When discussing the creation of artifacts, the anthropologist André Leroi-Gourhan once called the materially grounded series of skill-driven events a *chaîne opératoire* (an operational chain).[39] The creators of prehistoric hand axes, for instance, did not begin the process by immediately sharpening the edge of the blade. Instead, they proceeded through a series, a chain, of events. Axe makers, for instance, had to first search for an appropriate stone. They then had to knap the stone into an oblong shape. Only after the searching and the knapping could the tip of a stone be chipped into a blade. These acts reveal that the materiality of the stone influenced the order in which the skills of searching, knapping, and chipping occurred.

A similar kind of materially grounded situation existed for students learning to keep notebooks. They did not begin by binding quires. They first had to write, draw, or paint a page of script. Prior to that, they had to fold sheets of paper into bifolia that could be bound. In short, the materiality of the components influenced the order in which the skills of writing, drawing, painting, folding, and binding occurred as stages in a process.

I do not wish to suggest that I am the first to use a stage-based historiography to approach modern texts. Print historians, for instance, use what the historian Robert Darnton once called the "communications circuit" to study the evolution of books through time. They seek to trace the evolution of the book through a circuit than ran "from the author to the publisher . . . the printer, the shipper, the bookseller, and the reader" back to the author. In this exercise, the text, the printed or inscribed content of the book, serves as an object that is traced through its encounters with compositors, pressmen, warehousemen, suppliers (of paper, ink, type, and labor), shippers (agents, smugglers, entre-

---

38. The script of the radio program was published as MacGregor (2011).
39. Leroi-Gourhan (1993).

pot keepers, and waggoners), booksellers (wholesalers, retailers, peddlers, and binders), and readers (purchasers and borrowers in clubs or libraries).[40]

In many respects, Darnton's desire to foreground the stages of book making is similar to my desire to identify the stages that existed within the process that notekeepers used to create a notebook. Indeed, the literature surrounding his communications circuit influenced my early thoughts on the notebook as an object of enquiry.[41] But as I studied the notebooks, I realized that the anthropological notion of a stage-based framework made it easier for me to trace the material and visual development of different parts of the notebook as independent components in their own right that were created by an artificer who possessed a specific set of artificing skills. This realization was subsequently reinforced when I encountered Ludmilla Jordanova's work on the ways in which paintings, models, and statues are created through stages, each of which adds a new layer of meaning to the artifact.[42]

A point that I wish to underscore throughout the book is that notebooks provide evidence that speaks to the skills, the forms of artifice, that shaped how notekeepers learned to create components. I want to show how students, as artificers, used the material and spatial thinking involved in notekeeping to understand the subjects they were studying. I develop a historical account of this process as it operated in the homes and classrooms of Scotland on a daily basis. In my account of notebooks, each of their components functions as a stage in a notekeeping process that operated within the larger community of notekeepers spread throughout household and educational communities that operated at local, regional, and, as we will see in the final chapters, global levels.[43]

Reconsider the leather-bound Perth Notebook featured in figures 1.1 and 1.2. A simple component of each bifolium, for example, was the fold used to bisect it. The skill required to make it was folding. When enacted, the skill transformed the notekeeper into a folder, a person able to create a fold. A more involved notebook component was a hand-drawn diagram. The skill required to render it on a page was diagramming. When enacted, the skill transformed the notekeeper into a diagrammer, someone who knew how to create or replicate a diagram. Many other examples could be cited in relation to the

---

40. Darnton (1982a, quotation on 67). A similar stage-based approach was advocated by other book historians at the time, including McKenzie (1985; 2002). Darnton's notion of the communications circuit has since been expanded in a way that gives more attention to the material culture of predigital script and print. Calhoun (2020, 5–7).

41. Of particular note are the many essays of Jonathon R. Topham, especially Topham (1998), as well as Secord (2003) and Fyfe (2004; 2012).

42. Jordanova (1993; 2012).

43. The importance of balancing normativity with the network (or even meshwork) of skills provided both at local and regional levels is addressed in Knappett (2011).

multiplicity of components used by notekeepers to craft a notebook. But the overarching point is that student notekeepers did things with notebook components that enabled them to learn more efficiently and to see knowledge as a rearrangeable entity.

The process of acquiring notekeeping skills involved reams of paper, boxes of graphite, cases of pens, and bundles of string. Student notekeepers had to do many things with these materials. This means that in addition to learning facts relevant to the subjects they were studying, they also learned techniques that allowed them to manipulate facts on paper in a way that extended their own reasoning skills. At a time when Europe and its colonies were experiencing the largest explosion of print since the invention of the hand press, notekeeping skills were indispensable for those who sought to understand how to filter and order information in a meaningful way.[44]

Notekeeping was an active form of cognitive engagement, a collection of skills through which knowledge was transformed into understanding on the page. This connection between notekeeping and knowledge formation was very much on the minds of the students, teachers, and parents that we will encounter in this book, many of whom maintained that there was an inextricable skill-based link between learning and understanding. The connection was clearly voiced in Perry's popular *Royal Standard Dictionary*. Drawing from his considerable experience as a tutor and lecturer at the Edinburgh Academy during the mid-eighteenth century, he defined knowledge as "learning, skill, understanding." The words of the definition are listed in the order in which he saw them play out in his classroom and, like other dictionaries of the time, they reveal how skill was built into the notion that reason was an activity.[45]

When seen through the eyes of an educator like Perry, the components of school notebooks become a repository of evidence that can be used to historicize the skills that students used for the acquisition and maintenance of reason as it was perceived at the time. To make these skills more visible, each chapter of this book includes photographs of notebooks and their assorted components. Some of the figures are presented through the time-honored tradition of displaying the pages of manuscripts as if they were paintings. This approach presents the page of an inscribed or printed book as if the viewer is looking straight at it, as if she were standing in front of a well-lit painting in an art gallery.

---

44. For the explosion of print, see the chapters that address the eighteenth century in Cavallo and Chartier (2003).
45. Perry (1775, 206). For Latin, *peritia* was defined, in the words of John Mair's widely used *The Tyro's Dictionary, Latin and English* (1760, 330), as "skill, knowledge."

But for most of the images that I included in this book, I photographed notebook components in a way that depicted how they were fashioned from a specific kind of material, how an instrument was used to craft them, or how they were experienced in realtime. I will have more to say about the realtime capacities of notebook components later in the book, particularly in chapters 5, 6, and 7. But here it is important to note that, since I wanted my images to capture the everyday conditions experienced by someone interfacing with manuscripts at a desk or table, almost all of them were taken on an iPhone as I was conducting my research; that is, I sought to capture notebook components in situ, in a way that offered insight into how they were experienced by notekeepers over space and time.[46] My pictures also feature the traditional tools, including my own fingers, that I used to interface with the manuscript when I took the photograph. The result is a collection of images that depict the components of notebooks as kinesthetic learning tools in a way that makes it easier to understand the kinds of skills of artifice that students employed to design and use them.

Notekeepers as Artificers

At every stage and on every notebook page, students learned to hone a style of interface that was, at one level, bespoke to their own needs and, at another level, consistent with the larger mnemotechnic norms employed in the notes of their peers. Keeping a notebook inculcated scores of skills and routines that were in turn infused with a sense of purpose, a sense that notekeeping itself was just as important as the material notebook that they produced. At several points within the process, students even functioned as manuscript compositors, and within university and domestic contexts, they played the role of an editor as well. By inhabiting these different roles, they were able to craft manuscript books that helped them to remember things better after their studies.

Focusing on the components of notebooks to gain deeper insight into the skills of notekeeping lends itself to questions about who Scotland's student notekeepers were and why they would be motivated to work so hard to transform script and paper into a codex. When discussing the lives and ambitions of student notekeepers at conferences I sometimes introduce this topic with an invitation for listeners to think back to their childhood. How did they use writing, drawing, or even doodling to learn, organize, and remember informa-

---

46. For the presence of the iPhone as an essential archival research tool for historians doing archive-based research, see Madrigal (2020).

tion? What kinds of techniques worked better for them and why was that the case? I ask similar questions in everyday conversation with friends, students, and colleagues.

For some, the modes through which they learned to write fitted well with the abilities they possessed. For others, it was difficult, especially for those with learning disabilities. It was the same for the students we will meet in this book. As noted by George Chapman, the master of Dumfries Grammar School in the middle of the eighteenth century, the minds of some students were "late in opening." They struggled to acquire the proficiency with language that was required "during that period which is commonly assigned for school-education."[47] The notebooks they left behind show us that all learners were not the same, revealing that the individual strengths of students affected their ability to write and to keep notes.

Many of the students covered in this book struggled to internalize the skills required to draw a simple table or to replicate a teacher's diagram in a manner that they found satisfactory. Additionally, some students preserved only their rewritten, final copy of a notebook. Rather than being the starting point of a skill-focused analysis, such notebooks need to be considered in light of the fact that they sometimes represent the expression of notekeeping skills that took years to learn. In this sense, the notebook was a form and the notekeeper was a kind of skill-based agent who created the form. This is why I treat note-keeping as a kind of artifice, a mode of *form-giving*, that allowed a notekeeper to be an artificer capable of enacting a powerful media technology.[48]

Though a number of notekeeping skills were extensions of scholarly information management practices developed by humanists during the Renaissance, the components created in Scotland's homes and classrooms were learned in a British context that put them to different uses and placed different values upon them.[49] As intimated at the beginning of this chapter, the form and function of notebook components as artifacts can be used to conceptualize every notekeeper as an *artificer*. Prior to the nineteenth century, this term was sometimes used to describe someone capable of creating a graphic artifact. In the popular school textbook *Memoria Technica*, for instance, Richard Grey used the term "artificer" in the section that addresses the kinds of instruments that can be used to draw shapes.

---

47. Chapman (1784, 209–210).
48. For the relationship between movement and materials involved in "form-giving," see Ingold (2011) and Klee (1961; 1973).
49. The wider presence of humanism in eighteenth-century Scotland is explained in Allan (1993), McLean (2009), and Hammet (1985).

In his discussion of drawing tools, Grey mentions a "mechanical instrument" called a trammel that commonly was "used by artificers to describe ellipses."[50] This was a familiar tool for those who had studied practical mathematics or who had learned to draw geometric shapes. Grey's observation about the trammel is noteworthy because he uses the noun "artificer" alongside the verb "to describe." In addition to referring to verbal accounts of an object, this verb was used to capture the act of drawing a figure, which, as we will learn later in this book, was an activity pursued by many student notekeepers. Like my understanding of an artificer used throughout this book, the notion of an artificer for Grey and others during the long eighteenth century included someone capable of making and using a graphic artifact.

To present students in a historically sensitive manner that is grounded on the skills that they acquired when they made and used notebooks, I emphasize that they operated within notekeeping communities that existed across the homes, schools, and academies of Scotland. These communities provide the historical context that is essential for interpreting the efforts of notekeepers and for understanding the developmental advantages that their teachers attributed to notebooks. A key element of my analysis is the attention that I draw to how scholars can use notebooks to identify the attributes of different kinds of notekeepers that operated within a shared community.

Excavating the material and visual skills employed to construct and use notebooks will require us to meet individuals whom time has forgotten. The pages of this book transform hitherto nameless students and teachers into significant historical actors and underscore the importance of treating seemingly ordinary educational objects as extraordinary artifacts which offer new and refreshing insight into the factors that underpinned the success of modern forms of graphic representation. My investigation will touch on a number of topics relevant to how students learned skills that enabled them to be agents involved in their own education. One of my aims is to more clearly articulate the skills-based capabilities that underpinned the kinds of agency that modern students possessed as notekeepers.

That said, student notekeepers are often hard to identify. The identities of the notekeepers that I discuss were difficult to pinpoint because little biographical evidence remains. Add to this the fact that, though many learners were girls or young women, the visual and material evidence of student notekeeping is often masculine in nature. The extant depictions of learners which occur in the frontispieces of writing, bookkeeping, and ciphering man-

---

50. Grey (1737, 155).

FIGURE 1.3. "The Representative," George Bickham, *The Universal Penman* (London: Overton, 1743), frontispiece. The scene depicts a student writing on a quire. Image courtesy of the open access Internet Archive.

uals, for example, usually portray students as anonymous male notekeepers (figure 1.3).

In some cases, visual evidence of young female writers can be gleaned from the frontispieces of novels. Perhaps the most famous of these depictions is the frontispiece of the fifteen-year-old Pamela Andrews, a maidservant and the heroine of the eponymous novel written in 1740 by the English author Samuel Richardson (figure 1.4). The image presents the scene of Pamela writing a letter on a bifolium.[51] She holds a quill and the open lid of her writing box is turned away, leaving the viewer to imagine the usual supplies such as ink, black lead pencils, Indian rubber, and penknives contained therein. As I will show in later chapters, like Pamela and other young correspondents, students learned at home and in school to write their notebooks with the same supplies. This means that the frontispiece gives a good idea of what a scene of domestic notekeeping might have looked like for a number of the female notekeepers we will encounter later in the book.[52]

Setting aside visual depictions, there are material factors that obscure the identities of student notekeepers. Within Scotland, for example, there was a tradition of making a title page that stated a notebook's subject matter and then the identity of the notekeeper. Since it was placed near the front, the title page was often one of the first casualties of neglectful storage practices or overuse. This explains why many Scottish student notebooks are presently anonymous, including those created in university settings that contain rough or complete self-portraits of students or their friends (figure 1.5). There are some cases, however, where the identities of notekeepers have been preserved or are stated elsewhere in related institutional or familial ephemera. Yet even in cases where the identity of the notekeeper is known, little biographical evidence remains.

One of the great pleasures of researching this book was reassembling the identities of young notekeepers whose existence had long been undervalued and then forgotten by historians. They came from different kinds of families, followed different educational trajectories, and were exposed to a variety of notekeeping tools. Most of the students who made the notebooks covered in this book came from families associated with the rising merchant, professional, and artisanal classes. All of them reacted to the sociomaterial world of notekeeping in ways that allowed them to organize knowledge in a manner that worked with their own abilities, their wider ambitions, and their developing sense of human nature. In the many cases where the evidence regarding the identities or education of notekeepers covered in this book was lacking, I used the various attributes of the components presented in their notebooks

---

51. The epistolary conventions employed by Pamela Andrews in Richardson's novel are identified and examined in Curran (2016, 19–50).
52. The most popular letter-writing manual in Scotland at this time was Darling (1768). It gives a sample of the kinds of letters Scottish girls aspired to write.

FIGURE 1.4. The fifteen-year-old Pamela Andrews writing a letter. The scene is the first plate in a series of twelve engraved for Samuel Richardson's *Pamela* by L. Truchy and A. Benoist in London in 1745. The descriptive text reads: "Pamela is represented in this first Piece, writing in her Lady's dressing room, her history being known only by her letters. She is here surprised by Mr B who improves this occasion to further his designs." Wiki Commons, Typ 705.45.452, Houghton Library, Harvard University.

FIGURE 1.5. Student contemplating the pressure inside a cask. Anonymous, *Perth Academy Notebook*, vol. 3 (1787), Bound MS, NLS MS 14296, f. 8v. Reproduced with permission of the National Library of Scotland.

as an entry point. Like the artifacts employed by archaeologists to understand people who lived in the past, the components and associated skills presented by student notekeepers provided a helpful baseline of evidence from which I could then seek further information about the identity of the notekeeper.

My favorite student discussed in this book is Margaret Monro, a schoolgirl who lived in Edinburgh during the middle of the eighteenth century. When she was twelve years old, she used her notekeeping skills to create an enormous codex that addressed female conduct. Save for a few brief letters, her manuscript is the only extant evidence of her life. But what a manuscript it is. It reveals that she formatted all of its pages as a matrix, a table, into which she meticulously composited her script. She classified the paragraphs according to a binomial nomenclature of headings and subheadings, and she annotated and corrected the text. All of these activities indicate that she was a skilled "tabler," compositor, classifier, annotator, and editor. Her father placed a high value on these skills, praising her as a scholar whose intellect he greatly valued.[53]

Another fascinating notekeeper we will encounter later in the book is Thomas Cochrane, a medical student at Edinburgh University. His notebooks are the only evidence that we have of his life. But in addition to revealing how a university student used notekeeping to create a realtime learning device, his notebooks shed light on his prior education and, more broadly, they offer clues that speak to his goals as an aspiring medical student. The pages disclose, for example, that he diagrammed formulae, transcribed his notes, and interleaved blank sheets of paper for annotations. He also drew pictograms, figures of experiments, and profiles of his professors. These notekeeping activities show that he was a diagrammer, a transcriber, an interleaver, a pictogrammer, a figurer, and a profiler. The skills upon which these roles were based offered vital clues about his prior education and ambitions that made it easier for me to understand him as a notekeeper and as a significant historical personality.[54]

The examples of Monro and Cochrane reveal that much work remains to be done on identifying the student notekeepers discussed in this book and, more generally, those who lived in predigital Europe and colonial settings. But such research is necessary because it sheds light on how, why, and where students learned to become notekeepers.

Thought as a Realtime Activity

Since notekeepers used notebooks to filter, organize, and judge a remarkably large amount of data, the connection between notekeeping and thinking was woven into the cultural fabric of the time. The pedagogy and developmental psychology of Scotland was significantly influenced by a communal desire for student notekeepers, both female and male, to acquire reason. As explained in the first edition of the *Encyclopaedia Britannica*, reason was "a faculty or power of the mind, whereby it distinguishes good from evil, truth from falsehood." This definition reveals that reason, like a notebook, was not understood to be a static entity. Reason was an activity.[55] In the words of William

---

53. Monro Primus and Monro Bound MS (1739). The context of the manuscript is addressed in Monro Primus (1996/1738).

54. Black Bound MSS (1767–1768), Thomas Cochrane (notekeeper). Cochrane's notes were eventually published as Black (1966/1767).

55. The active nature of reason in Scotland is emphasized in Barfoot (1983). For the importance of human experience, broadly construed as an activity of engaging with the world through reason, see the introduction and readings in Friday (2004). See also Garrett and Heydt (2015).

Barron, the professor of logic, rhetoric, and metaphysics at the University of St. Andrews during the last decades of the century, "Practice is the chief mean of obtaining eminence in reasoning as in all other arts." In his view, reason had to be exercised regularly because "no rules of logic can make a mechanical reasoner."[56]

Put more clearly, reason was a collection of skills that had to be learned and performed simultaneously in realtime. Such skills were acquired and maintained through daily acts of reasoning, which Scots also called ratiocination. The *Encyclopaedia Britannica* jointly defined reasoning and ratiocination as "an act or operation of the mind, deducing some unknown proposition from the previous ones that are evident and known."[57] One point that I highlight in this book is that notekeepers were learning to be reasoners who could order knowledge on paper in a meaningful or coherent way that aided ratiocination.

The skills of reasoning were learned over time within the lives of individuals who were doing things with their minds and bodies in, around, and through notebooks on a daily basis. When seen from this angle, reason was closely tied to activities that enabled one to know how to know. As such, it was deeply embedded in the processes of social reproduction that enabled a person to recognize the kinds of skills, materials, and resources required to craft and re-craft a premise or to create and re-create a category. It was this sense of ratiocination that John Locke had in mind when he wrote, "I have seldom or never observed any one to get the skill of reasoning well, or speaking handsomely, by studying those rules which pretend to teach it."[58]

Until quite recently scholars writing about the reciprocality of notekeeping and reasoning at the dawn of the modern period framed the relationship in disembodied terms, as something that took place mainly in the heads of adult men. In recent years this focus has changed, with the attention being shifted to other kinds of historical actors who shaped their minds with the media technologies provided by manuscript culture. I see this book as part of that shift.[59] Focusing on the notebooks of both male and female students enables me to tell a more holistic story of the historical preconditions of reason, a story that speaks to the developmental context that underpinned the way in which reasoners learned to conceptualize ratiocination through the acts required to

---

56. Barron (1790, 158, 203).
57. The definitions for reason, reasoning and ratiocination are given in Smellie (1771, 3: 528–529).
58. Locke (1751, 83).
59. Classic studies that used manuscripts to explore the everyday lives of those forgotten by history are Ginzburg (1980) and Davis (1984; 1995). More recent studies that influenced my thought are Baggerman and Dekker (2009), Daybell (2012), Yale (2016), and Leong (2018a).

make and use a notebook, that is, through interfacing with the components of a codex that featured organized facts, observations, and rules.

Georgian Britain was fascinated with the relationship between inscription and ratiocination. Many believed that a disordered page could lead to a disordered mind (figure 1.6) or was the product of a deranged mind (figure 1.7). This view came part and parcel with the influential developmental psychology of John Locke and his *tabula rasa* metaphor. When attempting to describe an unsettled mind "disturbed with any passion," for example, he went so far as to compare it to a shaking piece of paper.[60]

We will see in later chapters that students learned to perceive notekeeping as a form of thinking through writing, as a process of ordering the mind through media. It was believed at the time that there were simultaneous cognitive relationships between acts of learning experienced through writing and acts of understanding strengthened through thinking. When seen from this angle, the handwritten page's role as an interactive nexus between the mind and body serves to remind us of how both the materials and the skills that underpinned the *tabula rasa* metaphor played an important role in shaping the value attached to notebooks as paper machines that could potentially help notekeepers distinguish between the many truths and falsehoods that existed in the world.

Put another way, in an age when the ability to make distinctions between true and false knowledge served as the hallmark of reason,[61] the materiality of notekeeping provided a reference point for those who used the *tabula rasa* metaphor to understand cognition as an embodied developmental process. At one level, the metaphor treated the mind as a blank sheet of paper, and words as signs of ideas.[62] But at another level, the metaphor was based fundamentally on the material culture of foliation, inscription, and codection that I have outlined in previous sections. As this book unfolds, it becomes clearer that the cognitive model provided by the *tabula rasa* metaphor was based firmly on the skills required to craft the fabric of the components of notebooks and other related forms of manuscript media.

Since student notebooks were artifacts that were made and used over time, they offer a wealth of evidence that can help historians identify the kinds of skills that students used to internalize and pursue reason as an unfolding

60. Locke (1752, 246–247).
61. See again the definition of reason in Smellie (1771, 3: 528–529).
62. Pasanek (2015). For corresponding metaphors drawn between books, writing and memory, see Draaisma (2000). The Scottish context of the *tabula rasa* is addressed in Eddy (2016b). For the use of other material metaphors that modern authors used to discuss the mind, see Silver (2015).

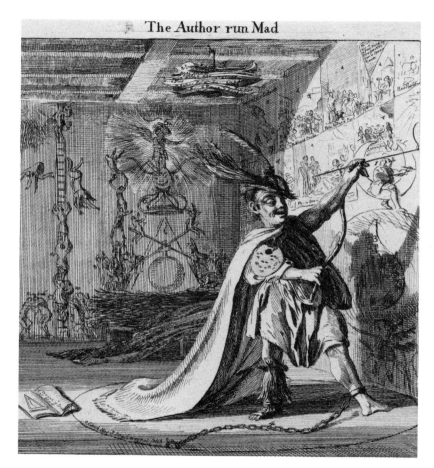

FIGURE 1.6. The relationship between ratiocination and inscription, which included writing, drawing, and painting, as visualized by Paul Sandby in *The Author Run Mad* (1753). Etching, from the New York Public Library, object number 107476, public domain. During the long eighteenth century, words written along ruled lines as grammatically correct sentences were analogous to ideas that flowed across the mind's eye as well-ordered trains of thought. Since writing words evoked images, inscribing cluttered words or figures was believed to cause madness. Sandby's print plays on this linear metaphor by filling the walls of the room (which represents a destabilized mind) with a confusing montage of images that visually explain why the mad author's "Words are loose as Heaps of Sand." The message conveyed is that, even though the author was standing in a protective "Circle that ye Devil may not fetch him," dabbling in disorderly forms of linear representation caused insanity. This link between ordered inscription and ordered thought played an important role in the theories of knowledge that underpinned eighteenth-century notekeeping.

FIGURE 1.7. James Kay, "A Letter to Sir L[awrenc]e Dundas," in James Kay, *A Series of Original Portraits and Caricature Etchings, with Biographical Sketches and Illustrative Anecdotes, Vol. II, Part II* (Edinburgh: Paton, Carver and Gilder, 1838), 28. Reproduced with permission of the Huntington Library, San Marino, CA. This satiric letter is addressed to Sir Lawrence Dundas, an unpopular Scottish member of parliament during the late eighteenth century. Kay uses pictograms as stenographic symbols to communicate that the writer, who signs off as the Devil, is suffering from a form of madness.

activity. In this sense every page of a notebook is a snapshot of knowledge in motion. Its very fabric presents us with evidence of the capabilities of self-organized knowledge that operated in conjunction with the facts, opinions, biases, and principles that were simultaneously utilized and problematized in a rich variety of educational, domestic, professional, and commercial contexts. Identifying such components and skills of notekeeping offers a fresh way of understanding the material basis of the cognitive metaphors employed by the literate world from the seventeenth to the nineteenth century.

By the time many students finished their education, they were artificers who knew how to keep notes in a way that operated in realtime within a mnemotechnic device they had personally created. This being the case, a student's role as a foliater, inscriber, and codexer offers opportunities to investigate what it meant to be a materially engaged reasoner, a thinker who knew how to craft reason with media technologies that improved mental operations such as perception, attention, and recollection in a way that made sense within the contemporary models used to understand the mind as an interactive entity. When considered from this angle, the fabric of student notebooks provides a way to reflect on how reason was understood to be an artifact made by a reasoner through reasoning.

Scholars writing about the social and cultural foundations of reason in Scotland at this time tend to frame their thoughts in reference to the "Scottish Enlightenment," a term traditionally used to denote the central role played by the life of the mind and its eighteenth-century relationship with progressive principles.[63] It is now widely acknowledged that the arc of the Scottish Enlightenment as a historical event is significantly larger than hitherto realized and that scholars must strive to engage more responsibly with gender, race, class, and other important intersectional factors that have been overlooked.[64] I accept this call for diversity and inclusivity, and this book works toward this goal by turning its attention to students; that is, boys, girls, and young adults who made and used art, literature, science, accountancy, and mathematics notebooks as they studied in homes, schools, shops, and universities. By concentrating on the experiences and artifacts of such actors, the book extends intersectional themes raised by scholars such as Christopher J. Berry, Silvia Sebastiani, Rosalind Carr, Jane Rendall, and Katherine Glover, who emphasize the importance of reconstructing the forms of social reproduction that facilitated the principles of utility, sociability, equality, and liberty that underpinned the cultural conditions that historians now collectively associate with

---

63. Wood (1993), Broadie (2001), Sher (2008; 2015), Emerson (2008; 2016), Sebastiani (2013), Carr (2014), and Ahnert (2015).
64. Sebastiani (2013) and Carr (2014) have made the strongest case for the importance of race and gender. The latter also plays an important role in Allan (2008), Glover (2011), Rothschild (2011), and Towsey (2010). These works extended earlier critiques of the traditional canon of the Scottish Enlightenment raised in Withers (2001), Withers and Wood (2002), and Wood (1993; 2000). The need to broaden the role played by different religious perspectives is raised in Jackson-Williams (2020). For a helpful overview of the historiography of the Scottish Enlightenment, see Dunyach and Thomson (2015, 1–19).

the Scottish Enlightenment as a process that evolved out of the late seventeenth century and lasted until the early decades of the nineteenth century.[65]

Framing student notebooks as sources that young learners themselves crafted makes it possible to search for their involvement in a capability-building process that enabled them to understand how the exercise of reason often depended on the organization of knowledge on and across a large amount of paper. My interest in such *self-made* artifacts crafted by hitherto overlooked agents seeks to understand the life of the mind at the ground level, where it was conducted by hands that oftentimes were not elite or destined for greatness. In this sense I am offering a long overdue chapter in the larger story of Scotland's popular enlightenment,[66] one that helps to more fully explain why the science of mind was based on cognitive metaphors drawn from the everyday practices of reasoning on paper.[67]

Here it is perhaps worth noting that the time-consuming self-organizational activities of student notekeepers coming from Scottish professional and merchant families stands in contrast to the model of the relatively unconstrained approach to childhood learning advocated by the philosopher Jean-Jacques Rousseau and other fashionable pedagogues who used his ideas during the latter half of the eighteenth century to further their careers as private tutors or as authors of educational treatises. This is not to say that the Rousseauvian model was unknown by the students and teachers under discussion in this book, or that teachers were somehow uninterested in nurturing the capabilities of students. Nor do I wish to suggest that notekeeping lacked the forms of expressive creativity advocated by Rousseau.[68] But when it came to the kinds of day-to-day forms of labor through which students learned to make and use notebooks in Scotland, and in many other countries for that matter, Rousseau's influence outside elite contexts was minimal.

Becoming an organized notekeeper and, by extension, a reasoner required an immense amount of labor that is implicitly marginalized by the *laissez-faire* model of human development articulated by Rousseau and his followers. As

---

65. Berry (1997; 2013), Rendall (1978; 2005; 2012), Sebastiani (2013), Carr (2014), and Glover (2011). See also the essays in Wallace and Rendall (2021).

66. I am using the term "popular enlightenment," also called *Volksaufklärung* by German historians, to refer to the ways in which liberal principles were used to reform and improve educational institutions in Europe and its former colonies during the long eighteenth century. Lerner (2000), Bartlett (2006) and Kashatus (1994). The term has been applied to Scotland, but primarily to the period falling between the 1780s and 1820s. Smith (1983). As several chapters in this book suggest, it is arguably the case that the origins of the popular enlightenment reach back to the late seventeenth century.

67. Some of these practices have been flagged in passing by historians of philosophy and literature. See for example the treatment of ink, paper, and writing throughout Frasca-Spada (2002).

68. Schaeffer (2014) gives a helpful revaluation of the kinds of conditioning that Jean-Jacques Rousseau used to understand the activities required to shape the faculty of judgment.

I will show, when Scottish schoolmasters or parents explicitly commented on matters relevant to developmental psychology, their model of learning was more strongly based on the forms of active guidance promoted by John Locke, and on the emphasis that classical authors such as Cicero and Quintilian placed upon writing and thinking as interlinked forms of developmental labor. In this model of learning, the skills of bookkeeping were just as important as those employed to organize a moral philosophy notebook.[69]

The various parts of this book that address the interface between cognition and manuscript media emphasize the advantages of treating notebooks as paper machines, as artifacts that offer insight into how thought was understood to be a realtime activity during the early years of the modern world. What, for example, would reason look like for someone who could not organize facts or observations on paper through writing? What would it look like for someone who could not arrange a table or draw a figure? Or what would it look like for someone who was unable to preserve inscriptions against the perils of time in a codex? These questions reveal that the acquisition and exercise of reason would be something very different for someone who did not possess the capabilities afforded by notekeeping.

One of the things that I underscore throughout this book is that, when it comes to identifying the everyday activities through which reason historically was acquired and valued, we risk losing a major part of the story if we fail to engage with how students learned to make and use notebooks. In the end, we will see that the act of notekeeping was a mode of recrafting reason that created capabilities for students as they progressed through educational settings and as they made their way in the world once they completed their education.

Science as a System

A final, but no less important, dimension of notebooks that is addressed in this book is the role that they played in the ways in which students learned to understand the relationship between science and media technologies. I am interested in this facet because notekeepers and instructors living at the time firmly believed that reason, science, and notebooks were closely connected.

In order to see this connection, we must note that prior to the 1830s the term *science* had a much broader meaning and the word *scientist* did not ex-

---

69. The emphasis that humanism placed upon the relationship between learning and various forms of useful inscription predated John Locke and was not unique to Scotland. For discussion of this matter, see Gianoutsos (2019).

ist.[70] The meaning of *science* more closely resembled the Latin term *scientia*, which referred to systems of knowledge that were based on authoritative facts and created according to a set of principles and related categories. The 1771 edition of *Encyclopaedia Britannica* defined science as "any doctrine, deduced from self-evident and certain principles, by a regular demonstration."[71] The reference to deduction here is important, because it reveals that collecting facts and observations into one place was not enough to create a science. The arrangement had to be ordered in a logical, coherent manner.

From the seventeenth century to the early nineteenth century, the word *science* was used in educational settings to refer to a specific subject, geometry for instance, that had been ordered into a system. As explained by the *Encyclopaedia Britannica*, a system was "an assemblage or chain of principles and conclusions, or the whole of any doctrine, the several parts whereof are bound together and follow or depend on each other; in which sense we say, a system of philosophy, a system of divinity, &c."[72] This definition, which mirrors that given in other dictionaries and encyclopedias at the time, explains why the Scottish students and educators examined in this book used the word *science* to refer to a body of knowledge, such as philosophy and divinity, that had been ordered into a system.[73]

Any subject could be turned into a science as long as it was arranged systematically. Subjects as different as chemistry and rhetoric, for instance, were regularly called sciences in educational, commercial, and domestic settings. The *Encyclopaedia Britannica*, for instance, referred to subjects ranging from mythology to mathematics as sciences. Throughout the century the term was increasingly applied in a more specific sense to emerging subjects as well. The philosopher Dugald Stewart, whose publications were read across Europe and the British Empire, used the word *science* not only to refer to subjects like mathematics, astronomy, anatomy, and botany, but also to newer, more specific, subjects such as the philosophy of the human mind.[74]

Notekeeping mediated the limitations of memory and facilitated the possibilities of representation possessed by every notekeeper. To efficiently iterate or invent a science, notekeepers, young and old, needed to order their note-

---

70. Yeo (2003).
71. Smellie (1771, 3: 570). The *Encyclopaedia Britannica's* definition of "science" was taken verbatim from the eponymous entry in Croker (1765, vol. 2). Notably, prior to listing the definition of science, Croker gave the word *scientia* as a synonym.
72. Smellie (1771, 3: 882).
73. Smellie (1771, 3: 882). The *Encyclopaedia Britannica's* definition of "system" is he same as the eponymous entry in Croker, (1765, vol. 2). For the use of "system" in Scottish educational settings, see Eddy (2008, 1–15) and Barfoot (1993).
74. Stewart (1792, 2, 17).

books according to clear and coherent categories so that they could remember the system that they had created; this was especially the case when a science was robust enough to fill an entire notebook. The end result of this kind of realtime interface was a vast number of systems committed to paper. Here student notebooks were again very important. Most were arranged according to sections labeled with headings that represented the main categories of a given subject. Such an organization meant that each student notebook presented a small scientific system when it was finished. Since making a notebook allowed students to learn scientifically relevant organization skills, several chapters of this book investigate the many ways in which they used notekeeping to condition the operations of their minds in a manner that was conducive to crafting a system.

The emphasis that educators placed on systems was linked to a changing perception of pedagogy that was emerging in modern Britain. Within Scotland, prior to the eighteenth century, one of the primary purposes of educational institutions, particularly universities, was to train clergy. But this began to change after Scotland joined England, Ireland, and Wales in 1707 to form the United Kingdom of Great Britain.[75] Scotland's economy slowly improved, creating the need for better resourced schools, academies, and homes. From the midcentury forward, specialty day schools and academies were founded in many cities. Likewise, grammar schools and "ladies" schools extended the curriculum to include a larger number of subjects. The increasing demand for practical knowledge led educators to offer courses that were packed with useful facts and organized into compact systems.

Notebooks enabled students to learn how to recognize and create accessible systems that functioned as sciences. Even at the university level, professors believed that a course of lectures should be presented via a system that broke down information into easily accessible categories.[76] When speaking about the complexity of the syllabi offered in Scotland's universities, for example, the Edinburgh medical professor Andrew Duncan noted that even "the most extraordinary memory cannot retain every thing."[77] A similar situation existed for the subjects that students studied in homes, schools, and academies. As we will see throughout this book, student notekeeping partially solved the mnemonic problem identified by Duncan because it was a mode of capturing well-ordered systems presented to students by educators. It pro-

---

75. The economic, social, and intellectual impacts of the union on Scots is explained in Colley (2005).
76. The relationship between systematics and teaching in Cullen's thought is explained in Barfoot (1993) and Eddy (2008, 53–68).
77. Duncan (1776, 12).

vided a kinesthetic mode of learning that connected the mind to the body as it interfaced with media technologies required to make and use a notebook as a scientifically organized artifact.

Book Outline

There are many methods that one might use to organize a historical study of the role played by notebooks as paper machines. The contents of the student notebooks that I consider in the following chapters were arranged under headings of important topics or according to rules relevant to the subject being learned. The headings functioned as categories that helped students understand how the process of notekeeping could be organized around freestanding terms or phrases that made both the content and layout of the script more intelligible. This being the case, one way I could have organized this book would have been to structure the chapters around the subject matter stated in the headings offered in the notebooks.

Another option would have been to treat notebooks as commentaries on the ideas being advocated by a specific educator or institution. Scholars using this approach to study the history of student notekeeping have organized their thoughts according to questions that speak to what a notebook has to say about the ideas of the teachers, professors, or authors who inspired it. Alongside this tradition, other scholars have employed descriptive approaches seeking to identify what kinds of disciplinary facts might have been learned by students. Alternatively, the processes of notekeeping associated with components such as indices and commonplaces have functioned as focal points for scholars interested in how student notebooks operated as information management tools.

Over the decade that I wrote this book, I used a number of the foregoing structuring mechanisms to organize the chapters of the evolving manuscript. In the end, I realized that my interest in the material and visual basis of the kinesthetic and perceptual activities used by students to transform notebooks into paper machines would be best served by organizing the chapters around the overarching skills of artifice that were learned via the process of notekeeping.

The book is divided into three parts, each of which contains three chapters. Each chapter addresses one superskill, that is, a multifaceted skill based on a collection of other more specific skills. Part I explores the skills of writing (chapter 2), codexing (chapter 3), and annotating (chapter 4) that students learned early in their educational journeys. Part II examines the skills of categorizing (chapter 5), drawing (chapter 6), and mapping (chapter 7) that were learned by slightly older students who attended burgh schools and academies, or who were taught at home by tutors or family members. Part III turns to the skills of systemizing (chapter 8), diagramming (chapter 9), and circulating (chapter 10) learned by university students. The final chapter offers a reflection on the historiographic advantages of treating notebooks as artifacts, notekeepers as artificers, and notekeeping as artificing. It revisits these themes with a view to highlighting how the notekeeping skills associated with inscription, foliation, and codection offer a helpful starting point for those interested in studying manuscripts as realtime artifacts.

In each chapter I use the components of notebooks to articulate and interpret notekeeping skills in reference to how and why they were being used by students operating in Scotland's educational communities, and in similar contexts elsewhere in Britain and Europe more generally. My aim is not to give an exhaustive explication of each skill, mainly because such a project would require inventories and secondary studies that have yet to be written. Instead, based on the hundreds of notebooks that I have seen over the past two decades, I am offering a guide to the kinds of skills repeatedly evinced in the manuscripts. I concentrate on those kept by young learners who were tutored at home or who attended Scotland's hospitals, parish schools, burgh schools, academies, and universities. In several places, I introduce insights based on similar scribal skills presented in domestic, mercantile, and industrial contexts that are relevant to how students and their families valued advanced forms of notekeeping.

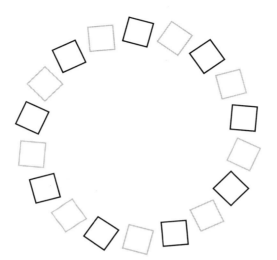

PART I

INSIDE THE *TABULA RASA*

In part I of this book, we will go inside the *tabula rasa* metaphor to explore the ways in which students gained an understanding of paper-based forms of knowledge creation and management. I focus on notebooks and ephemera that were made and used from the late seventeenth century to the early nineteenth century. The chapters concentrate on writing, codexing, and annotating, that is, three core skills that helped aspiring students craft the basic components that transformed their notebooks into sophisticated paper machines.

Starting in homes and then moving into schools and academies, I trace notekeeping skills in reference to how students learned to use them over time. My aim is to show how students used instruments such as pens, pencils, rulers, and tracers to craft materials such as paper, ink, and graphite into a notebook that operated as a handheld media technology. As the chapters develop, I reveal that working with the form and fabric of the manuscript was a multisensory exercise that implicitly conferred a visual and kinesthetic awareness of how the material culture of manuscripts could be employed to build a basic knowledge system on and around paper.

One of the points that I emphasize is that, in addition to using textbooks to establish what students were supposed to be learning, it is equally important to examine notebooks with a view to reconstructing a more accurate picture of what students were actually doing while they learned. Indeed, making notebooks helped students to think more efficiently and, more broadly, to understand the potential of the media technologies that existed around them at the time. The fact that the multisensory skills afforded by notekeeping were being experienced at such an early age reveals that the cognitive development of students was directly impacted by the forms of interface provided by the notebook in its capacity as a learning technology. I suggest that, when viewed from a social and cultural perspective, notekeeping skills underpinned essential realtime modes of information management that made it possible to represent and judge knowledge on paper during the modern period more generally. Part I's chapters offer a clearer indication of how student notebooks played a core role in the conceptualization and transmission of such skills.

Throughout the chapters I pause in several places to consider how student notekeeping skills offer a richer picture of the relationship between thinking and writing traditionally ascribed to the *tabula rasa* metaphor. When seen through the dynamic skills that students used to manipulate blank sheets of paper into a fully functioning codex, the meaning of the metaphor broadens. Instead of simply comparing words on the page to ideas in the mind, the metaphor is *activated* in a way that compares acts of inscribing to modes of reasoning. This aspect of the metaphor, I suggest, is easier to recognize when the components of notebooks are treated as enactive, purposeful, and user-friendly artifacts. By employing notebooks as evidence for what students were learning in realtime, I reveal the sociomaterial significance of paper for young learners who internalized various skills that operated in conjunction with the acquisition of knowledge and the exercise of judgment.

CHAPTER 2

WRITING

Writing as a Knowledge-Creating Tool

Sometime in 1738 the twelve-year-old schoolgirl Margaret Monro (1727–1802) received a letter from her father. Though she spent most of her early years at home, she was most likely attending a boarding school located outside Edinburgh in the Lowlands of Scotland. Written in a congenial and familiar manner, her father's letter outlined a plan that, if executed diligently, would improve her reason, advance her virtue, and garner the highest esteem from her acquaintances and relations. The plan was to send her a series of letters that explained how to make rational decisions in a variety of domestic, social, and commercial contexts.

In addition to reading her father's letters, Monro copied them out. As her father explained in his initial letter, the act of writing served jointly to strengthen her mind and to preserve useful knowledge for future reference. More specifically, the "art of writing" was "an assistant to memory." To learn it, she needed to write for at least "one hour per day."[1] The result would be a single codex that they could edit together at home. Monro's father viewed the codex as a collaborative project, one in which they were both participating in creating a shared artifact of learning. Not only did the codex preserve a collection of observations and facts, it also helped Monro to learn the many ways in which different kinds of writing could serve as a learning tool, as a kinesthetic mode of interface that helped her both remember and organize knowledge.[2]

Monro's father was Alexander Monro Primus (1697–1767), a professor in Edinburgh University's medical school and one of the most celebrated anat-

---

1. Monro Primus (1996/1738, 7, 9, 10).
2. Eddy (2019).

omists of the eighteenth century. Her mother was Isabella Monro (née Macdonald, 1694–1774), the accomplished daughter of a Highland clan chief. Both parents were highly literate. Primus had written or edited hundreds of manuscripts during his career and Isabella oversaw the family's financial ledgers, which required her to sort and order the data provided by the many scraps, slips, and receipts supplied to her by family members and household staff. Together, their abilities to extend their own minds on paper were substantial. It was for this reason that they recognized the value of sending Margaret and her three brothers to schools in the Edinburgh area that specialized in teaching students how to use script in a way that allowed them to order a large amount of information into a codex.[3]

True to her father's wishes, Monro assiduously enacted his graphic suggestions, creating a beautifully written 386-page codex, a notebook of sorts, that she finished as a twelve-year-old in 1739. Titled *Essay on Female Conduct*, it was later bound in red leather with a characteristically Scottish lozenge and vine edging tooled in gold. Today it is housed in the National Library of Scotland (figure 2.1).[4] The codex is a noteworthy historical document because it shows how a girl learned writing skills that were perceived as being rational in a manner that taught her how to filter and order a sprawling topic like female conduct on and through multiple genres of script. Not only does it afford insight into how a girl of her social status learned to use pen and paper, it also serves as a salient example of intergenerational knowledge in transit, or, as I like to put it, knowledge in motion.[5]

But how did Monro and other students learn to create the handcrafted scribal components, the script, of codices? What were the material and visual skills that underpinned such artifacts? How might we conceive the ways in which students learned to value script as a knowledge-creating tool that transformed them into writers capable of organizing knowledge on a page? In this chapter I suggest that many helpful answers to these questions lie in the relatively unexamined cache of Scottish school notebooks housed in the National Library of Scotland and in various other manuscript collections located across the globe. These notebooks tell a story of how students learned to conceive *writing* as an empowering skill, as a collection of related abilities such as compositing, copying, tabling, numbering, listing, and composing that often overlapped and had to be performed in realtime.

---

3. The biographies of Isabella and Alexander Monro Primus are given in Wright-St. Clair (1964).
4. Monro Primus and Monro Bound MS (1739). This manuscript was Margaret Monro's transcribed, corrected, edited, and expanded version of Monro Primus Bound MS (1738).
5. Secord (2004).

FIGURE 2.1. Top: Margaret Monro's handwritten nameplate. Bottom: The red leather and gold-tooled cover of Monro's manuscript. Alexander Monro Primus and Margaret Monro (transcriber and editor), *An Essay on Female Conduct* (1739), Bound MS, NLS MS 6659. Reproduced with permission of the National Library of Scotland.

In order to see the central role played by school notebooks as enactive artifacts, we will need to expand our understanding of writing as an object of enquiry. Ever since the nineteenth century, historians have traditionally treated writing as a fixed object on paper. Even today the word *writing* is used as a gerund, as a noun that refers to a published piece of work or the printed corpus of an author. But what about the performative aspect—the *writtenness*—of writing offered in the material traces of a manuscript? What were the skills that guided the modes through which the pen was put to paper in the past? How can they be used to understand the ways in which writers were taught to conceptualize writing as they transformed the *tabula rasa* into a *tabula verba*, a written page in a notebook?[6]

In many respects, the written components of student notebooks functioned as *paper tools*, that is to say, representations on paper that can be used as thinking devices.[7] In the early years of the modern period, numbers and lines were widely viewed as perceptual learning tools that could be used to shape the impressionable minds of students. One of the sources of this viewpoint was the developmental psychology offered in John Locke's *Some Thoughts Concerning Education* (1693), *Essay Concerning Human Understanding* (1690), and *On the Conduct of Understanding* (1706). In these works, Locke repeatedly compared words on the page to ideas in the mind. Likewise, he compared the bodily processes of inscription such as transcribing and composing to the mental processes, the operations, of ratiocination such as recollection, association, and judgment.[8]

Locke's views were influential in Scotland, where educators drew a close relationship between the movement of the hand and the operations of the mind in a way that implicitly created an enactive metaphor, that is to say, a metaphor that compared processes, particularly those that came into existence through various kinds of actions. These comparisons, moreover, introduced an embodied facet into the *tabula rasa* metaphor that tacitly assumed the importance of recognizing the centrality of the material culture of inscription.[9]

School notebooks provide helpful examples of how students learned the kinds of skills that enabled them to use the words, numbers, and lines of script as paper tools when they became adults. It was in this sense that the act of writing carried meanings relevant to the subjects that students were learning, *and* it served as a vehicle that students learned to manipulate as a paper tool through which different kinds of knowledge could be visually and kinesthetically generated. It is this interactive aspect of writing as a handwritten, real-

---

6. Brindle (2019) gives a useful overview and discussion of the historiographic parameters of handwriting as a form of material trace. The importance of noting the links between the materiality of the page and the script formed thereupon is addressed throughout Kreider (2010). Walter Ong also noted the dangers of conflating "language" with "writing," going so far as to state: "But to say that language *is* writing is, at best, uninformed. It provides egregious evidence of the unreflective chirographic and/or typographic squint that haunts us all." See Ong (1985, 27).

7. For the use of paper tools as material thinking devices see Klein (2003; 2001; 2019) and Damerow and Lefèvre (1996).

8. The influence of Lockean developmental psychology in Britain is examined in Richardson (1994), O'Malley (2003), and Winter (2011). For Locke's influence on Scottish educational institutions, see Eddy (2016b).

9. Gallagher and Lindgren (2015).

time facilitator of knowledge creation that informs my approach to the notebooks in this chapter.

Writing was an activity that occupied much of the daily life of literate Scottish children. It functioned as a mode of recording information and as a mode of learning in its own right. The cultural historian Michel de Certeau once underscored how authors living in the predigital era conceived this jointly developmental and performative facet of writing when he observed: "In front of his blank page, every child is already put in the position . . . of having to manage a space that is his own and distinct from all others and in which he can exercise his own will."[10] In what follows I reveal how children learned to use the act of writing to interface with the blank page in a way that offers insight into how they were able to create the scribal components of a notebook and why they valued such efforts as being a moral and hence worthwhile activity.

The Place of Writing within Literacy

There were many kinds of Scottish schools where students learned the kinds of writing skills that would eventually enable them to create a notebook. Throughout the century the pedagogy and curriculum of schools were guided by a number of factors, with the Church of Scotland playing a central role.[11] Basic literacy and numeracy were often offered at home. The foundations of organized writing were laid when children first learned to recognize the letters and numbers printed in hornbooks, ABCs, and the Shorter Catechism of the Church of Scotland.[12] The common act of learning letters with a paperback ABC at home during the 1760s was recounted in the autobiographical memoir written by the poet and dramatist Joanna Baillie. The daughter of a minister, her earliest memory was sitting on the stairs of Bothwell Manse near Hamilton, Scotland. Her sister held "a paper on which was marked the large letters of the ABC" and Baillie loudly repeated the alphabet.[13]

As intimated by Baillie's account and by Scottish instruction manuals, chil-

---

10. Certeau (2011, 134).
11. For key works that discuss the either the pedagogy or curriculum of Scottish schools, see Home (Kames) (1781), Bannerman (1773), Chapman (1773), Williams (1774), Turnbull (1742), Todd (1748), Barclay (1743), and Fordyce (1745). For English and other European authors whose thoughts influenced Scottish teachers, see Gerdil (1765), Anonymous (1772), Whitchurch (1772), Williams (1774), and Priestley (1768).
12. The Westminster Shorter Catechism was set in 1648 by the General Assembly of the Church of Scotland. It was published in various forms throughout the eighteenth century to promote piety and literacy. See General Assembly of the Church of Scotland (1736; 1757; 1737). All of these texts went through multiple editions. For more on Scottish catechisms, ABCs, and introductory penmanship primers, see Fox (2020, 26–35).
13. Slagle (2002, 47).

dren learned to say and recognize the letters first and then learned to write them out later. Baillie, for example, began learning the alphabet with her sister at the age of three but read in a "very imperfect manner" until she was sent to a day school in Hamilton when she was eight or nine. But encountering the pages of an ABC at such a young age imparted skills that would help children write in the future. Looking at the layout and letters of an ABC exposed them to a simple graphic format and shaped their ability to perceive the spatialization of words on the page. Put more simply, printed or written ABC charts allowed family members or tutors to train children to recognize and individuate different kinds of letters and blank spacings that were formatted on horizontal lines running across a rectilinear grid.

The next step on the road to scribal literacy was formal writing instruction from a schoolteacher or, for affluent families, a private tutor.[14] For orphans or poor children, there were workhouses, charity schools, hospitals, and Sabbath schools.[15] These institutions were coeducational and the level of writing instruction seems to have been modest. Some charity schools were able to offer higher-quality instruction because they had endowments and could award bursaries and hire trained staff. Such was the case for Hutchesons' Hospital in Glasgow as well as for the George Heriot Hospital, George Watson Hospital, and James Gillespie Hospital in Edinburgh (figure 2.2). For girls, there were also finishing schools and institutions such as Edinburgh's Trades Maiden Hospital and the Merchant Maiden Hospital.[16]

Boys and girls of modest means could attend village or parish schools if they lived in the countryside and burgh schools if they lived in a city.[17] The Church of Scotland usually played a role in running or supporting these institutions. For larger cities like Edinburgh, Glasgow, and Aberdeen, there were burgh schools run by city councils where the level of instruction was higher. Students learned spelling, mathematics, and basic penmanship by writing round, mixt, or secretary's hand on slate boards or loose-leaf paper (figure 2.3).[18]

14. A helpful account of how elite girls learned to write is given in Glover (2011, 27–29).
15. For more on Scottish household education, charity schools, and hospitals, see: Plant (1952, chap. 1), Mason (1954), Vance (2001, esp. 321–323), Cosh (2003, 41–47, 223–229), Graham (1899, 149–181), and Lochhead (1948, 224–239).
16. I was unable to locate extant notebooks from these institutions. For the history of the hospitals, see Hill (1881), George Watson's Hospital (1724; 1755), and Merchant Maiden Hospital (1776; 1783). *Merchant Maiden Hospital's Administrational Papers* (n.d.). Engravings and summaries of Edinburgh's hospitals are included in Stark (1821, 280–291).
17. A helpful reconstruction of the eighteenth-century curriculum taught in Fyfe's parish and burgh schools is given in Beale (1983, 122–177).
18. For the mathematical topics that children learned at this time in Scotland, see Cocker (1678), Paterson (1685), and Wilson (1935).

FIGURE 2.2. Scottish schools. Scottish children learned to write and keep notes in homes and schools. There were many kinds of schools, including parish schools, burgh schools, hospitals, and grammar schools. Top left: Merchant Maiden Hospital, Edinburgh, J. Stark, *Picture of Edinburgh: Containing a Description of the City and Its Environs* (Edinburgh: Fairbairn and Anderson, 1821), plate 10. Reproduced with permission of the Huntington Library, San Marino, CA. Top right: Perth Academy, William Morison (ed.), *Memorabilia of the City of Perth* (Perth: Morison, 1806), featured on the map affixed as a page occurring before the title page of the book. Reproduced with permission of the Huntington Library, San Marino, CA. Bottom left: Edinburgh High School. William Steven, *The History of the High School of Edinburgh* (Edinburgh: Maclachlan & Stewart, 1849), 123. Internet Archive. Bottom right: Hutchesons' Hospital, Glasgow, William H. Hill, *History of the Hospital and School in Glasgow Founded by George and Thomas Hutcheson of Lambhill, A.D. 1639–41* (Glasgow: Royal Incorporation of Hutchesons' Hospital, 1881), 83–84. Internet Archive Creative Commons License, original held by the National Library of Scotland.

Girls were often taught the same subjects as boys,[19] a situation that was increasingly illustrated from midcentury forward in primer frontispieces such as the ones in the Glasgow edition of Colin Buchanan's *The Writing-Master and Accountant's Assistant* and in the Scottish editions of George Fisher's *The Instructor* (figure 2.4).[20] Aside from a few copybooks housed in the National

19. For Sabbath schools, see Brown (1981). The coeducational history of Scottish burgh schools is addressed in Grant (1876, 535–537), Moore (2013, esp. 100), Vance (2001, 318–319, 323–324), Grant (1876, 527–531), and Bain (1965, 131).
20. Buchanan (1798, frontispiece).

FIGURE 2.3. Orthographic tables of cursive hands used in Scotland during the early and later parts of the eighteenth century. Top: Alexander Broadie, *The Penn's Practice, A New Copy Book* (Edinburgh: 1696). Reproduced with permission of the Huntington Library, San Marino, CA, No. 376772. Middle and bottom: "The Italian Hand" and "Secretary Hand," respectively, in George Fisher, *The Instructor: Or Young Man's Best Companion* (London: 1773). Reproduced with permission of the Wellcome Library, London. The notebooks written by Scottish students were written in Italian hand and secretary hand, or various combinations of the two. The neatest tend to be written in the Italian hand.

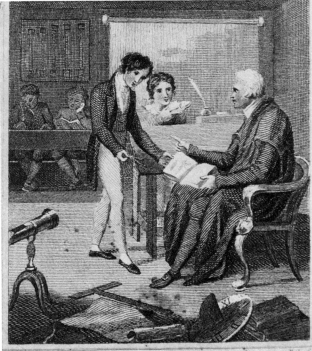

FIGURE 2.4. Coeducational vignettes of student writers in a classroom. Top: Colin Buchanan, *The Writing-Master and Accountant's Assistant* (Glasgow: Chapman, 1798). Image courtesy of Hamish Riley-Smith Rare Books and Manuscripts. Note the female teacher and students (wearing white blouses). Bottom: George Fisher, *The Instructor: or young man's best companion* (London: Johnson, 1806), frontispiece. Reproduced with permission of the Wellcome Library, London.

Library of Scotland, evidence for student notekeeping in these settings is minimal.

For students of means or for those who were given bursaries, there were several options when they finished their burgh school studies. For boys seeking a classical education that served as a gateway to university studies, there were grammar schools, many of which had been established for centuries. Grammar schools taught Latin, Greek, and classical history.[21] The premier grammar school in the country was Edinburgh High School.[22] It taught male students. There were also newer schools founded by enterprising masters who taught Latin, foreign languages, and other humanities subjects to boys and girls. James Mundell's avant-garde school in Edinburgh, for instance, taught these subjects to over 560 boys and just under 100 girls between 1736 and 1761.[23]

Learning Latin translation and English composition required a significant amount of writing that aided students when they made a school notebook. The bulk of the extant notebooks created by Edinburgh High School students, for example, are made up of Latin translations, called *lines* or *versions*, which required a significant amount of attention and time to make. Many grammar school masters were firmly committed to using this kind of writing as a mode of learning. George Chapman, the master of Dumfries Grammar School, for instance, required his students to spend "a portion of their time every day in writing."[24] His older students were required "twice or thrice every week" to "write a Latin version" of a translation of an English passage.[25]

Grammar school students and learners based in households were asked to write out translations taken from Latin authors such as Sallust, Livy, Terence, Cicero, and Pliny.[26] Chapman recommended the following method: "to prescribe them, every other day, passages from the said classics, to be translated into English, so that the work could be presented at the same meeting of school." Students were asked to correct inaccuracies of spelling, style, or idiom that occurred in what they had written. Like other masters, Chapman was influenced by the pedagogical practices of Thomas Ruddiman, the humanist author of Latin textbooks, who encouraged teachers to use the act of

---

21. Popular textbooks included Ruddiman (1779; 1790), Barclay (1758), and Adam (1772). Numerous editions of these primers were printed.

22. Steven (1849).

23. Anonymous (1789). No notebooks from his school seem to have survived.

24. Chapman (1773, 54).

25. Chapman (1784, 212–214).

26. In other schools, students translated Virgil's *Pastorals*, Ovid's *Metamorphoses*, Gerardus Joannes Vossius's *Compendium of Rhetoric* and the letters and orations of Cicero. For the Latin primers used in Scottish grammar schools, see Law (1965, chap. 3) and Bain (1965, 128). The curriculum of Glasgow Grammar School is addressed in McCallum (2014, 59–64).

writing as a learning tool to help students acquire ancient languages more efficiently.[27]

The atmosphere of Scotland's grammar schools was generally oriented toward producing gentleman scholars. But there was a second option for boys seeking to learn technical and scientific topics. For these students there were academies. From midcentury forward, Scottish academies grew in number, spreading as far north as Inverness by the end of the century.[28] The notebooks created by academy students display sophisticated orthography and draftsmanship techniques. Students made them to learn various inscription skills and to create a manuscript textbook that they could use as a reference work after their studies. Creating this kind of notebook was an important part of the curriculum in many European contexts and across the Atlantic world.[29]

The technoscientific subjects taught by academies resonated with the goals and ambitions of the rising mercantile, industrial, and professional classes. By the middle of the century, the classical curriculum of grammar schools was increasingly shunned by upwardly mobile parents who had concluded that academies provided the best value for money. The experimentalist Thomas Garnett appealed to this sentiment during the 1790s in the lectures that he gave at Glasgow's newly founded Anderson's Institution. He observed: "The inhabitants of North Britain have much juster and more liberal ideas of education than my English countrymen; and I cannot but express a wish, that many of the large schools in England, which are so nobly endowed, but in which the dead languages only are taught, were modeled according to the plan of the Scotch academies and universities."[30]

The standard of mathematical instruction in a number of academies was high, rivalling that of English military academies.[31] Student notebooks as well as the manuscript teaching notes kept by the Perth Academy rector Robert Hamilton, for instance, indicate that academy students learned topics such as accounting, geography, and practical mathematics and geometry.[32] Some-

---

27. Oral methods were also used to learn Latin. Law (1988, 69).

28. Inverness Academy was founded in 1790. Garnett (1811, 2–4). Other academies existed in Perth, Dumfries, Fortrose, Dundee, Banff, and elsewhere. Morison (1806) and Moore (2015). For a helpful comparative study of (primarily English) academies, see Hans (1966).

29. For eighteenth-century manuscript textbooks copied by students, see Knoles, Kennedy, and Knoles (2003), which includes several helpful photographs of copied textbooks. For a survey of manuscript textbooks made in North America prior to the 1850s, see Mulhern (1938).

30. Garnett (1811, 5)

31. An indication of the subjects that appear in the notebooks kept by students attending English academies is given in Forbes Bound MS (1781) and Lochee (1776).

32. Key accounting primers at this time were Macghie (1718), Lundin (1718), Malcolm (1718), and Colinson (1683). For practical mathematics and geometry there were Panton (1771), Ewing (1771; 1773), Wilson (1773), and Langley and Langley (1768). For geographical instruction, see Withers (2000, 72–74).

times academies taught natural philosophy, with the latter including chemistry and physics (especially mechanics).[33] The Perth Academy enjoyed a particularly strong reputation,[34] with several of its masters and tutors going on to professorships in Scottish universities.[35] While some academy subjects such as draftsmanship and surveying were taught primarily to boys, there is evidence that girls attended academies as well. Robert Nichol's Commercial Academy in Glasgow, for instance, offered classes on orthography, arithmetic, and geography to young ladies, that is to say, teenage girls who wished to progress from primary school subjects to more advanced forms of writing and calculating.[36]

The various skills of scribal literacy imparted through keeping notebooks were a key to the success of academies. This was the case in Scotland as well as in England, where students attending military and dissenting academies kept detailed notebooks as well.[37] Learning to create an academy notebook that integrated figures and narrative was a skill required for students who wished to become surveyors, architects, levelers, or gaugers, and for students wishing to enter the Royal Navy, where creating graphically coherent notebooks was required for cadets (midshipmen) wishing to be promoted.[38] Overall, the success of Scotland's academies led many Scottish grammar schools, including Edinburgh High School, to introduce more subjects oriented toward mathematics and the technosciences from midcentury onward.

A third educational opportunity existed for girls coming from professional, mercantile, or landed families of means who wished for their daughters to improve their writing skills through keeping advanced accounts, composing different kinds of letters, or learning several kinds of handwriting. Modern scholars have called these institutions finishing schools, but they were usually called a "school for young ladies" at the time. The schools existed in many Scottish towns from the late seventeenth century forward, and they played an important role in educating Scottish women. Though the role of notekeeping

---

33. Sutherland (1938, 86–92).
34. The Perth Academy's curriculum and its commitment to fine writing are summarized in Morrison (1806, 345–349) and Garnett (1811, 100–103).
35. Robert Hamilton (1743–1829), professor of natural philosophy and mathematics at the University of Aberdeen (Marischal College), served as Perth Academy's rector before his university appointment. John West (1756–1817), who taught as an assistant to Professor Nicolas Vilant (1737–1807), the chair of mathematics at St. Andrews University, unsuccessfully tried to secure the mathematics mastership at Perth Academy early in his career. See Craik and Roberts (2009, esp. 226) and Hart (1989).
36. "Education," *The Caledonian Mercury*, 26 August 1786.
37. The role played by notebooks in preserving the ideas of dissenting tutors is treated throughout Whitehouse (2015).
38. Cavell (2010, 102–105). For an example of the writing and drawing skills evinced in midshipmen's notebooks, see Pape Bound MS (1799). For more on the role of manuscript and print culture in the navigational arts at this time, see Schotte (2019).

in these schools has yet to be studied in detail, visitors to Edinburgh such as the English journalist and playwright Edward Topham held the Scottish education of "every boarding school Miss" in high regard, and the writing skills demonstrated by the letters sent home by students such as Amelie Murray during the 1740s and Mary Moore during the 1780s indicate that penmanship and various models of formal composition were being taught.[39]

From midcentury onward, Edinburgh's schools for young ladies attracted students from across the Highlands and Lowlands. Based on advertisements placed in newspapers such as *The Edinburgh Advertiser* and *The Caledonian Mercury*, boarding schools for young ladies were popular and taught writing skills alongside subjects such as French, English, geography, arithmetic, drawing, and bookkeeping.[40] Mrs. Kennedy's school in Musselburgh, instance, taught writing alongside "every branch of education," while Mrs. Denoon's school in Edinburgh incorporated writing within a portfolio of "useful and ornamental" skills that nurtured "morals, improvement, and happiness."[41] Margaret Monro, whom we met in the introduction of the chapter, attended one of these Edinburgh-area schools. As we will learn later in this book, the layout and script of her codex speak to the high quality of the writing skills of girls who learned to work with manuscript culture in such settings.

A fourth and equally popular educational option for boys and girls was to receive advanced instruction from a private tutor, who could be hired or who could come in the form of a family member. Private tutors specialized in a variety of areas such as accounting, drawing, orthography, foreign languages, or cookery. There were also teachers, tutors, governesses, instructresses, professors, and masters who taught groups of students at a school they ran out of their house or in rented rooms. Susanna Maciver, for instance, ran this kind of cookery school in Edinburgh for many years.[42]

The 1795 Edinburgh directory lists scores of writing masters, some of whom frequently circulated between schools, academies, and homes. George Paton, for instance, is listed as a "writing master and accomptant to the high school and rector to the commercial academy."[43] The curricula of private tutors and teachers remain relatively unknown and there are only a handful

---

39. Topham (1780, 110). For the correspondence of Mary Moore and Amelie Murray, see Moore Unbound MS (1786–1790); Murray Bound MS (1744–1745). The context of Amelie Murray's schooling in Edinburgh is addressed in Glover (2011, 33–36).
40. Houston (1993, esp. 386–388).
41. "A Boarding School for Young Ladies," *The Caledonian Mercury*, 12 June 1784, and "Boarding School & Education for Young Ladies," *The Caledonian Mercury*, 23 March 1799.
42. Maciver (1787). Mrs. Frazer took over the school when Maciver passed away. Frazer (1791).
43. Aitchison (1795, 145).

of extant student notebooks. The most intact specimens are two sets kept by students who attended the Edinburgh school run by the blind tutor and poet Rev. Dr. Thomas Blacklock, who operated in Edinburgh from the 1760s to the 1780s. Titled *Kalokagathia*, their content and format suggest that Blacklock dictated them as students wrote them out into a manuscript textbook of sorts.[44]

In addition to keeping notes in educational institutions, some students learned to make notebooks from their parents. The late seventeenth-century Scottish jurist Sir John Hope, for instance, took his sons as legal apprentices and, to help them with their careers, he dictated a treatise on Scottish law to them "for their Instruction, in Mornings while he was dressing."[45] He emphasized the importance of placing signatures and seals, using the front and back of official papers, leaving blank spaces on the page for future use, recopying and expanding scrolled notes, and using of headings to structure a page, to name but a few of the scribal skills he covered.[46] Several decades later, Alexander Monro Primus, whom we already encountered at the beginning of this chapter, asked his children to write out manuscripts that he had made for them with a view to improving their writing skills. He also encouraged them to a keep a communal *calendarium* notebook into which they copied Latin compositions.[47]

In addition to notebooks related to school subjects, students living in literate households learned to write a variety of manuscript codices such as diaries, commonplace books, recipe books, poetry books, and accounting ledgers.[48] As evinced in the elegantly written commonplace notebook of the seventeen-year-old John Greig, domestic notebooks sometimes reflect the interests of their keepers. A merchant's son who would go on to a successful publishing career, Greig crafted his notebook to include hand-drawn tables in pen and ink of the tides, the phases of the moon, and "The Cycle of the Sun," suggesting agricultural or maritime interests.[49] The early nineteenth-century commonplace books of Kath Sym (1788–1838) and the future antiquarian Elizabeth Caroline Johnstone Gray (1801–1887), both written when they were

---

44. Blacklock (anonymous notetaker) Bound MS (n.d.), EUL Dc.3.45; Blacklock (anonymous notetaker) Bound MS (n.d.), EUL La.III.80.
45. Hope (1726, vii).
46. Hope (1726) addresses signatures (218), seals (235), front-back (89, 237), blank spaces (183), recopying (172), and headings (174).
47. Monro Primus Bound MS (1740–1741). The role played by Primus in educating his children is summarized in Wright-St. Clair (1964, 50–58).
48. Eddy (2016a).
49. Greig Bound MS (1762–1764).

teenagers, feature literary extracts (including poems) that addressed behavioral and emotional topics relevant to moral conduct.[50]

Gray's commonplace book reminds us that Scottish students used note-keeping as a learning tool in a variety of contexts. They kept different kinds of notebooks at home and school with a view to learning writing skills alongside the official subjects of the curriculum. The importance of writing "lines" and keeping notebooks was also highly valued by parents and schoolmasters. In addition to the written and oral components of their examinations, students submitted specimens of their handwritten work. Not only were these specimens reviewed by inspectors, some schools placed them on public display and mentioned them in newspaper advertisements as evidence of superior instruction. An excellent example of this practice occurs in the 30 August 1786 edition of Edinburgh's *Caledonian Mercury*. In addition to praising the outgoing master's ability to teach English, Latin, Greek, and French, the managers of the school in Alloa noted the "Many beautiful specimens of Writing" that the students had created under his supervision. As we will see below, many of these "beautiful specimens" were in fact notebooks or sheets that were later bound to make a singular codex.[51]

Script and Observational Learning

The writing skills that underpinned the construction of school notebooks were conceptualized as being part of a larger regime of observational learning that enabled students to actively engage with the world around them. Learning to observe one's own writing as a notekeeper energized the kinds of skills that served as the starting point for understanding how script on the page could be mindfully structured to expedite different kinds of information. This means that school notekeeping influenced the way that learners perceived the modes through which the written page could operate as a verbal picture, a *tabula verba*, in motion.

The legibility of the script in a notebook depended on the level of orthographic instruction the student received. At the time, orthography was,

---

50. Sym Bound MS (1805–1841); Johnstone Bound MS (1816–1826). Johnstone even created a shorthand code in her notebook for key words. Some of the symbols included: ")" (opportunity), "G" (great), "A" (accordingly), "t" (triumph) and "tx" (trouble). She is now known by her married name: Elizabeth Caroline Hamilton Gray. See Williams (2009).
51. The writing specimens of Mr Bell's students are mentioned in Anonymous, "Alloa, August 24. 1786," *Caledonian Mercury*, 30 August 1786. When writing masters applied for jobs they had to submit specimens to school managers. "A Writing-Master and Precenter Wanted," *Caledonian Mercury*, 9 November 1782.

in the words of Hugh Christie, the rector of Montrose Grammar School, "That part of grammar which treats letters, and their several affections or properties."[52] Notably, it was not a subject that was included in the core curriculum of most schools. It was learned by paying an additional fee to a professional writing master, a situation that existed throughout the Atlantic world.[53] Parents or guardians also had to purchase engraved exemplars.[54] The lessons took place at school or at home. Perhaps the most well-known Scottish writing master of the late eighteenth century was Edmund Butterworth, who famously taught the young Sir Walter Scott and other Edinburgh High School students.[55] In addition to publishing popular orthographic primers and classroom posters, he taught many young "ladies and gentlemen" from the city in his rooms at the high school and at his own house in Brown Square.[56]

But here we need to make a distinction between theory and practice that will help us better understand the kinds of script that students used to craft the pages of their notebooks. In many respects, orthography was the study of what handwriting should look like. It was theoretical in the sense that it promoted aesthetically pleasing letters formed according to geometric principles. But as the pages of student notebooks so readily indicate, this ideal was seldom achieved. The reality was that the handwriting of students reflected their training and the visual decisions that they made in terms of layout, spacing, letter size and neatness. The *art of writing* required to shape these elements of script was sometimes called *chirography*. In many respects chirography was for script what typography was for print.[57]

Students of writing masters or household tutors spent hour after hour observing orthographic exemplars and then writing their own specimens. Once they were confident with their abilities, they were ready to create their own type specimens that were then bound into a small copybook that served as evidence of their newly learned skills.[58] Depending on their abilities and re-

---

52. Christie (1758, 1).
53. Copybooks of the Atlantic world are explored in Monaghan (2005, 286–292), and Douglas (2001).
54. An early printed copybook published in Scotland is Broadie (1696). A rare copy is housed at the Huntington Library, San Marino, CA, no. 376772. The classic text on printed copybooks, mainly for masters living in England, is Heal (1931). For a wider European perspective, see Becker (1997).
55. McKinstry and Fletcher (2002, 64).
56. Butterworth (1778; 1785). A good example of a folio-sized orthographic poster appears as the second sheet of Butterworth (1778). Butterworth promoted his desire to give extramural instruction to young "ladies and gentlemen" in the advertisement section of the *Caledonian Mercury* on 29 November 1780.
57. The term "chirography" or "chirographer" appears in many primers published by writing masters. Gough (1760), for instance, defined it simply as "the art of writing."
58. Hardly any handwritten Scottish copybooks have been preserved. For two good examples, see Greig Bound MS (1763) and Richardson Bound MS (1778).

sources, students learned to enact different kinds of orthographic hands at different speeds.

Learning to write required students to source and use basic instruments, some of which included "black lead" graphite pencils for ruling sentence lines and margins, gum for rubbing out graphite, razors for scraping off words written in ink, and powder for drying (pouncing) the ink.[59] After acquiring the skills of ruling, rubbing, scraping, and pouncing, some students learned to design the space of the page through the use of a lead pen. Also called a tracer or leaden plummet, this was a pen with a hard, smooth metallic tip. It was a tool that had been used since ancient times to impress a line into the fabric of parchment, paper, and other malleable writing surfaces.[60] This tradition was continued in many school notebooks, where students used lead pens to impress a grid of ruled lines into their pages.

A helpful account of how writing masters taught students orthography at this time was published in 1763 by William Massey in *The Origin and Progress of Letters*. It reveals that masters sometimes communicated the goal of "fair" writing through poems. Massey reproduced early English poems of this nature, including one written around 1570 by a master only known as "E.B." It ran as follows:

> HOW TO WRITE FAIR.
> To write very fair, your pen let be new,
> Dish, dash, long-time flie; writing eschew;
> Neatly and clearly your hand for to frame,
> Strong stalked pen use, best of raven;
> And comely to write, and give a good grace,
> Leave between each word small letter's space,
> That fair, and seemly, your hand may be read,
> Keep even your letters, at foot and at head;
> With distance alike, between letter and letter,
> One out of others shews much the better.[61]

The writing implements used by students prior to the early twentieth century required more attention than today's disposable pens and pencils, in terms of

---

59. The shape and composition of a black lead pencil at this time is described in *Saturday Magazine*, 24 March 1838, 109–111.
60. Tracers are described and depicted in Barrow (1792). Fisher (1763), 29, uses the term "leaden plummet." For the ancient use of lead pens, see the "plumatus" entry in Rich (1860, 512).
61. "E.B.," c. 1570, reprinted in Massey (1763, 21–22).

both making them and using them. To master different hands, young writers learned to observe their own use of an assortment of pens and quills. There were, for instance, heavier, expensive fountain, Italian, and round hand pens that had metallic shafts and nibs, the materials of which were durable, but could not be easily adjusted with a penknife. There were also lighter, affordable goose, crow, and swan quills. Their feather shafts were less durable, but could be shaped and adjusted with a penknife to suit the needs of a writer. Understanding the form and function of the nib was important because its shape and parts affected the thinness and thickness that could be achieved with a hair stroke or a bold stroke, respectively. Put another way, as intimated by the names of the pens mentioned above, certain kinds of nibs were used for certain kinds of hands. Once students chose a nib and shaft, they then had to learn how to hold the tip at different angles to further refine the strokes that they wanted to learn.[62]

Quills sold in shops did not have nibs and had to be shaped from scratch. Students began the process by "dutching" the nib to make it more durable.[63] Since the tips of homemade and store-bought quills came unsharpened, students had to learn to use penknives. If the nib was misshapen, the ink could disfigure the page through blotting or it could distort the lines of the letters, making them thick and hard to read. Cutting the vents, shoulders, tines, slit, and tip of the quill's nib was not easy and had to take the curvature and thickness of the shaft into consideration (figure 2.5). Some writing manuals gave a robust account of quill cutting, but the finer details were effectively trade secrets that writing masters divulged only through face-to-face instruction.[64] Some masters, Lowlands-based Colin Buchanan for instance, offered diagrams on how to select different sized feathers. For "strong paper," he recommended using goose and swan feathers to craft a "large text pen" and a "square hand pen." For smaller "very nice writing" he recommended using crow and swallow feathers to make a "fine pen."[65]

Learning how to make and use ink also was also necessary. Though was

---

62. Instructions on how to hold a quill or pen are given in Morton (1720), Lawrie (1779, 43), and Butterworth (1785). See also Bickham (1750, 3).
63. A helpful summary of quill making, including the dutching process, is given in the "pens" entry in Sandford, Thomson, and Cunningham (1836, 732–735). See also *Saturday Magazine*, 13 January 1838, 14–16.
64. For further examples of quill-cutting instructions, see Lawrie (1779, 46) and Bickham (1750, 3). Two of the most thorough discussions of quill cutting are given in Wilkes (1799) and Lewis (1825). Students also had to consider the advantages of different penknives. See Savigny (1786).
65. Buchanan (1797, 2 and plate I).

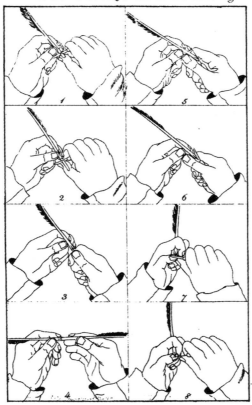

FIGURE 2.5. How to make a quill nib. James Henry Lewis, *The Royal Lewisian System of Penmanship: Or, New Method of Rapid Writing . . . by which Any Person, Though But Little Acquainted with the Subject, Can Detect and Easily Remove the Various Impediments which Retard His Progress in the Art of Writing* (London: Printed for the author, 1816). Unnumbered plate appearing between pages 74 and 75. Google Books, original from New York Public Library, public domain.

it possible to buy ink, students often had to make it on their own,[66] a practice that was versified by E.B. as well:

TO MAKE INK.
To make common ink, of wine take a quart,
Two ounces of gumme, let that be part;

---

66. Instructions for making and preserving red and black ink are given in T.H. (1796, 58–60). Instructions for making ink and pens were also included in books meant for older children. See Fisher (1763, 31, 37, 45).

Five ounces of galls, of cop'res take thee,

Long standing doth make it better to be;

If wine ye do want, raine water is best,

And then as much stuffe as above at the least,

If ink be too thick, put vinegar in,

For water doth make the colour more dimme.[67]

Students who used different kinds of ink learned that some kinds faded with time and that others were absorbed more readily by paper. Add to this the fact that different pens and quills required thicker or thinner ink. Various household substances and liquids could be used to dilute or thicken ink in a way that suited a notekeeper's needs.[68]

The relationship between self-observation and the material culture of writing in educational settings also involved elements of posture. Students attending schools usually did not learn to write on the kind of individual desks that have been used in Western classrooms since the mid-nineteenth century. Instead, they often stood or sat (on benches) side by side, facing a raised, slanted writing board that was either mounted on a wall or on a table-like structure (figure 2.6).[69] The board's slanted position sometimes prevented them from setting inkpots on it. To overcome this problem, some students tied the pot onto a rope necklace so that it hung in front of them. This innovation made the pot ready to hand and ensured that they did not spill or lose their ink.[70] The physicality of writing on tables or desks was noted by many authors, even those who today are known primarily for their philosophical or literary accomplishments. John Locke, for instance, observed that a pen needed to be held "between the Thumb and the Fore-finger alone" and that it was advisable to employ a writing master if possible. He also felt it was important to teach students how to "lay" the paper and how to place the "Arm and Body to it."[71]

Alongside all the foregoing media and instruments, there were pedagog-

---

67. Massey (1763, 22).

68. A popular method for making ink was developed by Joseph Black, Edinburgh's professor of chemistry. See Black (1966/1767, 143–144). The ink recipe of "Dr Black" is also given in Wilkes (1799, 45–46). A helpful account of the kinds of household trials that could be conducted with ink appears in Lewis (1763, 377–401).

69. An eighteenth-century Scottish school table desk is depicted in James Kay, "Modern Moderation Strikingly Displayed; or, A Ministerial Visitation of a Sabbath Evening School," in Kay (1837–1838), opposite page 356.

70. A slanted school desk affixed to a wall appears in Sir George Harvey's painting "The Schule Skailin" housed by the National Galleries of Scotland in Edinburgh. The image depicts a schoolmaster and several students in a one-room schoolhouse with slanted desk boards and benches fixed to two of the walls.

71. Locke (1752, 234).

FIGURE 2.6. Slanted writing desks. James Kay, "Modern Moderation Strikingly Displayed; or, A Ministerial Visitation of a Sabbath Evening School," in James Maidment (ed.), *A Series of Original Portraits and Caricature Etchings, with Biographical Sketches and Illustrative Anecdotes, Vol. I* (Edinburgh: Paton, 1837–1838), opposite page 356. Reproduced with permission of the Wellcome Library, London.

ical theories that engendered different approaches to writing posture and to holding the pen.[72] The instrumental skills necessitated by writing rendered it an activity in which students became mindful choreographers of the haptics that influenced their own scribal performance on paper. It would have been very difficult, for instance, to form letters, plot headings, draw sentence lines,

---

[72]. For an overview of the physiological theories surrounding the use of body braces, see Sheldrake (1783).

demarcate margins, align columns, or cross-align rows without observing how various kinds of impressing affected the movement of the hand across the page. Nor would it have been possible to sharpen quills, pounce pages, scratch out mistakes, or even make (or select) ink without being an observer of one's own writing. It was for this reason that instructors encouraged students to treat the act of writing as a form of self-observation.

Tutors, schoolteachers, and householders played an important role in showing students how to be observers of their own writing. The 1779 teaching handbook used by Edinburgh's Merchant Maiden Hospital for girls, for example, gave the following advice to its writing instructors: "Here you copy three or more letters, but be at great pains to cause them [to] keep the distance equal, and to *observe* the shape of the letters." Likewise, when demonstrating how to make a pen from a quill, the school's teachers were to ensure that the girls knew how to "*observe* the method of one who makes a pen well, and endeavor to imitate it."[73] Penmanship textbooks made similar connections between the act of writing and the act of observing.[74] In sum, inscription was often a process of observation.[75]

From midcentury forward, many educators in Scotland and other parts of Britain interpreted the performativity of writing via a Lockean developmental psychology that drew strong links between visual and mental order.[76] Dugald Stewart, whose works were widely read during the late Scottish Enlightenment, articulated this longstanding view in the 1790s moral philosophy lectures that he gave at Edinburgh University. Like many educators of the time, he conceptualized the page of a notebook as a picture that both visualized and influenced how ideas were organized in the mind.

As Stewart's comments on commonplace notekeeping indicate, ordering a notebook through writing was hardly an inconsequential act. In his words, "A COMMON-PLACE [note]book, conducted without any method is an exact picture of the memory of a man whose inquiries are not directed by philosophy. And the advantages of order in treasuring up our ideas in the mind, are perfectly analogous to its effects when they are recorded in writing."[77] Since everyone's mind was different, notekeepers had to judge for themselves which

---

73. Both quotations are taken from Lawrie (1779, 44–46). Emphasis added.

74. The first page of Fisher's *Instructor*, a popular teaching compendium in Scotland, makes a direct connection between the act of observation and the modes through which a child read the letters of the alphabet. Fisher (1763, 1). Observation and literacy are linked on pages 5 and 61 as well.

75. Here I am treating observation as a learned process. My thoughts on the matter were influenced by Bleichmar (2012), which addresses the modes through which making and using visual culture reflexively influenced how observers interfaced with the natural world.

76. Richardson (1994, 127–142), O'Malley (2003, 86–101), Winter (2011, chap. 1).

77. Stewart (1792, 423).

form of commonplace ordering worked best based on "their own peculiar habits of association and arrangement."[78]

Stewart's lectures were given to teenage boys who had learned to write and rewrite notebooks in Scotland's schools and academies. There they had extended their writing skills, the "methods" in Stewart's terminology, that enabled them to create a useful school notebook. Perhaps the most important skills were those that enabled them to structure the layout of the page into a format that helped them to observe what they were writing as they wrote, an act that transformed the page into a visual object, a verbal picture, that was easy to use.

In addition to functioning as a mode of mental organization, it was widely believed that an ordered writing posture helped to more firmly instill ordered thoughts as well. Erect posture was a key element taught to some Scottish students, the idea being that orthography was a performance that required bodily alignment if script was to be executed well. As recalled by Mary (née Fairfax) Somerville in her reflections on her early education during the 1790s at Miss Primrose's boarding school in Musselburgh, some writing instructors went so far as to force students to wear a brace that held their back and head in an upright position. A straight rod forced her chin up, while the ribs of the brace pulled her back and torso down. It was in this constrained condition that she, along with the other young girls of the school, was asked to complete writing and reading lessons (figure 2.7).[79]

Somerville's uncomfortable experience with a writing brace illustrates how upright posture was sometimes seen as being an embodied representation of an ordered mind. The desire to achieve this kind of erectness, therefore, was reflected in visualizations of the acts of writing and drawing. Depictions of hand posture in orthographic manuals and portraits such as Paul Sandby's undated drawing *Portrait of a Lady at a Drawing Table* (figure 2.8) underscored the importance of manual comportment.[80] Sandby's drawing depicts a teenage girl, or possibly a young woman, of means, sitting erect in front of a notebook positioned on a luxurious drawing table. Her face, and hence her mind, is calm. Her writing arm is poised in a manner similar to Apollo's right arm in the famous belvedere pose used by contemporary artists in sculptures and paintings. The symbolism is clear: like the arm of Apollo that fired arrows,

---

78. Stewart (1792, 441). The order of the written page captured the attention of many European educators, including Immanuel Kant, the philosopher and professor at the University of Königsberg. See his 1768 essay "On the first ground of the distinction of regions in space." Kant (1991, esp. 29).

79. Somerville (1873, 21–22). A visualization of a posture brace appears in Feutrie (1772, plate 2).

80. The relationship between education, behavior, and orthography is addressed throughout Thornton (1998).

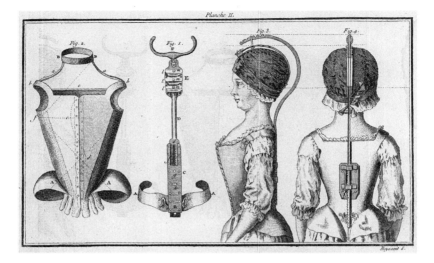

FIGURE 2.7. Posture machine invented by Thomas Levacher de la Feutrie and depicted in his *Traité du Rakitis, ou L'art de Redresser les Enfants Contrefaits* (Paris: Lacombe, 1772), plate 2. Reproduced with permission of the Wellcome Library, London. A ventral version of this kind of machine was used to teach the teenage Mary Somerville to write in an erect position during the 1780s. Instead of holding the head with a cradle, the machine had a chin rest that pressed the head up.

the girl's arm is poised and ready to direct the ink in an ordered manner onto the pages of her notebook.[81]

Grids and Verbal Pictures

Students were often surrounded by educators and literate family members who valued the ability to shape the material face of the page into a visual artifact, a *tabula verba*. In order to create such a verbal picture, they had to organize the space of the page through a skill called *ranging*. At the time the verb *to range* meant to order things by putting them in a line or dividing them into categories. When used in reference to the act of writing and sewing, *ranging* referred to an act of linearizing or patterning that took place when the hand worked with the eye to visually align ink or thread across the flat, square surface of a sheet of paper or cloth.[82]

81. Sandby, *Portrait of a Lady at a Drawing* (n.d.). See also Sandby, *A Lady Copying at a Drawing Table* (c. 1765). For the design and fashionable status of drawing tables, see Sheraton (1802, 49–54).
82. *OED*, *to range*, eighteenth-century usages: to order in a line (II.9.a); to lay or set down in a line (II.9.e); to categorize (II.11).

FIGURE 2.8. Paul Sandby, *Portrait of a Lady at a Drawing Table* (n.d.), graphite and brown wash on medium, slightly textured, cream laid paper, Yale Center for British Art, Paul Mellon Collection, B1975.4.717. Public domain.

Put more clearly, ranging captured the process of writing words, numbers, or symbols in a straight line, particularly within the rows or columns used to structure the layout of a blank page. A good example of this meaning occurs in William Perry's textbook *The Man of Business*. A successful academy master in Edinburgh, Perry described the columns of a handwritten ledger as places "wherein the several articles belonging to every different person or subject, which are dispersed in the journal, are collected together and ranged in their natural order."[83] Here ranging was a transitive act, one that illustrates why the *tabula rasa* functioned as an enactive metaphor within Lockean theories of mind.[84]

Students sometimes made their first attempts to range information by writing inside the premade columns or rows that had been drawn by a tutor or printed in works like the *Ladies Pocket Books*.[85] Used by both boys and girls, the pocket books contained weekly calendars ranged as a blank table of boxes that represented the days of the week. Each day was given a blank box that invited young readers to write "Appointments, Memorandums, and Observations."[86] Aside from developing a student's sense of observation and organization, the skills required to insert information into this kind of box helped young writers to understand and value the kinds of ranging skills that could be used to structure the space of the page into columns that were divided into boxes that served as informatic containers.

Since eighteenth-century paper was not ruled, students wishing to clearly and efficiently order script had to first learn how to draw a graphite grid of even, parallel lines running horizontally and vertically across the page. The grid served as the visual framework for ranging words, numbers, and figures. It had to be made in its entirety before the script was written or drawn. In addition to using the medium of graphite to craft a grid, students could, in the words of Todd's *The School-Boy and Young Gentleman's Assistant*, "trace with a lead Pen the Space for regulating the Head and Feet of Letters."[87]

Writing instructors often drew a sample grid that children learned to replicate. Edinburgh's Merchant Maiden Hospital School teaching handbook used by the writing mistresses and masters made the following suggestion: "After you have ruled their [a student's] book, copy it on the left side with a pen and ink, drawing straight scores all the way down as on the margin: Cause them [to] keep these strokes at an equal distance from each other, point the first line to them."[88] As the verbs "draw" and "stroke" indicate, crafting the written page was conceived to be a visual activity of interface with the page.[89]

---

83. Perry (1777, 210). See also Gordon (1763, 166). Ainsworth (1746) uses the phrase "Ranged in their natural order" to describe the words listed in the columns of a dictionary.
84. The psychological importance of ranging a visually accessible layout is still underscored today. See Wright (1999).
85. The use of ladies pocket books as information management devices is addressed throughout Vickery (1998).
86. *The Ladies Complete Pocket-Book* (1780).
87. Todd (1748, 67).
88. Lawrie (1779, 43–44).
89. For the pictorial aspects of the typography in British primers, particularly those that addressed geographical topics, see Mayhew (2007). A helpful overview of the visual elements of textbooks is given in Hartley and Harris (2001).

As indicated by Todd's reference to a lead pen, some students used impressions to create a grid on the paper. Margaret Monro used this subtle technique. She employed a lead pen to vertically divide each page into two columns into which she could write her script. The lead pen was a graphic design tool included in the cases of writing instruments sold in bookshops. Since the linen-based paper of the day was softer than today's, the dull, pointed metal tip of a lead pen was used to impress faint lines that helped to demarcate columns or to make geometric schemata over which diagrams were drawn. Some students pushed so hard that their impressions went through the paper to the surface on which they were drawing.[90]

Since the skills required to make grids were not subject specific, they were transferable and could be employed across the pages of all the notebooks kept by a student. In other words, grids formed the visual basis of many of the handmade graphic patterns that ran through most school notebooks. These patterns, moreover, formed a fundamentally rectilinear mode of representation that implicitly framed students' efforts to observe the patterns of their own script in realtime as they wrote. This means that they were learning to observe by hand via graphic conventions that connected to the priority that Lockean developmental psychology implicitly assigned to linear modes and forms of representation.[91]

The most common structure that students used to order script on the grids they designed was what we might call the standard module. At the most basic visual level, the module consisted of a heading, usually centered, followed by a block of narrative formatted as a paragraph (figure 2.9). In this layout, the heading and the paragraph functioned as two visual units.[92] The standard module could be used to structure one page, or it could also be designed to run across two facing pages.[93] It had emerged as an important information management structure in European medieval codices after scribes began inserting blank spaces between words.[94] The fact that it was still being used in eighteenth-century school notebooks indicates that it was appropriate and useful to the everyday function it needed to fulfil, both in terms of making and then using it. Its usage over such a long period of time shows that it was a

90. A good example of a secondary impression occurs on the endleaf of Purdie Bound MS (1823).
91. As shown by Omar Nasim (2013), the use of hand-drawn patterns was a core skill required to represent scientific ideas in the eighteenth and nineteenth centuries. The epistemological advantages and challenges of repeated or stylized patterns at this time are traced out in Benjamin (1968) and Crary (1992).
92. For more on the headings and paragraphs as units of information, see, respectively, Eddy (2010a) and Slauter (2012).
93. James Fowler of Strathpeffer, for instance, turned both the verso and recto pages into one unified module. Fowler (1780, 69v–70r).
94. Saenger (1997).

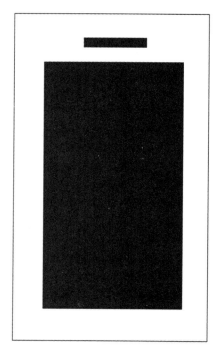 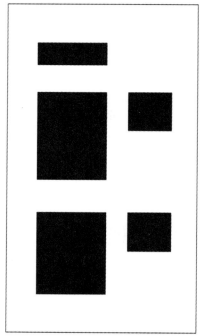

FIGURE 2.9. Stylized modules of the layouts students used to structure words on school notebook pages. Left: The standard notebook module. © Matthew Daniel Eddy. It was formed of one column and contained a heading (sometimes centered), a block of text, and a white frame created by margins. The modules could be short or long, on one page or running over several pages. Right: Expanded notebook module. © Matthew Daniel Eddy. Older school students sometimes added a second column for subheadings that described the content of the adjacent text. The visual affordances of notebook modules transformed them into simple gestalt images.

stable media technology and that it was firmly a part of the mnemonic culture surrounding the design of manuscripts.[95]

Designing a word module involved a number of ranging skills and constituted a mode of interface with what would eventually become a well-structured page. Working with the graphite grids that they themselves had created, students had to learn to judge the spacing required to plot headings in different locations around the grid. They also had to decide the kinds of indentations to use at the beginning of paragraphs and which kind of handwriting worked best for the keywords they wanted to highlight. There were additional techniques required if they chose to include figures or tables.

[95]. Norman (2013) holds that there is a direct, and oftentimes overlooked, relationship between "the psychopathology of everyday things" (see chapter 1) and "the psychology of everyday actions" (see chapter 2).

Students could also use the grid to divide their pages into columns. The importance of this usage was emphasized in the letters received by Margaret Monro from her father. He asked her to use loose-leaf sheets and to range each page so that it had "large Margins," an act that effectively divided the page into large and small columns.[96] As he explained in subsequent letters, this layout, which was created through "the Act of ranging," provided a way for her to materially enact visual order on the page.[97] Monro implemented the layout on every page (figure 2.10). The addition of a second column significantly expanded the page's potential as a sophisticated information management platform. It enabled her to use the larger left column for the main headings and the bulk of the script, and the smaller right column for topical headings.

The double column format occurs in other student manuscripts as well. Take for instance the handwritten codex made by Amelie Keir (1780–1857) during the 1790s when she was a teenager. Like Monro, Keir's father, the Scottish inventor and scientist James Keir, had learned to keep notebooks as a student. He too wanted his daughter to learn advanced scribal skills through transcribing a manuscript codex that he had created. The codex was titled *Dialogues on Chemistry between a Father and His Daughter*. The original version, which might have been a collection of rough notes, has not survived. But Keir's transcribed copy reveals that she used two columns to organize the script. Each page offered a large column reserved for the main text and a marginal column that allowed her to more efficiently insert corrections into the script at a later date.[98]

The use of columns to divide the space of a page was a common technique in bookkeeping, a skill that students such as Margaret Monro had studied prior to making her codex.[99] Bookkeeping was a subject that was learned by both boys and girls in Scotland from the late seventeenth century forward.[100] As noted by Monro's father, "the art of ranging" columns was an essential skill for those who wanted to organize "[ledger] books in the most distinct method and order."[101] In addition to the instruction she received from her ledger-writing master, Monro would have witnessed pages ranged into columns at home in the household accounts kept by her mother. Isabella Monro was

---

96. Monro Primus (1996/1738, 7–8).
97. Monro Primus (1996/1738, 10, 68).
98. Keir Bound MS (1801). Earlier versions of the manuscript are lost. This extant version is written in Amelie Keir's hand, bears her annotations and features her diagrams. Moilliet (1964). For more on the cultural milieu surrounding the Keir family, see Schrantz (2014). Notably, girls and women learned chemistry during the eighteenth century. See Fulhame (1794) and Eddy (2022).
99. Monro Primus (1996/1738, 9, 10, 68, 106–107, 112, 121) addresses bookkeeping.
100. Eddy (2016a). For the role of women in family bookkeeping in England, see Vickery (1998).
101. Monro Primus (1996/1738, 18).

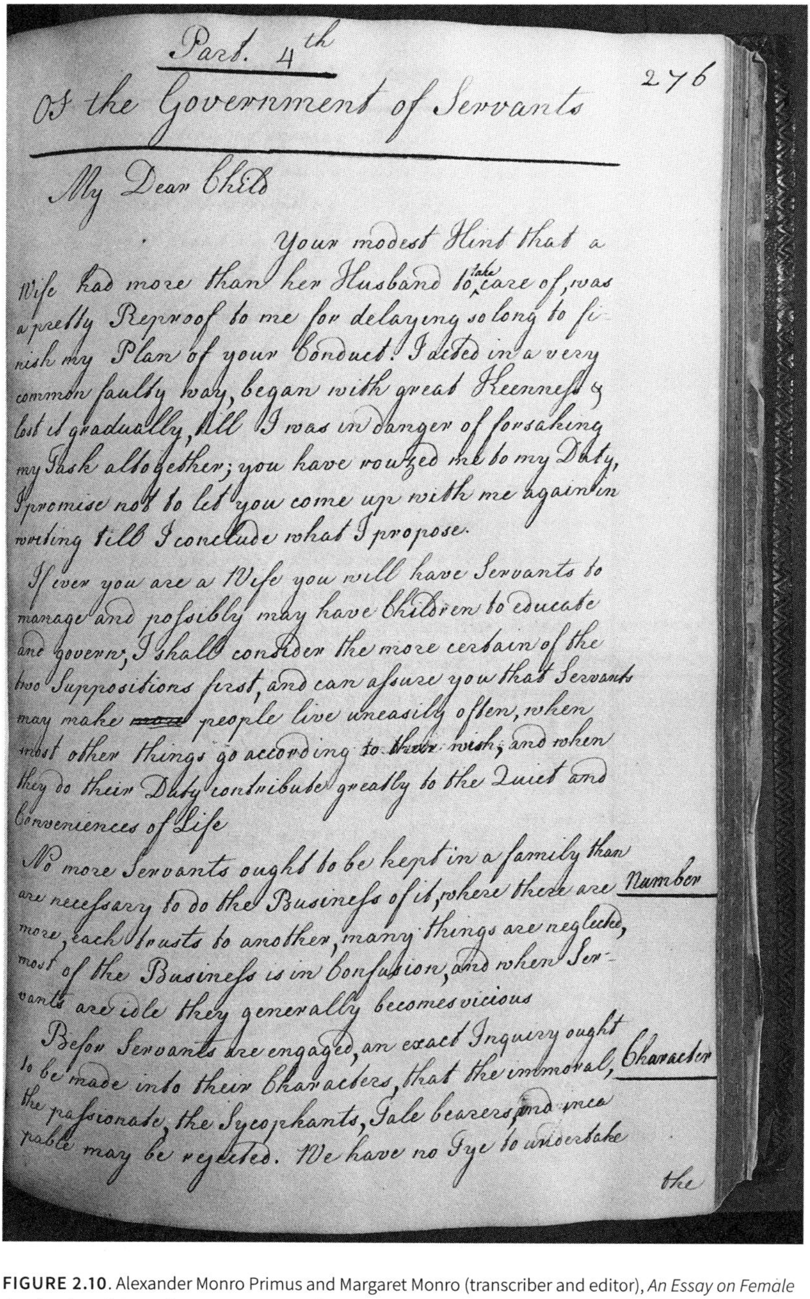
Part. 4th

Of the Government of Servants 276

My Dear Child

Your modest Hint that a Wife had more than her Husband to take care of, was a pretty Reproof to me for delaying so long to finish my Plan of your Conduct. I acted in a very common faulty way, began with great Keenness & lost it gradually, till I was in danger of forsaking my Task altogether; you have rouzed me to my Duty, I promise not to let you come up with me again in writing till I conclude what I propose.

If ever you are a Wife you will have Servants to manage and possibly may have Children to educate and govern; I shall consider the more certain of the two Suppositions first, and can assure you that Servants may make people live uneasily often, when most other things go according to ~~their~~ wish; and when they do their Duty contribute greatly to the Quiet and Conveniences of Life

No more Servants ought to be kept in a family than are necessary to do the Business of it, where there are more, each trusts to another, many things are neglected, most of the Business is in Confusion, and when Servants are idle they generally becomes vicious Number

Befor Servants are engaged, an exact Inquiry ought to be made into their Characters, that the immoral, the passionate, the Sycophants, Tale bearers and incapable may be rejected. We have no Tye to undertake Character

the

FIGURE 2.10. Alexander Monro Primus and Margaret Monro (transcriber and editor), *An Essay on Female Conduct* (1739), Bound MS, NLS MS 6659, f. 276. Reproduced with permission of the National Library of Scotland. The original right column was larger when it was ranged. Its size was reduced when the manuscript was rebound.

highly literate and oversaw the collection of the household receipts into the accounts kept in the "Family Book," which was a variety of what was sometimes called a "household book" at the time.[102] Like other numerate mothers of professional households, Isabella most likely showed her daughter how to write the entries.[103] Thus, even before her father gave her advice on how to structure a page, Monro had been familiarized with ranging.

The skills of rendering script inside the columns and modules created the capability to order and sort knowledge in an efficient way that could be applied to numerous mercantile, academic, and personal contexts experienced by students when they became older. This in part explains why Dugald Stewart's comments about notekeeping entailed the assumption that the writing skills used to structure blank notebook pages were just as important as the factual content that they preserved. Once created, modules became graphic containers, visual sorting tools, inside of which students efficiently processed and stored information.

In addition to designing modules, students learned to create combinatorial tables. As shown in the work of the anthropologist Jack Goody, tables of words, numbers, and other symbols played a central role in the development of Western culture, especially those that were structured as "a matrix of vertical columns and horizontal rows."[104] Ever since medieval times, the matrix had served both as an information management technology and as a mode of interface between users and texts.[105] This means that the student notekeepers who learned to create word and number tables in Scottish classrooms were mastering a stable *mise-en-page* technology that had been successfully user-tested over the past one thousand years. Yet though the matrix was a long-standing technology, it had to be learned, adapted, and valued anew by every student who wished to make or use one in a notebook.

Students were of course exposed to matrices via their regular appearance in textbooks.[106] Printed compendia featured tables of comparative weights, measures, and prices.[107] Practical mathematics primers gave tables of "pow-

---

102. Household books were, in Marion Lochhead's words, "a treasury of homecraft" that also contained menus, recipes, and other necessary information required to successfully run a home. Lochhead (1948, 34–40, quotation on 34). For a printed example of a household book, see Baillie (1911).
103. Monro Primus (1954, 93) praised Isabella Monro's wise "Management" skills, which included overseeing household accounts.
104. Goody (1977, 53). Goody's interest in tables as mediators of writing systems also influenced his work on the anthropology of inscription in Goody (1987; 1986).
105. Higgins (2009).
106. Scotland's textbooks are reviewed in Law (1965), Wilson (1935), and Michael (1987).
107. Weights and measures conversion tables are given in the many Edinburgh editions of George Fisher's *The Instructor: Or, Young Man's Best Companion*. See Fisher (1799, 70–80). For another example, see Ramsey (1750).

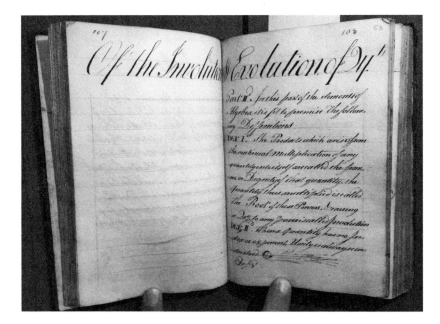

FIGURE 2.11. Recto and verso pages being used to create one module with two columns. James Fowler, *Schoolbook of James Fowler* (1780), Bound MS, NLS MS 14284, ff. 52v–53r. Reproduced with permission of the National Library of Scotland.

ers" (squared, cubed, and biquadrated) and logarithms used by surveyors, levelers, gaugers, and navigators.[108] Grammars featured vocabulary, conjugation, and declension tables, and geography primers presented tables of countries and cities.[109] The rows and columns of the tables required vertical and horizontal reading skills, a mode of interface that Lorraine Daston has called "right-angle reading."[110] Such skills did not come easily to some children. Likewise, though it might have been easy for some to read a matrix in a printed book, ranging one in a notebook was much more difficult and required a host of different scribal skills. This was especially the case for younger students who were learning penmanship and bookkeeping.

Students adapted matrices into visual forms that suited their own needs. As shown in figure 2.11, the Strathpeffer schoolboy James Fowler fused two open-faced notebook pages (verso and recto) into a singular module that fell

108. Fisher (1763), 323–327, and throughout Hoppus (1799), Ewing (1771), and Panton (1771).
109. Grammatical declension schemes occur throughout Adam (1786) and Ruddiman (1779). Good examples of geographic tables occur in Fisher (1763, 265–296), and throughout Mair (1775). All three texts were popular in Scotland. At least ten editions of *The Instructor* were published between 1750 and 1810 by printers in Edinburgh and Glasgow alone.
110. Daston (2015, 205).

under one heading. He then used each page as a column, which effectively created a matrix. Throughout his notebook both verso and recto pages contain script; however, I have chosen the specimen in figure 2.11 because the empty verso page makes it easier to see the graphite grid that he first drew to create a unified structure across two facing pages. Once the grid was in place, he then used his impressive chirographic skills to make the headings, the sub-headings, and the prose.[111]

Some students found that ranging a table by hand was a challenge. This was especially the case for those who experienced developmental reading or writing difficulties. Needless to say, the process often ended in frustration. A striking case of scribal irritation is evinced in a matrix of shorthand symbols drawn by the adolescent James Dunbar in his school notebook during the early eighteenth century. He drew the lines of his columns so unevenly that many of them simply ran off the page, rendering the right side of his table use-less. Dunbar became so annoyed with his unsuccessful effort that he wrote the following in the space of a malformed (and hence, unusable) column: "I am angry that I left a blank here and wrote filthy Scribble Scrabble on the side and that I did not contrive it better."[112] Here we can see that right-angle writing was an acquired skill (figure 2.12).

Dunbar persevered and the next matrix in his notebook was designed bet-ter. This illustrates the fact that simply attempting to structure the space of the written page as a verbal picture, a *tabula verba*, was a capability-building exercise. Ranging served as a form of visual training that strengthened the kinesthetic relationship between mind and hand. It reinforced the utility of information being arranged into symmetric, perpendicular, rectilinear, an-gular, and parallel patterns. It enabled students to learn how to adapt visual formats to fit their needs and to reposition headings, keywords, paragraphs, rows, and columns in a manner that instilled the ability to create a combina-torial word scheme on their own.[113] The rich variety of the skills required to cre-ate such schemes indicate that students were not simply *writers*. In the end, once students learned how to range grids, modules, and word tables, they became what might be called *gridders*, *modulators*, and *tablers*, that is, note-keepers capable of compositing script on paper. The skills involved in creat-ing these structures were powerful elements of graphic design that opened

---

111. Fowler Bound MS (1780), ff. 52v–53r.
112. Dunbar Bound MS (1710). For further treatment of the Dunbar notebook and its visual context, see Eddy (2013, 215–245).
113. For examples of the kinds of word-based combinatorics available prior to the nineteenth century, see Yates (1966), Schmidt-Biggemann (1983), and Grafton (2012, 25).

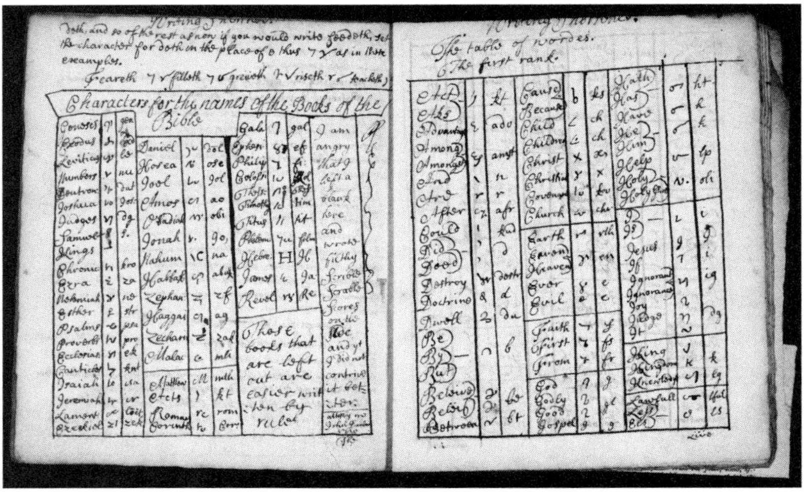

FIGURE 2.12. James Dunbar's Scribble Scrabble written in a notebook table titled "Characters for the names of the Books of the Bible," James Dunbar, *A Volume Completed by James Dunbar* (1710), Bound MS, NLS MS Acc 5706/11. Reproduced with permission of the National Library of Scotland. In the far right column of the left page, Dunbar wrote, "I am angry that I left a blank here and wrote filthy Scribble Scrabble Scores on the Side."

up new learning possibilities as students encountered novel or unfamiliar subjects.

Copies and the Exercise of Memory

Today a *copy* is seen as an object that requires little time or labor to create. This perception is reinforced every time we *copy and paste* data in word processors and *carbon copy* emails to friends and colleagues. But in the world of predigital notekeeping, this flat, relatively disembodied understanding of a copy would have been a bit perplexing. As we have seen in this chapter, crafting a written copy involved a variety of skills and required a significant amount of labor that was believed to have a remarkable impact on a learner's mind. The cognitive importance of copying explains why many students endeavored to range and rewrite their preliminary class notes into neater versions that eventually could be bound. This kind of copying was practiced throughout modern manuscript culture. It was common, for example, to rewrite data and observations in commonplace or bookkeeping notebooks.

In Scotland, a handheld corpus of preliminary notes was called a scroll book. In Scots English, it was called a *scrow buik*.[114] Since students tended to reuse the paper of their scroll books for other purposes when they finished school, the extant evidence for this kind of writing, which might be called "scrolling," is thin, consisting of a few bifolia specimens.[115] Scrolling was learned through recording information in the classroom and, perhaps even more frequently, through writing the drafts of exercises, translations, and essays that students then rewrote and submitted for examination. It is this kind of rewritten copy, which was called a transcript, that often has been preserved under the general label of school notebooks, copybooks, or ephemera in special collections libraries. This means that many extant school notebooks are in fact copies, transcriptions of the many forms of handwritten genres that students used to learn and organize knowledge as script on paper.

There were many words used to refer to the process of making the kinds of handwritten copies that students created in the pages of their notebooks. Within school settings, teachers used words such as *repetition, correction, transliteration, translation, transposition, exemplification,* and *composition* to refer to the various modes of copying carried out by their students. Some of these words, *translation* for instance, were used to describe the final product as well. The terms used to describe such processes also give a helpful indication of what students thought they were doing while they were copying things. Instead of *copying,* their notebook pages represented the acts of repeating, correcting, transliterating, translating, transposing, exemplifying, and composing. Put more clearly, student notekeepers who wrote copies were not merely copiers. Rather, they were repeaters, correctors, transliterators, translators, transposers, exemplifiers, and composers. In short, a copier was an interfacer, a writer who actively engaged with the page.

The various modes of copying lent themselves to a number of terms unique to eighteenth-century manuscript culture. A female translator, for instance, was a *translatrix*. Alternatively, a person who somehow egregiously miscopied something was a *transcribbler*. The rich variety of these and other terms associated with the different kinds of copying reveals that the iterative nature of student notebooks presents an excellent opportunity to examine one of the most prevalent forms of writing practiced during the modern era. In order to treat copying as such a multifaceted, effective media technology, we will need to set aside the traditional notion that copying was solely a rote

---

114. Robinson (1985, 592).

115. The specimens are bound with rewritten notes in Fowler Bound MS (1780) and Purdie Bound MS (1823).

exercise, something that involved little cognitive effort and had minimal historical value or impact. In this conventional interpretation, copying is framed as a matter of cut and paste, with attention being given to the accuracy of the copy when it is compared to a printed exemplar.

But the copied pages of school notebooks were tangible artifacts bursting with material and visual evidence of the hands and minds that made them. The anthropologist Tim Ingold once wrote that "there is more movement in a single trace of handwriting than in a whole page of printed text."[116] When seen from this perspective—from the perspective of a notebook's status as a hand-made artifact—it becomes easier to recognize that students were using copying as a dynamic mode of knowledge acquisition and cognitive regulation. As shown by Peter Damerow's work on ancient writing systems, the longstanding importance of inscription as this kind of developmental media technology lays in the fact that it gave inscribers the ability to learn that the control of behavior can itself become "the subject of controlled acts of behavior." In this process, "the mediated control of behavior acquires the characteristics of self-reference," that is to say, it becomes a conscious form of regulation.[117]

Since the pages of student notebooks were often copies, they instantiated the power of iterative writing recognized by many educators.[118] Across modern Europe and its former colonies at this time, students were increasingly treated as interactive learners who could use copying to train and shape the structure and content of their own minds.[119] Within Scotland, the value of the copied page was underpinned by the model of the learning mind provided by the *tabula rasa* metaphor as expressed by Locke and his Scottish followers such as the cleric and moral philosopher Frances Hutcheson.[120] In the words of the Scottish savant and pedagogue Lord Kames, "But the mind of a child is white paper, ready to receive any impression, good or bad."[121]

In many respects, ranging and rewriting notebook pages fostered ideas and skills that were widely associated with a sound sense of judgment. The prerequisite of judgment in Lockean psychology was the capability to see imprinted or inscribed words as "sensible marks," signs, that corresponded to ideas in the mind. A person who possessed judgment was able to collect related words, either through thinking or writing, into a group and then to cate-

---

116. Ingold (2007, 93).
117. Damerow (1996, 3).
118. Douglas (2017), Thornton (1998), and Howell (2015).
119. Benzaquén (2004), Ezell (1983), and Brown (2006). For Scotland, see Eddy (2010b). For children's literature, see Bartine (1989), Cohen (1977), and Pickering (1981).
120. Scott (1900). Hutcheson's use of the *tabula rasa* is addressed in Scott (1900, 173, 196, 255).
121. Home (Kames) (1781, 282–283).

gorize them with an appropriate name. As a consequence, Locke averred that the act of writing was a mode of "copying our thoughts."[122] This standpoint gave rise to the notion that, in the words of Lockean scholar Hannah Dawson, "Our embodied minds yearn for sensible marks with which to think."[123] Since thinking with such marks involved simultaneous mental operations such as recollection, perception, attention, and association, there was a sense that acts of writing, which included the forms of copying practiced by student notekeepers, acted as a mode of realtime training.

In the years preceding the creation of a school notebook, children learned to associate the act of copying with the act of shaping their own sense of volition. The 1763 Edinburgh edition of George Fisher's *The Instructor* emphasized this interface between the mind and hand with the following poem:

> A penknife razor-metal, quills good store;
> Gum sandrick powder to pounce the paper o'er;
> Ink shining black; paper more white than snow,
> Round and flat rulers on yourself bestow:
> With willing mind, these industrious hands,
> Will make this art your servant at command.[124]

Fisher's poem occurs near the section of the book that lists scores of maxims, most of which were moral in tenor. Students were meant to copy them over and over with a view to improving both their handwriting and their behavior. Once perfected, pristine versions of the maxims were rewritten on slips or sheets of paper and bound as a small copybook.[125] The multiple acts of copying required to create a copybook reveal that Fisher's reference to the "willing mind" is inextricably linked to the writing hand. His reference to the will communicates that copying was a formative activity, an everyday performance in which inscription and volition reinforced each other. This linkage between rewriting words and rethinking ideas was a core assumption that underpinned the broader developmental psychology that educators teaching in Scotland and in British dissenting academies used to understand the process of childhood and adolescent learning. The relationship between copying and volition in these settings was often understood in moral terms, thereby creating a con-

---

122. Locke (1781, 160).
123. Dawson (2007, 250).
124. Fisher (1763, 29). The poem occurs in the section titled "Directions to Beginners in Writing." The section that follows it is "To hold the Pen." Pryce-Jones and Parker (2014, 9–10).
125. Greig Bound MS (1763).

text in which transcribing the pages of a school notebook was treated as a virtuous activity.

The moral and rational utility of copying served to reinforce the notion that training one's sense of judgment could be accomplished through writing. The burgh school in Kirkcaldy, for instance, placed a high value on copying as a tool that helped develop students' rational capabilities. The school's minutes state that students were to be trained to "write a version [copy] to exercise their judgements to teach by degrees to spell rightly, to write good style, good sense and good language."[126] Here we can see that copying was treated as an activity that significantly aided a student's capacity for sound judgment and "good sense," which, at the time, were essential elements of reason.

The link between writing and judgment transformed copying into a central learning tool that led to the creation of a student notebook. Copying Latin or English sentences, which were commonly called version or lines, was a standard element of Scottish education. Masters sometimes referred to this process as "exemplification."[127] Copying scroll books into a neater school notebook was an activity that jointly inculcated writing skills and created a reference work that could be used in the future. Bannerman explained this process in the following manner: "[O]ur pupil is trained to writing, and a proper experiment made of his genius for numbers. In the former he makes considerable proficiency, after his hand is formed, if he shall, from a fair copy written by his master, transcribe his system of divine and moral truth, of geography, and of history, into [note]books, which he ought to preserve and peruse so long as he lives."[128]

The many forms of copying outlined above both bestowed and preserved useful information in the mind and on the page. Perhaps one of the clearest advocates of the psychological power associated with copying was the Scottish jurist and savant Lord Kames. He saw copying as a realtime act of learning that trained the flow of ideas in a student's developing mind. As both a humanist and a firm devotee of John Locke, his popular *Loose Hints upon Education* underscored the value of employing the act of copying as a learning tool, with the goal being the retention of ideas and the training of the mind's operations in a way that created what he called "a facility of writing." He recommended the ordered copying of extracts because the act of writing impressed

---

126. Fay (1956, 50–52).

127. John Mair refers directly to exemplification in the title and preface of Mair (1790).

128. Bannerman (1773, 26–27). Both Thomas Reid and Dugald Stewart recommended *Memoria Technica* to their university students as well. Reid (2004, 65, f. 155).

ideas into the memory and strengthened the copier's skills of attention and recollection in a way that imparted a stronger sense of judgment.[129]

Like many Scottish educators and literati, Kames's commitment to the cognitive advantages of copying was deeply influenced by his humanism, particularly the links drawn between thinking and writing voiced by classical orators such as Cicero and Quintilian. By the eighteenth century, this linkage spread far beyond Scotland's grammar schools into the vernacular curricula of academies and schools. The humanist tradition of using writing as a learning tool influenced the way that many teachers viewed copying and its relationship to the kinds of writing enacted in a school notebook. Richard Grey's introductory comments in his *Memoria Technica*, a popular compendium used in Scotland, explained how the general modes of instruction of this tradition could be used to inculcate keywords as well as phrases and verses. After discussing the pedagogical views of the Roman orator Quintilian, Grey asserted that "the frequent Repetition of . . . Memorial Lines would certainly answer this End; and if I might also recommend, as he does, the Writing of them too, in order to make a deeper Impression."[130]

Kames encouraged students to rearrange the words in the sentences of their copied excerpts as well. Manipulating and improving the meaning or clarity of a text was essentially a realtime form of writing, thinking, and correcting, a collective skill that had been prized within European humanism since the Renaissance.[131] Such a practice bestowed a "facility" of arrangement, one that enabled students to order an extract "better than by the author himself."[132] Grammar school students learning to translate Latin into English "versions" or "trials" also practiced the copying and rewriting of prose in this way. They were asked to reposition the words in Latin sentences and English translations in a way that made them clearer or more poetic. Such translations and rearrangements were done on paper so that the students remembered what they were thinking when they read them out in the next lesson.[133] They were then bound as a notebook, a codex, at the end of a student's studies.

The practice of reworking excerpts and translations effectively hybridized elements of transcription and composition in a way that students could apply to a myriad of literate contexts that they encountered when they completed

---

129. Home (Kames) (1781, 128–129).
130. Grey (1756, xi).
131. Grafton (2001).
132. Home (Kames) (1781, 237–238).
133. Chapman (1784, 208–215; 1790, 66–67, 163–64).

their education. In addition to the intertwined acts of thinking and writing required by this activity, the reworking process took place across multiple sheets of paper, all of which contributed to the final copy that eventually became part of a student's notebook. The necessity of so many sheets and the need to assemble them into a coherent collation draws our attention to the fact that writing was not the only skill that students needed to learn if they wanted to create and preserve a notebook. When they learned to write they also learned how to source, shape, order, and bind different kinds of sheets in a way that transformed them into useful paper machines. The next chapter explores this point in more detail. As we will see, learning to craft a codex not only enabled students to design and assemble a sophisticated mnemotechnic device, it also enhanced their abilities to make powerful media technologies.

CHAPTER 3

CODEXING

Paper Machines as Material Artifacts

On the 8th of October 1790, the Edinburgh High School student and future lord chancellor of Great Britain Henry Brougham was in need of a piece of paper. His tutor had given him an English sentence to translate into Latin. In the end, he did not manage to find a new, clean, or unwrinkled sheet. Instead, the one that he found was an oddly sized scrap of paper that had most likely been used as a letter envelope. Its edges were partially ripped and its lightly soiled surface was interrupted by well-worn folds that divided its front and back faces into several rectangular sections. Undeterred by the sheet's seasoned surface, Brougham used three of the square sections to write an English sentence and its translation into Latin. He left the other sections blank. In so doing he completed his homework and was free to enjoy the rest of the day.[1]

Brougham's homework is a kind of juvenile ephemera seldom studied by historians. It does not present an earth-shattering political proclamation. It does not foreshadow a revolutionary scientific discovery. Nor does it signal the beginning of a social movement. It is simply a normal, everyday specimen of manuscript culture that was created by many schoolchildren during the modern period. But what if we were not looking for extraordinary specimens? What if we were looking for the beginnings of an everyday artifact that was an essential knowledge management tool? If that is what we seek, then Brougham's scribbled exercises are noteworthy because they represent the beginnings of a codex, that is, one of the most important paper machines that a student at the time could learn to create.

---

1. Brougham Loose-leaf MS (1790). See also the translations of Cullen Loose-Leaf MS (c. 1800) that consist of two quarto sheets folded into an octavo booklet of four pages. For the curriculum of Edinburgh High School, see Steven (1849) and Law (1965).

Brougham was participating in the time-honored process of making an exercise notebook. It began with sourcing scraps, slips, and sheets and then writing translations and calculations that had been set by a teacher. Once all was corrected, students copied their work onto fresh bifolia, which they then stacked or sewed. The end result was a small notebook, a codex, that served as a reference work while they studied over the coming days and weeks. Others took the process even further by gathering sets of codices together into a leather-bound volume that served as a reminder of their studies and which they or their family members could use as a reference tool for many years to come.

In addition to strengthening their knowledge of literary and numerical subjects, making small and large codices endowed students with a host of skills that could be deployed to create a variety of media technologies relevant to the data circulating in commercial, domestic, literary, and technoscientific contexts. In many respects the skills of codexing were just as significant as the skills of writing or drawing. One of the aims of this chapter is to show the importance of recognizing that notekeepers were also codexers and that the process of creating a codex, which I call *codection*, influenced how students learned to conceive the malleability of knowledge.

The codex was a crucial technology during much of the modern period. Though forgotten today, codexing skills formed the manual and material bedrock of the *tabula rasa* metaphor, especially as it was formulated by John Locke in his *Some Thoughts Concerning Education*. In the book's conclusion, Locke reinforced the material basis of the metaphor by using verbs such as *mould* and *fashion* to explain how the mind is structured like a piece of "white paper, or wax, to be moulded and fashioned as one pleases."[2] This reference to material manipulation was firmly grounded on the many physical uses of paper that permeated manuscript culture. His metaphor took for granted that students commonly learned to mold paper into a communications media platform that was appropriate for the kinds of notebooks they wished to make. In other words, before there could be a *tabula rasa*, there had to be a *tabula folia*, a quadriform, flexible artifact that preserved what students learned in the classroom and served as a form of user training for the many kinds of paper documents they would design or encounter as adults.[3]

The informatic potential of paper as a flexible material went far beyond a student's ability to create a notebook page with it. Since many notebooks

---

2. Locke (1752, 324).
3. For the historical importance of paper as a form of information management, see Gitelman (2014).

were hundreds of pages long, making one from scratch allowed students to see the value of creating a large, well-designed, and interactive artifact. But what were the skills that students had to learn in order to transform paper sheets into components that could be assembled into a codex, a manuscript book that functioned as a three-dimensional learning and retrieval device?

A helpful way to tackle this question is to recognize that students designed their notebooks to operate as material artifacts, that is, as instruments that used the properties of paper to create or manage knowledge. Historians who approach paper from this perspective treat it as an epistemic object, as an entity that can carry various kinds of meaning based on how it was experienced or manipulated. Within cultural history, the term *paper technology* is used by scholars who seek to explore how the material properties of paper or paper-based forms of organization such as indices, archives, and commonplaces function as tools that can be used to remember, sort, and order knowledge in ways that correspond with the values and beliefs of different communities.

An early advocate of treating the notebook as an interactive paper technology was the historian Anke te Heesen. Her work explored how the materiality of notebooks influenced the creation and transmission of knowledge systems that existed both inside and outside of science broadly construed.[4] Other scholars have since used the paper technology concept to understand the epistemic function of paper within a variety of knowledge communities, particularly those involving ephemera relating to modern archives, homes, correspondents, libraries, and classrooms.[5] As evinced in Heesen's work, the notebook offered a number of unique paper-based affordances. Rather than being a simple collection of slips, scraps, and sheets spread across shelves, classrooms, drawers, and bags, notekeepers often crafted paper with a specific view to creating a codex, that is to say, a manuscript book that functioned as a specific kind of informatic artifact with parts designed to work with each other.

Put more clearly, a notebook, especially a student notebook, was a specific kind of media technology, one that possessed interchangeable paper components and one that could be assembled and used with combinatorial patterns designed to impart and organize knowledge. It was, in fact, a paper machine. As shown in Markus Krajewski's work on the history of index cards

---

4. Heesen (2005).

5. The paper technology concept is also considered throughout the essays in Bittel, Leong, and Oertzen (2019). See also Leong (2018b), 87–103; Hess and Mendelsohn (2010); and Kassell (2016). For the advantages of using the material history paper to understand modern print culture, see Heesen (2014), Calhoun (2020), and Senchyne (2019).

and Anthony Grafton's work on scholarly notebooks, paper machines were part and parcel of the modern literate world. When individual paper storage devices such as cards or pages were manipulated in concert over time, explains Krajewski, "The force taking effect is the user's hand." The movement of the sheets in this manner produced "a mechanical work taking place under particular conditions."[6] In this chapter, I would like to apply Krajewski's insight to the paper components of a modern student notebook, with a view to utilizing their material properties as evidence of artifice. I will focus on the roles played by the general properties of paper, quires, and bindings. I reveal that, like a pile of index cards, the malleable and moveable pages of a student notebook functioned collectively as a paper machine with distinctive characteristics both before and after it was bound.

At first glance, the adaptability of a notebook page as a material object might not be that obvious to the twenty-first-century observer. This is mainly because the screen-based texts that we use today inhabit a different kind of materiality than that which existed in the predigital modern world—a flat materiality that makes it difficult to see that paper can be an epistemic object, that is, a dynamic artifact that helps create knowledge.[7] If left unrecognized, this flatness makes it difficult for historians to see that the material or visual properties of paper often operated as important learning or thinking tools.[8] In the case of predigital notebooks, it overlooks the "paperness" of their components and it obscures the fact that they were temporal objects that were oftentimes assembled from different kinds of paper over a period of time. Add to this the fact that extant notebooks that have survived in libraries usually are packaged as leather-bound volumes, giving the impression that the content and order of their pages was fixed at the time of composition and subsequent usage.

But for many school notebooks nothing could be further from the truth. The notebooks created by Scottish schoolchildren oftentimes did not begin as notebooks at all. As intimated in the case of Henry Brougham's Latin translations mentioned at the start of this chapter, notebooks often began as blank sheets of paper that could be ordered and reordered through folding, bending, creasing, gluing, piercing (with pins), gathering, stacking, shuffling, turn-

---

6. Krajewski (2011, 7). For the use of slips or sheets as information management devices, see also Krajewski (2015) and Müller-Wille and Charmantier (2012a). Grafton (2012) treats notebooks as information machines.
7. Here I am extending Juhani Pallasmaa's notion of flat materiality, the idea that the late modern world's understanding of the word *material* is based on a significantly weakened conceptualization of the relationship between matter and the five senses. Pallasmaa (2008, 31–34).
8. Heesen (2014).

ing, ripping, scratching (a mode of erasure), sewing, and binding, all according to the notekeeper's needs as a learner.[9] All of these skills added up to the larger ability to conceive and implement a codex, a skill that I call codexing.

Paper as an Informatic Medium

Predigital notekeepers lived in an age when paper was regularly used to make a variety of ingenious machines that were powered by the human hand. It is worth noting this context so that this chapter's discussion of the codexing skills learned by school notekeepers might be better understood in reference to other paper-based capabilities that existed at the time. Contemporary paper machines ranged from instruments fashioned from only a few slips, scraps, sheets, and scrolls to those comprising multiple components that were placed within a framework of wood, metal, or leather. The rotating frames, shelves, and joints of the enormous book wheels used by Renaissance humanists, for instance, were made of wood and housed books made of paper (or vellum) and other materials such as glue, leather, and thread. By the eighteenth century, there was a proliferation of small and large paper machines that could be cut out of books or purchased from bookshops. The fashionable paper fan, for instance, was called a "little gay fluttering machine."[10] There were also paper calculators, the smallest of which were made from movable slips. Perhaps the most impressive was Napier's Box of Tables. A precursor of the modern computer, it was based on the logarithmic tables devised by the Scottish polymath John Napier. As evinced in the fine specimen housed in the National Museum of Scotland, it consisted of handwritten tables and cylinders made from paper and fitted into a book-sized wooden box.[11]

Students coming from British households that could afford to buy the fashionable learning tools produced by the London book market learned to shuffle the images on paper featured in picture books, geographical puzzles, and flashcards that contained images of noteworthy personalities. At a more general level, school and academy students used primers that presented opportunities to work with paper that was already bound into a codex. Figures and

---

9. A helpful source that outlines some of the materials and processes required to manipulate different kinds of paper into codices is the collection of essays titled "The Renaissance Collage" edited by Juliet Fleming in 2015 for the *The Journal of Medieval and Early Modern Studies*. The essays address the skills involved in collaging, signcutting, signsewing, cutting, pasting, sticking, pricking, and clipping. For a summary of the collection's themes, see Fleming (2015).
10. Stabile (2004, 155–157).
11. Anonymous, (n.d.), *Napier's Box of Tables*. For the mathematical importance of Napier's tables to the history of mathematics, see Whiteside (2014).

tables, for instance, were often gathered into plates inserted at the back or middle of the book. In these cases, students learned how to shuffle between pages with a view to connecting related information that was presented in different parts of the book. Such an activity taught them how to think about the potential placement of information in a codex.

Some of the metrological instruments featured in primers circulating in Britain and its former colonies were designed so that they could be cut out and pasted on a hard surface. Students using primers such as Fisher's *The Instructor* and David Gregory's *A Treatise of Practical Geometry*, both published numerous times in Scotland during the long eighteenth century, could learn to shape paper into useful instruments by cutting out quadrants, rulers, and dials that could be pasted or sewn together.[12] Such activity showed students that it was possible to create a simple and accurate paper-based instrument if they found themselves in need of one when they were away from home or at sea. Though the ephemeral nature of paper instruments makes it difficult to find extant specimens, those that do exist reveal that students who made them from scratch learned how to master cutting, folding, gridding, pinning, knotting, and stitching skills that also could be used to design and create the various components of a notebook.

Take for instance the components of the paper quadrants that students used to measure the placement of the sun, moon, and stars in the sky. Fisher's *Instructor* features one that could be cut out, pasted to cardboard, and then given a movable hand by pinning a piece of string with a plummet to it (figure 3.1).[13] Once the instrument was finished, a student needed only to align the base with an object in the sky, let the plumb line hang straight, and then take a degree reading at the point where the string passed the face of the dial. In this scenario, the frame of the quadrant was powered by the human hand and the plumb line was powered by gravity pulling it downward. As evinced by the circa 1800 paper quadrant tucked in the Huntington Library's copy of the 1773 Edinburgh edition of William Wilson's *Elements of Navigation*, save for a joint made of string tied into knots, it was possible to make all the instrument's parts from paper (figure 3.2).[14] It is unclear who crafted the quadrant, but the physical features of its various components speak to the kinds of paper-based skills that many students learned before they set off on a sea voyage. Pinprick-

---

12. Gregory (1761) and Fisher (1763).

13. For paper quadrants in printed primers, see Gregory (1761, plate I, following p. 20). Fisher (1806, 279). For a paper quadrant made by a student in a school notebook, see Anonymous Bound MS (1804, 6).

14. Wilson (1773). The Huntington Library copy of the book (call no. 492548) belonged to a William Lauriston. Whilst examining the book I discovered the quadrant tucked between the book's cartographic tables. Its construction is similar to the instruction given in Gregory's *Practical Geometry*.

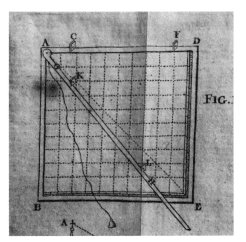
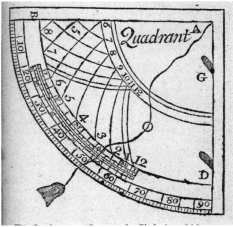
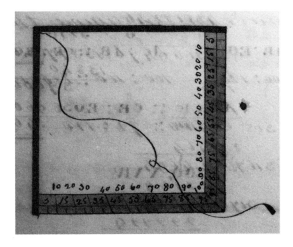

FIGURE 3.1. Paper quadrants, geometrical squares, and grids with plummets appear in printed primers and handwritten student notebooks. They were sometimes cut out of textbooks and pasted to cardboard. Left: David Gregory, *A Treatise of Practical Geometry*, 4th ed. (Edinburgh: Hamilton, Balfour, Neill, 1761), plate I, following page 20. Reproduced with permission of the Huntington Library, San Marino, CA. Right: George Fisher, *The Instructor: Or Young Man's Best Companion* (London: Johnson 1806), 279. Reproduced with permission of the Wellcome Library, London. Bottom: Anonymous, *Practical Mathematics* (1804), Bound MS, NLS 14285, f. 45. Reproduced with permission of the National Library of Scotland.

ing the outline of the frame (the perforated evidence for which occurs on the back), drawing the lines of quadrant, cutting out hands, and tying the joint together with string familiarized students with skills and materials required to make a codex and, crucially, helped them to more easily conceptualize a school notebook as a paper machine.

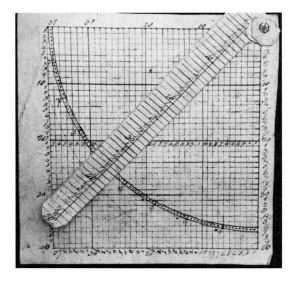

FIGURE 3.2. Handmade paper quadrant with a string joint. Its construction is similar to the instruction given in Gregory's *Practical Geometry*. Found by the author tucked inside William Wilson, *Elements of Navigation: Or the Practical Rules of the Art, Plainly Laid Down, and Clearly Demonstrated from their Principles; with Suitable Examples of These Rules. To Which Are Annexed All the Necessary* (Edinburgh: Wilson and Donaldson, 1773), Huntington Library, San Marino, CA, call no. 492548.

Within Scottish homes there were a number of sophisticated handmade paper machines that functioned as learning aids. Lady Charlotte Murray, for instance, made numerous sets of handwritten flashcards that she used to teach her nieces and nephews topics ranging from history to botany. Each card featured a topic on the front and a list of key terms on the back. The children were shown the front and then asked to repeat the key terms from memory. The terms and classifications on the botanical flashcards were based on the expert knowledge that she would later demonstrate to the reading public in her *The British Garden: Descriptive Catalogue of Hardy Plants, Indigenous Or Cultivated in the Climate of Great Britain*, which was published in 1799.[15] To further enhance the children's learning experience, she devised paper volvelles, chronographs, and drawings as well as nature walks, all of which were meant to visually and kinesthetically reinforce the information on the cards. To ensure that all these paper components could be operated as one machine, she housed them in two bespoke mahogany boxes.[16]

---

15. The third edition had appeared by 1808 as Murray (1808).
16. Noltie (2019). A private collector holds Lady Charlotte's box of manuscript learning devices. The tradition of making boxes to house paper-based educational material existed in other European settings as well. See Heesen (2002).

The foregoing examples reveal that, far from being a unique form of interface with paper, student notebooks operated in a rich world of paper machines that existed in both homes and schools. Despite the presence of so many machines, the nature of their construction and usage by students, or even adults, is relatively unexplored. This means that student notebooks present a welcome opportunity to investigate how children and adolescents in modern Britain learned to make, use, repurpose, and value paper devices as media technologies.

When it came to making the components of a notebook, students first had to find and select appropriate sheets of paper. Though paper was a ubiquitous medium of communication during the predigital era, we are only just beginning to learn about its cultural importance as an epistemic material in Scotland.[17] We do know, however, that Scotland had a healthy paper industry, and near the end of the century Edinburgh became a publishing powerhouse.[18] Understanding the difference between the many kinds and formats of paper was vital to understanding how it could be adapted into different kinds of scribal media. It was for this reason that primers included broad summaries of the kinds of paper that students could use for their studies. The 1767 Glasgow edition of George Fisher's *Arithmetic in the Plainest and Most Concise Methods* gave the following summary for "paper and parchment": "1 Bale is to 10 Ream; 1 Ream 20 Quires, 1 Quire 24 (or 25) sheets; 1 Roll of Parchment, 5 Dozen; 1 Dozen, 12 Skins."[19]

As most students came to learn over the course of their education, the foregoing sizes, though often used by stationers when they sold paper, could be manipulated when it came to tailoring the components of a notebook to their own specific learning needs. The sizes indicate that there were many choices to be made regarding the different formats and brands of paper that might be used. Some student notekeepers selected quarto sheets with the intent of leaving little white space around their inscriptions and drawings. Others used folio sheets so that they could space out the content. These choices were influenced by a student's finances and remind us that economic restrictions often played a role in the kinds of materials that notekeepers could use to make the pages of a notebook.

---

17. Paper devices were often mentioned by pedagogical authors influenced by John Locke's writings. A prominent Scottish voice on this topic was Home (1781, 61–64, 234–35). For the cultural importance of paper in Scotland at this time, see Friend (2016). For the uses of different kinds of paper within print culture, see Fox (2020).
18. The Scottish paper and publication industries are addressed in Waterston (1949) and Sher (2008), respectively.
19. Fisher (1767, 39).

The fact that prebound notebooks were made up of discrete, rearrangeable sheets of paper introduced an element of adaptability into the learning routines that influenced how learners conceived and used a school notebook before and after it was bound. From the very start of the notekeeping process, even before students began to write, the paper itself was formatted as an orderly material. It was sold as a quadriform sheet, that is, a symmetric geometric shape. Students then created pages by folding sheets in half into bifolia. Like the act of reading words, this act of folding paper imbued various kinds of meaning that made sense to the *foldmaker*. The haptics of the technique facilitated an act of visual order that enabled students to begin to think about the material strictures and possibilities presented by a blank sheet of paper, transforming the page into a device intended for organizational activities.[20] In this sense the *tabula rasa* was also a *tabula folia*, a material artifact shaped from the very beginning to function as an adaptable communications medium.

As mentioned in chapter 2, many of Scotland's extant school notebooks are recopied manuscripts based on a student's scroll books, that is, rough notes or draft copies that were then rewritten as a neater set of notes. Scroll books could be made from any shape or size of paper. They could be loose or bound. They could be kept on scraps or blank memoranda books sold at book shops. Working with the pages of a scroll book allowed students to experiment with paper as an adaptable medium. The writing of scroll books occurred at every level of literate society and, as can be seen in those produced by the clerks recording the minutes of Edinburgh City Council, they sometimes existed within a vast system of scribal culture that historians have only begun to understand.[21]

Students had to learn how to materially interface with paper through the act of writing as well. Moving the tip of the pen across paper was not a straightforward task, especially when it came to inscribing neat, curved letters. The task also presented difficulties for students who had learning disabilities with symptoms of distractibility and inattention, or who struggled to understand handwritten words or the layout of the page. The nuanced relationship between the material qualities of paper and the evolving skills of a notekeeper reminds us that paper at this time was, in the words of Caroline Fowler, a "made-thing."[22] Crafted from pulped linen, it was thicker than

---

20. The meaningfulness of folding is explored in Derrida (2003, 11–13). Helpful reflections on the haptic and kinaesthetic advantages of folds and folding are given in Friedman and Schäffner (2016); see also Friedman (2018) and Livingston (2008, 89–107).
21. Edinburgh City Council, *Scroll Council Minutes*, Bound MSS (c. 1682–1875) and *Final Council Minutes*, Bound MSS (c. 1682–1875). The minutes are housed in the Edinburgh City Archives.
22. Fowler (2019, 120).

modern paper and prone to grabbing the tip of the pen as a student drew the curved lines of letters and figures. Add to this the fact that linen pulp was strained and dried into paper inside gridded sieves that created parallel microgrooves called chain lines.

The foregoing considerations reveal that a blank piece of paper, the bedrock of the *tabula rasa* metaphor, was not completely blank or smooth. Put another way, the material structure of a piece of paper affected how students used the pen to interface with its surface. Depending on how the paper of a notebook was cut by a stationer, the microgrooved lines ran across the entire surface of the page in either an up-and-down or right-and-left pattern. In some cases, the rectilinear structure of the microgrooves made it easier to draw a grid, or to attempt to write straight sentences freehand or to form the edges of margins.[23] But the microgrooves also could present challenges to inexperienced notekeepers. Hitting them the wrong way with the tip of a pen, for example, could easily jolt a student's stroke, making the line of a letter or figure uneven or filled with small deviations. This jointly material and visual aspect of the blank page as both *tabula rasa* and *tabula folia* governed how student notekeepers interfaced with the blankness of paper, especially when they endeavored to make a *tabula verba*.

The evenness of the script in many notebooks shows that students were remarkably adept at making well-formed letters across the textured surfaces inside the space of columns, rows, and paragraphs. Some students even accomplished this feat when writing on poorly bonded paper or on the rough paper covers that protected some paper books. The Edinburgh High School student William Erskine managed to write his Latin translations in a good hand over the coarse surface of his cardboard notebook cover when he ran out of paper.[24] As evinced in the scraps of notes kept by the Scottish naturalist Robert Brown in Australia and in the letters written by wives of Scottish merchants and diplomats living in India, the capability to write well when presented with poor materials was essential for those who needed to record or organize information outside the comfort of the home.[25]

Students also supplemented their ability to work with paper by writing in chalk on rectangular slate tablets. This practice was summed up in 1748 in James Todd's *The School-Boy and Young Gentleman's Assistant*: "I say he [a

---

23. A good example of how a student used horizontal chain lines as the basis of a page's grid are the ruled lines drawn across several blank pages of Fowler Bound MS (1780, 28–30).
24. Erskine Bound MS (1784). The notebook consists of several quires sewn together inside a paper cover.
25. Moore and Beasley (1997). The letter writing skills of Scottish women abroad are addressed throughout Rothschild (2011) and Jasanoff (2006).

student] should have one [geometry book] by him while he Works on his Slate, or his paper, as he has his Grammar by him, for culturing in parsing, &c. of Language."[26] Learning to interface with multiple kinds of media such as slates, scraps, slips, sheets, bifolia, thread, string, and cardboard in a way that transformed a *tabula folia* into a fully functioning paper machine required a considerable amount of labor.

The mobility and manipulability of paper in the hands of schoolchildren created a situation in which they were learning to see the prebound components of their notebooks as the building blocks of a powerful information management device. By the time they had finished school, they had learned to conceptualize the pages of their notebooks as indispensable media technologies that could be manipulated by scores of adaptable selecting, ripping, folding, and shuffling skills.

Quires and Knowledge Management

When viewed through the material culture of student notekeeping, the main structural component of the notebook as a codex was the *bifolium*, a sheet of paper that was folded in half to make four pages, that is to say, a booklet of a front page, two inside pages (left and right) and a back page.[27] When several bifolia were folded together into a booklet, it was called a *quire* (some scholars use the term *gathering*). Within manuscript studies, the term *quiring* is used to describe the ability to mold sheets of paper into a booklet of bifolia conducive to being part of a codex. It was this kind of interchangeable looseleaf quire that students used to construct school notebooks. Though most students made and used unbound quires, it was also possible to buy those which were sewn together.

The prices of the blank paper that students used to make quires can be gleaned from the accounts of booksellers and stationers. The terms used to describe the formats were *scrap paper*, *quire*, *paper book*, *memorandum book*, and *note book*. The formats came in different sizes, most commonly octavo, quarto, and folio. A helpful indicator of the costs of the foregoing formats in Edinburgh during the late eighteenth century is the accounts ledger of Charles Elliot, the popular bookseller whose shop was conveniently located in Edin-

---

26. Todd (1748, 70). Though the word *gentleman* is used in Todd's title, the facts and skills communicated in the book were studied by both girls and boys.

27. Student letter writers used the bifolia folding technique as well. They stacked bifoliated letters into collections called "letter books." For an example containing letters written and received by a teenage girl living in eighteenth-century Scotland, see Moore Unbound MS (1786–1790).

TABLE 3.1. Kinds of notepaper used by students

Student	Date	Number	Price
Alexander Bartram	17 November 1776	1 paper book	2 shillings
	2 August 1777	4 quires of scrap paper	4 shillings
	15 August 1777	2 paper books	2 shillings
John Harsky	2 October 1776	2 quires of paper	1 shilling
	2 October 1776	2 paper books	2 shillings
	23 October 1776	3 paper books	3 shillings
	15 December 1776	6 memorandum books	2 shillings 6 pence
Mr. Piggot	17 June 1777	thick notebook	2 shillings

Data extracted from Elliot Bound MS (1771–1777, 429, 430, 437, 446). Special thanks to Warren McDougal for providing this information.

burgh's Parliament Close. These reveal that there was a healthy demand for different kinds of blank paper. A representative example of an accounts page from the mid-1770s is given in table 3.1.

Elliot's shop was a short walk from many schools based in the city. His ledger entries show that both paper books and quires sold for around one shilling. Other forms of notepaper were slightly more expensive.

After students had written and drawn all the loose or bound quires for a given school subject, they then had to bind them. I will have more to say about the binding process below, but at this point it might be helpful to refer to figure 3.3, which summarizes the codexing process used by students when they assembled their notebooks. In the figure we can see that students started by practicing their writing and ranging skills on slates, sheets, and scraps of paper. They then moved on to scroll a bifolium. Older students sometimes skipped the preliminary steps and began the process with scrolling. Next, students copied the scrolled bifolium onto a single blank bifolium or onto bifolia that were part of a quire that they had either bought or made. Since the quires were often unbound, it was possible to remove a bifolium and lay it flat on a table while the writing or drawing process occurred. Once all the quires were complete, students then bound them as a codex. To save money, many students wrote on the front and back of their bifolia. Since eighteenth-century paper was thicker than the forms of wood pulp paper introduced in the early nineteenth century, script usually did not bleed through in a way that was disruptive (figure 3.4).

A helpful visualization of the process of quire-based scrolling in a school was captured in a satirical print made by the Scottish artist and cultural commentator James Kay in the 1790s. In addition to revealing the energetic envi-

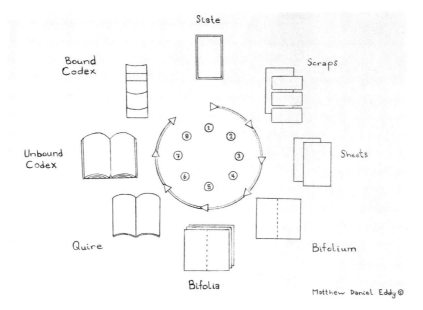

FIGURE 3.3. The paper components of a student codex. © Matthew Daniel Eddy. Students used a number of paper-based components to create notebooks that operated as paper machines. Young notekeepers started by (1) interfacing with the rectilinear surfaces of slate boards and then moved on to use (2) scraps and (3) sheets. As students grew older, they learned to fold a blank sheet in half to create (4) a bifolium, which was a small booklet with a front page, two inner pages (left and right), and a back page. Students also learned to fold two or three sheets in half to create (5) bifolia, slightly larger unbound booklets. They used the bifolium and bifolia formats to keep rough notes (scroll books). Once these were complete, students rewrote their rough notes on (6) quires, which they made by folding around a dozen or more sheets in half to create a large unbound booklet. They then collected and (re)ordered all the quires into a stack that functioned as (7) an unbound codex. The process ended when students took their quires to a bookshop to have them stitched into (8) a bound codex with boards and a leather spine and cover.

ronment of a Sabbath school, it depicts students standing at desks. In front of each student is a blank quire waiting to be used (see again figure 2.6).[28] It was this kind of quire that formed the basis of a scroll book, that is, a collective set of rough notes. Making the copied version of a notebook from these quires gave young notekeepers the opportunity to learn how to select and manipulate the kinds of paper required to organize information in different kinds of notebooks.

Shaping a manuscript book via the act of quiring was an established technique reaching back to the codices of the Middle Ages. As noted by codicol-

---

28. James Kay, "Modern Moderation Strikingly Displayed or a Ministerial Visitation of a Sabbath School," in Kay (1837–1838, opposite page 356).

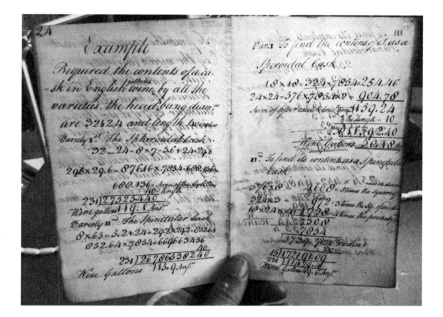

FIGURE 3.4. Bifolium held up to a light to reveal writing on all four pages, tacket holes in the spinefold, and a watermark (running across the top, center of the page over the spinefold). Anonymous, *Perth Academy Notebook* (1790), Bound MS, NLS 14291 f. 103. Reproduced with permission of the National Library of Scotland.

ogists and ethnologists alike, transforming paper (or vellum) into efficiently designed quires required time, labor, and skill.[29] The experience of shaping paper in this manner was different from the kinds of folding used to make the printed pages of a book (figure 3.5). As explained in hand press manuals, compositors intentionally arranged printed pages out of order on a sheet. Once printed, the pages on the sheet were folded in a manner that placed them in proper numeric order.[30] When compared to codex quiring, the practice of placing several pages onto a printed sheet made it harder to replace individual pages when mistakes were made.

The number of bifolia in a school notebook quire depended on the skills and resources of the student and on the subject matter being addressed. A student might, for instance, settle on quires comprising four bifolia only to find that there was a bit more material that still needed to be written when

---

[29]. For the kinds of medieval and modern scribal decisions required to shape a piece of paper see, respectively, Thaisen (2008) and May and Wolfe (2010). For the epistemological issues surrounding the use of paper as a thinking tool, see Livingston (2008, 89–107).
[30]. Luckombe (1771, 410–438).

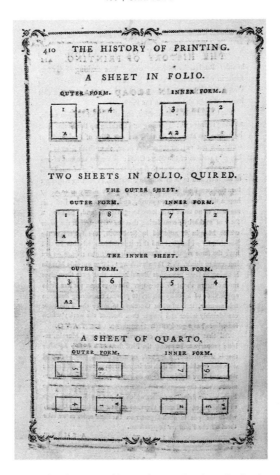

FIGURE 3.5. Philip Luckombe, *The History of the Art of Printing* (London: Adlard and Browne, 1771), 410. Reproduced with permission of the Huntington Library, San Marino, CA.

the pages of the quire were filled.[31] Since inserting more pages into the quire would disrupt pagination, a simple solution was to write the information on one bifolium and then to bind it at the end of the four-page quire. The overarching point to note is that quires were an adaptable medium. The quires of notes made at Edinburgh High School, for instance, vary between eight and sixteen pages.[32]

---

31. Examples of quires that differed in length and subject matter occur throughout Fowler Bound MS (1780).
32. Specimens of prebound quires made by students are housed in the Edinburgh High School collection in Edinburgh City Archives. See, for example, Archibald Cullen's exercises, Cullen Loose-Leaf MS (c. 1800). It consists of several quarto sheets folded into an octavo quire.

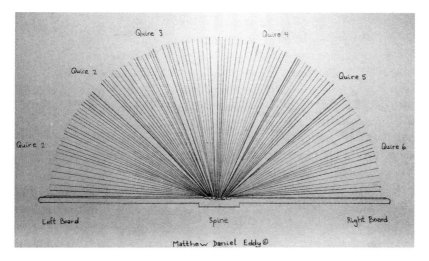

FIGURE 3.6. The architecture of a bound student notebook composed of six quires. © Matthew Daniel Eddy. Most student notebooks contained many quires, the lengths of which differed. The diagram depicts a codex based on features displayed in student notebooks made in Scotland during the long eighteenth century. Each quire is made from a gathering of bifolia, the number of which varied. Quires 1 and 6, for instance, are longer, while quires 3 and 5 are shorter.

When working with a bound student notebook created during the predigital era, a helpful way to see the material order and structure of its quires is to look at the bottom of the spine in a manner similar to the arrangement depicted in figure 3.6. Carefully trace the path of each page from the outer edge down to its base at the spine. If the codex is not tightly bound, a slightly larger gap will occur in the spine when you move from one quire to another. Since the spines of eighteenth-century school notebooks are usually fragile, this exercise must be done carefully so that it will not damage the bindings.

The ability to fold and order paper into a quire was an essential technique that had to be learned at some stage by many people who made a modern notebook.[33] Learning to quire constituted a mode of internalizing a jointly linear and flexible conceptualization of ordered ideas even before they were represented as inscribed facts on paper. The iterative routines required to fold a piece of paper reinforced, for instance, the utility of gestalt principles such as symmetry, rectilinearity, and parallelity that were presented in the pages of each bifolium. Using loose-leaf bifolia also made it easier in the prebound stage to remove pages that contained mistakes or to shuffle pages into a dif-

---

33. Folding was used to make interactive paper devices in the early Royal Society as well. See Hunter (2013, 73–78).

ferent order before they were bound. When students shaped and shuffled paper in this manner, they learned that ordering knowledge was an adaptable exercise.[34]

As mentioned above, in addition to making quires, some schoolchildren purchased blank paper books. These were quires of bifolia sewn together in the crease. Though they are seldom studied as an interactive medium that facilitated the management of predigital knowledge, paper books were sold by most early stationers and booksellers because adults frequently used them to manage domestic, commercial, or artistic information rendered in ink, graphite, or watercolors.[35] The sewn binding prevented the pages from becoming disordered. Sometimes stationers sewed on a piece of marbled endpaper or rough brown paper as a cover. As we will see later in this chapter, student notekeepers followed this practice as well.

Rather than recopying their notes all at once from start to finish in a bound, blank volume such as a paper book, most students assembled all of their handwritten quires into a bound compendium. It is this kind of compendium that most special collections today call a student notebook. The techniques of assembly were contingent upon the learning needs of the notekeeper. At one level, the order of the pages (and the subsequent content) was relatively set for the quires they had retained for each school subject. But at another level, students could arrange the different subjects in whatever order they deemed appropriate when the time came to bind all their completed quires into one leather-bound volume. Such compendia could be collated immediately after a student's studies were completed or at a later date.

Put more clearly, close inspection of the watermarks, pagination, and bindings of the paper inside bound school notebooks reveals that most were composites, that is, collocations of several sets of copied quires kept at different times, often in different courses. As evinced in the different quires bound in the *Schoolbook* of the student James Fowler during the 1780s, notes were not always collocated in the order that they were written. Fowler lived in the Strathpeffer area of northwest Scotland and attended school in Fodderty. As shown in table 3.2, although the school book is a single leather-bound volume today, it was originally four different sets of quires that were kept over a period of four years.[36]

---

34. For the gestalt principles of visual culture, see Arnheim (2004) and Tversky (2011).

35. The uses and prices of Scottish paper books remain relatively understudied. As intimated in table 3.1, a good indication of their price and popularity are the sales ledgers of the Edinburgh stationer Charles Elliot: see Elliot Bound MS (1771–1777). See also Brown (2011).

36. Fowler Bound MS (1780). For a set of bound notebooks that were also assembled from a collection of smaller notebooks, see Anonymous Bound MSS (1787), I.M. (notekeeper).

TABLE 3.2. The four sets of notes in James Fowler's *Schoolbook* of 1780

Year written	Order in the notebook	Subjects	Original pagination	Secondary pagination	Tertiary Pagination
1780	4th	composition and translation	1–47	none	137r–159v
1782/3	1st	arithmetic and algebra	1–136	none	1r–66v
c. 1782/3	2nd	advanced algebra	1–98	none	67r–107v
1784	3rd	trigonometry	5–52	99–158	108r–136v

Once Fowler's quires were bound together into a collocated codex, they ceased to be individual manuscript units and became sections. The erstwhile status of the quires was all but erased when the binder trimmed their page edges to form one unified volume. The process of gathering together and ordering notebook quires was arguably the most common method used to assemble the contents of a leather-bound student notebook. It was used to assemble notes on various subjects as well as exercises that featured translations, maps, and figures. Alexander Kincaid's *Latin Exercise Book*, kept at Edinburgh High School during the 1760s, for instance, was created from exercises recopied on individual quires that were eventually bound together.[37]

The transformation of quires into sections of larger books was a crucial knowledge management technique at the time employed to organize large systems of printed and inscribed knowledge, including those used to teach university students.[38] The material structure of Fowler's notebook shows that schoolchildren were learning to practice this ordering skill as well, albeit on a smaller scale. Fowler chose to order his smaller notebooks topically, placing the mathematics sections first and the composition section last (see again table 3.2). This kind of ordering, which was essentially an extension of what Walter Ong calls topical logic, was a flexible process and could have been done differently had Fowler decided that the composition section was the most important.[39] Overall, shuffling quires in a topical manner was an essential manuscript management technique for students wishing to order their bound school notebooks in a manner that they found most useful.

The presence of so many quires, some of which consisted of only one or

---

37. Kincaid Bound MS (1764).
38. Eddy (2010a), 227–252.
39. Ong held that topical logic was a mode of classifying that grouped things according to convenient topics (Greek: *topoi*; Latin: *loci*) and not necessarily according to natural kinds. Ong (2004).

two bifolia, serves to remind us that students learned a host of codexing skills long before the leather binding process occurred. Indeed, they effectively used the bound and unbound quires of a given subject as small paper machines while they learned at home or in school. They did this as they wrote and rewrote their notes and, as we will see in later chapters, as they sketched and redrew their maps and figures. Using the pages inside the quires in this manner happened countless times before they finally decided to fix all the quires into a single leather-bound notebook that operated as a powerful media technology, one that served as evidence of their codexing skills and as a memory aid when they became adults. I will return to such binding skills in the next section. But I would first like to highlight an important set of codexing skills that students learned from the affordances offered by the flat pieces of paper available in a prebound quire.

A quire that contained under a dozen or so bifolia easily could be laid flat on a hard surface like a desk or table. This property of *flatness* became harder for students to create as they added more bifolia to quires. The problem was that the flatness of the page was inversely proportional to the thickness of the quire. Put more simply, adding more bifolia to the quire made it harder to press the page flat while writing or drawing, the reason being that the collective thickness of the paper caused the pages to bend upward at the center. The problem became more pronounced when quires were bound. The surface area of a page in a hardbound quire was especially bent because it was attached to a spine and boards. The pages of thick, thread-bound quires exhibited a similar bent surface, though to a lesser degree. This meant that children seeking to write neat script or to draw precise images tacitly learned to recognize flatness as an important property of paper that had to be observed while working with a page in realtime.

When the surface of a quire page was bent, it was difficult for a student notekeeper to render a straight line with a ruler or a curved line with a compass. Using single sheets or thin quires before they were bound eliminated this problem. Though prebound school notebook quires were sometimes sewn together with a few stiches of thread, students seeking to inscribe neat words or figures learned to keep them thin, between ten and twenty bifolia. Since such quires were a single set of folded sheets, they naturally sat together. All of these characteristics made it easier to press the quire open so that it lay flat, thereby making it possible to draw or write accurately. A helpful visualization of a student who had drawn a figure on the page of an unbound quire is depicted in the frontispiece of the 1786 Glasgow edition of George Fisher's *The Instructor* (figure 3.7). It features a student holding an open quire. The flexible

FIGURE 3.7. George Fisher, *The Instructor; or, Young Man's Best Companion. Containing Spelling, Reading, Writing, and Arithmetic, in an Easier Way Than Any Yet Published* (Glasgow: 1786), frontispiece. Reproduced with permission of the Huntington Library, San Marino, CA, No. 65022.

pages bend over his left hand as he gives it to his teacher for inspection. Both pages feature a symmetric and well-proportioned geometric figure, indicating that he pressed the page flat when he used his ruler and compass during the drawing process.[40] The frontispiece, therefore, depicts what a student could learn to do with a flat page that was within a thin quire. It gestures to the many skills and tacit acts of paper-based observation that took place when student notekeepers worked with quires on a daily basis. Indeed, working with quires

40. Fisher (1786).

in this manner enabled them to see that creating an accurate geometrical image was best achieved when the material affordances of paper were taken into consideration.

Books and Customized Packaging

Once student notekeepers had transcribed their notes onto the pages of quires, they had to learn how to collate and fix their work into a codex that was designed to operate as an interactive paper machine. The student notebook's status as a codex, as a manuscript book designed to facilitate interface, draws our attention to the fact that the skills required to construct one constitute a fascinating chapter in book history. Framing notebooks and their components as tangible, manipulatable artifacts is in many respects a departure from the way that books were conceptualized after the nineteenth century. In recent years, however, historians of publishing, authorship and readership have increasingly emphasized that a predigital *book* was an extraordinarily diverse material and visual artifact, one that blurs the traditional distinctions between print and manuscript culture. In this reading, the book is conceptualized, in the words of Amaranth Borsuk, as object, as content, as idea, and as interface.[41]

When students participated in the process of making a codex, they often had to learn how to craft a binding and a cover. Notekeepers who wanted to successfully make these components had to repeatedly practice a variety of skills such as stitching, tacketing, folding, bending, pasting, knotting, and cutting. Practicing these skills through iteration enabled students to understand how they might source materials such as string, thread, glue, paper, and boards, or locate appropriate instruments such as knives, scissors, and compasses with sharp points that served as hole punchers. In learning such skills, students became stitchers, tacketers, pasters, and more, in a way that transformed them into codexers who knew how to materially package knowledge into a manuscript book.

Let us first turn to bindings. Student notebooks in Scotland were bound in a variety of formats that ranged from simple stitches of string to elaborately bound leather books.[42] Once students had gathered together their loose-leaf

---

41. Borsuk (2018). Core studies on the nature of the book and the interface between print and scribal culture are Febvre and Martin (1976), Love (1998), Chartier (1994), Johns (1998), Jackson (2001), McKitterick (2003), Sherman (2009), and Blair (2010b).
42. Records of Scottish bookbinders from this period have not survived. Zachs (2011). For England, see Hill (1999) and Middleton (1963).

or quired notes and exercises, they were then in the position to consider the kind of binding that suited their skills and resources. The simplest way to keep a handful of sheets together was to use the spinefolds of bifolia as rudimentary bindings. Sometimes students, such as the future novelist Sir Walter Scott, used this technique to keep their loose-leaf bifolia exercises together, leaving behind helpful evidence of how students designed the kinds of quires that potentially could be made into a notebook.[43] Some students created bindings by looping or weaving string or thread through holes that they had punched with pins, scissors, or compass points in the spinefolds of their bifolia. This process of making spinefold holes in quires reached at least as far back as the Middle Ages, when the skill of tacketing was used to pierce tackets, slits for binding string, into parchment.[44]

A good example of the improvised string bindings created by students is the 1780s Latin exercise notebook crafted by the future orientalist William Erskine.[45] During his student years at Edinburgh High School, he sewed a number of his Latin translations into a small notebook with string in an effort to keep them together (figure 3.8).[46] Unlike the professional binders with their elaborate stitching patterns, he did not thread one continuous piece of string through several holes punched in the spinefold. Instead, he punched two holes and bound the bifolia together with a central stitch fashioned as a loop. More specifically, he looped a piece of string through the two holes, tied it in a knot, and then cut off the excess. The spinefold of his notebook also features a set of empty stitch holes, indicating that his looped stitches were in fact a second attempt to bind some of the pages. The empty holes further indicate that some, but not all, of the pages had been previously bound and then unbound in another smaller quire. Erskine's old stitch holes reveal that the material culture of binding allowed notekeepers to actively arrange and rearrange knowledge.

Students used a variety of notebook covers, all of which familiarized them with the different ways that knowledge could be packaged into a durable, long-lasting notebook. Some simply used an extra sheet of bonded paper as cover. Others used slightly thicker paper. The cover of the 1815 *Maps* notebook of the schoolgirl Jemima Arrow, for instance, is a simple thick sheet of marbled paper. She customized it by writing the title on a piece of bonded paper

43. Scott Unbound MS (c. 1800). See also Cullen Loose-Leaf, MS (c. 1800).
44. Kwakkel and Thomson (2018, 20).
45. Erskine Bound MS (1784).
46. Erskine Bound MS (1784).

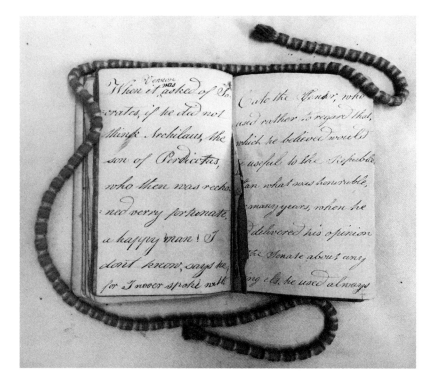

FIGURE 3.8. William Erskine, *Latin Exercise Book* (1784), Bound MS, ECA SL137/9/38. Reproduced with permission of the Edinburgh City Archives, City of Edinburgh Council.

cut out to resemble an oblong octagon. She then pasted the paper on the center of the front cover (figure 3.9).[47]

Other notebooks feature covers made from thick, rough brown paper that stationers sometimes used to make covers for the blank paper books that they sold in their shops. A good example of a notebook with this kind of stationer-made cover is the 1778 copybook of Robert Richardson. Kept when he was a student at the Perth Academy, the front and back faces of the cover are filled with strikethroughs, swirls, and other scribbles.[48] Other students used similar kinds of rough grey or brown paper to create their own covers. Such was the case for the teenager and future geologist Hugh Miller. In 1815 he created a cover for a notebook of his recopied poems from thick, rough

47. Arrow Bound MS (1815).
48. Richardson Bound MS (1778).

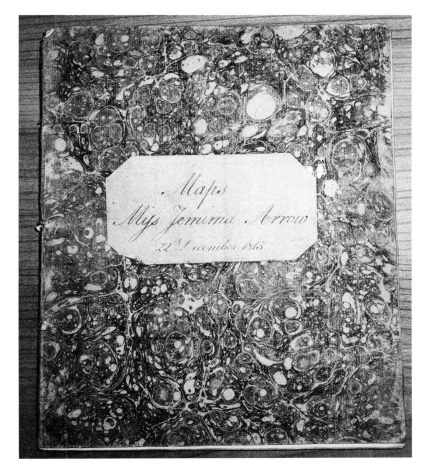

FIGURE 3.9. Jemima Arrow, *Maps* (1815), Bound MS, NLS MS 14100, cover. Reproduced with permission of the National Library of Scotland.

paper on which he painted a green, red, and yellow design of circles, squiggly lines, and dots that resembled Native American art (figure 3.10).[49]

The stitches and covers of student notebooks often speak to the improvised nature of bindings and covers and how they were crafted from the materials at hand. Students used different kinds of string, thread, and paper. The latter was perhaps one of the most recycled materials in schools and literate households, making it a multipurpose information management medium.[50] The

49. Miller Bound MS (1815–1826).
50. For the early modern recycling of paper and other household items, see Stabile (2004), Strasser (2000), and Werrett (2019).

FIGURE 3.10. Hugh Miller, *Juvenile Poems* (1815–1818), Bound MS NLS MS 7520, cover. Reproduced with permission of the National Library of Scotland.

cover of Erskine's notebook, for instance, is marbled cardboard that he either bought from a stationer or repurposed from a disused book or notebook. The sections of his notebook were demarcated by a thicker and rougher kind of brown paper that he most likely repurposed from the cover of a paper book. Such covers were often detached from paper books and then reused to hold other kinds of manuscripts. Some student notebooks also feature improvised covers from recycled newspaper, a practice used by Peter Purdie, or perhaps a member of his family, to protect his student notebook from the elements.[51]

The skills required to manipulate and mold paper bindings and covers into a codex were just as important as those required to write script. Some educators recognized the importance of such materials and skills and integrated them into their pedagogical systems. The Scottish philosopher, educator, and novelist Elizabeth Hamilton, for example, held that the material facets of books and notebooks were themselves learning tools. In her *Letters on the Elementary Principles of Education*, she observed that:

---

51. Purdie Bound MS (1823). Since the articles in the newspaper are dated 1850, the cover was added long after Purdie was a student.

the leather binding of books, the paper which forms the leaves, the thread on which these leaves are strung, and the characters that are printed on them, may be made instrumental in invigorating the conceptions: and I am persuaded, that habits of attention thus acquired, would be found of greater use in developing the faculties, than any lessons which the poor ignorant children could be made to read, or get by heart.[52]

At the time, "conception" and the notion of a mental "faculty" were central components of the Lockean developmental psychology promoted by many Scottish educators.[53] Hamilton used them to frame her entire pedagogical system.[54] From her standpoint, the material management of paper was an activity that shaped a student's memory and created skills that facilitated reason.

Hamilton thought that interfacing with the folds and stitches of a binding was a manual mode of performance that ordered the mind. The links between thinking and everyday forms of artifice like sewing, needlepoint, and embroidery had been part of the Western cultural history since the ancient Greeks.[55] Writing in the first decade of the nineteenth century, Hamilton joined a number of female authors who sought to draw out the links between writing and other modes of bodily movement such as sewing, exercising, cleaning, and caring for children. The specific overlap between the skills of writing and sewing (which was replicated in binding), was increasingly visualized in prints during the 1810s (figure 3.11). Notably, such figures sometimes depict sewing and writing utensils together, signaling the conceptual overlap between writing and sewing when they were used to represent the process of thinking.

The practical and interactive nature of the notebook as an evolving codex continued when students chose to bind all their notes into a single leather-bound volume. This was a more expensive option in which students or parents took the sheets or quires to a binder. But these bindings still speak to the pragmatic nature of the role played by notebooks as paper learning machines. The relative absence of wheel, lozenge, and fishbone bindings, for instance, shows that students or their families often chose not to use the elaborate gold embossing or tooling available in Scotland at the time.[56] Nor do we see very

---

52. Hamilton (1810, 251–252).
53. The centrality attributed to "conception" at the time is summarized in the philosophy of mind lectures given by Professor Dugald Stewart at Edinburgh University during the 1790s. See Stewart (1792), 132–150, for his discussion of conception.
54. Gokcekus (2019) and Price (2002).
55. Bergren (2008).
56. The patterns of Scottish bindings are discussed in Morris (1987), Sommerlad (1967), Prideaux (1903, 22–25), and Quaritch (1889, 34, plates 90–93).

FIGURE 3.11. A woman sewing next to a table with writing instruments and books. Anonymous, *The Female Instructor, or Young Woman's Companion: Being a Guide to All the Accomplishments which Adorn the Female Character, Either as a Useful Member of Society* (Liverpool: Nuttall, Fisher and Dixon, 1811), frontispiece. Wellcome Library, London, Digital Collections, public domain.

many that have overly decorated endpapers. Instead, most leather-bound school notebooks have plain or marbled endpapers. For those with an embossed or tooled cover, the designs are usually simple, occurring only on the spine or as thin vines or lines running as frames around the edges of the faces of the front and back covers. When such frames are present, they are often worn away from repeated usage. Such is the case for the tooling on the cover of Fowler's notebook, suggesting that it was consulted frequently after it was bound.[57] Additionally, other than the letters of the title, many student notebooks have no tooling on the spine at all.[58]

Some notebook keepers anticipated future usage by including blank pages at the front or end for annotations.[59] Some leather-bound notebooks were used so often that they needed to be rebound. Evidence of this practice appears in the form of the odd sizes and closely cropped pages evinced in many school notebooks. Perhaps the most striking evidence of rebinding appears as a worn leather patch pasted on the endpaper glued to the inside front cover of the first volume of an anonymous set of 1787 Perth Academy notebooks. The patch is the size of a modern fifty pence coin and is embossed with letters reading "PERTH ACADEMY." The gold inlay of the letters is worn away. The title reveals that the patch is in fact a piece cut from the spine of a previous binding of the notebook. Judging by the cracked leather running across the patch, it is likely that the previous spine was worn down from overuse.[60]

The bindings discussed above reveal that school notebooks continued to function as paper machines well after they were bound. This means that when students needed to find information in their notebooks after they finished school, they did not necessarily have to move through the pages one after another from start to finish. As the makers of their own notebooks, they were already familiar with the order of the subjects. They could dip about and jump from one part or page to the next as necessary, creating their own order within the bound pages whenever they used them. From this perspective, the fixed order of the pages corresponded to a more flexible understanding of the notebook's architecture that was preserved in a student's memory. Put more clearly, since students created the material structures of their notebooks and organized the order of the subjects therein, they could move efficiently through the content. Thus, even after they were bound, school notebooks could be used by their creators as organized, but adaptable, collections of knowledge. Like manuscript commonplace books, student notebooks were, in the words of Anthony Grafton, efficient "information-retrieval machines."[61]

---

57. Fowler Bound MS (1780), spine, front and back covers.
58. For representative examples of relatively plain academy notebook spines and covers, see Anonymous Bound MSS (1787).
59. The technique of inserting blank pages in modern books and notebooks was commonly practiced by adults. See Gibson (2010).
60. Anonymous Bound MSS (1787), vol. 1.
61. Grafton (2012).

CHAPTER 4

ANNOTATING

Revisibilia Made through Annotation

How many shillings are in a pound? How many pennies are in a shilling? These kinds of calculations were carried out hundreds of times a day by students learning to keep a ledger notebook prior to the nineteenth century. Every time they entered an expense in their ledger, they had to remember that one pound was worth twenty shillings and that a shilling was worth twelve pennies. Whereas basic bookkeeping was commonly learned by Scottish girls and boys, students being trained to manage large companies or households were given advanced instruction by accomptants, that is to say, writing masters who specialized in the art of keeping ledgers.

Bookkeeping required its practitioners to process a large amount of information. Even the data preserved in a relatively straightforward household ledger far exceeded the memory of its keeper. Data and their corresponding labels had to be corrected in one notebook and then copied into another. It was for this reason that students learned to treat an accounting ledger as an interactive annotation platform, as a place where they could practice transferring scribbled numbers and words from bills into hand-drawn tables that tracked credit and debits over time. Indeed, bookkeeping was in many respects a never-ending exercise in annotation that only ended when companies or households folded. In this sense, annotation was a skill that enabled a writer to organize and reorganize the names and numbers of predigital data by hand on paper.[1]

To learn the forms of sophisticated annotation used by bookkeepers, students wrote and rewrote loose-leaf sets of accounts from exemplars. In most

---

[1]. Müller-Wille (2017).

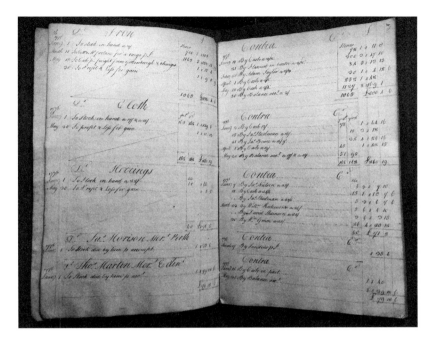

FIGURE 4.1. Robert Richardson of Pitfour, *School Exercise Ledger* (1776), Bound MS, NLS MS 20985, ff. 3v–4r. Reproduced with permission of the National Library of Scotland.

cases, the rewritten sets are the only material trace left of the long process that students employed when they were learning how to annotate and transfer bookkeeping data. Once students had finished their sets, they bound them as a notebook that served as a reference point in future endeavors. Two superbly crafted notebooks of this nature were made in the mid-1770s by the Perth Academy student Robert Richardson (figure 4.1). As the nephew of the successful merchant John Richardson of Pitfour, Robert was preparing for a career in the Richardson company, which traded in port cities located in the Baltic and the Low Countries, as well as France, Spain, and other parts of the Mediterranean. Every page of his neatly written labels and numbers reveals that the management of data on paper required a wide range of mutually dependent conceptual and scribal skills.[2]

Richardson's recopied sets are a sophisticated example of the bookkeep-

---

[2]. The history of John Richardson and his company is given in Haldane (1981). Robert Richardson's notebooks are housed in the NLS's Richardson Family Collection. Robert was identified as John Richardson's nephew in a letter written by Dr. James Wood to the renowned physician William Cullen. See Wood, MS Letter (1789). Robert had become ill, and both Wood and Cullen recommended that he summer in Portugal.

ing feats that could be accomplished by students who had learned how to combine annotating skills with the larger writing and codexing skills of notekeeping. Though the sets address primarily bills and commodities, their visual organization speaks to the rich variety of ranging skills that many Scottish schoolgirls and schoolboys placed in the service of annotation. But perhaps more importantly for the purposes of this chapter, Richardson's advanced ability to structure quantitative and qualitative data on the page presents an instructive example of the hybrid nature of student notebook annotation. In one sense, the script could be a transcription, that is, material that was copied from an exemplar. But in another sense, some of the script could be a composition, that is, numbers, labels, and corrections that a student such as Richardson created on his own to serve as annotations that increased the script's accuracy and accessibility. This kind of hybridity was by no means unique to Richardson's educational journey. Indeed, transcriptions interwoven with annotations regularly appear in many school notebooks and in other codices made by students while they lived at home.

As we will see in this chapter, students learned many important notekeeping skills through making a variety of annotative marginalia, paratexts, and ciphers as they participated in the process of transforming graphite, ink, and paper into a notebook, that is to say, a useful artifact that was both a learning device and reference tool. The journey to becoming an annotator was not immediate and required the guidance of teachers ranging from professional masters to family members. It took place in schools and households. Instead of following an explicit curriculum, it was learned along the way, frequently intersecting with the rich world manuscript culture that thrived through the creation of diaries, memoranda books, and even ephemeral paper slips. Put another way, within the larger world of student notekeeping, annotation was not solely a matter of scribbling on the pages of a recopied notebook.

Since the skill of annotating involved elements of composition *and* transcription, it served as a mode of revision, by which I mean an intentional alteration that constituted a handwritten textual innovation such as a "deliberate deletion, addition, insertion or rephrasing" that operated as part of the text after the change was made.[3] Here the modes of annotating being learned or exercised by student notekeepers drove the process of revision via the insertion of revisibilia, things revised, into the script. Examining such revisibilia makes it easier to see the many functions of annotations, including the role

---

3. Beal (2008, 344–346). Beal attributes a revision to the "author" of the text. In the world of predigital notekeeping, the term "reviser" is more suitable, as it affords agency to whoever was doing the writing.

that they played as paratextual tools that, when used collectively, allowed student notekeepers to improve or clarify what they were asked to copy or what they already had written.

In what follows I examine the collection of annotating skills that students learned while making or using revisibilia. Since the material culture of student, or even childhood, annotation is a sizeable and relatively unmapped world of ephemera, I concentrate on skills that offer insight into how young Scottish annotators were learning to create notebooks. I focus on marginalia, paratexts, and ciphers, that is to say, revisibilia that occur often in notebooks kept by students. In following this path, I wish to present a clearer picture of how young notekeepers learned the skills that allowed them to become adept marginalists, cipherers, bookkeepers, and editors. Throughout the chapter, I take care to highlight the ways in which the material and kinesthetic aspects of inventing and inserting revisibilia ensured that student notebooks were dynamic artifacts of learning. We will see that writing and ranging revisibilia constituted a transformative mode of interface that treated notebooks as fluid artifacts that could be rewritten and reshaped in realtime as they were made and used.

Marginalia as Scribal Interface

Students who kept school notebooks sometimes lived with families that maintained modest libraries or attended churches that kept collections of religious books like hymnals, bibles, psalters, and sermons. In some cases, even before they entered school, students were tempted to doodle in these books and, in so doing, began to learn how to use script as an annotation tool that fitted their own interests or needs. The end result was a meaningful collection of marginalia that functioned as annotations and accordingly transformed a printed book into a kind of rough or improvised notebook.

As argued by the historian of literature M. O. Grenby, marginalia potentially offer direct, yet limited, access to the ordinary lives of children because they are more spontaneous and sometimes occur in unsupervised settings.[4] The marginalia inscribed by Scottish girls and boys from the seventeenth century onward appeared in places that ranged from court books to presbytery reports. It was common for children to write in the same book over a period of time, inscribing things like signatures, poems, scripture verses, or maxims they had memorized. The schoolgirl Jeane Masson, for example, staked her

4. Grenby (2011, 226).

claim on a seventeenth-century copy of Ellon Kirk session minutes by writing *"hic Liber ad me pertinet"* and the young John Greig wrote financial calculations in a copy of the *Edinburgh Almanack* during the 1750s and 1760s.[5]

While historians sometimes note the importance of children's marginalia, hardly any of it has been collected or catalogued. One of the richest Scottish collections of extant marginalia is located in the Dunimarle Library collection kept at Duff House near Banff. The collection consists of hundreds of inscriptions made from the 1770s to the 1790s by the children of Lieutenant-General Sir William Erskine of Torrie. Erskine had received a baronetcy for his outstanding service to the military. Since he was abroad during many of the years the children made their marginalia, especially during the 1770s when he served as quarter-master general for Lord Cornwallis during the American War of Independence, the education of the children was overseen by his second wife, Frances (née Moray) Erskine.[6]

Though the collection is now kept at Duff House, the children originally made the marginalia in the books of the family library at Torrie House in Fyfe. Aside from being catalogued, the collection has yet to receive sustained attention from historians.[7] But the collection is a treasure trove of information about the annotation habits of children and is a helpful indicator of the kinds of marginalia that can be used to investigate the forms of graphic order being spontaneously internalized by potential student notekeepers. It also points to the need for further studies on the kinds of childhood writing and drawing practices that were used in settings like libraries and homes.

Examining the Erskine marginalia as evidence of artifice will require us to slightly adjust our gaze so that we can properly appreciate their significance. Studies that address adult marginalia that were written during the modern period have traditionally used annotations to determine how readers reacted to the content of printed books.[8] This approach sometimes gives implicit priority to the information being conveyed through print. When exploring how children used marginalia, we need to treat this kind of prioritization with care. Young writers often inserted marginalia on topics that were on their minds,

---

5. Jeane Masson is quoted in Simpson (1942, 206–207). The teenage John Greig wrote financial calculations on the 1756 and 1767 editions of the *Edinburgh Almanack*. The volumes are housed in the NLS Greig Papers, Dep. 190, Box 4. For seventeenth-century scribbles, see Williamson (2002).
6. Stephens and Stearn (2004); "Erskine, Sir William, 2nd Bt. (1770–1813), of Torrie, Fife" in Thorne (1986).
7. Cumming [n. d.]. There are also marginalia in the educational books housed in the NLS's Castle Fraser Collection, but most of it seems to have been inscribed by adults. For example, the underscoring in John Locke's *Some Thoughts Concerning Education, Fourteenth Edition* (1772) is made on passages in which Locke is commenting on how to teach headstrong children.
8. Jackson (2001; 2008).

thereby creating a situation in which their script was not responding directly to the content of the printed page on which they wrote it.

Sometimes young writers used the flyleaves, margins, and other blank areas of printed books as scrap paper, as places to write thoughts that were tangential or even irrelevant to the book's subject matter. Based on the current evidence, it seems that the Erskine children, as well as other Scottish children writing in religiously oriented books, were sometimes prone to using the blank spaces of printed books as a kind of *ad hoc* notebook for thoughts relevant to what they were thinking at that moment. Put more clearly, though the Erskine children occasionally reacted to the content of a book, most of their marginalia addressed other topics that they found interesting or which were relevant to other subjects they were learning at home and in school.

As observed by Seth Lerer, children's marginalia "are sometimes barely legible, tantalizingly irrelevant to the texts before them, evasive, duplicitous, or just plain weird." He goes on to mention that these qualities make them exciting historical sources because they "change, irrevocably, the status of the annotated book as artefact."[9] I concur with this assessment and, as we will see below, when Lerer's observations are placed in conversation with school notekeeping practices, it becomes easier to see how young writers used marginalia to transform printed books into *ad hoc* notebooks. I suggest that the Erskine marginalia are annotations in the sense that they, firstly, added meaning to the page that was not there prior to their arrival and, secondly, functioned as a training exercise through which children learned to see the printed page as an inscribable entity.

Most of the Erskine annotations were made in ink, graphite, and watercolors by Magdalene, Henrietta, Elizabeth, John, James, and William Erskine (figure 4.2). They wrote signatures and a variety of marginalia in the library's primers and children's literature books, and other annotations in books used more often by adults.[10] They ranged tables, wrote words, aligned letters in palindromes, calculated equations, sketched polygons, and drew pictures. All of these marks shed insight into their interests and ambitions. The oldest son would be given the title and wealth of the father; all the younger siblings, like many children of the lower ranks of the Scottish aristocracy, would most likely have had to make their way in the world through marriage, industry, trade or a profession.

---

9. Lerer (2012, 127).

10. The ages of the Erskine children are established in Cumming [n. d.]. They are pictured in David Allan's *Sir William Erskine of Torrie and his family*. The painting also depicts Torrie House, the original site of the Erskine library, in the background. Wright, Gordon, and Smith (2006, 63).

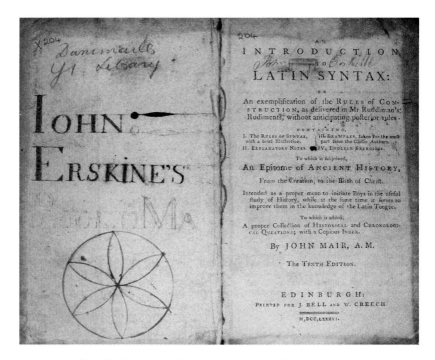

FIGURE 4.2. John Erskine's inscription of his name and a geometric flower in John Mair's popular *An introduction to Latin syntax... Tenth Edition* (Edinburgh: J. Bell and W. Creech, 1786), DH LIB 204. Reproduced with permission of the Dunimarle Library, Duff House, Historic Environment Scotland.

The Erskine children expanded the family's library books into their own bespoke notebooks by adding annotations about topics relevant to natural history and other juvenile interests. They also inserted lists that named notable professions and honed mathematical skills that would help them keep track of accounts in the future.[11] Notably, these marginalia were not part of their official lessons and reveal that the children were thinking about their various school subjects in their own way via different kinds of media, library books in this case, that they found conducive to the abilities that they possessed.

Perhaps one of the clearest examples of self-directed learning occurs in the mathematical marginalia most likely penned by the teenage John Erskine in the library's copy of Eutropius's *Historiae romanae breviarium* (1779).[12] Erskine wrote a neatly aligned multiplication table (figure 4.3). Writing the table both

---

11. For inscriptions about birds, arithmetic calculations, professions, and drawings, see the following books in the Dunimarle Library collection: Moir (1775, DLDH No. 2130); Boyer (1782, DLDH No. 986); Anonymous [17—], DLDH No. 247); Mair (1779, DLDH No. 1791).
12. Eutropius (1779, DLDH No. 362).

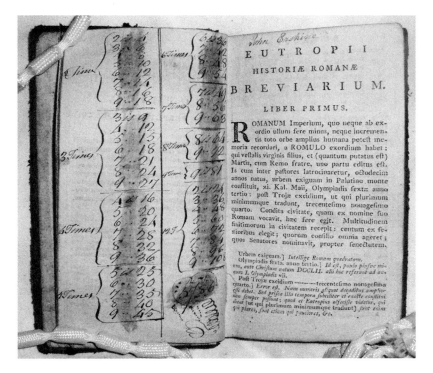

FIGURE 4.3. John Erskine's inscription of a multiplication table in Eutropius, *Eutropii historiae omanae breviarium, ab urbe condita usque ad Valentinianum et Valentum Augustos . . . In usum scholarum* (Edinburgh: Gordon, Gray, Dickson et Creech, 1779), DH LIB 362. Reproduced with permission of the Dunimarle Library, Duff House, Historic Environment Scotland.

improved his memory and created a tabular learning device that he could use to further memorize multiplication products via a graphic artifact designed to meet his own needs. To refresh his multiplication skills, he only needed to cover the part of the table that he wanted to memorize with a piece of paper and then see if he could remember the answers.

The self-organized nature of the Erskine marginalia stands in contrast to paper calculating instruments used by other students at the time. Rather than memorizing the multiplication table, some students used the matrices printed in books, while others used a Napier box, a device that contained rotating paper cylinders labeled with numbers turned in various ways to get an answer. Alternatively, there was even a "Multiplication Table" song devised in 1795 by Preston & Son of London. Printed across three sheets of paper, the melody was "adapted for Juvenile Improvement as a lesson for the piano forte." The lyrics related all the combinations and answers of the times table up to twelve times twelve. As intimated in 1798 by the wry book reviewer of *The*

Monthly Magazine, the music was more suited for learning to play the piano than to learning multiplication tables.[13] Rather than using such complicated modes of paper-based interface, Erskine opted to learn his multiplication tables through writing his own inscriptions in a library book. His table is but one example from the Dunimarle marginalia that illustrates how the children transformed the blank spaces of library books into makeshift notebooks that allowed them to learn graphic structures that could be used to organize qualitative and quantitive data.

Another example of scribal interface in the Erskine marginalia is the graphite grid drawn by "Miss Erkine," one of John's younger sisters. Occurring across two facing endpapers in the library's copy of Robert Lowth's popular *A Short Introduction to English Grammar*, Miss Erkine's wobbly, slightly uneven lines suggest that she drew it freehand, or with a straight edge made from folded paper.[14] The purpose of the grid seems to have been to keep track of data associated with a list of thirty-seven surnames such as "Bell," "Graeme," and, indeed "Erskine." On the verso page, she created twenty-eight rows and eleven columns which took up the entire space of the page. On the recto page she ranged nine rows and eleven columns, which took up the top half of the page. On both pages, she placed one surname at the start of each row. The end result was a word table in which each row began with a surname that was followed by a series of boxes that could be marked or left blank. The columns are unlabeled, so their meaning is not known. After she finished designing the table, she put a plus sign and a circle next to five of the names. She left the rest of the grid blank.

The gridded table is a snapshot of graphic knowledge in action. It strongly suggests that Miss Erskine found herself in a situation, most likely a game consisting of eleven steps or rounds, in which she needed to order a cache of data that she understood would exceed her working memory. The wobbly lines show she was in a rush to draw the grid quickly in order to record the information then and there, because making better lines would have required her to spend time finding the rulers and compasses that were used by her siblings to make well-formed geometric marginalia in other library books. The ephemeral nature of the need to order the data is further evinced by the fact that she only completed the first column.

Though Miss Erskine's desire to order the data had for some reason ended abruptly, perhaps by the quick conclusion of a game, the grid's importance is

---

13. Preston (1795) and Phillips (1798, 50).
14. Lowth (1783, DLDH No. 205).

not diminished by its blankness. Indeed, her inscriptions reveal that, when she found herself in need of a data management device to help with an everyday task, the first thing she thought of was a grid. Not only did she think of it, but she knew how to draw it and how to arrange data inside of it, thereby indicating prior training in the mnemotechnics of paper-based forms of information management. Additionally, the presence of the grid speaks to the ways in which she, and, more broadly, children with access to family or institutional libraries, could use marginalia to repurpose printed books into makeshift notebooks that helped them order the data of everyday life.

The Erskine library contained a number of primers and dictionaries in which the children drew or wrote. As indicated by insertions in primers like Alexandre Scot's *Nouveau Recueil*, the children learned modern languages that would help them pursue international business interests. James Erskine's marginalia, for instance, indicate that he studied French while learning other practical subjects at the Perth Academy during the 1780s—a fact memorialized when he wrote the following inscription in Scot's book: *"James Erskine Dec. 28th 1784 Perth Academy, 8 o'clock in the morning at French."*[15] John Erskine also signed his name, "John Erskine his book 1789," and doodled in the library's copy of Abel Boyer's *The Complete French Master for Ladies and Gentlemen*. Likewise, in the 1790s Henriette Erskine signed her name in the front of Boyer's *The Royal Dictionary Abridged* and Thomas Deletanville's *A new French Dictionary*.[16] Such inscriptions reveal how library books were used as graphic learning platforms by the Erskine children.

The playful and pragmatically oriented nature of the Erskine inscriptions both bear witness to what Seth Lerer has called the "culture and motives" of the children who wrote and drew marrginalia.[17] They reveal how the power of marginalizing, and self-directed annotating more generally, was used by children to reinforce the strong utiliarian facets of Scottish developmental psychology in a way that was similar to the pragmatic outlook evinced in the notebooks made in schools and academies. This developmental context in turn shaped, and was shaped by, an educational ideology that, even at the privileged level inhabited by the Erskines, drew strong links between the act of writing and self-improvement. This suggests that annotating, even when pursued in leisure as scribbles in library books, was indirectly a form of self-assessment, a mode of internalizing information that children associated

---

15. Scot (1784, DLDH No. 293).
16. Boyer (1720, DLDH No. 334); Deletanville (1794, DLDH No. 328).
17. Lerer (2012, 127).

Paratexts and Editorial Training

Within the pages of student notebooks and other related kinds of ephemera, annotations sometimes functioned as paratexts that made the final version of the script more accessible. There were many kinds of paratexts in the modern world of print and script, including title pages, prefaces, postscripts, indices, footnotes, sidenotes, catchwords, and other kinds of apparatus that augmented or changed the meaning of a text. Within manuscript culture there were also marginalia, such as carets, underscoring, and quotation marks that operated as diacritics. The importance of paratexts has received more attention in recent years, particularly by scholars interested in the material and visual culture of handwritten and printed books,[19] and more widely by scholars who study classical and prehistoric writing systems.[20] In this section I would like to highlight how paratexts played two important annotative functions within the wider culture of notekeeping that surrounded Scottish children keeping notebooks. In the first role, they functioned as tools that helped younger writers see how a handwritten text could be corrected. In the second role, they helped young annotators use the paratexts on their own or with guidance to make revisions. In both cases, the young writers were learning important scribal annotation skills that could be applied to a variety of domestic, commercial, or academic documents when they became older.

Students learning to use marginalia usually wrote with the expectation of surveillance or intervention on the part of a tutor. As indicated by the extant ephemera of Edinburgh High School, grammar school students presented their written work to be read by their Latin teachers as well, most likely for prizes.[21] There was also the larger expectation that parents would read the final version of a school notebook. After attending Lewis Lochee's Royal Military Academy in Chelsea, for instance, the sixteen-year-old James Ochoncar Forbes recopied his notes into a 1781 notebook titled *A New System of Practical Geometry*. As indicated in the notebook's title, Forbes fully anticipated that his father, the Scottish Admiral and Lord James Forbes, would read it: "presented

---

18. For further information about the Erskines and their books, see Cumming (2017).

19. Genette (1997), Grafton (1999), Sherman (2009), Smith and Wilson (2011), and Tweed and Scott (2018).

20. Jansen (2014), Tversky (2001; 2019, 226–227), and Faulkner (2005). The cognitive framework of symbolic representation in ancient writing systems is addressed in Damerow (1996).

21. Anonymous Bound MS (1742–46); Anonymous Bound MS (1744–1787). For Edinburgh High School, see the EHS collection housed by the Edinburgh City Library.

to Admiral Forbes by his most obedient humble Servant James Forbes."[22] Add to this the fact that teachers working in fee-paying schools anticipated parental oversight and gave students material relevant to the moral and utilitarian topics promoted by enlightened adults.[23]

Students learned early in their education to recognize the utility of having their script corrected with marginalia written by an instructor. The first stage of the process of correction involved a tutor or family member reading a young writer's work and then correcting the spelling and grammatical mistakes with underscoring, strikethroughs, and insertions. Learning how to recognize and use such corrections often began at home through writing copy sentences and sometimes through writing a diary. As shown in the work of Arianne Baggerman and Rudolf Dekker, this practice was not unique to Scotland, or even Britain and its former colonies.[24] Some students continued to receive spelling corrections from their parents on the letters they sent home from school. During the 1780s, for instance, the sixteen-year-old Mary Moore received grammatical advice from her father, Dr. Robert Moore of Blairston, on the letters that she sent him while attending a boarding school in Edinburgh.[25]

Family members and teachers corrected the spelling and grammar of children's diaries with a view to improving style and clarity. But crucially, such corrections also showed young learners how to annotate a text.[26] Good examples of the kinds of corrective annotations witnessed by a young Scottish writer occur in the diary kept at home by the eleven-year-old Marjory Fleming. Made up of unbound quires, Fleming's diary features the spelling corrections written by Isabella Keith, Fleming's first cousin and governess. Keith's diligence and Fleming's perspicuity are evinced throughout the script, inspiring nineteenth-century artists such as Warwick Brookes to visualize their scribal efforts as a collaborative performance (figure 4.4). An example of the kinds of corrective marginalia Keith used to annotate Fleming's script can be seen in the following excerpt from the diary:

[f. 16v]
To Day I aff-
ronted myself be-

---

22. Forbes Bound MS (1781). For further information on James Ochoncar Forbes, see his *ODNB* entry.
23. The pedagogical and mnemonic advantages of copied textbooks are underscored in Knoles, Kennedy, and Knoles (2003).
24. Baggerman and Dekker (2008; 2009) and Baggerman (2008).
25. Moore Unbound MS (1786–1790). Mary Moore mentions her father's spelling corrections in letter 7, dated 5 December 1786.
26. See again Baggerman and Dekker (2009) as well as Rossignol (2015).

FIGURE 4.4. A nineteenth-century depiction of the child diarist Marjory Fleming and her tutor Isabella Keith by Warwick Brookes in L. Macbean, *The Story of Pet Marjory (Marjory Fleming) with Her Journals, Now First Published, 2nd ed.* (London: Simpkin, Marshall, Hamilton, Kent & Co., 1905), opposite page 192. Author's copy.

ANNOTATING | 127

fore Miss Margret & Miss Isa Cer[?]-
ford Mrs Craford
& Miss Kermical
which was very
nauty but I
hope that there
will be no more
ever in all my
Journal,

[f. 17r]
~~To Day~~
To Day is Saturday
& I sauntered
about the <u>woulds</u>
& by the burn sid
I dirtied myself<u>e</u>
which puts me in
mind of a song my
mother con[m]p[o]sed it was
that she was out & I. . . .[27]

Keith corrected the script by underlining incorrect spellings ("woulds" and
the "e" in "myselfe") and inserting an "m" and "o" above "conpsed" to make
"con[m]p[o]sed" (figure 4.5). These and many similar annotations written by
Keith in the diary illustrate the fact that young learners wrote in the expectation that their writing would be subjected to scrutiny, familiarizing them
with the culture of correction that permeated the material world of manuscripts. The situation transformed the handwritten page into a media platform through which young notekeepers and diarists could communicate their
thoughts, concerns, observations, and emotions to the adults who they knew
would read their script. Whereas the relationship between media and the
mind was ostensibly guided by the watchful eye of an adult, children and adolescents found creative ways to subvert the process. Fleming, for instance, offered apologies and thinly-veiled criticisms, thereby cleverly communicating

27. Fleming Bound MS (1810–1811) MS 1097, 16v–17r. Fleming's diary was published as Fleming (1934).
Her manuscripts in the NLS are catalogued under Marjory Fleming, *Papers of Majorie Fleming* (1810–1811),
Bound MS, NLS Acc. MSS.1096–1100.

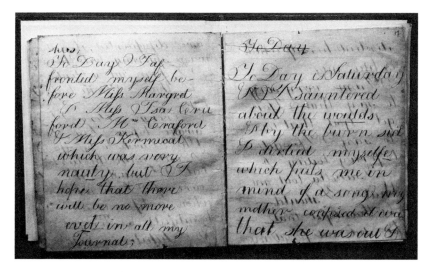

FIGURE 4.5. Annotations of Isabella Keith in Marjory Fleming, *Diary* (1810–1811), Bound MS, NLS MS 1097, 16v–17r. Reproduced with permission of the National Library of Scotland.

her views to Keith, her instructor and mentor, and any other member of the family who might decide to read the diary.

There are only a handful of extant diaries that were kept by Scottish children and adolescents, but they all feature various kinds of corrective marginalia.[28] Observing or inscribing such revisions helped young learners more readily understand the editorial potential of marginalia when they became older and wanted to correct the script of their school notebooks, particularly at the scrolling stage.

Aside from minor corrections, a persistent form of annotation presented in school notebooks is the headings, that is to say, words or phrases that served as labels for narrative or figures. A good example of this practice occurs in Peter Purdie's 1823 *Mathematics* notebook, which features a large number of doodles and was written in an unsteady hand, indicating that it might possibly be the rough version of material that he partially copied and calculated on his own (figure 4.6).[29] A number of Purdie's calculations addressed problems in which the numbers represented metrological or monetary units. When Purdie first drafted the manuscript, he did not write these units for some of his an-

28. See Sinclair Bound MS (1809), Campbell Bound MS (1789), Bogle Bound MS (n.d.), Sandy Bound MS (1788), printed as Sandy (1942). Summaries of the content of the Sandy and Bogle diaries are given respectively in Matthews (1984, 92, 130).
29. Purdie Bound MS (1823).

ANNOTATING | 129

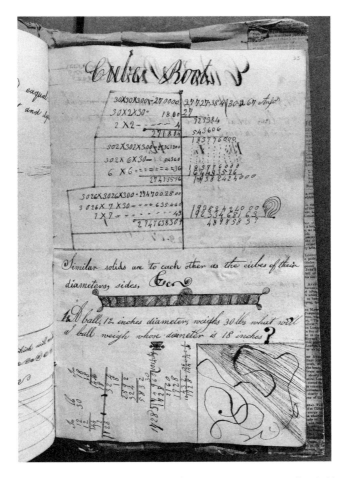

FIGURE 4.6. Peter Purdie, *Mathematics* (1823), Bound MS, NLS MS 14288, 23r. Reproduced with permission of the National Library of Scotland.

swers to the set problems. At some point after he finished the draft, he went back and added the names of the units, "guinea" or "mile" for instance, in darker ink and a slightly larger font.[30] The addition of such labels, either composed or copied, was an indispensable mode of visual and conceptual categorization that I will explore more fully in later chapters.

Once students had learned to use paratexts to correct or label what they had written, they were in a better position to see that annotations were a form of revisibilia that allowed them to shape a notebook into an artifact that fit-

---

30. See Purdie Bound MS (1823). Good examples occur across ff. 5–6.

ted their personal learning or organizational needs. One of the best examples of this kind of enhanced paratexting occurs in the conduct codex created by the twelve-year-old Margaret Monro, which we encountered in previous chapters. We will recall that her father, Alexander Monro Primus, wrote out the original script on bifolia over the course of several months. Monro then copied his script to make a beautiful codex of her own. According to the developmental psychology of the day, the many writing and ranging skills honed by such an exercise significantly improved her rational abilities. Her efforts, therefore, were not a mindless form of replication, but a vital interactive mode of thinking that increased her capabilities in way that prepared her to use writing as a form of realtime thinking when she became an adult.

In addition to working with the script Primus had written, Monro's codex reveals that she used paratexts to learn how to significantly revise his text. Her use of paratexting in this manner can be studied by comparing her manuscript to the one he made.[31] In other words, the fabric and layout of her manuscript presents evidence that helps us understand how she changed elements of her father's draft through omissions and insertions that reflect her own knowledge. All of these transformed her codex into an artifact that made more sense to her and was easier for her to use in the future.

Primus organized the content of his manuscript into seven main headings. They were as follows: "Part I: Of the Education of Girls"; "Part II: Of the General Conduct of Life"; "Part III: Of the Commerce with Men"; "Part IV: Of the Government of Servants"; "Part V: Of the Management of Children"; "Part VI: A Summary of Religion"; "Part VII: Of the Origine of Government." He then divided the content of the parts into subheadings.[32] It is in the placement, alteration, and removal of these paratextual headings and subheadings where we can best see how Margaret Monro exerted her budding editorial prowess in her version of the script.

To fully appreciate the nature of her agency, however, we first need to note several important features of her father's manuscript. Primus wrote his script on the front and back of single sheets. To show Monro how to lay out her script, he ranged the face of each side so that it had wide margins that provided the space for each part's subheadings. He utilized different sizes of paper, most likely because he was sourcing whatever was readily available at home. The sizes were similar to folio and quarto measurements, but they seem to have

---

31. Monro Primus Bound MS (1738).

32. Eddy (2019) discusses the structure and themes of the headings and subheadings found in Monro Primus Bound MS (1738) and Monro Primus and Monro Bound MS (1739).

been cut slightly smaller for reasons that are unclear. Once he finished the first draft of his script, it seems he edited it three more times. I would first like to summarize the edits and then turn to the ways in which Monro used her handwritten paratexts to change the script so that it suited her needs and interests.

In the first edit, Monro's father used the wide margins to add more subheadings. He also inserted minor corrections consisting mainly of crossed out mistakes and brief clarifications here and there. In the second edit, he added further marginal insertions that were in slightly darker ink and usually around a sentence long. In the third edit, he placed each written sheet inside a blank bifolium. This created a format in which each side of each sheet had a new blank page facing it. These fresh pages offered a much larger area for annotations. He used the additional space to write longer insertions that often ran to one or two paragraphs.[33]

When considered collectively, Primus's sheets and the blank pages added up to a large handwritten book being edited in over time. The first and second set of his edits (and possibly the third) almost certainly would have been visible on the script when Monro rewrote her own version. Though she copied all of the first edits, she did not copy all of the second and third edits. When it comes to understanding the enactive notekeeping skills that she was learning, the important point here is that interfacing with her father's manuscript showed her how to use the layout of the page to make corrections and how to insert and stack folios in a way that was conducive to both ordering and editing a manuscript book in realtime. It also showed her how to navigate between the pages of a manuscript that consisted of several different sizes of paper.

When Monro's manuscript is compared to that written by her father, it can be seen that she made significant changes at the paratextual level, resolving what (and what not) to underscore, indent, capitalize, italicize, and insert. Each of these decisions impacted the presentation of words differently and affected how she interacted with each page while she was writing it. The indentions that she inserted at the beginning of paragraphs, for instance, injected blank space into her script in a way that made it easier for future readers to skim through the pages. There are also several pages in her manuscript where she broke her father's longish paragraphs into two and placed a new indention at the beginning of the new paragraph. Her father seems to have encouraged the development of this editorial skill. In the editorial comments

---

33. Good examples of darker ink insertions and the use of new folio pages occur across Monro Primus Bound MS (1738, esp. ff. 32v–33r).

written in his manuscript, Primus suggested how she might split several of his long paragraphs. In these cases, he wrote "a new line" in the margin and an "x" at the end of the sentence where Monro might want to make the break.[34] She implemented most of her father's suggestions, which enabled her to edit her text in her own way.

Monro introduced new forms of underscoring to her script as well. Her underscoring functioned as paratextual apparatuses because they drew attention to keywords and headings, making the script more accessible. Her thin and thick lines denoted lesser and greater forms of emphasis. Drawing these lines under words required her to understand the thickness of the paper and the miscibility of the no fewer than three different kinds of ink that she used. With this knowledge, she made thin lines by pressing softly and changing the angle of her nib (or quill). She made thick lines by pressing harder, moving slower and re-angling her nib.

Monro's manipulation of the manuscript reveals that she understood how to interpret and deploy annotations in a way that enabled her to innovatively craft her thoughts in ink on paper. Her thick underscored lines, most of which do not occur in her father's script, are placed strategically on words that she wanted to highlight. Her additions arguably made her codex easier for her and others to use.[35] As can be seen in the section of the page that is depicted in figure 4.7, she added a thick line to separate the main headings from the script's prose. Her father did not use this technique and, consequently, the main headings in his manuscript are harder to spot.[36]

Another informative example of Monro's underscoring occurs in the subheadings that she wrote in the margins. She used thick strokes to underline all of them, a practice that can be seen in the "Judgement of Persons" subheading written in the right column of the page depicted in figure 4.7. Though her father wrote his subheadings in his marginal columns as well, his were formatted differently. He added some and demarcated others by drawing boxes around them. But Monro's underscoring was her own; here, and for the running headings positioned at the top of the pages, she shaped the script in a way that made sense to her. She enacted this kind of paratexting throughout the text, adding underscoring and other markings or spacings to make the text more visually accessible.

---

34. Examples where Primus suggested a break occur in Monro Primus Bound MS (1738, 17, 29).

35. A number of cancellations presented in Margaret Monro's codex suggest that she, or perhaps another person, read it at a later date with a view to transcribing or publishing it.

36. For Margaret Monro's underscored main headings, see Monro Primus and Monro Bound MS (1739, ff. 5, 54, 207, 276, 289, 308). For Primus's main headings, see Monro Primus Bound MS (1738, ff. 3, 14, 84, 125, 113, 144).

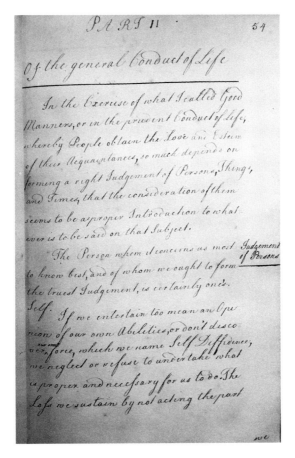

FIGURE 4.7. Paratexting in action. The twelve-year-old Margaret Monro learned to shape the script in her own way by underscoring the main heading ("Part II" and "Of the general Conduct of Life") and the subheading ("Judgement of Persons"). Alexander Monro Primus and Margaret Monro (transcriber and editor), *An Essay on Female Conduct* (1739), Bound MS, NLS MS 6659, f. 54. Reproduced with permission of the National Library of Scotland.

In addition to making the foregoing paratextual changes, Monro made further revisions in her script by altering or omitting several of the subheadings that her father originally had used to classify and label the subsections in his letters. His script contains fewer subheadings than Monro's. It is possible that the first version of his script contained hardly any at all. Closer inspection of Monro's subheadings reveals that she invented and inserted a notable number of them into her script by herself. In Part II of her manuscript, for instance, she omitted her father's subheadings "Surlyness" and "Petting" and instead introduced a new one for "Experience." Such changes indicate that she categorized and recategorized the script as she went along. This activity was facilitated by the fact that she had formatted the margins of the page as wide columns into which she could write her own subheadings.[37]

Monro's revisions of the subheadings are present from the very start of the manuscript. Though she faithfully copied her father's prose in the first five paragraphs of Part I, she did not elect to use the original subheading that he suggested for this part of the script. He had only included one, "Mercating," which flagged a passing reference in the second paragraph to bookkeeping. Monro, however, omitted this subheading in her codex and inserted new subheadings for "Reading" and "Writing," that is, topics discussed in the third and fifth paragraphs.[38]

In many respects Monro's subheadings more accurately represented the overarching themes discussed in the opening sections of the manuscript, demonstrating how she sharpened her skills of categorization through revision. Her sense of revising allowed her to see that the "Mercating" subheading was better suited for a later section of the script that more thoroughly addressed the larger importance of household management and the necessity "for a Girl to know how to manage rightly the Things that are purchased." Monro accordingly moved the "Mercating" subheading to a later section, revealing that she understood how to revise categories thematically and how to design the script in a way that was clearly laid out on and across paper.[39]

Perhaps the most impressive example of Monro's paratextual skills is displayed in the three-page table of contents that she crafted and then bound into the front of the codex (figure 4.8).[40] Without this crucial piece of paratextual apparatus, the content of the finished manuscript, which ran for several hundred folio-sized pages, would have been difficult for her to access quickly. Like all tables of contents, hers was an index, a list of well-selected and ordered entries. The completed version provided a way for her to overview the content of the entire manuscript at a glance. Her father's manuscript does not contain an index, nor does it give instruction on how to make one. The columns, indentions, and chirography of Monro's index were, therefore, her own.

Indices were organizational tools that enhanced a student's learning experience. As shown masterfully by Ann Blair, within the wider world of manu-

---

37. Margaret Monro's subheadings for Part II, "Of the General Conduct of Life," occur across Monro Primus and Monro Bound MS (1739, ff. 54–207). Primus's subheadings for part II occur across Monro Primus Bound MS (1738, ff. 14–83).

38. Monro Primus and Monro Bound MS (1739, ff. 5–8); Monro Primus Bound MS (1738, f. 2).

39. Monro Primus and Monro Bound MS (1739, ff. 38–40); Monro Primus Bound MS (1738, f. 10).

40. Monro Primus and Monro Bound MS (1739, iv recto, iv verso, and v recto).

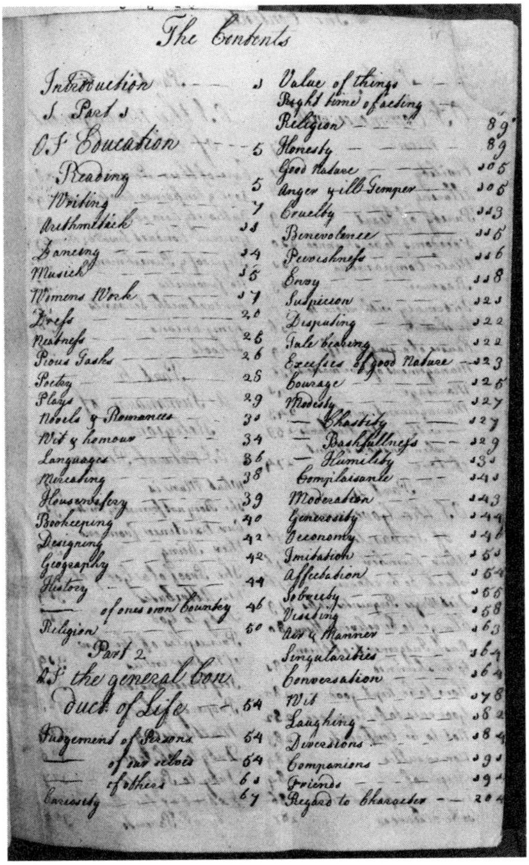

The Contents

Introduction — 1
1 Part 1
Of Education — 5
Reading — 5
Writing — 7
Arithmetick — 11
Dancing — 14
Musick — 15
Womens Work — 17
Dress — 20
Neatness — 25
Pious Tasks — 26
Poetry — 28
Plays — 29
Novels & Romances — 31
Wit & humour — 34
Languages — 36
Mercating — 38
Housewifery — 39
Bookkeeping — 40
Designing — 42
Geography — 42
History — 44
— of ones own Country — 46
Religion — 50
Part 2
Of the general Conduct of Life — 54
Judgement of Persons — 54
— of our selves — 54
— of others — 61
Curiosity — 67

Value of things —
Right time of acting —
Religion — 89
Honesty — 89
Good Nature — 105
Anger & ill Temper — 108
Cruelty — 113
Benevolence — 115
Peevishness — 116
Envy — 118
Suspicion — 121
Disputing — 122
Tale bearing — 122
Excesses of good Nature — 123
Courage — 125
Modesty — 127
— Chastity — 127
— Bashfullness — 129
— Humility — 131
Complaisance — 141
Moderation — 143
Generosity — 144
Oeconomy — 148
Imitation — 150
Affectation — 154
Sobriety — 155
Visiting — 158
Air & Manner — 163
Singularities — 164
Conversation — 164
Wit — 178
Laughing — 182
Diversions — 184
Companions — 191
Friends — 194
Regard to Character — 204

FIGURE 4.8. Margaret Monro's index. Alexander Monro Primus and Margaret Monro (transcriber and editor), *An Essay on Female Conduct* (1739), Bound MS, NLS MS 6659, f. iv. Reproduced with permission of the National Library of Scotland.

script culture, indices were arguably a leading finding device used to order codices from the Middle Ages forward.[41] Designing one was predicated upon the ability to assemble lists of ideas or facts extracted from print, script, the world, and even the mind. By the eighteenth century, this humanist principle of purposeful and organized extracting and list-making functioned as an essential organizational tool for those revising and transcribing notebooks in households and schools alike. Its use as an everyday organizational tool in domestic settings, for instance, underpinned the assemblage of everything from recipes

41. Blair (2010b), 132–152.

to natural history specimens.[42] The centrality of list making as a scribal information management tool also appeared regularly in eighteenth-century novels. The character Bridgetina Botherim in the 1800 feminist novel *Memoirs of Modern Philosophers* by the Scottish author Elizabeth Hamilton, for instance, created an index to help her tally the potential advantages of a relationship with the dashing Henry Sydney.[43]

Monro's exquisite index represents the final stage of a revision process that utilized list-making capabilities that she already possessed but were expanded exponentially while she was collecting and organizing her subheadings into preliminary lists that she then transformed into the index. Some of the subheadings, as we learned above, began originally as her annotations. Put another way, Monro's comprehensive index is an artifact that speaks to the ways in which she learned to create a thematically ordered knowledge system made up of headings and subheadings. Since she made the index after she had written and numbered the other pages of the manuscript, the act of creating it reinforced the modes of organization she had experienced while writing the main body of the script.

Ciphers and the Acquisition of Numeracy

The presence of calculations in the Erskine marginalia and Purdie's notebook reminds us that students used annotations to improve their numeracy skills. As observed many years ago by Keith Thomas, the history of numeracy is often overshadowed by the history of literacy. He astutely observed that numeracy "is of equal importance to anyone concerned to reconstruct the mental life and cognitive apparatus of the past, for numbers and number systems are among the most basic of all mental categories."[44] As we have seen in recent chapters, part of that apparatus was the ability to write and organize numbers on paper and across pages. In this section I wish to investigate this process in a

---

42. The overarching relationship between humanism and the creation of indices for manuscripts in modern Britain is addressed in Yeo (2014, 113–123). For natural history lists, see Cooper (2007, 72–80), and Müller-Wille and Charmantier (2012b). Leong (2018a) addresses the relationship between list making in recipes and health. Many of the manuscripts in the collection of modern Scottish cookbooks held by the National Library of Scotland feature indices. For an example of a manuscript Scottish cookbook index, see Malcolm Bound MS (1790, 1–8).
43. Bridgetina Botherim presents "A speech which had long been conned, twice written over in a fair hand" and was ordered on paper as a list of headings that were "a sort of index taken of the contents." Hamilton (1804, 99–100). For lists and indices in other eighteenth-century literary works, see Barchas (2003, 173–213).
44. Thomas (1987, 103).

way that makes it easier to see how students enhanced the function played by their notebooks as media technologies.

At the dawn of the modern era, arithmetic was understood to be "the science of numbers, the art of computation by figures."[45] The characters associated with arithmetic, bookkeeping, and other forms of advanced calculation were often called *ciphers* (or *cyphers*) and they usually were written as numbers, or as letters and symbols in equations with variables. Like letters, numbers were signs to which meanings were attached. In the words of the Edinburgh schoolmaster Thomas Bruce, "A Cypher depending upon it self is nothing in Value, but by its filling up of Places, hath Dependency upon some other Figure, to point, or speak out its Number or Value."[46] The act of writing and manipulating such ciphers was called *ciphering* (or *cyphering*). It was an essential skill for anyone seeking to keep track of data expressed as numbers across the many sheets, slips, receipts, ledgers, and memoranda notebooks that existed within manuscript culture.[47]

The efficient management of households, plantations, churches, businesses, governments, and even laboratories hinged on the presence of competent cipherers who knew how to write, correct, and evaluate the numbers flowing in and out of ledgers. The kinds of ciphering required for these tasks depended on a number of notekeeping skills, many of which were best learned under the watchful eye of an experienced tutor, master, or family member who could correct mistakes and make suggestions in realtime.[48] Outside bookkeeping, ciphering was conducted in many subjects, including basic mathematics, geometry, algebra, gauging, leveling, and surveying. Students ciphered numeric tables either as standalone entries or alongside relevant text and images. The "Mensuration of Planes" section depicted in figure 4.9, for instance, is a visually unified table in which the left page serves as the left column and the right page serves as the right column.

Ciphering numbers in scroll books and recopied notebooks was oftentimes a mode of annotating that sat on the border between transcription, composition, and revision. This connection led many schoolmasters and governesses

---

45. See the "arithmetic" entry in the *OED*.

46. Bruce (1724), viii.

47. The larger context of British mathematical instruction in which ciphering took place is addressed in Froide (2015), Harding (1972), Otis (2017), and Wardhaugh (2012, 63–90). For the British colonies, see Clements and Ellerton (2014), Dauben and Parshall (2014), and Denniss (2008).

48. Much of the research on ciphering and ciphering notebooks addresses colonial North America and the early American republic. Notable studies include Crackel, Rickey, and Silverberg (2014; 2015), Denniss (2012), Ellerton, Nerida and Clements (2012, 2014), Monroe (1917), Hertel (2016), and Richeson (1939).

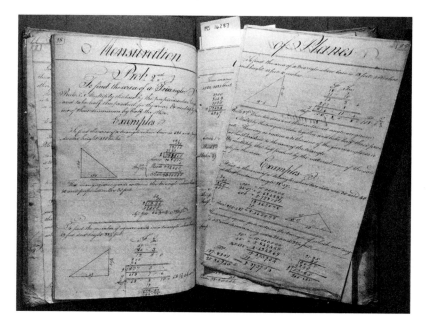

FIGURE 4.9. Anonymous, *Pure Mathematics* (1816), Bound MS, NLS MS 14287, 13v–14r. Reproduced with permission of the National Library of Scotland.

to underscore the link between the art of bookkeeping and the art of note-keeping that was fostered by Scotland's humanist tradition. The title page of William Gordon's *The Universal Account* (1763) is a good example of this relationship. It features the following Latin quotation from Cicero: *Quid munus republicae majus meliusue affere possumus, quam si juventutem bene erudiamus?* Taken from the second book of *De Divinatione*, its translation is: "For what greater or better service can I render to the commonwealth than to instruct and train the youth?"[49] Those aware of the significant influence of Cicero's works within eighteenth-century Scotland would have seen the link that Gordon was drawing between civic humanism and various forms of practical education.[50] Gordon was a master at an academy in Glasgow and, like many authors of primers, he saw the title page as a form of advertising. The quote

---

49. The Cicero quotation appears in the first edition of Gordon (1763). It remained in the 1765 second edition and the 1770 revised edition. The wording of the quotation, which appears on the title page of several primers published by Scots from the 1770s to 1810s, is slightly different from that used today: *Quod enim munus rei publicae adferre maius meliusve possumus, quam si docemus atque erudimus iuventutem?* For this Latin version and its English translation, see Cicero (1989, 374–375). The larger connections between humanism, bookkeeping, and morality are also addressed in Aho (2005).
50. For Scottish Ciceronianism, see Packham (2013).

was meant to appeal to parents or guardians who valued the way in which Scottish humanism fused morality with scholarly activity.[51]

Since ciphering was the key ingredient of becoming a successful householder, merchant, or professional, bookkeeping was often associated with commercial and moral order. A striking visualization of this sentiment appears in William Hogarth's *Industry and Idleness*, a series of prints made in the middle of the century. Hogarth's prints were popular in Scotland, and Scottish pedagogues such as Lord Kames advocated their use because the scenes reinforced morality through association.[52] Kames's favorite print series was *Industry and Idleness*, which he called "Hogarth's *good Apprentice*." He thought that the prints exhibited "an excellent moral for children." He also noted that the series was "too complex for beginners," which signals that he thought the images were appropriate for students who had moved beyond basic writing, ciphering, and reading skills.[53]

Hogarth's *Industry and Idleness* tells the story of Francis Goodchild, a virtuous and industrious apprentice who uses his notekeeping abilities to build a successful career. The tools and skills of ciphering are relevant to many of the plates, but they are perhaps most clearly represented in plate 4, titled *The INDUSTRIOUS 'PRENTICE, a Favourite, and entrusted by his Master* (figure 4.10). In the scene Goodchild stands next to an open cabinet, the shelves of which are filled with large, folio-sized ledger books. Beneath the cabinet there is a foldout writing desk, a traditional piece of furniture for cipherers working in shops and households.

A writing quill and an inkpot sit on the desk. Goodchild holds an open daybook, which was a kind of bookkeeping ledger that would have contained a variety of annotations, and stares at the folded, handwritten bill of sale affixed to the top of the lumber being carried by a worker. The ledger cabinet is guarded with a lock and the apprentice holds the keys in his right hand. The overarching message communicated by the scene, as well as others in the series, is that ciphering and other associated notekeeping skills literally gave Goodchild the key to knowledge and success.

The value that artists like Hogarth assigned to ciphering was shared by

---

51. The moral gaze that emerged from the interface between Scottish Calvinism and humanism is addressed throughout Allan (1993). See also Ahnert (2015).

52. Prints were increasingly used as teaching tools from the seventeenth century onward, particularly after the publication of Johannes Comenius's internationally successful Latin primer *Orbis Sensualium Pictus* (1658). See Capkova (1970). John Locke famously advocated their use as well. Locke (1779, §156). The use of prints to teach children in the eighteenth century is addressed in Nenadic (1997). Some adolescents collected and traded prints as well. Sandy (1942).

53. Home (1781, 77). The longstanding relationship between modern prints, morality and education in Europe is explored in Heesen (2002).

FIGURE 4.10. William Hogarth, *Industry and Idleness*, plate 4. *THE INDUSTRIOUS 'PRENTICE a Favourite, and entrusted by his Master*, 30 September 1747, etching and engraving on paper. Image courtesy of Andrew Edmunds, London.

Scottish parents and educators. Since ciphering involved both the transfer and calculation of numbers, it was a form of annotation that included elements of composition and revision. It was a core skill for students keeping school notebooks that addressed mathematical subjects. The layout of the pages in figure 4.9 reveals that, like writing, ciphering required notekeepers to range the layout of a page in a way that enabled them to write information in straight rows and to align different kinds of ciphers into columns. Within such rows and columns, different operations required students to order ciphers in a specific way so that a correct answer could be calculated.

Ciphering numbers into ledgers, which was sometimes called *journalizing*,[54] has largely escaped the notice of historians of literacy.[55] But keeping a ledger was perhaps one of the most advanced graphic skills attained by middle-class students, many of whom learned to range the page into sophisti-

---

54. For an example of the term *journalize*, see the "Book-Keeping by Double Entry" sections at the end of Perry (1774).
55. An exception to this trend is Monaghan (2005, 293–296), who underscores the direct link between writing and mathematics in books used in the British Empire.

cated numeric matrices. The ability to cipher complex accounts in this manner was one of the key factors that contributed to the rise of fiscal literacy. Economic historians have shown how this new form of calculability made it easier for the middle class to track the commodities that flowed through Britain's imperial markets.[56]

Many young cipherers learned from family members to keep simple accounts of their own expenses. But transcribing and revising the multiple entries written in a formal accounting ledger was more involved. The Italian method, which is now called double-entry accounting, was not something that most children could learn to cipher by simply reading a general instruction manual such as Todd's *School-Boy*, or even popular accounting textbooks like John Mair's *Book-Keeping Methodized* or William Webster's *An Essay on Book-Keeping According to the Italian Method*. Boys and girls not only had to learn how to transpose ciphers from slips, bills, and memoranda notebooks into ledger book entries, they also had to learn how to range the many lines and columns that were used to structure the space of a typical ledger book page.[57]

The process of writing and rewriting ciphers included many kinds of correction, laying the foundation for the conceptual importance of financial annotations. Students usually learned to cipher in columns by first practicing on large slate boards. A visual representation of this kind of preliminary slate ciphering appears in the oval vignette featured on the title page of Colin Buchanan's *The Writing-Master and Accountant's Assistant* (see again figure 2.4) and in the frontispiece of the 1765 edition of William Gordon's *The Universal Accountant* (figure 4.11). Buchanan's vignette features a boy proudly handing his practice slate to his teacher—a scene that undoubtedly occurred frequently in schools that taught writing and ciphering. Gordon's frontispiece depicts a teacher who is holding a slate of numbers ciphered as columns.[58] Dressed as a goddess, the teacher instructs a bookkeeping student as she cradles the ciphers in her left hand. Her right hand points to a collection of scales, packages, barrels, and, further off, ships in the harbor. The message is similar to that communicated in Hogarth's *INDUSTRIOUS 'PRENTICE*. Commerce depended on accurate ciphering skills, not least because it required various forms of annotation to ensure that the numbers used to represent the volume, weight, and price of

---

56. Macve and Hoskin (1994). For the historical context of accounting and information management, see Soll (2014).

57. From the late seventeenth-century Scottish girls were taught accounting and then went on to become merchants. Dingwall (1999). For evidence of accounting columns being practiced by a girl in a notebook, see Montgomery Bound MS (1787). Montgomery married Alexander Walker of Bowland in 1811.

58. Buchanan (1798, frontispiece).

FIGURE 4.11. William Gordon, *The Universal Accountant and Complete Merchant*, 2nd ed., vol. 1 (Edinburgh: Donaldson, 1765), frontispiece, NLS ABS.2.89.3. Reproduced with permission of the National Library of Scotland.

goods traveled safely across the columns and rows kept by accountants, merchants, and quartermasters.

Bookkeeping textbooks included instructions on how to range a variety of columns into which numbers could be ciphered. By the end of the century the instructions became longer and more detailed, with book 3, "On Ledgers" in the 1797 seventh edition of John Mair's *Book-Keeping Moderniz'd* serving as an excellent example of the kind of detail on offer. But despite the growth in details, learning how to craft a ledger page based solely on instructions printed in a book was not an easy task.[59]

The best way to learn ledger writing was to seek the assistance of an instructor who specialized in ranging and ciphering. An informative source that sheds light on how students learned the scribal skills of advanced accounting is the aforementioned set of school ledger books kept by Robert Richardson while he was a student at Perth Academy during the 1770s (see again figure 4.1).[60] Ciphered in a neat hand into columns made from graphite lines drawn on a grid printed in red ink, each book contains sets, exercises, of detailed accounts. Richardson's sets reveal that he had to cipher specimens from across the business world, with different kinds of accounts requiring slightly specialized columns and entries. He followed the common practice of ciphering the accounts across two columns, one for incoming credits and the other for outgoing debits.

An important point to note about the numbers featured in the multicolumned pages of student ledgers is that they served as the nexus of other notebooks. The ciphers of Richardson's ledger, for instance, were extracted from "fieldbooks," "cornbooks," and other kinds of waste books used to keep track of all the facts and figures relevant to running a solvent business or farm. Students also learned to see the ledger as the central repository, a database for the handwritten information preserved on the many bills of exchange that recorded the individual transactions that companies had with clients. Such bills played a crucial role in the day-to-day business of trading companies based in Scottish port cities such as Glasgow and Leith, as well as those as far north as Inverness and as far east as Peterhead. The bills underpinned the circulation of the data generated by Scottish merchants who traded with Baltic ports, Lowland Countries, France, and even Italian states such as Venice. Students learning to make ledger books, therefore, were being prepared to enter the sophisticated scribal world that tracked bills of exchange through a sizeable trading network.[61] Learning to make and annotate a ledger allowed them to see how the paper systems that underpinned the accounts of a successful company might be constructed and managed.

When students like Richardson annotated a master ledger, they were do-

---

59. Out of the 466 pages of the National Library of Scotland's copy of Gordon (1770), only page 281 features graphite annotations.
60. Richardson Bound MSS (1776; 1777).
61. The impressive scope of Scottish trading companies during the last half of the eighteenth century is given throughout Haldane (1981).

ing it within columns and rows that carried specific economic meanings. Assigning different meanings to ledger columns and their respective rows created what Jonathan Gibson has called "significant space," that is, a place on the manuscript page that carried a specific kind of meaning.[62] He argued that the space on the page assigned to the margins, salutations, paragraphs, and postscripts of handwritten letters were distinct components of a collective image. The space inhabited by each component carried a meaning that worked alongside the other meanings that writers assigned to the words that they wrote. In order to understand these meanings, he showed that historians need to ask questions about the education and intentions of the writer and the wider scribal context of the time. In the case of ledger pages, the columns could represent debt or credit, or they might represent different kinds of commodities, such as wool, fish, or coal. Organizing accounting notebooks in this manner enabled students to know where to look for specific kinds of data and, more generally, allowed them to practice categorizing the space of a page in a way that could be applied to ledgers and other kinds of manual script when they left school.

The skill of designing the space of a ledger page was emphasized in many compendia, including Todd's *School-Boy*. After extolling the virtues of Locke's pedagogical methods and then discussing genteel matters such as civil history and religion, Todd noted the benefits of keeping ledgers and using the double-entry method of accounting in which ciphers were arranged into two columns on the page.[63]

The importance of using a ledger as a central database for information collected from several notebooks was underscored by Robert Hamilton, who taught accounting and mathematics at the Perth Academy, where he also served as rector. Hamilton eventually went on to become the professor of natural philosophy at the University of Aberdeen, at which time he published his Perth teaching material as a school mathematics and accounting textbook titled *An Introduction to Merchandise*.[64] Like other authors of bookkeeping primers, Hamilton's book described how to extract numbers and various facts from bills of parcels or exchange, promissory notes, "book debts," receipts, orders for stock, certificates, letters to factors, and different kinds of notebooks received from clients. If his students went on to become accountants working with farmers, for instance, he explained that they would have to contend with "field books" or "corn books."

---

62. Gibson (1997).
63. Todd (1748, 71).
64. Hamilton (1788).

Once the data was extracted, students then had to learn how to cipher the numbers into the bespoke categories and columns that they created in their ledger books. To make matters even more nuanced, there were different chirographic conventions that could be used for different kinds of data across the table.[65] Extracting information from so many different kinds of manuscripts required considerable annotative skills. Add to this the fact that many farming account books, especially those kept for large estates, were highly sophisticated information management devices in their own right.[66]

Hamilton's guidelines for keeping and using ledgers and other accounting notebooks are instructive, as they communicate the graphic elements (columns) and conceptual skills (assigning categorical meanings to the columns) being learned by students through the combined skills of ciphering and annotating accounts. When discussing field books, he took care to remind his students that:

> A FIELD-BOOK, [is a notebook] where several pages are allowed for each field, and ruled with two columns, for extending the sum of the expenses, and the produce. The rent, seed, manure, ploughing, reaping, and the number of sheaves, quantity of corn, and value, are entered from the journal of work, or the cornbook. The register of the same field, in successive years, is continued from page to page, that success of the different manures, rotations, and methods of treatment, may be easily compared, and a judgment formed of the whole, when continued for a sufficient length of time.[67]

After the foregoing instructions, Hamilton gives a specimen of how the ledger page should look, thereby supplying a visual example of what a student might want to cipher under the guidance of a teacher.[68] The overarching point is that learning to keep ledger notebooks, which were effectively elaborately structured tables, required ranging, ciphering, and other kinds of annotation skills that took time, effort, and money to learn. Notably, the process of keeping a master ledger made it possible for students to exercise their sense of judgment, that is, a mental ability widely associated with rationality at the time.

---

65. For examples, see Champion (1759). A well-preserved edition of this book is held by the Huntington Library, Call No. 622325.

66. The family collections in the National Library of Scotland contain a plethora of manuscript account books relevant to farming, industry, and merchandise. For an example of the kinds of accounts kept on Highland farms, see Grant (1924).

67. Hamilton (1788, 492). Some of the skills used by adult accountants to extract numbers from different sources in eighteenth-century Britain are recounted in Edwards (2014).

68. For further examples of printed ledger pages, see the many examples given in Scruton (1777).

When Richardson's ciphers are considered alongside the Erskine marginalia, Monro's additions, and the other forms of correction and revision mentioned in this chapter, it can be seen that students used annotation to develop skills of artifice that allowed them to exert power over paper within the manuscripts or library books that they treated as notebooks. Their understanding of paper-based forms of knowledge creation and management bordered both scholarly and domestic forms of manuscript usage. Through correcting and reworking the form and fabric of notebooks and other kinds of codices, students learned how to build and, crucially, creatively adapt knowledge systems on and around paper. I have shown that, by employing their script as evidence for what they were learning, oftentimes in realtime, it becomes easier to understand how students internalized and externalized annotation skills that enabled them to actively interface with a notebook, in its capacity as media a technology.

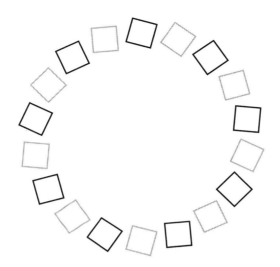

PART II

AROUND THE *TABULA RASA*

In part I we learned that the page of every student notebook began as a *tabula rasa*, as a collection of blank sheets of paper. Students then used writing, codexing, and annotating skills to further shape the sheets and other materials into a fully functioning notebook. In this part of the book, we will take a tour around the space of the page fashioned from a *tabula rasa*, with a view to gaining deeper insight into how students learned to transform its blank field into an accessible and efficient mnemotechnic device that operated within the notebook in its capacity as a paper machine. I concentrate on the core skills of categorizing, drawing, and mapping that enabled students to organize words, numbers, and lines on the page in way that transformed it into a picture. Learning such skills further expanded a student's awareness of the kinds of material, visual, and kinesthetic activities that could be used to enhance the function of the notebook page as a media technology.

Through expanding the *tabula rasa* metaphor, I further extend the notion that notekeeping went beyond the rote transfer of words and figures because it required students to engage with the notebook as a material artifact. It was not a simple matter of input and output. Notekeepers were constantly doing things with the materials as the notebook took shape. They built on skills that they had learned in the past only to learn more skills as they continued to keep notebooks. Likewise, we will see that students modified their skills to craft the page of a notebook as an artifact that would be accessible to them in the future. Some of the visual elements students used to structure each page were similar to those that compositors employed to lay out textbooks, thereby linking the subtle interplay between the forms of graphic design used in the script and print of educational texts. But unlike their experience with the fixed formats of the printed books cited by their teachers, students used their own graphic intelligence to choose which kinds of layouts and figures suited them best in their notes. In making such important visual decisions, they were learning how to more efficiently manage knowledge on paper.

Throughout the chapters, I underscore that, when the *tabula rasa* metaphor is viewed in light of the developmental richness evinced on the pages of student notebooks, it can be seen how its meaning far exceeds the notion of blankness that scholars have frequently attributed to it. Instead, it had the capacity to reflect a much wider meaning, one that drew from the considerable amount of time, effort, materials, and concentration inherent in the skills required to make the different components of a manuscript page. In this meaning, the *tabula rasa* was a longitudinal entity. It provided a fertile way to frame thinking and notekeeping as conjoined skills of artifice that students applied in stages over time in way that actively enabled their abilities to judge information in realtime on and across paper.

CHAPTER 5

CATEGORIZING

Headings as Realtime Categories

When the young David Hume decided to reevaluate his understanding of the foundations of Western philosophy, he embarked upon an ambitious reading program that included keeping notes on an impressive collection of authors. To organize his thoughts, he wrote them as lists of epitomes in a notebook that he had made from a handful of bifolia sheets. The word used at the time for such entries was *head*. Today the word is *heading*, and, as we learned in previous chapters, it was a core organizational tool in the manuscript culture of Scottish schools, businesses, and homes. Each of Hume's headings was a brief summary, usually a line or two, or observation of a concept based on what he had read. After he finished composing his headings, he then used them as recollection and reflection tools while writing his early books (figure 5.1).[1]

In his 1739 *Treatise of Human Nature*, for instance, Hume used the heading "Of Laws of Nations" to label section XI of part II in book III in a way that provided a focal point for his ensuing reflections. He explained this organizational strategy in the following manner: "Under this *head* we comprize the sacredness of the persons of ambassadors, the declaration of war, the abstaining from poison'd arms, with other duties of that kind, which are evidently calculated for the commerce, that is peculiar to different societies."[2] These topics were a list, written horizontally in a sentence, of subheadings that he employed to further organize his analysis. He used headings and subheadings in a similar manner in the rest of his publications to name the sections, parts, chapters, and books of his treatises.

---

1. Hume Unbound MS (1729–1740). A transcribed edition of the memoranda was published as Mossner (1948b).
2. Hume (1740, 188–193, quotation from page 189), emphasis added.

152 | CHAPTER 5

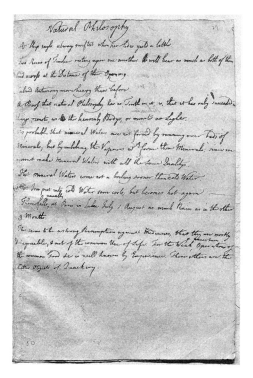

FIGURE 5.1. The earliest extant set of notes written by the philosopher David Hume. It reveals that he used headings to preserve his early ideas. He wrote a main title, "Natural Philosophy," at the top of the page. He then listed headings, wording them as epitomes or observations on specific topics. David Hume, *Collection of Memoranda* (1729–1740), Loose MSS, NLS MS.23159, No.14m ff. 28–29. Reproduced with permission of the Royal Society of Edinburgh.

Like many literati educated in Scotland at the time, Hume had learned the art of creating headings at some point in his early education. From that point forward they functioned as mnemonic and kinesthetic sorting devices. This means that many of the headings that appear in his publications are in fact time capsules, artifacts, that gesture to the processes on paper that he used to categorize his thoughts before he published his books. Yet while it is clear that headings played a notable role in Hume's thought process, it is far less clear how he and other Scots learned to use them in different ways as students or, in the case of other authors, before they became published or even famous. How did the young Hume and the host of students learning to keep notes in Scotland's homes, schools, and academies learn to use the humble heading as a knowledge management tool?

The foregoing question could be asked of many personalities who lived not only in Britain, but in Europe and colonial contexts as well. It gestures to our relatively limited knowledge of how modern schoolchildren learned to think with pen and paper in realtime. In this chapter, I explore this topic by using the handwritten headings of Scotland's student notebooks as an entry point into the fascinating history of childhood learning and predigital data management. I wish to explore how writing and rewriting something so seemingly simple as a heading served as a crucial mode of material and visual training through which students learned to label chunks of information, demarcate important topics, and structure the script into a format that was easier to scan. Approaching headings from this perspective, I suggest, restores their status as powerful informatic learning tools that students used to strengthen their literacy skills in a way that enabled them to actively interface in realtime with script and print.

In previous chapters, I showed that handwritten headings played an important structural role in student notebooks and other forms of codex-related ephemera. In this chapter, I further extend my investigation by suggesting that the creation and subsequent use of every notebook heading was effectively a multifaceted act of realtime categorization. As verbal categories, headings were words that acted as mnemonic labels for large chunks of information. As visual categories, headings were gestalt cues, technic focal points, that stood out to the eye. As vectoral categories, headings were coordinates for paths, sightlines, over which students could run their gaze across the page in a way that made it more accessible to scan. When students wrote headings on the page, they were creating sites of realtime categorization, both at the time they were written and then later when they were read. As such, writing and reading headings strengthened the literacy skills already possessed by students, while at the same time enabling them to actively acquire new skills of interface.

In many respects the skills evinced in Hume's headings were commonplace in literate society across Europe. Indeed, middle-class British readers of his books used headings as epitomes to record their reactions to his ideas in their own notebooks.[3] But like the childhood notes that Hume kept before he began his memoranda book, the early notes of many such writers have been lost to the sands of time. Thankfully, since handwritten headings were so prevalent in the Scottish educational system, extant school notebooks can be used to gain deeper insight into the skills required to transform the headings of Hume's day into categories that could be used to efficiently organize knowledge on the handwritten page.

One of the overarching points that I wish to emphasize is that each school

---

3. Towsey (2010, 262–292; 2008).

notebook page was a material and visual artifact that transformed John Locke's *tabula rasa* metaphor into a three-dimensional cognitive model based firmly on the embodied skills performed everyday by notekeepers.[4] Accordingly, I wish to turn the focus to the notebook page itself with a view to showing how students learned to make and use it as an artifact that unfolded kinesthetically over time and space. In following this path, I employ two axioms that provide a framework for understanding the nature of realtime learning during the predigital era. One addresses space, the other addresses time.

The first realtime axiom relates to graphic space. More specifically, it is the notion that the acts of writing, codexing, and annotating surrounding the mindful creation and usage of a modern handcrafted page were often conducted across space in realtime, that is, they were "performed or occurring in response to a process or event and virtually simultaneously with it."[5] In the case of student notebooks, the jointly material and kinesthetic process surrounding the creation of the page, foliation, is the object of enquiry. Though this axiom is not explicitly articulated by historians when they discuss topics relevant to the many kinds of media presented by print and manuscript culture, it increasingly informs studies that seek to understand the historical relationship between material culture and the practices, routines, and performances that shaped the arts, humanities, and sciences. Such studies range from Anthony Grafton's learned and entertaining account of the fifteenth-century "Monty-Pythonish fire-farting rabbit" designed by the Paduan engineer Giovanni Fontana during the Renaissance to Matthew Reason's fascinating account of the relationship between documentation and live performance during the late twentieth century.[6]

My second realtime axiom relates to graphic time. Hal Sackman voiced it on the eve of the digital revolution during the mid-twentieth century. He had realized that there would always be differences in information systems developed at different times. He intimated that "real time" interface was by no means restricted to the modern era or to specific kinds of digital information systems. In his words, "Real time essentially refers to events—their appearance and duration, their passage and succession, and the hypothesized interrelations of events as empirically tested and demonstrated in any reference system and its environment."[7] Sackman's understanding of realtime activity reveals that

---

4. Eddy (2018).
5. See *Oxford English Dictionary Online*, s.v. "real time."
6. Grafton (2002, 24) and Reason (2006). Historians of photography also emphasise the relationship between material cultural and simultaneous acts of moving and thinking. Sassoon (1998) and Edwards and Hart (2004).
7. Sackman (1968, 1493).

it operates over time within an ordered system. In the case of notebooks, or even other forms of predigital manuscript and print culture, such systems were simultaneously conceptual, visual, and material. Though the word *real-time* was not used in the predigital era, there was a distinct awareness that repeated acts of interface with manuscript and print media could be used to train the mind to simultaneously sense, perceive, recollect, and abstract facts and ideas in a way that instantaneously facilitated judgment. I will have more to say about this cognitive facet of realtime knowledge in later chapters when I discuss the manual and mental skills that students learned while making university lecture notebooks.

In what follows, I use the appearance, duration, and succession of headings in their capacity as components on the handwritten notebook page as the basis for an examination of realtime activities that involved the hand and the mind of a student. Over the course of the discussion, I identify a number of ancillary predigital data management skills that were included in the larger process of student notekeeping. Like scholars who study the many screen-based forms of human-computer interface, I readily treat the visual groupings of words, symbols, and figures as tools that can be used to think through complex ideas and to locate information.[8]

For reasons that will become apparent in later sections, I occasionally find it necessary to approach the pages of texts as artifacts that can be read like maps,[9] and I treat the meaningful patterns that run across the headings of a page as important elements of visual communication.[10] By taking this approach, I reveal a hitherto neglected aspect of the kinesthetic foundations of Scotland's knowledge economy and, at a more general level, I help to clarify several aspects of the nature of the predigital history of layouts that we see regularly on the page or the screen today.

Headings as Mnemonic Labels

Let us first investigate the role played by student notebook headings as mnemonic labels, as conceptual categories that encapsulated a clump of thematically related data. Historians have compared the informatic role played by the modern heading to a box into which related facts, observations, epitomes,

---

8. Teevan et al. (2009).

9. For the map-like qualities of the written and printed page, see Tversky (1981; 1993) and Tversky and Schiano (1989).

10. The map-like usage of graphic space occurs throughout human history, including in prehistoric times. Tversky (2019, 194–201).

and other kinds of *describabilia* are placed.[11] The ability to use a heading as a category to encapsulate pieces of information was a longstanding skill that had existed since antiquity, particularly in scholarly contexts influenced by humanism.[12]

During the Renaissance, merchants, clerics, professors, and other professionals educated in the arts of commonplacing and bookkeeping used headings, which they sometimes called *topoi* (Greek) or *loci* (Latin), as both sorting and learning devices.[13] The usefulness of headings as everyday information management tools in manuscript culture ensured that the process of selecting them, that is, the skill of encapsulating, operated via rules of thumb developed by individuals or communities from the sixteenth century forward.[14] By the seventeenth century, compilers and professionals such as physicians and lawyers regularly used headings as categories to arrange the diverse entries of manuscript miscellanies and notebooks into organized schemes.[15]

By the eighteenth century, handwritten headings were commonly used by scholars as diverse as the English theologian William Paley and the Swedish naturalist Carl Linnaeus to manage the flow of information through their notebooks.[16] This use of headings as informatic capsules occurred as students and adults alike interfaced with the handwritten page. When viewed in reference to the wide reach of manuscript culture, headings, particularly in their function as everyday reading and writing tools, were part of the quotidian world that surrounded contemporary data management practices. As such, they are relevant to discussions held today by scholars concerning the nature of, in Sackman's words, "historical interpretations of time, and by extension, 'real time,' [which] have assumed that varied forms of the conceptual containers into which notions of time, like a liquid, have been poured."[17]

The organizing principles behind the pedagogical use of headings as labels in scholarly educational settings were usually thematic, or, in the words of Walter Ong, they operated according to a form of topical logic that was sometimes underpinned by dichotomies.[18] Ong and other scholars working on the

---

11. For the historical role played by the box (*arca* in Latin) as a metaphor for a collection of related memories, see Carruthers (2008, 42–44). The role played by headings as labels that served as metaphorical boxes is addressed in Müller-Wille (2018, 209–210).
12. Vickers (1989).
13. Blair (2010b, 62–74).
14. For use of headings as teaching and organizing tools, see Eddy (2010a) and Ong (2004).
15. Vine (2017). The organization and layout of headings in early modern medical notebooks is explained in Timmermann (2008).
16. For Linnaeus, see Eddy (2010a). For Paley, see Eddy (2004).
17. Sackman (1968, 1492).
18. Ong (2004, chaps. 5, 8, 9) discusses topical logic, which he also called place logic.

humanist art of memory and its direct effect on the scribal skills of common-placing have shown that, by the eighteenth century, the topical use of headings was part of the conceptual wallpaper that underpinned the organization of school and university textbooks throughout Europe. Within these works, headings served both as labels and as classification categories that teachers used to organize the pages of primers.[19]

In Britain, school primers sometimes stated explicitly that headings could be used to label and remember information. The instructions given underneath "The Numeration Table" in George Fisher's *Instructor*, a popular primer throughout Britain and its former colonies, gave student readers the following advice: "For easier Reading of any Number, first get the Words at the Head of the Table by Heart."[20] The larger importance of using headings as organizational capsules in this manner was recognized by many British schoolmasters and pedagogues. The scientist and Warrington Academy teacher Joseph Priestley, known today for discovering oxygen, underscored the importance of selecting headings that young learners could remember easily. His view of the pedagogical facet of encapsulation is perhaps most clearly evinced in the preface of his *Chart of History* (1770), where he explains that the selection of headings must take the needs of the intended user into account.[21] A similar sentiment is communicated in the widely read books of Isaac Watts that used keywords as mnemonic learning technologies.[22]

The general method that was employed to create a useful heading, in Scotland and beyond, was nicely summarized in the *Encyclopaedia Britannica* "common-place book" entry. To create a head, it instructed its readers to "consider to what head the thing you would enter is most naturally referred; and under which one would be led to look for such a thing."[23] This was precisely the kind of heading that Scottish students used at home and at school to label the different sections of their notebooks. They usually wrote simple and plainly worded headings that connected to the subject matter at hand.

Whereas adult commonplacers usually created their own headings, teachers and tutors made the task easier for younger student notekeepers by selecting headings for them to copy as part of their notes. Even though the students

---

19. Works that flag the important role played by headings include Yates (1966), Rossi (1968), Blair (2010b; 1997), and Moss (1996).
20. Fisher (1806, 29). "The Numeration Table" occurred in the "Arithmetic" section of most editions. See, for instance, the 1763, 1771, and 1799 editions. See also Mayhew (2007) on the typography and spacing in British schoolbooks.
21. Priestley (1770).
22. Watts (1786). For keywording, see Bartine (1989) and Ross (1988).
23. Smellie (1771, 2: 241). For further insight into the selection of headings in Britain at this time, see Dacome (2004) and Yeo (2001, 101–119).

were not selecting the headings per se, the act of writing them repeatedly in their notebooks reinforced the advantages of matching a simplified encapsulation to the textual material it was meant to label. Educators living at the time treated this kind of active copying as a mode that positively shaped the minds of young learners. Once students understood this important skill, they were ready to start creating headings of their own.[24]

As we learned in the discussion about Margaret Monro's codex in chapter 4, copying and creating headings was an effective method that trained young learners to categorize the text of a manuscript as they wrote it. Some educators intimated that inventing or writing headings facilitated logical naming and ordering skills that helped students understand how to judge the world around them. Throughout European literary culture, the goal of education in many places was the acquisition of a sense of judgment, which, as most conduct manuals indicated, served as the basis of moral decision making.[25] Within Scottish educational settings, logicians such as William Duncan, professor at Marischal College, Aberdeen, expressed the common view that ordering headings, which, as words, represented ideas, into multiple categories was a logical act if it was done according to a clearly articulated organizing principle.[26] He held this view because he believed that learning to order the world into cogent categories was a precondition for those wishing to make rational judgments.

In order to become an organizer of categories, a student needed to learn the skill of categorizing at the ground level. This was accomplished through seeing or writing a collection of relatively straightforward headings that represented the main topics of a given school subject. A simple organized list of such headings at this time was sometimes called a "scheme" and it served as a foundational media technology through which students became young categorizers. The importance of a curricular scheme of headings was noted in Scotland and beyond. Its developmental importance was flagged, for instance, by the aforementioned Isaac Watts. In addition to being an English nonconformist minister, Watts was a pedagogue whose views as a "hesitant Calvinist" enjoyed currency in Scotland and other religious communities spread across Britain and its North American colonies.[27] In *The Improvement of the Mind*, which addressed the education of children, Watts summarized the impor-

---

24. Howell (2015).
25. Sheehan and Wahrman (2015) explain the importance of self-organization and judgment in British literature at this time.
26. Duncan (1752).
27. For Watts's status as a "hesitant Calvinist," see Sweeney (2020, 34).

tance of a curricular scheme in the following manner: "The best way to learn any science is to begin with a regular scheme of that science, well drawn up into a narrow compass, omitting the deeper and more abstruse parts of it."[28]

Though not traditionally treated as learning devices by historians, curricular schemes served as powerful predigital tools during the long eighteenth century, particularly for students seeking to turn their loose-leaf, and potentially disorganized, school notes into fully functioning paper machines arranged according to the order offered in the headings. Learning to recognize the relationships among the categories of an organized system served as a goal in the educational processes enacted in many schools and literate households spread across Europe and its colonies. Students who did not learn according to this jointly interactive and systematic method were, according to Watts, condemned to live on "the Scraps of the Sciences."[29] Put more clearly, if chunks of information were not presented in the form of an organized scheme, they would amount to nothing more than scraps in the minds of students because they were not united into a system of categories that were somehow related to each other. This position united the media of script and print with the cognitive capabilities of the mind.

A sure way for school-aged children to learn to recognize, replicate, recollect, and order headings was to write them repeatedly across multiple sheets. Here paper served as a medium through which the skills of logic and, by extension, judgment, were internalized. Schoolchildren implicitly learned to replicate such skills when they kept organized school notebooks because teachers and parents encouraged them to encapsulate clumps of prose according to headings that served as categorical labels. Near the beginning of the century, the schoolboy James Dunbar, for instance, organized the arithmetic content of his notebook under the headings of "numeration," "addition," "subtraction," "division," "reduction," "proportion," etc., with the organizational principle of selection being, roughly, the main facts and operations one needed to know about the mathematical sciences.[30]

Scottish students continued to use headings to arrange the content of their schoolbooks throughout the eighteenth century. During the 1780s, for instance, the schoolboy Robert Jackson ordered the content of his *Geometry Notebook* under the following main headings: "Of a Circle," "Of divers Solids

---

28. Watts (1741, 316). For similar advice regarding the design of catechismal schemes, see Watts (1786).
29. Watts (1741, 317).
30. Dunbar Bound MS (1710). Headings used to label mathematical problems are featured throughout Fowler Bound MS (1780). For further examples of organized headings in student notebooks, see Anonymous Bound MS (1780s–90s); Anonymous Bound MSS (1787), I.M. (notekeeper).

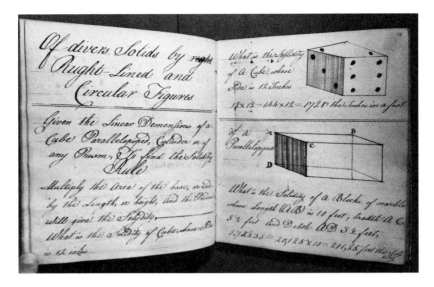

FIGURE 5.2. Robert Jackson, *Geometry Notebook of Robert Jackson, A Schoolboy* (1788), Bound MS, NLS MS 9156, f. 17v–18r. Reproduced with permission of the National Library of Scotland.

by Right Lined and Circular Figures," "of a Parallelopiped," "Of a Prism," "Of a Right Pyramid," "Of a Cone," "Of a Cylinder," "Of a Frustum," "Of a right Cone," "Of A Prismoid," "Of a Sphere and its Parts," "Of a Circular Spindle," "Of a Spheroid," and "Of a Paraboloid" (figure 5.2).[31] The headings served a mnemonic function in that they were conceptual categories of the geometric system he was both recording and learning through writing. The mnemonic facet of such headings was aided by the fact that the terms were relatively straightforward encapsulations of the content that appeared under them.

For student notekeepers, using headings as organizational categories in their notebooks was a heuristic practice of classification. Writing headings in this manner familiarized them with the fact that categories needed to make sense to the person who was using them. Outside of keeping notebooks, students were familiar with the organizational potential of headings because, as we learned above, they appeared in printed primers. But learning to create or use a heading in a notebook required different skills than those required to recognize one in a printed book. This was especially the case for school notebooks, where students sometimes added or omitted headings based on preference.

31. Jackson Bound MS (1788).

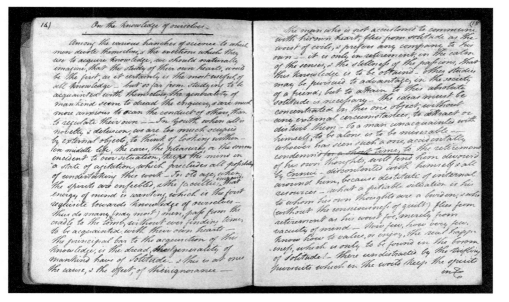

FIGURE 5.3. Exercise notebook pages with a heading titled "On the knowledge of ourselves." Williamina Belsches, *Exercises* (1795), Bound MS, msdep7—Personal papers, Box 22, no. VII/9, f. 14. Reproduced with permission of St. Andrews University Special Collections. Belsches wrote an essay that stated her position on the topic given in the heading. She structured the prose into brief sections separated by dashes that made the script easier to navigate. The essays were written during a time when the young philosopher James Mill tutored her.

The practice of using headings to label the contents of notebooks was employed by students receiving private tuition on more advanced topics as well. The 1795 *Exercises* notebook of the seventeen-year-old Williamina Belsches (1777–1810), for example, presents excellent specimens of this practice (figure 5.3). Belsches's education was overseen by her mother, Lady Jane Belsches, a respected scholar and aesthete in her own right. During the 1790s, Lady Jane hired the young philosopher James Mill to tutor Williamina on topics relevant to moral philosophy and the philosophy of mind. Williamina's notebook was written around the time she was Mill's student. It contains a series of perspicuous essays on topics that were most likely agreed upon by her and Mill.[32]

The titles of Belsches's essays were centered, placed at the top of the pages, and used as headings. Like David Hume, she employed the headings to structure her thoughts. Her headings were as follows: "On the regard to be shown

---

32. Belsches Bound MS (1795). For a brief account of Williamina Belsches's contact with James Mill, see Cumming (1962, 158).

to the opinions of the world," "On Steadyness," "On acknowledging a fault," "On the duties we owe to others," "On the knowledge of ourselves," "On Happiness," "On Adversity," "On Friendship," "On Solitude," "On Death," "On the employment of the present," and "On Benevolence." In addition to serving as focal topics for her essays, the presence of the headings made her notebook's content easier to navigate and, when viewed collectively, they amounted to a compact system of moral conduct.[33]

Overall, the headings featured in Scottish student notebooks provide a helpful window onto the way that young learners categorized information as they wrote. This form of interactive writing implicitly depended upon equally interactive modes of reading their own script in a manner that allowed them initially to see what they were writing, and then to proofread what they had written after they had finished writing. The result was a simple handwritten system that was easy to understand, access, replicate, and remember.

Headings as Visual Cues

In addition to using headings as informatic capsules, student notekeepers used them as focal points, visual cues, that made the script of a page easier to navigate. In order to fully understand this visual role, it is worth observing that, as in printed books, the script on every notebook page was usually organized according to a modular format. We will recall from chapter 2 that school notebook modules often treated the entire face of the page as a column in which students wrote a freestanding heading that was followed by a block of text (see figure 2.9). The use of this modular format ensured that main headings appeared in the same place, at the center or left margin, on every page of the notebook, thereby making the text easier to navigate.

To make the headings easier to spot and to denote different kinds of information, students learned a number of chirographic skills that allowed them to further transform headings into bespoke focal points.[34] The use of headings in this manner was effectively a locational scheme, that is, a collection of visual cues on the pages of manuscripts that made information easier to memorize. Such schemes were employed within European manuscript culture as early as the twelfth century.[35]

---

33. The headings occur across Belsches Bound MS (1795).
34. The visual principles that underpin the creation and usage of focal points on both word- and figure-based images are explained in Bradley (2018).
35. Carruthers (1998; 2002; 2008).

Before we begin to consider the crucial role that headings played in student notebooks as visual cues, it is worth pausing briefly to remember that historians tend to overlook the layouts used to structure words, numbers, and symbols on a page or even on a screen. They simply take such things for granted until they encounter unexpected variations in layout or typography that disrupt the way that they experience the in- formation. This point has been underscored in recent years by the use of foveal dots such as the one embedded in this paragraph. Such figures illus- trate how the mind uses per- ception to mediate foveal vis- ion, which is precise, but lim- ited to a small area, and periph- eral vision, which is less precise but helps us to see a larger area. Staring at the dot in the center of the circle reduces the sentences of the paragraph to a grey, shimmering block of text, transforming the blank space inside the circle into a calm white field.[36] The end result is a figure and ground gestalt image that operates under the principles of symmetry and closure. Importantly, the structure of the foveal figure draws the eye away from the shimmering block to the white space of the circle. From there, the eye is then drawn directly to the dot in the center.[37]

The perceptual facets of the foveal figure embedded in the foregoing para- graph help us gain insight into the spatial functions played by the headings presented in student notebooks and on modern handwritten pages more generally. When seen from this angle, headings functioned as visual cues in a manner similar to that of the dot in the foveal figure. Headings in a notebook, like the dot, attracted the eye. In other words, they had the capacity to imme- diately attract a viewer's attention. Scholars working in different disciplines sometimes call such cues *focal points* or *preattentives*.[38] Every time students transformed a heading into this kind of a cue, they were learning the skill of *cuing*. Perhaps more accurately, they were learning the skill of *focalpointing*, the ability to create focal points within a handwritten page.

Students learning to craft headings into focal points were aided by various early modern layouts in books that combined the dark and light areas of script in a way that transformed the words and space of a page into an artifact that

---

36. The foveal dot in this paragraph was inspired by the one featured in Monaco (2009, 174–175). For a brief explanation of the differences between foveal and peripheral vision, see Lauwereyns (2012, 5–7). **37.** The relationship between reading and foveal vision is summarized in Jordan, McGowan, and Pater- son (2012). **38.** The relationship between preattentives and attentive processing as we understand it today is ad- dressed in Friedenberg (2013, 63–66).

was easier to understand and use.[39] Some printers, Philip Luckombe for instance, obviated the visual impact of different layouts by creating elaborate typographical schemes such as the one depicted in figure 5.4.[40] As evinced by the starburst effect running through the words, such schemes were often gestalt images. The schemes by Scottish students in their notebooks were of course much simpler than Luckombe's. The most common was in fact the standard module itself. Its heading and block of words functioned collectively as a visual unit that occurred on one page after another. Within this format, students frequently focalpointed their headings into effective visual cues by altering their alignment or chirography.

Let us first examine alignment. More specifically, I wish to examine how students aligned headings in a way that effectively created a gestalt image in which a slight alteration to a symmetric pattern functioned as a focal point. To illustrate how this kind of gestalt image operated, it would perhaps be helpful to consider the focal point created by the two circles set above the three rows of six circles in figure 5.5. The pattern is symmetric, but the open space above and beside the two top circles draws the eye to the top of the structure. The overarching visual effect of the pattern is one in which the circles at the top operate as a focal point, as a visual cue that attracts the eye.

Instead of using shapes like circles, students keeping notebooks aligned the words of their headings in patterns similar to that depicted in figure 5.5. To implement this technique, they positioned their headings in the center of the page and plotted open space above and beside them. The headings made by Robert Jackson in figure 5.2 and by Williamina Belsches in figure 5.3 are good examples of this kind of positioning and plotting. At the same time, they justified the left and right margins of the following block of script. The right margin was hard to align. But Jackson and Belsches, like so many Scottish students keeping notebooks, demonstrate a keen resolve to monitor the spacing of the letters and words of each sentence so that the right margin is relatively justified. The overarching effect of positioning, plotting, and justifying the heading and the block is a vertically symmetric image, one in which the placement of the heading at the top and the surrounding white space attracts the attention of the eye.

In addition to alignment techniques, student notekeepers learned to use

---

39. Henri-Jean Martin and Bruno Delmas once called the expansion of open textual space during the eighteenth century a triumph of white over black. See Martin and Delmas (1988, 295–299). Chartier and González (1992, 54), discuss Martin and Delmas's views.

40. See the typographic starburst and the related discussion in Luckombe (1771, 218). It also appears in Smith (1755, 153). For the "fashionable" mode of inserting extra space into texts in late-eighteenth-century Scotland, see Bell (1802).

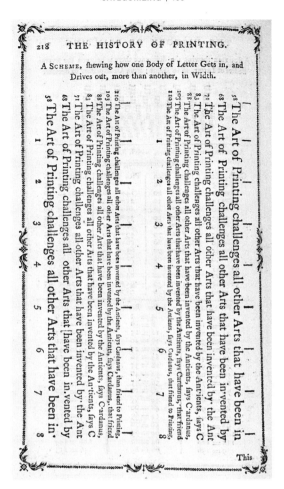

FIGURE 5.4. A typographic "scheme." Philip Luckombe, *The History of the Art of Printing* (London: Adlard and Browne, 1771), 218. Reproduced with permission of Glasgow University Library, Department of Special Collections, Sp Coll Hunterian Q.6.1. The scheme consists of two expertly designed columns of text in which the white space between the words presents a starburst effect. Luckombe described the scheme as an image that reveals "how one Body of Letter Gets in, and Drives out, more than another, in Width." The same typographic scheme appears in John Smith, *The Printer's Grammar* (London: Smith, 1755), 153.

chirographic alterations to transform headings into cues that made them stand out to the eye. In making these changes they learned how to manipulate their handwritten letters in ways that utilized gestalt principles of obliquity (orientation), intensity (tone), scalability (size), and plasticity (shape). The scribal techniques that they employed to enact the principles enabled them to transform the letters in their headings into cues that were noticeably differ-

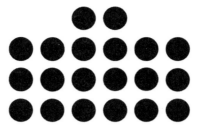

FIGURE 5.5. Focal point created by two circles over a series of circles. © Matthew Daniel Eddy.

FIGURE 5.6. Focal point created by two squares over a series of circles. © Matthew Daniel Eddy.

ent from the other words that they wrote elsewhere on the page in the block of script.[41]

To get a better idea of the gestalt principles exhibited in student notebooks, it would perhaps be helpful to step away for a moment from handwritten letters to simple shapes. Consider, for instance, the focal point created by the two rectangles in figure 5.6. At one level, the two rectangles stand out because, like the two circles of figure 5.5, they occur at the top center of the image and they are surrounded by extra white space. But they stand out at another level on account of their tilted orientation, lighter tone, larger size, and different shape.

The changes that students used to transform the letters in their headings into striking cues operated according to the same gestalt principles as the rectangles in figure 5.6. The "Prob. IX" heading that appears at the top of the anonymous *Practical Mathematics* notebook page depicted in figure 5.7

---

[41]. Smith-Gratto and Fisher (1999) and Chang, Dooley, and Tuovinen (2002) explain the larger relationship between graphic design and gestalt principles such as proximity and symmetry.

serves as a helpful example of how students enacted the gestalt principles required to transform the altered letters of words into visual cues. Though we do not know who made the page, paying attention to the principles offers insight into the focalpointing skills that the anonymous student notekeeper was learning to enact in the script. The page reveals that the skills involved in altering the orientation, tone, size, and shape of the letters were expressed through techniques such as italicization, emboldification, minusculization, majusculization, sublinearization, and capitalization. The final result was letters with altered chirography that drew attention to the heading.

Gestalt-oriented alterations were by no means unique to the headings created in Scottish student notebooks. Primers printed across Britain and more broadly in Europe from the Renaissance onward used typographic alteration to demarcate headings as well.[42] Like the compositors who used typographic alteration to structure the space of the page in primers, students used chirographic alteration to transform the letters of their headings into cues. From a student notekeeper's perspective, the handwritten headings were purposeful components of the page's layout. The act of creating them functioned as an important mode of learning how to structure the space of the page into an information artifact.

The heading in figure 5.7 serves as a helpful example of the kinds of visual decisions that students made when they used handwritten script to interface with the module that they had crafted on the page. To create a contrast between the heading and the prose in the block of script, the student selected different hands, an act that required a significant amount of prior training. The sentences of the block are written in what seems to be a cross between the Italian and running hands. The letters are tilted to the right in a sharp oblique orientation. They are smaller and the strokes are thinner.

The student who wrote the "Prob. IX" heading in figure 5.7 further differentiated the letters by using two different hands. The letters of "Prob." are in Old English hand. They are bigger and rectilinear in orientation. The letters of "IX" are in Italic print hand.[43] They are tilted to the right in a slight oblique orientation that is less slanted than the hand in the block of prose below. The letters of the Roman numerals are larger as well, with thicker strokes. The resulting

---

42. Fisher (1763), for example, put capitalized keywords inside his centered and italicized headings to cue topics that he felt most children would want to know. Adam (1786) employed a system of large capitalized headings for overarching main headings and smaller capitalized subheadings. Mayhew (2007) treats the uses of typography in other British primers.
43. Samples of Italian, running, Old English, and Italian print hands are given in Champion (1750).

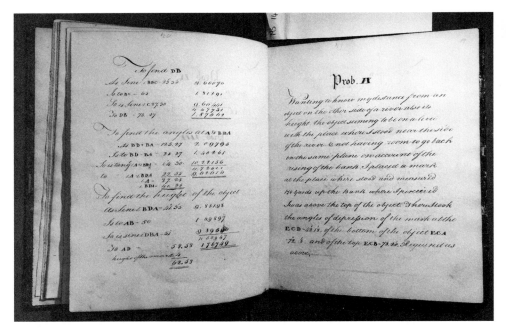

FIGURE 5.7. Open notebook with a heading on the right page. The heading is easier to see because it is majusculed and rendered in altered typography. Anonymous, *Practical Mathematics* (1804), NLS MS 14285 ff, 20–21. Reproduced with permission of the National Library of Scotland.

effect is a collection of contrasts between oblique and rectilinear orientations, intense and mild strokes, and large and small letter shapes. These contrasts collectively make the letters of the heading stand out as an eye-grabbing visual cue, and they occur across many of the main headings that the student wrote in the notebook.

Students employed a plethora of chirographic alterations based on the kinds of cuing that they had already learned or wished to improve. One way to examine the variety of cues is to compare the letters that appear in the headings of notebooks made by different students. One kind of cue, for example, that appears in many notebooks is majuscule lettering, that is to say, letters that students enlarged so that they were bigger than the others on the page. We have seen this cue already in the *Practical Mathematics* notebook discussed above (figure 5.7). Its majusculed headings were written in a hand that was different from the ensuing block of script.

Another approach to majusculing was to use the same hand, but to make the letters of the heading significantly larger than the size of the letters in a page's block of text. Robert Jackson used this technique in his *Geometry Note-*

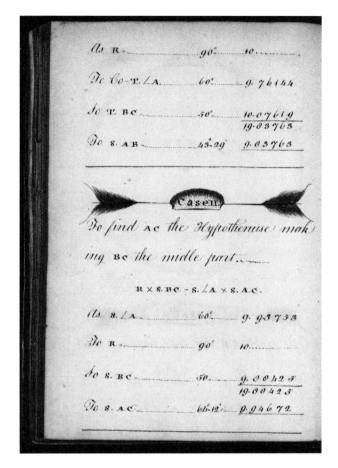

FIGURE 5.8. Slightly majusculed heading surrounded by a target with embedded arrows. Anonymous, *Perth Academy Notebooks*, vol. 1 (1787), NLS MS 14294, f. 39v. Reproduced with permission of the National Library of Scotland.

book (figure 5.2). Yet another option was to use majuscules that were only slightly larger than the block. For example, the main headings designed by the anonymous student notekeeper who crafted a set of *Perth Academy Notebooks* in 1787 were minimally majusculed. But this minimalization required the student to add extra cues to make the headings easier to see. More specifically, he added shapes, shading, and arrows embedded in the heading as if it were a target (figure 5.8).[44] Arrows at this time were used in the figures that accompanied primers on mechanics, surveying, and optics. But their use in

44. Anonymous Bound MS (1787). For arrow headings, see vol. 1, 39v–40r.

the *Perth Academy Notebooks* effectively transforms the modestly majusculed headings into arresting cues.[45]

The different options for focalpointing reveal that there were many graphic decisions that students had to make in realtime when crafting the placement of headings on a notebook page. In addition to learning how to select visual cues, the notebooks also show that students clearly understood that increasing the number of focal points on a page created a form of repetition blindness.[46] In other words, they understood that adding too many focal points effectively made headings less attractive to the eye and, hence, less effective as visual cues.

Space does not permit a discussion of all the visual cues that students used to adjust the baselines (upper and lower), spacing, ruling (especially Scotch ruling), and serifs of letters. Nor does space permit a discussion of the anatomical elements of handwritten letters such as the apex, arm, bowl, ear, crossbar, leg, ligature (link), counter, chin and crossbar. It will suffice to say that all of these elements provided a means to adapt the shape and placement of the letters in headings in a way that transformed them into cues, spatial categories that made the page more understandable on an individual basis.[47]

Headings as Coordinates for Scanpaths and Sightlines

The role played by headings as mnemonic capsules and technic cues was complemented by a third, equally useful, function as coordinates, that is to say, as points on the page that facilitated kinesthetic modes of realtime interface along a *scanpath*. Put more plainly, headings could be used to scan the script of a notebook page in a way that offered quicker access to the information. Though this kind of dynamic scanning regularly is used today in the visual layouts created by designers who structure information on the screens of digital devices, its presence in predigital manuscript culture remains relatively unexamined.

The importance of paying attention to the visual, oftentimes vectoral, movement of knowledge across the page was increasingly recognized during the long eighteenth century. William Playfair, the Scottish engineer and political economist widely credited as the inventor of line graphs during the 1780s, summed up this recognition in the following way: "The giving [of] form and

---

45. Arrows in headings and diagrams occur across Anonymous Bound MS (1787).
46. Friedenberg (2013, 98–99).
47. Richard Rubenstein offers a helpful discussion of the typographic anatomy of letters and spaces. Rubenstein (1988, 153–194). See also Ambrose and Harris (2010) and Hartley and Harris (2014).

shape, to what otherwise would only have been an abstract idea, has, in many cases, been attended with much advantage; it has often rendered easy and accurate a conception that was in itself imperfect, and acquired with difficulty."[48] Like the gridded framework surrounding the lines running across Playfair's graphs, the visual components, the layout, of notebooks played an important role in the relationship between the form and meaning of headings rendered in script.

As shown by Lorraine Daston, the headings and keywords in modern lists and tables required their viewers to possess "super-vision," a set of vectoral skills that went beyond simply reading or writing words in a sentence.[49] It is this kind of super-vision that helped even the most articulate authors to process and filter data through their notebooks. David Hume's memoranda, depicted in figure 5.1, for instance, were ordered as a list of longish headings, each of which was neatly separated by a thin blank line that made it easy for him to jump vertically from one entry to another. As shown by James Harris, these memoranda were most likely a recopied, neater version of an earlier set of notes.[50] This situation reveals that Hume valued the visual utility of ordering and copying headings into linear patterns on the page, an act that allowed him to become more familiar with what he had already written, and which created an accessible vertical layout.

How did Scottish students educated at the time of Hume learn such note-keeping skills? To answer this question, we first might wish to reflect on the kinesthetic framework offered to students by the evolving structure of a notebook page or, more generally, by a page of script or print. Scholars working on the history and theory of graphic design have noted the emergence of what they call the Gutenberg diagram, that is, a pattern that was repeatedly used from the Renaissance onward to structure the flow of information across the pages of handwritten and printed documents. Since the pattern offers a helpful way to think about how students laid out the flow of information in their notebooks, I would like to note its main features.

The diagram is named after Johannes Gutenberg, the fifteenth-century inventor and printer who introduced movable type to Europe. It offers a way of thinking about how the movement of information on a page was increasingly facilitated in early modern texts by several reoccurring elements of paginal layout. It also offers the opportunity to reflect upon how a relatively ordinary

---

48. Playfair (1786, 3). Funkhouser (1937) historicizes Playfair's visualizations. See also Costigan-Eaves and Macdonald-Ross (1990).
49. Daston (2015).
50. Harris (2015, 146).

student notebook page had the extraordinary capacity to operate as an architectural mnemonic. As shown by Mary Carruthers in her groundbreaking research on word pictures in medieval manuscripts, architectural mnemonics provided visual structures that facilitated memory systems.[51] Using the Gutenberg diagram, I wish to make a similar point about the mnemonic structure of the student notebooks under examination in this chapter.

The Gutenberg diagram is a heuristic representation in that it provides a way to think about the general flow of information, the axis of orientation, across the words on a page (figure 5.9).[52] It visualizes the areas of the page that are more likely to attract the attention of a viewer. The first element is the primary optical area. It is located at the top and left side of the page and serves as the eye's entry point. The second element is the strong fallow area that pulls the eye toward the top or right side of the page. It is "strong" because it is the part of the page toward which a viewer's gaze is more likely to move after it has entered or moved down the structure. The third element is the weak fallow area that pulls the eye toward the bottom or left side of the page. It is "weak" because when a person casts her gaze from the right to the left side of the page, she is not seeking to read the information; rather, she is merely going back to the left margin so that she can read the words from left to right. The final element is the terminal area located at the bottom right of the page. It is "terminal" because it is where the eye usually leaves the page.[53]

The flow of information created by headings as focal points on the pages of notebooks designed by Scotland's students, like so many texts of the modern period, operated according to the Gutenberg diagram. Students further enhanced the function played by headings as coordinates by using various forms of chirographic differentiation, an act that made the headings even easier to see when they wanted to move quickly through their completed notes. Since students did not avoid implementing this kind of customized chirography (and this would have certainly been the easier option), a noteworthy aspect of notebook layout lies in the fact that they spent much time crafting headings that operated according to the Gutenberg diagram, thereby making their pages more accessible.

The skills required to craft headings as coordinates are clearly exhibited in the way that students plotted and chirographically altered headings in a man-

---

51. Carruthers (1998, 89–98).
52. The origin of the term "Gutenberg diagram" is attributed to the mid twentieth-century typographer Edmund Arnold. Lidwell, Holden, and Butler (2003, 118–119). Sometimes the term "reading gravity" is used instead of "axis of orientation."
53. Lidwell, Holden, and Butler (2003, 100–101), Eldesouky (2013, 152–153).

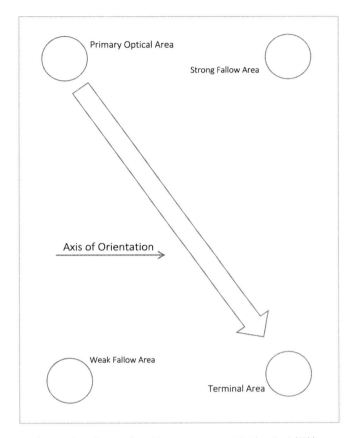

FIGURE 5.9. The Gutenberg diagram of graphic space on a page. © Matthew Daniel Eddy.

ner that made it relatively easy to jump from one heading to another across the downward flow of information on a page. The end result was a *sightline*, a visual axis that drew the eye from one cued heading to another.[54] When followed, sightlines offered paths on which students could track "the scent of information" across the page.[55] Using headings as coordinates plotted according to the information flow depicted in the Gutenberg diagram, student notekeepers could craft sightlines that created what today's graphic designers

---

54. Within the modern field of visual studies, a sightline is also called a "line of sight" or a "visual axis." Historical examples if sightlines are given in Meadows (2003) and Elkins (2000, 133–145). See also the discussion of "fields of force" and "positional enhancement" in Gombrich (1979, 155–153), and the explication of "perceptual force" in Arnheim (2004, 16–19).
55. The phrase "tracking the scent of information" is employed in Johnson (2010, 97–107), to describe the process of using graphic cues to find information. The basic paths taken by the eye across different kinds of graphic fields are explained in Dondis (1973, 64–66).

call a zig-zag flow pattern.[56] The path of this pattern ran from left to right down the page between focal points in a way that resembled a series of repeated z's.

A relatively simple example of a zig-zag sightline can be seen in figure 5.10. Based on the headings presented in Fisher's *Instructor*, it presents a stylized account of how a young reader might have interfaced with the page along sightlines running between headings functioning as visual cues. As such, figure 5.10 uses the flow of information captured in the Gutenberg diagram to visualize a mode of textual interface from the perspective of a student wishing to create a scanpath in realtime through the page to find the heading of a topic that she would like to read about. The dashed line represents the act of scanning and the solid line represents the act of reading.[57] She enters the page at point A and scans the text to point B. She then reads the heading that appears between points B and C. Since she does not find the topic she seeks, she scans further between points C, D, and E. She reads the second heading that appears between points E and F. Since she again does not find relevant information, she continues to scan the text between points F and G. She leaves the page at point H to continue the process on the next page.

Though specific names were not given to different kinds of sightlines at the time, their value was implicitly identified by a number of teachers operating in Britain, including Richard Grey and Joseph Priestley, both of whom were influential among Scottish educators. Grey called sightlines "technical lines,"[58] and Priestley praised the pedagogical efficacy of headings because they helped a student to "carry his eye *horizontally*."[59] Add to this the fact that printers laid out the headings and keywords of primers in structures that afforded ocular navigation. Even if educators did not explicitly identify sightlines, they readily acknowledged the fact that words, especially headings, could be patterned on the page so that they functioned as markers on a map of text.[60]

Though students were familiar with the zig-zagging sightlines running between the headings featured in printed primers, scholars seldom consider how they learned to make and use such spatial scanning patterns in handwrit-

---

56. In addition to the zig-zag-pattern, graphic designers today structure words and images to create a z-pattern, a golden triangle Pattern, and an F-pattern. Lidwell, Holden, and Butler (2003), Eldesouky (2013, 153–157), and Bradley (2018).
57. The classic paper on scanpaths is Noton and Stark (1971).
58. Grey (1799, 48). Grey explained how the technical line ran between typographic markers that he had developed for the word matrices in his book.
59. Priestley (1786, 12–13).
60. The relationship between typography, layout, and paginal navigation in the headings of textbooks is raised throughout Eddy (2010a) and Mayhew (2007). The use of typographic variation and layout to frame information in eighteenth-century paratexts is treated throughout Barchas (2003). For the kinds of conventions that guided the usage of capitalization and italics, see Moxon (1963 [originally published in 1683–1684], 211–219) and Luckombe (1771, 379–390).

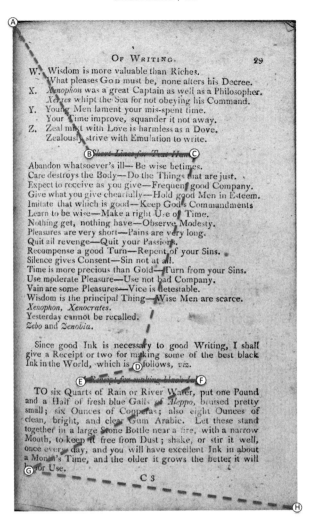

FIGURE 5.10. Zig-zag sightline running across the page of a primer. © Matthew Daniel Eddy George Fisher, *The Instructor: Or Young Man's Best Companion*, 29th ed. (London: Johnson, 1806), 29. Original image of Fisher courtesy of the Wellcome Library, London.

ten documents. This means that the zig-zag sightlines created by Scotland's students in their notebooks present an entry point into this relatively unexplored chapter of cognitive history.[61] A key point to note in reference to the frequency of zig-zag patterns presented in student notebooks is that, unlike

---

61. The absence of historical case studies on how children acquire spatial awareness is noted in Uttal and Chiong (2004, 126).

a printed primer, students themselves designed the spaces of their notebook pages. Crafting the zig-zag pattern would have taken a significant amount of time and effort, namely because it required the skills of coordinated gridding, ranging, plotting, and writing.

A particularly striking example of headings working as visual cues oriented to guide sightlines appears in the 1787 *Perth Academy Notebook* we encountered above.[62] The headings are formed and positioned in a manner calculated to catch the attention of a viewer. The first heading is depicted inside a semipictograph of a target featuring arrows embedded on the right and left. The second heading, placed near the bottom of the script, is surrounded by outward flowing contour lines that give the impression of a radiating sun. These and other chirographic features make both headings stand out, transforming them into coordinates that viewers could use to scan the page with a zig-zag sightline.

To make the utility of the sightlines easier to see, figure 5.11 presents a scenario of a student seeking to locate a "PROBLEM" heading that signals a block of text addressing the calculation of angles in a spherical trigonometry diagram. The dashed lines represent the act of scanning and the solid line represents the act of reading the heading to ascertain its meaning and, hence, relevance to the interface at hand. Following the conventions of the flow of information depicted in the Gutenberg diagram, the student's gaze enters the page at the top left and moves to the center to read the "Case III" heading. Since this is not a "problem" heading, the student continues to scan down the center of the script to read the "PROBLEM IV" heading. To ascertain the nature of the problem, the student then reads the subheading. The subject matter addresses: "The Hypothenuse and a leg given to find the other Parts," which means it is not the information that the student is seeking. Having read the heading and subheading of "PROBLEM IV," the student's gaze exits at the bottom right of the page. The student then repeats the process several times, since the relevant information occurs several pages later.

The navigational structure of the headings presented in figure 5.11's depiction of a *Perth Academy Notebook* page is evinced in many of the student notebooks created in schools, homes and academies. The presence of the zig-zag sightlines in figure 5.11 also reveals that many student notekeepers wrote with the expectation that they would be strategically scanning headings once the page was finished, or once they had completed school and needed to use the notebook as a reference resource. This meant that ranging headings was po-

---

62. Anonymous Bound MS (1787, 1:39v–40r).

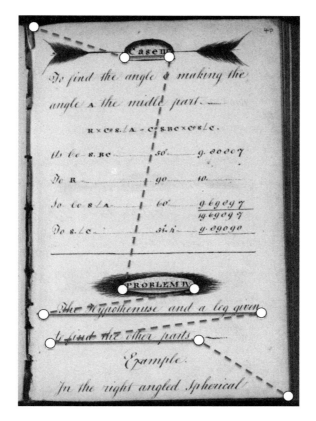

FIGURE 5.11. Sightlines running between headings in a student notebook. © Matthew Daniel Eddy. Anonymous, *Perth Academy Notebooks, Vol. I*, Bound MS Notebook (1787), NLS MS 14294, f. 40r. Reproduced with permission of the National Library of Scotland. Each white circle represents a change in the eye's movement across the page.

tentially a proactive enterprise, one in which students had to think ahead to how they might interface with the pages of their notebook in the future, or perhaps to how others in their family or school would read the notebook as evidence of their notekeeping abilities.

In order to maintain a regular pattern of sightlines through an entire notebook, students had to enact and consistently implement a graphic nomenclature of sorts, a chirographic code that enabled them to shape the internal space of the page in a systematic manner. Every time they mapped out the framework of the pattern with graphite or lead pen impressions, and every time they attempted to write according to the codes they had selected, they were effectively transforming the pages of their notebooks into interactive visual platforms as they wrote.

Designing pages of notebooks as informatic maps reinforced skills that students needed to learn through repetition so that they became automatic and did not tax the working memory, that is, the short-term memory skills that enabled them to quickly process or perceive information.[63] Indeed, the use of sightlines as navigation cues depended upon this ability to do something automatically without having to think too much about how it was being done. Such graphic automaticity fostered what the art historian Martin Kemp once called "structural intuitions," the ability to iterate or recognize visual patterns in images. Once learned, children used such skills and other related intuitions to access the visual affordances present in other kinds of textual, or even figural, pictures.[64] This kind of enhanced writing, so crucial to designing a school notebook in Scotland, operated in realtime and was a form of graphic interface that provided skills that students could use across different subjects and extend when they became adults who wanted to create more sophisticated manuscript layouts.

The sightlines presented in school notebooks trained students to structure and track information across lines in a quick manner, giving them the capability to understand that qualitative and quantitative data could be arranged according to zig-zagging lines running back and forth across the page. The striking nature of this pattern can be seen more easily if we remove the script of the student notebook presented in figure 5.11 and retain only the zig-zag pattern of the sightline. Figure 5.12 shows the pattern turned on its side with a view to further demonstrating how the automaticity that came with the familiarity of this kind of structured eye movement arguably made it easier for students to understand both the established and emergent genres of modern graphic representation more generally.

The ability to follow information in a zig-zag pattern across a graphic structure could potentially applied other texts. The pattern presented in William Playfair's innovative line graphs during the 1780s, for instance, consisted of a line moving back and forth across points plotted on a grid in a manner very similar to the sightlines used by students to create and scan the headings in their notebooks (figure 5.13). Whereas the inspiration for Playfair's graph is a matter of dispute among historians, it would not be going too far to say that its structure would have been much easier to understand for students familiar with the kinds of sightlines that operated between the headings plotted on the grid of a notebook page.

---

63. Henry (2011).
64. Kemp (2000).

CATEGORIZING | 179

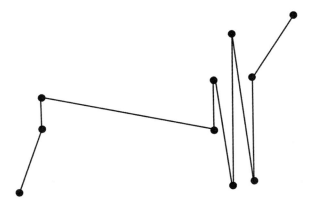

FIGURE 5.12. Zig-zag sightline extracted from the notebook page depicted in figure 5.11. © Matthew Daniel Eddy.

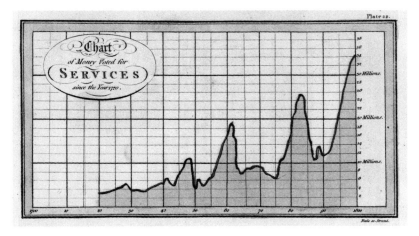

FIGURE 5.13. "Chart of Money Voted of Services since the Year 1720," William Playfair, *The Commercial and Political Atlas, Representing, by Means of Stained Copper-Plate Charts, the Progress of the Commerce, Revenues, Expenditures and Debts of England, during the Whole of the Eighteenth Century*, 3rd ed. (London: Burton, 1801), Plate 22, following page 88. Internet Archive, Public Domain Mark 1.0.

The formative and interactive aspects of handwritten school notebook headings remind us that they served as mnemonic, technic, and kinesthetic markers. They facilitated extremely useful modes of interface that taught students how to make and use different kinds of verbal, visual, and vectoral categories. Thus, when the young Hume opted to use headings in his notebook, he was creating a multivalent form of mnemotechnic representation, one that expedited his interpretation of what he was reading and preserved his thoughts in a pattern that was easier to navigate at a later date.

Transforming words into patterns via handwritten modes of interface was often a tacit enterprise for the students such as Jackson, Belsches, or those who attended academies. The success of their handcrafted designs depended on skills that they had to learn through much practice. This means that school notekeeping effectively trained students how to perceive and organize various kinds of data in realtime in a way that laid the foundation for accessing and creating a knowledge system. Crucially, making notebook headings taught them how to replicate and adapt various kinds of word patterns in a way that was conducive to guiding their gaze through any handwritten document that they might choose to create in the future.

CHAPTER 6

DRAWING

Description and Movement across a Page

Imagine yourself sitting in a cottage surrounded by the rolling foothills of northern Scotland. The only clock in your house has stopped and you need to reset it. How might you determine the correct time? This question could be easily answered today by consulting the myriad of digital devices that surround us. But three centuries ago, the inhabitants of remote Highland hamlets often had to walk for several hours should they wish to reach a town that had a publicly accessible clock. It was for this reason that many students had to learn how to set a clock by using a sundial.

But what kind of sundial might have been available to a student living in a Highland hamlet? And how might that student learn to use it? Fortunately, helpful answers to these questions are offered in the school notebook of James Fowler, a student who lived in the hamlet of Fodderty in the Strathpeffer Valley of the Ross and Cromarty region of the Highlands.[1] In one of the mathematical sections that he inscribed in the notebook, he crafted a sundial. At the top of the figure, he wrote his name, "1784," and "Fodderty LAT: 58° N" (figure 6.1).[2] The various ink, graphite, and chalk components of the dial reveal that he learned to use and understand its structure by drawing it.

Fowler was most likely in his early teens when he kept his school notebook. His dial is an example of the interwoven relationship that existed between writing and drawing technologies used in educational settings during the modern period. A clue to the relationship occurs in the prose that he wrote

---

1. Fodderty was located in the Strathpeffer Valley and was part of the Dingwell Presbytery in the region of Ross and Cromarty county. Mackenzie (1793, 410–415). Fodderty school is briefly mentioned in on page 414, where Mackenzie notes that it had moved into a new schoolhouse erected in 1779.
2. Fowler Bound MS (1780, 134r).

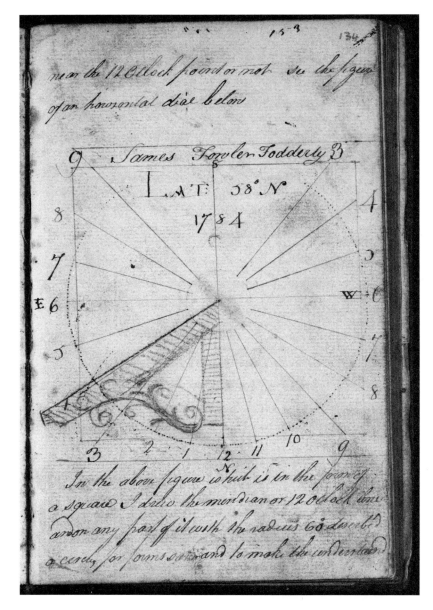

FIGURE 6.1. James Fowler, *Schoolbook of James Fowler, Strathpeffer* (1780), Bound MS, NLS MS 14284, f.134r. Used by permission of the National Library of Scotland.

to explain how he made the dial and what its components were meant to represent. There he uses the verb "describe" alongside the verb "draw" to explain how he had rendered the image. Under the dial he wrote: "In the above figure which is in the form of a square I drew the meridian or 12 Oclock line and on any part of it with the radius 60° described a circle."[3] Here and elsewhere in his notebook Fowler conceived the act of drawing the line of a figure as a mode of describing, which, like the description offered by a written sentence, required the movement of the hand across the page and of ideas inside the mind.

Fowler's use of the verb *describe* reveals that it had a wider meaning than it does today. At the time the verb still retained the meaning of its Latin root *describere*. The verb combined the prefix *de*, which meant *down, from,* or *off,* with the verb *scribere*, which meant *to inscribe*. The eighteenth-century entry for *description* in the *Oxford English Dictionary*, for instance, defines it as "the process of describing or delineating a geometrical figure" and "the tracing out of a given path or region by the motion of an object following a certain course." A *description* could also be "a representation in art; a picture, painting, sculpture, etc., representing a person or thing."[4] The visual understanding of description communicated in Fowler's dial was not unique. Similar usages of the verb *describe* appear in primers, treatises, essays, and dictionaries that addressed how to make and use the many figures required for technoscientific subjects. The 1763 Edinburgh edition of Fisher's *Instructor*, for instance, uses the verb when explaining how to draw geometric lines.[5]

The jointly manual and visual process of description evinced in Fowler's dial reveals how paying attention to the creation of a notebook's figures can help us further understand the material and visual foundations that underpinned the success of the *tabula rasa* metaphor and the way that it was used to understand the enactive potential of manuscripts as media technologies. More specifically, Fowler's dial offers an invitation to enter the fascinating multimedia world that made it possible for young learners to "describe" what could be called a *tabula figura*. In previous chapters we learned that a student notebook page began as a blank sheet, a *tabula rasa*, that students then transformed into a *tabula verba*, a word picture ranged and written by hand. A similar process existed for the creation of a page designed to function as a figure in a notebook.

This chapter reclaims the drawing skills that students learned while making and using a *tabula figura* as an interactive media technology. As in previous

---

3. Fowler Bound MS (1780, 134r).
4. For the uses of *description* within the visual arts, see the essays in Bender and Marrinan (2005).
5. Fisher (1763, 246–247).

chapters, I will treat notekeepers as artificers who crafted each page through a process that involved specific kinds of skills, materials, and instruments. To this goal, I will concentrate on how students learned to literally create a *tabula figura* from start to finish. This will allow me at times to reflect on how the material culture that underpinned the *tabula figura* helps further explain the contemporary power of the *tabula rasa* metaphor. In later chapters, I will return to these reflections when I discuss the relationship between media technologies and the formation of the skill-based understanding of the human mind that permeated developmental psychology in Scotland and Europe more broadly.

The drawing skills that notekeepers used to make a *tabula figura* enabled them, to paraphrase the historian and visual artist Nick Sousanis, simultaneously to orchestrate a conversation with themselves and to step outside themselves in a manner that allowed them to see visual relationships.[6] The multimodal facet of describing was a fundamental feature of the increasingly important role played by the acts of drawing, sketching, and tracing exhibited in scholarly, domestic, artistic, and commercial contexts throughout Europe at this time.[7]

Learning to Draw a Picture

Scottish civic humanism drew strong links between the visual arts and a liberal education.[8] This link was reinforced at a developmental level in popular pedagogical works published by authors such as John Locke. In *Some Thoughts Concerning Education* (1693), for instance, he averred that the "Skill in Drawing" needed to be learned by students because it helped them to create images that expressed "in a few Lines well put together, what a whole Sheet of Paper in Writing would not be able to represent and make intelligible."[9]

In addition to the links between drawing and humanism, it was common for girls and boys of means to learn advanced drawing skills and then to employ them as pastimes later in life. Princess Elizabeth (1770–1840), George III's daughter, for instance, learned to draw from the artist Peltro William Tomkins

---

6. Sousanis writes: "Drawing, as Suwa and Tversky suggest, is a means of orchestrating a conversation with ourselves. Putting thoughts down, allows us to step outside ourselves, and tap into our visual system to see in relation." Sousanis (2015, 79), citing Suwa and Tversky (2001). See also Tversky, Suwa, and Agrawala (2003). The process through which learners make sense of the world with pictures (or art) created by others is explored in Perkins (1994) and Gardner (1990).
7. Alpers (1983), Baxandall (1985), Rosand (2002), Nickelsen (2006), Ingold (2007), Daston and Galison (2007), and Alpers and Baxandall (1996).
8. The larger relationship between art, civic humanism and liberal education in Britain is treated throughout Barrell (1986).
9. Locke (1752, 234–235).

FIGURE 6.2. Paul Sandby, *Lady Francis Scott and Lady Elliot* (ca. 1770), watercolor and graphite on medium, cream, moderately textured laid paper, Yale Center for British Art, Paul Mellon Collection, B1977.14.4410. Public domain image.

and then used her skills when she became an adult.[10] It was a similar story for the children of other Scottish aristocrats. Lady Frances Douglas (1750–1817), daughter of the Earl of Dalkeith and stepdaughter of the English politician Charles Townsend, for instance, received drawing instruction as a child and went on to use a *camera obscura* to make drawings of Scottish landscapes (figure 6.2).[11]

Over the eighteenth century, an increasing number of middle-class students in the Scottish Lowlands learned to draw. As intimated in teaching manuals such as Hannah Robertson's *Young Ladies School of Arts* (1767), drawing and painting were increasingly being included in a larger repertoire of graphic skills associated with a proper middle-class education. Robertson (b. 1724) had run a school in Perth and the fact that her book was published in Edinburgh indicates that there was a bankable audience of eager Scottish parents

---

10. Marschner, Bindman, and Ford (2017). After teaching Princess Elizabeth to draw, Peltro William Tomkins engraved a number of her drawings; they were printed as *The Birth and Triumph of Cupid* (1795), *The Birth and Triumph of Love* (1796), and *The Power and Progress of Genius* (1806). For Tomkins's innovative engraving methods, see Tomkins (1814).
11. Lady Douglas's use of a *camera obscura* is depicted in Paul Sandby, *Lady Francis Scott and Lady Elliot* (ca. 1770), Yale Center for British Art, Paul Mellon Collection, B1977.14.4410.

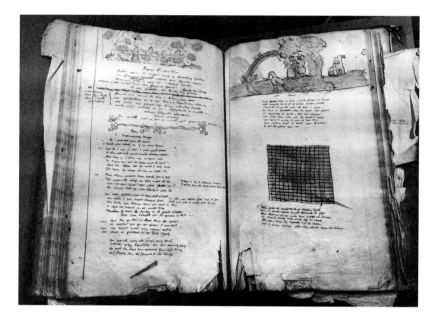

FIGURE 6.3. John Black, *Juvenile Poetic Works*, NLS MS 1797–1798 14233 ff. 26v–27r. Reproduced with permission of the National Library of Scotland.

seeking to enhance the drawing skills of their children.[12] Such skills could be practiced in school notebooks or in other kinds of notebooks or loose-leaf ephemera.

Learning to draw often provided an alternative or complementary mode of description for young writers who found words difficult to remember or spell. Even basic sketches could expand a young learner's creative potential. The *Juvenile Poetic Works* notebook of the poet John Black, which he wrote between the ages of fifteen and nineteen, for example, included watercolored illustrations of characters, crowds, landscapes, grids, and botanical naturalia (figure 6.3). Although the figures were not included in the parts of the manuscript that were eventually published as *The Falls of the Clyde* in 1806, it is clear that drawing them helped Black think through the visual imagery of the verses.[13]

Instead of drawing the scenes of a poem, the teenage George Sandy depicted vignettes from his own life in the diary he kept during his years as an apprentice lawyer in Edinburgh during the 1780s (figure 6.4).[14] Rendered in pen,

12. Robertson's birth date and other biographical details are given in Robertson (1792).
13. Black Bound MS (1797–1798, vol. 2). The notebook contains fragments of what was eventually published as Black (1806). Black's life and work are treated in Yoshino (2013, 126–161).
14. Sandy (1942) and Sandy Bound MSS (1788).

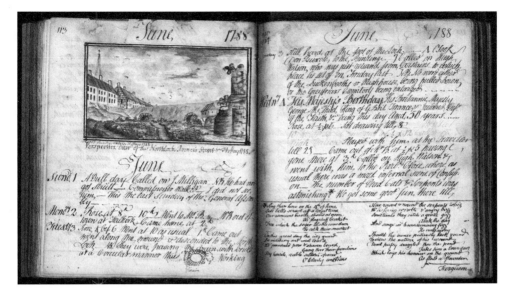

FIGURE 6.4. George Sandy, *Schoolboy Diary* (1788), ff 113–144. Reproduced with permission of the WS Society, Signet Library, Edinburgh.

ink, and watercolors, the vignettes depicted places and objects addressed in the narrative and, like Black's drawings, operated as realtime thinking tools that allowed Sandy to process his experiences as he sketched, inked, and colored the image. The rise of this kind of graphic intelligence in Scotland and across Britain had a positive impact on the abilities that students could use to execute technical drawings and to make or buy important media such as ink, paint, and drawing instruments.[15]

Keen to capitalize on the fine arts educational market, schools and academies began to include basic drawing instruction or endeavored to retain a drawing master whom students could hire for an extra fee. The Perth Academy's drawing master Mr. Maccomie, for instance, charged half a guinea per term.[16] Students learned basic drawing skills in geography and geometry and then moved on to "figuring," that is, the skills of rendering more advanced images such as diagrams or maps. The advertisement featured in the front matter of the 1770 edition of William Gordon's *The Universal Accountant and Merchant*, for example, stated that the Trongate Academy in Glasgow offered

---

15. The materials and skills required to make paints are explained throughout Boutet (1739), with page 107 being a particularly good example.
16. Anonymous, "Perth Academy," *Caledonian Mercury*, 25 Aug 1781, Anonymous, "Academy—Perth," *Caledonian Mercury*, 10 August 1785.

lessons on "Languages, ancient and modern; Writing, *Figuring*, Geography, Geometry, Algebra, and Merchants Accounts."[17]

Independent drawing masters were available to affluent households from the early part of the century. Most masters were male, but female masters, Mary Cummins for instance, emerged near the end of the century.[18] Masters advertised their expertise in postal directories and in local newspapers. To cite one of the many examples, during the winter of 1786 George Walker advertised lessons for "Ladies and Gentlemen" on "Drawing, Perspective, and Elements of Fortification" that he intended to teach at his "drawing academy" based at Foulis's Close on Edinburgh's High Street.[19] The presence of such masters, however, was haunted by the ever-present problem of quality control.

To ensure the presence of a high standard of artistic instruction, especially drawing, the Academy of St. Luke was founded in Edinburgh in 1729 by the artist Richard Cooper (1701–1764), members of the university, and the town council.[20] In 1754, Glasgow's Foulis Academy of Art was founded. Edinburgh's Trustees Academy of Art was organized in 1760. To maintain their links with patrons, the academies publicly exhibited oil-painted portraits either housed in their collections or created by their best students. As revealed by David Allan's majestic depiction of a Glasgow academy portrait exhibition staged outside in the quadrangle of the university to celebrate the coronation of King George III, visual records of such events were preserved in paint and print.[21]

When we look past the public exhibitions organized to stimulate the pocketbooks of patrons, a more practical purpose of the academies emerges, one that sheds light on the kinds of drawings that appear in school notebooks. The academy in Glasgow was founded by the Foulises, a family with strong ties to the printing industry and the university, who saw the history of the visual arts as an important part of a liberal education.[22] The Trustees Academy in Edinburgh, on the other hand, was founded and overseen by the Board of Trustees

---

17. Gordon (1770, advertisement in the front matter). Emphasis mine.

18. See the entry for "Mrs Robert Cummins, née Mary Forbes, (fl. Edinburgh 1797–1800)," in Jeffares, online edition at http://www.pastellists.com. Accessed 12 April 2020.

19. Anonymous, "Drawing Academy," *Caledonian Mercury*, 18 November 1786. George Walker was active in Edinburgh from circa 1781 to 1815. In addition to the techniques that he advertised in the newspaper, he also taught students how to draw landscapes with chalk crayons. His written account of his technique was published posthumously as Walker (1816).

20. Richard Cooper became the master of the academy. Laing (1871). Rock (2000) touches on the Academy of St. Luke and its connections to the arts and science. The Scottish artist Sir Robert Strange, who trained at the Academy of St. Luke, noted that "Mr Cooper encouraged me as much as possible in the study of drawing." Strange (1855, 22–29, 26).

21. Allan (c. 1753).

22. Gibson-Wood (2001–2003).

for Fisheries, Manufacturers and Improvements in Scotland.[23] Both the Foulis family and the Edinburgh Trustees wanted their academies to train adolescents to draw, paint, and stitch so that they could pursue careers as illustrators, painters, and graphic designers in the printing, woolen, and linen industries.[24]

The utilitarian aims of Scotland's drawing academies applied to both girls and boys. The trustees of the academy in Edinburgh offered prizes for drawings and "to the boy or girl under twenty years of age, who shall produce the best pattern of his or her invention for a Scots carpet, the pattern to be drawn upon design-paper, from which the carpet can be put into the loom." It offered further two guinea prizes for the best damask carpet, damask table linen, and a "flowered lawn."[25] The overarching aim of the trustees was to support the Edinburgh academy "with a view of promoting the Knowledge of Drawing, the Principles of Art, and of assisting those who aim at being Professional Artists."[26] This is why the academy admitted both girls and boys and offered additional instruction on damask weaving and embroidery (including tambour work). True to its improvement-minded aims, the academy waived the tuition fees of promising students who could not afford to pay.

The presence of esteemed masters in Glasgow's and Edinburgh's flagship drawing academies not only raised the standard and reputation of the visual arts, it also guaranteed the presence of qualified masters who could offer quality extracurricular lessons to students seeking to make drawings for their school notebooks or, as in the cases of John Black and George Sandy, for their personal notebooks. The Trustees Academy master John Graham, for instance, offered "Two extra Classes; one for Ladies, and the other for Gentlemen," during the late 1790s. His classes ran on Monday, Wednesday, and Thursday, with "the Ladies hour from eleven till one, and the gentlemen from one to three." As with many educationally oriented advertisements of the day, "ladies" and "gentlemen" included teenage girls and boys.

---

23. The masters of the Trustees Academy were William Delacour (d. 1767), Alexander Runciman (1736–1785), David Allan (1744–1796), and John Wood (late 1790s). Wood's tenure was short-lived and he was dismissed after the trustees learned that he could not paint and had submitted the paintings made by another artist when he applied for the academy's mastership. After his departure, the trustees appointed the painter John Graham (1754–1817) around 1798 and the academy began to flourish again. Campbell (1802, 256–257). Thomson, "John Graham (1754–1817)," *ODNB*. Wood tried to salvage his reputation in Wood (1799), the figures of which confirm that he knew more about optics than he did about painting.
24. Tromans (2007, 61–64). For the Foulis Academy, see Murray (1913), and the manuscript letters of the Foulis Press housed in the Murray Collection, GUL, MS Murray 506. The Trustees Academy served a thriving art market in and around Edinburgh. See Nenadic (1998, 177).
25. Laing (1871, 38–39).
26. Anonymous, "New Drawing Academy," *Caledonian Mercury*, 28 November 1799. The quotations in the next paragraph are taken from this advertisement as well.

But it is worth noting that the prices charged by skilled instructors were not cheap. Lessons for male or female students wishing to learn the "Drawing of Flowers and Landscapes" at Alexander D'asti's Edinburgh academy, for instance, cost £2 and 2s. Lessons for "Fortification, and drafting plans in colours" cost £3 and 3s, that is to say, roughly the same as the three guineas it cost to take one course of lectures for a full academic year at Edinburgh University.[27] Since three guineas was worth around £3, and considering that the annual incomes of male and female laborer's in Scotland were around £9 and £4, respectively, it can be seen that hiring a drawing master was an option primarily available to the professional, merchant, and landed classes.[28] This expense in part explains the presence of more affordable self-help drawing manuals and the popular practice of children learning to draw from family members.

A core part of the training for students learning to draw was keeping a sketchbook. One of the few known sketchbooks of a Scottish student is that of the portrait artist Allan Ramsay, part of which was made while he attended the Academy of St. Luke in Edinburgh during the 1730s.[29] It features drawings mainly of the human form, many of which were rendered in chalk crayons. Ramsay was an exceptional artist who had received superior training as a student. So we must be careful not to take the drawings in his sketchbook to be normative, especially for student notekeepers seeking to illustrate the subjects that they were learning in school. Nevertheless, Ramsay's sketchbook reveals that he was working with many of the same materials, graphite pencils and red chalk for instance, that other students used to create visualizations in their school notebooks (figure 6.5).

Further insight into how students learned to make sketchbooks is offered in David Allan's print *Interior of the Academy of the Fine Arts in the University of Glasgow*, which was based on one of his paintings and first published in the 1760s (figure 6.6). It depicts the students of the Foulis Academy taking lessons in the upper floor of the Glasgow University library. The presence of screens on the lower sections of the windows reminds us that light, as well as the drawing materials, in such professional studios was carefully managed.[30] The foreground of the scene features advanced students working on canvas. But

---

27. Anonymous, "Edinburgh, 11th August 1787," *Caledonian Mercury*, 16 August 1787.
28. The annual wages of Scottish laborers are given in Morgan (1971, 183).
29. Ramsay Bound MS (1730–1731). Ramsay's sketchbook and its relevance to the time that he studied at the Academy of St. Luke in Edinburgh are discussed in Brown (1984a; 1984b).
30. A helpful account of the kind of window screen depicted in Allan's print is given in Russell (1772, 11). "If the window is too near the ground, the effect cannot be pleasing, because the shadow will be distracted into too many parts: In this case, the lower part of the window will be darkened with something that will quite obstruct the light, about the height of six feet, which will throw the shades into very agreeable masses on the subject for imitation."

FIGURE 6.5. Pen and ink sketchbook drawings made by Allan Ramsay at St. Luke's Academy, Edinburgh, c. 1730. Left: *Female Figure Dancing with a Crotalum in Each Hand*, n.d. Right: *Drawings of Lions and Camels (after Stefano della Bella)*, 20 December 1730. It is likely that some of the drawings in the sketchbook were entered in the annual student competition hosted by the academy. Allan Ramsay, *Sketchbook Containing Copies of Old Master Prints and Drawings* (1730–1731), reproduced with permission of National Galleries Scotland. Bound MS, Accession Number D 5109, ff. 9 and 14.

the figures in the image that offer the most insight into how sketchbooks functioned as developmental media technologies are the five younger students drawing at the desk shelves under the windows. They are diligently practicing their sketching skills on individual sheets of paper. This kind of activity was repeated on a daily basis until they had curated a collection of their best sketches, which, like writing samples and notebooks offered for exhibition in schools, were sometimes called specimens. The collection, bound or unbound, became a sketchbook. The loose-leaf nature of a student's individual sketches can be seen in Ramsay's sketchbook, the binding of which was made after he had finished his studies (see again figure 6.5).

Like writing, drawing required a variety of media and instruments. These in turn were used to enact a plethora of skills, such as tracing, sketching, crosshatching, smudging, stippling, pricking, grooving (with trace pens), and watercoloring. As we will see below, the smoothness and regularity of the lines in notebook drawings indicate that students were constantly observing their strokes and scores to fit the kinds of media they sourced and the evenness of the lines they wanted to make in their figures. Add to this the fact that many

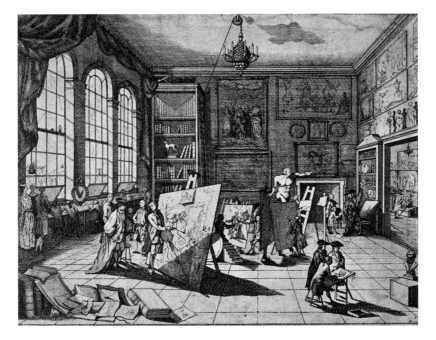

FIGURE 6.6. David Allen, *Interior of the Academy of the Fine Arts in the University of Glasgow*, David Murray, Robert & Andrew Foulis and the Glasgow Press (Glasgow: Maclehose, 1913), 68. Author's library.

students learned to draw on desks equipped with slanted surfaces or on easels and other kinds of transportable furniture that made it possible to control the movement of their hand when they were drawing and painting.

The precision and lack of deviations in the strokes in school notebooks indicate that student notekeepers learned to use various instruments of draftsmanship, the most common being the steel pen, the "black lead" (graphite) pencil, the ruler for straight lines, and the compass for circular lines. The steel pen was particularly important for precision drawing. The London-based lexicographer and instrument maker Benjamin Martin described it in the following manner: "The *Drawing Pen* is only the common Steel Pen at the End of a Brass Rod, or Shaft, of a convenient Length, to be held in the hand for drawing all kinds of *straight* Black Lines by the Edge of a Rule. The Shaft or Handle has a Screw in the middle Part; and, when unscrewed, there is a *fine round Steel Pin*, or *Point*, by which you make as nice a mark or Dot upon the Paper as you please, for terminating your Lines in curious Draughts."[31] For more advanced

---

31. Martin (1760, 3). The modern use of steel pens is treated throughout Bore (1890) and in Anonymous (1838b).

figures, there were protractors and sectors as well. Instrument makers sold these items individually or as part of a collection that was called a "pocket case" (figure 6.7).

School notebooks contain circles drawn first in graphite and then traced over in ink. This indicates that students were using compasses that had several interchangeable "points" for drawing or impressing different kinds of lines. More specifically, they used compasses equipped with a plain point for impressions, a pencil point for graphite lines, an ink point for solid ink lines, and a dotting point for dotted ink lines. These instruments were sometimes included in a pocket case as well. For more advanced students there was the magazine case, which, in addition to including a wider variety of the foregoing instruments, offered calipers and a watercolor set.[32] It is also possible that some students used a pantograph (also called a parallelogram), which the *Encyclopaedia Britannica* defined as "an instrument designed to draw figures in what proportion you please" (see again figure 6.7). It was made up of pieces that could be assembled in different ways that allowed students to replicate, enlarge, or reduce a figure. But it was difficult to make and use and often did not work, a point that was underscored in the *Encyclopaedia Britannica* as well: "Few will do any thing tolerably but straight lines; and many of them not even those."[33]

Students or their parents had to purchase their notebook drawing supplies from a variety of shops. In addition to selling maps and prints, for example, John Ainslie's shop at the head of Borthwick's Close in Edinburgh sold "Drawing and Writing Papers of all kinds," Indian ink, Indian rubbers, black lead pencils, hair pencils, chalk (black, white, and red), and boxes of watercolors. In addition to the items sold by Ainslie, the Edinburgh warehouse of Charles Esplin and Company sold carmine and "all other Colours, dry or prepared," "Sets of Superfine Prepared Water Colours, dry or prepared," "Crayons, single and in set," port crayons, "English, Dutch, swan-quill, and miniature hair pencils," and "lead eaters." Other media sold by Esplin included "Drawing books of all sorts, for teaching beginners" as well as transparent silk, drawing paper, and gum flower paper (in different colors). Those wishing to store their writing utensils in a fashionable container could buy a silver pencil case at Robert

---

32. Barrow (1792, 6–7).
33. For both quotations in this paragraph, see Bell and Macfarquhar (1797, 14:135). John Wood attempted to attract students after he was fired by Edinburgh's drawing academy by offering lessons on how to use a pentagraph. Wood (1799, 123–124). For a comprehensive list with prices of the various draftsmanship instruments, including a small pentagraph, that were sold during the late eighteenth century, see Martin (1797, 19, 20).

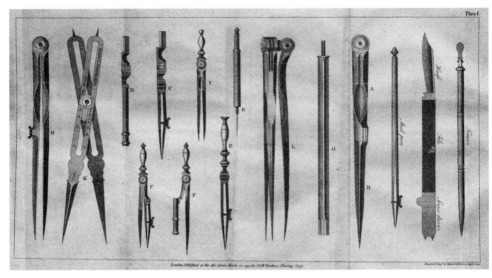

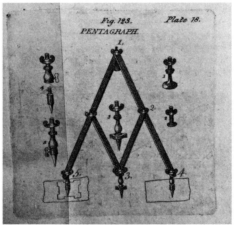

FIGURE 6.7. Drawing Instruments. Top: J. Barrow, *A Description of Pocket and Magazine Cases of Mathematical Drawing Instruments; in which Is Explained Use of Each Instrument* (London: Watkins, 1792, plate I). Reproduced with permission of the Thomas Fisher Rare Book Library, University of Toronto. The three instruments on the right side of the plate, starting from the edge and moving inward, are a tracer, a knife-file-screwdriver, and a steel pen. The remaining instruments are as follows: (A–B) plain compass, (C) ink point, (D) pencil point, (E, top) fine steel point, (E, bottom) drawing pen, (F) bow compass, (G) lengthening piece for a large compass, (H) hair compass, (I) small compass, (K) proportional compass, (L) triangular compass. Bottom: A pentagraph used for replicating shapes or maps. John Wood, *Elements of Perspective; Containing the Nature of Light and Colours, and the Theory and Practice of Perspective* (London: Cawthorn 1799; Edinburgh: Hill, 1799), plate 18, figure 123. Huntington Library, San Marino, No. 498643.

Johnston's toy shop, which was next door to the entrance of the Edinburgh exchange.[34]

But why would parents and guardians go to such lengths to pay for drawing lessons, media, and instruments? The traditional answer to this question, as intimated above, was that it was a skill valued by polite society. In one sense this answer is true, particularly for elite families living in metropolitan London or for aristocratic households with the resources that could be spent on a battalion of expensive, talented instructors. It is also true that the middling classes throughout Britain valued drawing as a skill that might help them gain access to the patronage networks that valued polite artistry. But even in an age when an Edinburgh property owner could rent out a house to a journeyman for £2 per annum, the aspirations for polite society do not fully explain why Scottish parents from merchant or professional households would pay over £3 for a child to learn advanced skills of botanical or architectural draftsmanship.[35] Nor does it explain why so many Scottish students were motivated to diligently draw figures in their notebooks.

Student notebooks show that boys who attended academies learned draftsmanship alongside other subjects such as accounting, gauging, surveying, and leveling, indicating that several kinds of drawing had concrete applications for those who wished to enter a number of professions or trades, particularly those of a technoscientific nature. Likewise, girls who learned to draw or sew floral patterns stood a good chance at gaining employment in the linen and decorative arts industries.[36] These examples show that families believed that drawing was a useful skill that could potentially provide financial remuneration as well as social advancement.

Figures as Developmental Tools

As pointed out by the Scottish educator and moral philosopher George Turnbull, learning to "draw lines and figures upon paper," no matter how simple, was a useful knowledge-creating skill. In his view, images made through the act

---

34. The various supplies mentioned in this paragraph are taken from the following advertisements: Anonymous, "Paper Hangings," *Caledonian Mercury*, 27 July 1782; Anonymous, "Toy Shop," *Caledonian Mercury*, 17 January 1781; Anonymous, "On Friday and Saturday next," *Caledonian Mercury*, 19 November 1783; Anonymous, "Maps," *Caledonian Mercury*, 12 December 1781.

35. Arnot (1779, 557), states that a journeyman with a £14 annual salary could expect to pay around £2 per year to rent a house in the city of Edinburgh.

36. For an excellent example of the kinds of decorative flower patterns drawn in Scotland at this time, see Montgomery Bound MS (1809). The notebook contains drawings at different stages in the design process, ranging from graphite sketches and grids to the pen and ink final versions.

of drawing could be used "as a means, and not as an end; when they lead us to something else, and not when they are rested in as the principal part of instruction."[37] In many cases, that "something else" for students crafting notebooks was the ability to recognize and replicate symmetry, similarity, perpendicularity, angularity, equality, and proportionality in geometric shapes and, more generally, in objects encountered at home, in daily life, or in professional settings.[38]

When seen from the perspective of a notekeeper who was learning to use drawing as a thinking tool, Turnbull's observation reminds us that, like the acts of ranging and writing words, the act of drawing a figure on a notebook page fostered the ability to make realtime judgments on, through, and around paper. It was a jointly manual and mental activity that helped learners internalize and spatially understand different kinds of knowledge. Turnbull's observation also raises the larger question of what kinds of figures, precisely, were being drawn in school notebooks. Despite the centrality of figures to the technoscientific subjects taught throughout Britain and its former colonies, surprisingly little research has been done on the topic. The purpose of this section is to address this lacuna in reference to the ways in which drawing figures in notebooks operated as a knowledge-creating skill for students.

Within the wider world of children's books in Europe at this time, publishers increasingly printed figural forms of visual culture such as board games, prints, and illustrated chapbook stories as the eighteenth century progressed.[39] Yet the picturesque beauty and charm featured in these relatively expensive images does not seem to have played a significant role in the figural forms of representation used to educate most Scottish schoolchildren keeping notebooks, particularly those of the working and middle classes who attended parish or burgh schools. In fact, painterly figures rarely appear in school notebooks. As shown in previous chapters, the predominant visual forms were word modules and matrices, and all the visual cues that they entailed.

Nevertheless, from midcentury onward a rising number of Scottish students learned to draw a notable assortment of schematic figures for school subjects relevant to mensuration, geography, geometry, and, in a handful of schools, the experimental technosciences included within natural philosophy. Consequently, when we speak of school notebook "figures" and the processes used to describe them through drawing, we are speaking about a specific kind

---

37. Turnbull (1742, 273).

38. Linear concepts such as proportionality, similarity, perpendicularity, parallelity, etc., are mentioned repeatedly in geometric textbooks, some of which included Douglas (1776), Simson (1781), Scott (1782), Vilant (1798), and Ingram (1799). For an informative look at how the concept of proportionality functioned as a tool in the building of geometric systems, see West (1784, vi–vii).

39. Heesen (2002) and Stafford (1996).

of schematic representation designed to help young learners understand how they might reduce the world to lines of measurement that, on the whole, could be used in trade, industry, interior design, land management, or various maritime or military contexts. This explains why most notebook figures were schematic in appearance. The practical facet of visualization was also supported by theories of the young mind promoted by authors writing about developmental psychology. As pointed out by pedagogues such as John Locke in England and August Hermann Franke in Germany, simple figures facilitated a child's powers of observation and apprehension. No matter what kind of knowledge was being represented, schematic simplicity was a virtue.[40]

School notebooks reveal that students learned basic drawing skills by sketching and resketching simple two-dimensional shapes such as circles, squares, and triangles. They then moved to three-dimensional shapes such as cubes and pyramids. As indicated in figures like the expertly crosshatched, but partially translucent, pyramid drawn in the 1788 *Geometry Notebook* of the schoolboy Robert Jackson, students taking advanced mathematics courses learned through figuring polyhedrons (figure 6.8).[41] The technique of progressing from simple to complex shapes was common in textbooks that addressed the expansive topic of practical geometry. But as with tables and matrices, simply seeing such a progression was not the same as using it to learn how to make shapes for oneself. Indeed, iterative acts of sketching were a crucial part of learning how to draw, which means that even smudged or ink-blotted sketches of shapes constitute important forms of visual evidence.

Take for instance the evidence of pen and ink skills presented in Jackson's pyramid. Like most students creating pen and ink drawings, he began by sketching the figure in graphite with a black lead pencil. As he sketched, he rubbed out unneeded lines, most likely with an Indian rubber. Switching to a metal-nibbed pen, he traced the contour lines in black ink and created shaded areas with hatching, giving a three-dimensional appearance to the pyramid. Elsewhere in the notebook, he used other drawing techniques as well. The edges of the contour lines and shading strokes are clean, indicating that he understood that pressing too hard on the pen released extra ink that made the edge of the line look jagged or smudged. The precision of Jackson's lines, strokes, and other visual elements presented in the figure as a hand-drawn artifact all speak to the skills that he had learned over the hours, days, weeks, and even years that preceded the creation of his geometry notebook.

---

40. Locke (1693, 184). Franke's use of pictures is discussed in Whitmer (2015, 56).
41. Jackson Bound MS (1788, 23).

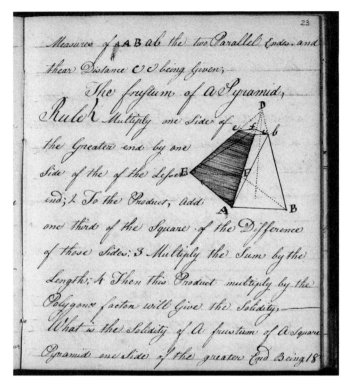

FIGURE 6.8. Robert Jackson, *Geometry Notebook of Robert Jackson, A Schoolboy* (1788), Bound MS, NLS MS 9156, f. 23. Used by permission of the National Library of Scotland.

In addition to being a successful mode of learning, the movement from simple to complex geometric shapes was valued by Scottish educators and parents because it conformed to the Lockean developmental model. The Scottish pedagogue Lord Kames summarized this relationship between drawing and learning with the following advice: "Begin not to teach Euclid, till he [a student] is well acquainted with the different figures. In that view, employ him to inscribe a circle in a square, a triangle in a circle, and so on. This manual operation will be an enticing amusement: and at the same time contribute to make the demonstrations more readily apprehended."[42] Kames was communicating a view that educators should treat the act of drawing as a developmental learning tool. Like other European pedagogues who accepted Locke's theory of mind, Kames placed a high value on the geometric figures that helped students transform a blank page into a *tabula figura*.

42. Home (Kames) (1781, 235).

According to Locke, "Diagrams drawn on paper, are copies of ideas, and not liable to the uncertainty that words carry in their signification."[43] The key words in Locke's formulation for the purposes of our interest in student notebooks are "drawn on paper." These reveal that he was not referring to figures that simply appeared in a printed book. Instead, he was referring to figures crafted across time and space by the human hand, which means that he thought something important happened in the mind as the result of drawing activities taking place through the media of ink and paper. His observation makes it easier to see that the act of drawing a diagram carried cognitive importance at the time. As such, the act influenced the ways in which students used the psychological relationship between drawing and thinking to rationalize their educational experience when they became adults. This relationship, as we will see in chapters 8 and 9, played a notable role in the way that Scottish university students and professors interpreted the proactive interface between cognition and manuscript media technologies.

The use of simple geometric shapes, especially circles, was widely seen as an efficient and uncomplicated way of learning how to understand the regularities of nature. It was based on the aesthetic axiom that nature was best represented through geometric symmetry. In addition to playing a significant role in the emergence of heliocentrism, especially in the work of mechanical philosophers such as Nicolas Copernicus, the axiom was expressed in Scottish primers as early as the seventeenth century.[44] George Sinclair, who called himself the "sometime Professor of Philosophy at the College of Glasgow," referred to pedagogical uses of circles in the subtitle of his *The Principles of Astronomy and Navigation*, published in 1688, which announces that the book will give "a clear, short, yet full explanation, of all the circles of the celestial, and terrestrial globes." Sinclair, like his Renaissance forebears, believed that simple shapes worked best for the teenage students who attended his lectures.[45]

Moving from the stars to everyday life, cooking manuals used circles to represent plates in diagrams designed to help young learners discuss and understand different kinds of place settings, and drawing manuals began by comparing houses to squares and roofs to triangles. Students learning technosciences at home or in school such as chemistry came to understand basic forms of crystallization via drawing simple geometric shapes. An example of this practice occurs in the figures made by the adolescent Amelie Keir in

---

43. Locke (1767, 164).
44. The relationship between nature and symmetry classical and modern thought is addressed in Spary (2004) and Franklin (2017).
45. Sinclair (1688).

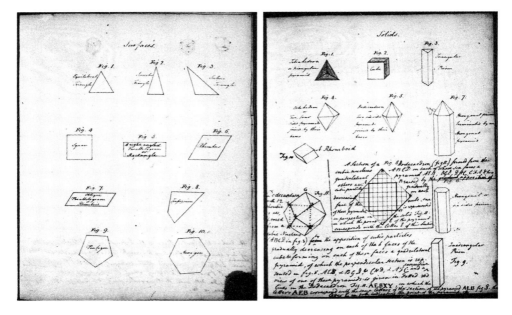

FIGURE 6.9. Chemical crystals represented as simple polygons and polyhedrons. Amelie Keir (later Moilliet), *Dialogues on Chemistry between a Father and His Daughter* (1801), reproduced with permission of the Assay Office Birmingham. Bound MS, James Keir Archive, Box 33, dialogues 19 and 20.

the manuscript chemistry primer she and her father designed as a series of dialogues during the 1790s (figure 6.9). On the first plate she drew polygons to communicate the names and two-dimensional shapes of crystalline surfaces. On the second plate she drew polyhedrons to communicate the names and three-dimensional forms of crystalline solids. Her use of the shapes in this manner reveals how simple polyhedrons and polygons were multivalent learning tools in that students or teachers could transfer them from one subject (geometry) into another (chemistry).[46]

As revealed in Keir's diagrams, once students had learned to recognize the representational potential of shapes, they then applied their new skills to creating more advanced visualizations by labeling the different parts of their geometric figures with alphanumeric headings. It was at this point that their shapes took on a more sophisticated diagrammatic function. Students keeping notes with figures learned to insert alphanumeric headings that labeled various straight lines, angles, and curves. Alongside or nearby such diagrams, students wrote the meanings of the labels.

46. Keir Bound MS (1801, dialogues 19 and 20).

Once the process of drawing and labeling was learned, students then went on to fashion more involved geometric diagrams relevant to the subject matter at hand. Those keeping notes on practical geometry, for instance, began by learning to label the points on lines, triangles, and squares and then moved on to combine these forms into more complex figures relevant to longimetry, the measurement of distances and lengths; planimetry, the measurement of plane surfaces; and altimetry, the measurement of altitude or height. Rather than being static entities, these figures functioned as calculation devices. For instance, in the longimetry exercises set by gauging and leveling instructors, students used the labeled figures to determine distances between objects situated in a landscape.

Some of the most visually impressive figures were rendered in notebooks kept by students attending Scottish academies during the last decades of the eighteenth century. They feature prominently in notebooks that contained "sailing" and "leveling" exercises.[47] A good example appears on the notebook page featured in figure 6.10. It depicts a leveling exercise in which the student is tasked with reducing the side of a mountain to a series of flat surfaces. The figure is similar to the visualization offered on plate III of Alexander Ewing's *Practical Mathematics* (1799). If we compare the landscape of Ewing's plate to one made by a student in an academy notebook (figure 6.10), it can be seen that attempting to use the former to create the latter was not a straightforward process. This is because the figure in the notebook is colorized, contains more detail, and is significantly larger than the figure presented on Ewing's plate. Such elements reveal that translating a figure from a printed plate onto a handcrafted notebook page required a host of additional, and interconnected, writing, sketching, drawing, and painting skills. In other words, a student notekeeper who created images such as the one presented in figure 6.10 would likely have had some sort of contact with a drawing instructor, either at the time it was made or in lessons that occurred prior to its composition.

Students drew the diagrams of the sailing and leveling figures in school and academy notebooks in a manner that demonstrated the metrological relations between several objects. Sailing diagrams were useful to students seeking to become ship navigators who could calculate everything from the height of coastline cliffs to the width of ports. Leveling diagrams appeared in notebook exercises relating to different heights of buildings and the ledges, shelves, and plains of mountains and valleys. As the name suggests, the measurements were conducted with a view to creating flat surfaces conducive to

---

47. Anonymous Bound MS (1787); Anonymous Bound MS (1790); Anonymous Bound MS (1804).

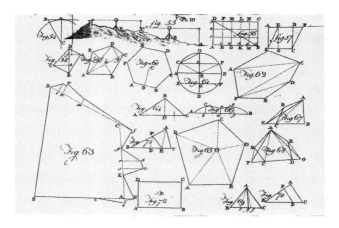

FIGURE 6.10. Top: Alexander Ewing, *A Synopsis of Practical Mathematics*, 4th ed. (Edinburgh: Fairbairn, 1799), plate III. Open access Internet Archive, original from the University of Michigan Library. Figure 55 (top row, left side), a depiction of a leveling diagram, does not appear next to the narrative that described it. Bottom: Anonymous, *Practical Mathematics* (1804), Bound MS, NLS 14285, f. 123. Reproduced by permission of the National Library of Scotland. In contrast figure 55 presented in Ewing's plate, the landscape diagrams in student notebooks were much larger and watercolored. They appeared on the same page as the narrative that described them.

farming, the erection of a building, or, on a larger scale, the creation of a canal system.

Since the early part of the century, British mensuration primers regularly featured simple renditions of practical geometry figures. John Love's popular *Geodaesia*, which was on its thirteenth edition by 1796, featured altimetry and longimetry scenes that involved landmarks such as castles and trees.[48] Other more expensive books, such as Adams's *Geometrical and Graphical Essays* (1791) or John Macgregor's *A Complete Treatise on Practical Mathematics* (1792) feature similar scenes.[49] More affordable mensuration textbooks such as Ewing's *Practical Mathematics* or books such as Ingram's *New Seaman's Guide* (1800), offered scenes as well, but they were fewer in number.[50] Although the diagrammatic scenes offered in such books helped students understand various principles of mensuration, they were awkward to use as learning tools because the images were seldom placed next to the text that they were meant to illustrate. Instead, the figures were situated on a plate elsewhere in the book, where they were depicted alongside a jungle of other visualizations.

A good example of the dislocation that often occurred between the placement of a figure and its descriptive text occurs with the landscape diagram featured on the top left of plate III in Ewing's popular primer *Synopsis of Practical Mathematics* (1799) (see again figure 6.10). The diagram was placed on a plate with twenty other figures that binders sometimes placed at the back of the book, or in some other place that was not anywhere near the text that described it. The dislocation between word and image was common in printed primers at the time, especially those the covered practical mathematics. It effectively forced students to flip between different parts of the book every time they encountered a reference to a diagram in the text. Such a practice of moving between different pages in a primer required an excellent working memory that was capable of holding a significant amount of qualitative and quantitative information. Today we know that this kind of memory is difficult to strengthen through training, especially in cases where the learner has a specific learning disorder.[51]

This being the case, it is worth noting that, unlike the diagrammatic scenes featured on the plates included in printed textbooks, the figure and prose of each mensuration diagram in most student notebooks occur in the same place, a convention that is evinced in the hand-drawn leveling diagram pre-

---

48. Love (1786, 181, 184).
49. Adams (1791, plate 28, figure 40); MacGregor (1792, plates IV and V).
50. Ewing (1799) and Ingram (1800).
51. Henry (2011).

sented in figure 6.10. The direct relationship between visual and verbal description eliminated the need to flip pages and allowed students to more easily concentrate on the meaning and purpose of the diagrammatic scene as they drew it and then when they or another viewer consulted it in the future. Including this kind of custom-made and user-friendly scene for every exercise made it easier to understand, and hence learn, how practical geometry could be applied to everyday situations.

The mensuration figures that students drew in their notebooks repeatedly demonstrate that the diagrams were designed to work in conjunction with handwritten verbal descriptions. The diagrams in the Perth Academy notebooks, for example, feature explicit labels and prose that explain the meaning and purpose of the visualization. In the trigonometric exercises section of an anonymous academy notebook of circa 1790, the notekeeper worded the headings as questions, a move that helped with the interpretation of the figures. One heading asks: "How far can a person see an object from an eminence 2 miles high on the level of the horizon[?]" (figure 6.11).[52] The angles between the objects depicted in the diagram served as a visual prompt that helped the student interpret the question and then calculate an answer of 126 miles.

Another way that student notekeepers learned by drawing figures was through depicting a variety of metrological instruments. Drawing a stylized version of an instrument helped them better understand the meaning of its parts and how they might be used. This explains why student notebooks feature scores of metrological instruments such as dials, compasses, rulers, and quadrants.[53] Many of these instruments were used across related subjects such as surveying, gauging, leveling, and sailing.[54] Notebooks on other subjects, particularly those that addressed technoscience, contain detailed iterations of air pumps, common pumps, thermometers, magic lanterns, compound microscopes, *camera obscuras*, and prisms.[55] Some renderings were executed with great skill. The air pump depicted in figure 6.12, which was drawn by the anonymous notekeeper of the 1787 *Perth Academy Notebook*, is

---

52. Anonymous Bound MS (1780s–1790s), f. 15v. The figure depicted on f. 15v was probably inspired by Gregory (1761, fig. 22).

53. Metrological instruments occur in many gauging and surveying notebooks. For representative specimens, see Anonymous Bound MS (1804), f. 6 (quadrant), f. 33 (mariner's compass), f. 45 (geometrical square).

54. Families of related metrological instruments used across different fields are discussed in Bennett (1987).

55. Exemplary specimens used in the Perth Academy natural philosophy course occur in Anonymous Bound MS (1787, vol. 3): f. 56v (air pump), f. 57r (common pumps), f. 57v (thermometers), f. 77v (magic lantern), f. 78r (compound microscope), f. 90r (quadrant), 90v (camera obscura), f. 91r (prism).

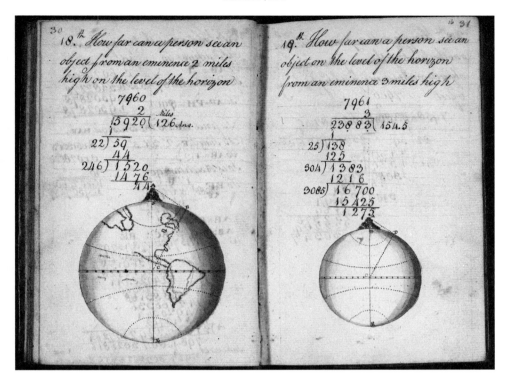

FIGURE 6.11. Student notebook exercises. The left exercise is titled "How far can a person see an object from an eminence 2 miles high on the level of the horizon?," Anonymous, *Perth Academy Notebook* (1790), Bound MS, NLS MS 14291, ff. 15v–16r. Reproduced with permission of the National Library of Scotland. The act of drawing the globe was a learning exercise that helped students internalize information. Placing the figure next to the question and the calculations made it much easier to see all the information at one time. Most printed textbooks placed the figures on plates located at another place in the book, making it more difficult to see the question, calculation, and figure all in the same field of vision.

particularly striking and displays the advanced draftsmanship skills of hatching, crosshatching, and contouring.[56]

Sometimes the figures of instruments in student notebooks resembled in one way or another those found on plates in mensuration or geometry primers.[57] The impressively executed figure of "The Universal Compound Mi-

---

56. Anonymous Bound MS (1787, 3: f. 56r). The image of the notebook's microscope was based on Ferguson (1760, plate XIV). The same plate with the same number remained in the republications of Ferguson's book until at least the tenth edition (1803).
57. Hoppus (1799), Ewing (1771), Panton (1771), Fisher (1763), and MacGregor (1792) offer representative specimens of instruments included on compendia plates. Some authors encouraged students to cut out the figures and paste them on pieces of wood so that they might be used in lieu of more expensive instruments made of metal. For more on the relevant context of eighteenth-century mathematical instruments, see Wess (2012).

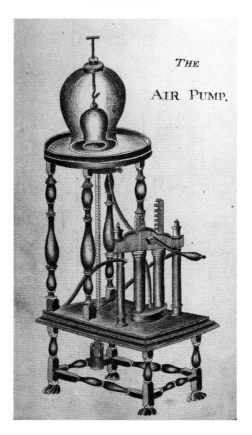

FIGURE 6.12. Pen and ink drawing of an air pump. Anonymous, *Perth Academy Notebook* (1787), vol. 3 NLS 14296, f 56r. Reproduced with permission of the National Library of Scotland.

croscope" rendered in an anonymous *Natural Philosophy* notebook kept at the Perth Academy in 1787, for instance, was based on a plate in Benjamin Martin's 1759 *Philosophia Britannica*.[58] The detail of the notebook's hand-drawn microscope, like the figures discussed above, indicates that some students worked very hard to understand an instrument through drawing it. The final version of such figures was in many respects a technical image, that is to say, a visual artifact that was "predominately instrument-based or the results of imaging procedures."[59] In this case, the figure was a discrete handmade artifact designed to function as a learning device.

The advertisements run by academies in newspapers indicate that courses

---

58. Anonymous Bound MS (1787, 3: f. 78); Martin (1759, vol. 3, appendix II, plate 1).
59. Bredekamp, Dünkel, and Schneider (2015, 1). For the essay in this collection that engages most directly with drawings, see Fischel (2015).

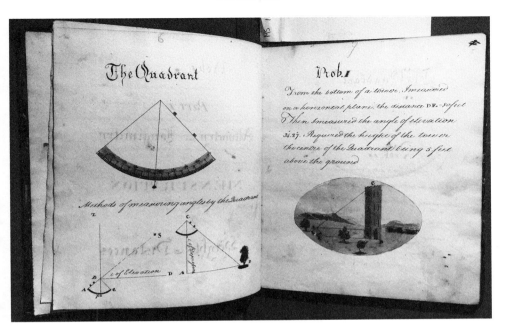

FIGURE 6.13. Pen and wash drawings of quadrants and a practical geometry exercise involving the height of a tower on a horizontal plane. Anonymous, *Practical Mathematics* (1804), Bound MS, NLS 14285, f. 6–7. Reproduced with permission of the National Library of Scotland.

on practical geometry frequently required students to use a variety of handheld devices. This means that drawing an instrument or a piece of apparatus was often reinforced by the hands-on experience of either holding it or using it. The large quadrant that occupies the top of the left page in the 1804 *Practical Mathematics* notebook featured in figure 6.13, for example, is part of a tableau designed to show how the instrument could be used to measure an angle set in a landscape.[60] Below the large quadrant, the notekeeper drew examples of how to read angles of elevation and depression by using a natural object such as a star in the sky or a tree in a landscape. The hand-drawn examples made the instrument more intelligible because they helped the student both understand and remember how to use it.[61]

A striking example of how students learned to experience their drawings as artifacts in motion is the "figure of an horizontal dial," a common sundial, in the dialing section of the James Fowler *Schoolbook* that we encountered at

---

60. Tableaux of this nature were an important mode of conveying of practical information in print culture at this time as well. Bender and Marrinan (2010, 19–52).
61. See Anonymous Bound MS (1804), f. 6. Further examples of notebooks that treat gauging include Anonymous Bound MS (1804) and Anonymous Bound MS (1790).

the beginning of this chapter.[62] As we can see from figure 6.1, Fowler did not finish it. Incomplete figures of this nature are rare occurrences in most school notebooks, mainly because drawings were often composed on loose leaves or bifolia that could be discarded if mistakes were made.[63] Its incomplete state, however, makes it easier to see the kinds of drawing skills that students employed to create a *tabula figura* from scratch. Dials were essential learning devices in modern homes and classrooms because they were used alongside the compasses, gridded globes, and maps that were employed in navigation. Accordingly, textbooks not only gave instructions on how to draw them, but also how to calibrate them to work in different latitudes.[64] Indeed, the latitude that Fowler penned on his dial (58 degrees north) reveals that he customized it to correspond to his own location in northwestern Scotland.

The form and fabric of Fowler's incomplete dial reveal how students learned to understand technical instruments through sketching, tracing, and shading them on a piece of paper that was destined to become a page in a notebook.[65] He began by penciling a square in graphite with a ruler. Using a small compass, he then drew a graphite circle inside the square. Next, he drew radiating lines in graphite through the circle and square. He then used ink to inscribe stippling dots (over the circle), a signature, the date, a latitude bearing, the four points of the compass, and numeric labels for the radiating lines. Finally, he drew a hand for the dial in the red chalk used by painters to make preliminary sketches. Had Fowler completed the dial, he most likely would have followed the common practice of washing its different parts with watercolors.[66] Each step shows how different figuring skills were required to composite knowledge on the page—literally—in layers. As he progressed through the stages of drawing the circle, the square, the radiating lines, and the labels, he added new meanings to the dial in terms of how he finally came to understand it as a singular artifact. Fowler's stages of composition reveal that the act of drawing the dial constituted a sophisticated form of mnemotechnic interface, both inside the image as well as between the image and the surrounding narrative that described the construction of the dial.[67]

---

62. Fowler Bound MS (1780, 134r). "Dialling" was included in many printed primers, including Fisher (1763, 313–319).

63. A few examples of incomplete diagrams occur near the end of Anonymous Bound MS (1787).

64. Fisher (1763, 314–315).

65. Fisher's dialling sections are accompanied by a foldout table of dials. Fisher (1763, 319–321).

66. Fisher advised his readers to cut his dials out, paste them to wood, and then to paint them. Further sections explain how to make paint from household items. Fisher (1763, 319–321).

67. It is worth noting here that the dial occurs at the end of several pages of narrative on various kinds of dialing. Fowler Bound MS (1780, 123r–133v). Here my notion of using the material layers, or parts, of an artifact to understand its meaning extends Jordanova (1993, 55–56; 2012, 70–72).

Like other students drawing figures in their notebooks, Fowler kinesthetically learned to understand the components of the dial as thinking tools by drawing and then labeling them. The dial, therefore, was not a static entity. The skills of artifice required to draw figures in this manner were a crucial part of the knowledge-making process through which students learned how to conceptualize instruments and other objects relevant to their educational experience. The skills gained from creating such figures endowed them with the capability to observe in realtime how rectilinear and curvilinear depictions could serve as thinking tools to which useful empirical or metrological information could be attached through inscription. This transformed the lines of the figures into different kinds of significant space such as inches and degrees.[68]

Put more clearly, the points, sketches, traces, and strokes of student notebook figures were all developmental learning tools that facilitated the acquisition, mediation, and transmission of knowledge. As intimated in previous chapters, particularly in my discussion of the work of Barbara Tversky and Tim Ingold, a developed ability to efficiently attach different kinds of meaning to a line in this manner is not something that exists a priori in the mind. It must be learned. This means that the lines that appear in the figures discussed above are more than what they may seem in that they were interactive thinking tools that functioned in a way very similar to that of the sightlines we observed in chapter 5. There I pointed out that sightlines were linear patterns that ran between notebook headings in a way that enabled students to establish conceptual and visual relationships across the page. Like sightlines, the lines of notebook figures were tools through and on which students learned to attach different kinds of information. The lines aided their ability to access the page in front of them in a way that enhanced their individual sense of judgment, as well as their understanding of the relationship between measurement and the natural world more generally.

When considered from the perspective of how drawing helped students learn, making even the simplest geometric figures played an important role in facilitating the internalization of spatial and mathematical capabilities that could be applied both on and beyond a piece of paper. The process helped students to acquire a graphically literate eye that knew how to match skills such as tracing, sketching, contouring, and hatching to what needed to be learned or understood. Within Scottish homes and schools, particularly those

---

68. For the cognitive links between using instruments and creating metrological knowledge, see Livingston (2008).

of middle-class students, this intelligence fostered a form of visual literacy that was infused with utility, a desire to treat drawing as both a means to an end and as an attempt to truthfully (or even accurately) represent objects and phenomena.

Scenes as Observational Training

The figures drawn by Scottish students in their school notebooks served as a practical mode of visual training, one that did not require the expensive oil-based paints, designer chalk crayons, large canvasses, or colored drawing papers used by visual artists operating in elite circles of patronage. Aside from profiles occasionally doodled in the margins or flyleaves, portraiture and other kinds of life drawing were absent as well, even though, as evinced in Allan Ramsay's student sketchbook, they were clearly being taught in southern Scotland at the time in the institutions run by the Foulis family, the Edinburgh Trustees, and independent drawing masters such as Mary Cumming or George Walker. But though student notebooks do not feature portraiture, they do, as we have already seen in figure 6.10, contain miniature landscape drawings. More specifically, notebooks that address practical geometry and its application to gauging, leveling, and sailing feature a variety of pen and watercolored scenes of objects such as houses, ships, and trees arranged strategically within seascapes, landscapes, riverscapes, and mountainscapes.

Such scenes served several pedagogical purposes, one of which was to learn how to use the act of drawing as an observational skill. The benefits of this practice were identified by John Locke in *Some Thoughts Concerning Education*, where he pointed out that learning to draw the elements of a landscape helped students observe important facets of the environment around them. It was especially useful when traveling through new landscapes at home or abroad. Locke underscored this point by rhetorically asking how the ideas of "many buildings," machines, and habits "would be easily retained and communicated by a little Skills in Drawing; which being committed to Words, are in danger to be lost, or at best but ill retained in the most exact Descriptions?"[69]

Locke's views were underpinned by the practical acknowledgment that drawing helped travelers remember what they had seen. It was this kind of acknowledgment, which was sometimes assumed in modern educational settings of the middling and upper classes, that further explains why Scottish stu-

---

69. Locke (1752, 235).

dents learned the skills required to make the landscapes presented in their notebooks. Put more clearly, drawing scenes helped students remember the many ways that practical geometry could be applied to the natural world. To this goal, students superimposed geometric diagrams on the scenes to illustrate the various kinds of planimetry, longimetry, and altimetry questions that they used to facilitate their sailing, gauging, leveling, and surveying calculations.

Let us consider how students learned to draw and watercolor their notebook scenes. The London market was flooded with handbooks that addressed the art of drawing with pencils and chalk crayons, and painting with watercolors and oil paints. Most of these works explained how to make a wide variety of inks, dyes, and adhesives from local materials as well. They also offered helpful diagrams on how "To make a convenient BOX to hold Colours" or how to design a palette for mixing colors.[70] It is unclear to what extent such books were used by students in Scotland; however, many of the masters who taught in Edinburgh, Glasgow, and Perth had spent time training in London or abroad and would have been able to teach their students the drawing, painting, and media-related techniques mentioned in such books. Add to this the fact that books published by Scottish presses, Hannah Robertson's *The Young Ladies School of Arts*, for instance, gave basic advice on drawing composition and the transformation of local materials into dyes that could be used for painting.

Like writing and sketching, creating a scene took practice. This might seem obvious today; however, during the eighteenth century there were artists who eschewed the use of prints or paintings as exemplars. Some believed that asking students to copy such visualizations damaged their ability to discern beauty or to eventually produce original drawings or paintings. The implicit division between learning and creating that underpinned this interpretation of copying was summarized by the London-based artist John Russell: "Some Artists reject the use of Drawings or Prints for the Student's imitation; 'it is, say they, a servile method of proceeding, because it cramps the ideas, and hence, Genius suffers too great a confinement.'"[71] We will soon see that Russell thought this view was unrealistic. But other drawing masters, Edinburgh's Alexander Nasmyth for example, resolutely rejected the use of exemplars as learning tools.

While the rejection of exemplars might have gained traction in the world of art connoisseurs and aesthetes who sought patronage, it was an impracti-

---

70. For a figure and instructions on how to make a box to hold painting supplies, see Bowles (1796, 63). For a figure and instructions on how to make a color palette, see Anonymous (1732, 67).
71. Russell (1772, 8).

cable principle when it came to teaching students how to draw. Mary Somerville, who studied at Nasmyth's academy when she was a teenager during the 1790s, sardonically summed up her teacher's method in her autobiography: "I was not taught to draw, but looked on while Nasmyth painted."[72]

The rejection of exemplars was also respectfully questioned by art theorists and teachers alike. George Turnbull's *A Treatise on Ancient Painting* (1740), which used the history of painting to discuss the nature of aesthetics, repeatedly emphasized the importance of using copies to understand art.[73] Russell, who was both a studio artist and a successful drawing instructor, pointed out that, notwithstanding the aesthetic arguments against copying, "the usual practice is to copy after Prints and Drawings at first, and I imagine experience has determined the advantages accruing from this method. To set the Drawing (in particular) of the most eminent Artists before a young beginner, at his first commencement, must be highly beneficial, as it most undoubtedly will prevent rudeness and inaccuracy, against which the most exact cannot be too much guarded."[74]

The reoccurrence of rough drawings in student scroll books, and the repetition of similar objects and landscapes in their school notebooks, indicates that Scottish students were learning their drawing, shading, washing, and watercoloring skills through copying. Most of their scenes were pen and wash and were based on exemplars, that is to say, images displayed in the classroom or passed around as handouts. Students began by sketching the figure or landscape with a black lead pencil. They then removed unwanted lines with an Indian rubber and traced the contour lines in black ink with a metal-nibbed pen. After letting the ink dry, they used freehand brush strokes to watercolor the sea and the sky. Then they "washed" the contoured objects such as buildings or trees, most of which were usually in the foreground, with watercolors, taking care to add the lighter shades first and darker shades last.

A good example of a watercolored mountainscape can be seen in figure 6.10. It appears in the leveling section of an anonymous student notebook titled *Practical Mathematics* and it visualizes the side of a mountain reduced to a series of right angles.[75] The contouring and washing techniques evinced in the scene speak to the prior training of the student. The brush stroking on the mountain, for instance, reveals that the student knew how to control the pressure and direction of the brush in a way that regulated the even flow of

---

72. Somerville (2001, 38).
73. Turnbull (1740).
74. Russell (1772, 8–9).
75. Anonymous Bound MS (1804), f. 123.

the paint across the paper. Like figures made in other notebooks, scenes of this nature were rendered on the same kind of paper that was being used by the student for the other handwritten pages of the notebook. The techniques and media used to create such scenes also required students to sketch, trace, and wash it on a flat sheet or unbound quire, which means that such paintings were made before the notebook was bound.

The level of precision evinced in the scenes featured in many academy notebooks suggest that students had already learned how to sketch basic scenes in graphite and how to lay them out according to rules of landscape composition commonly taught by drawing masters or discussed in manuals. Robertson's *School of Arts*, for example, offered the following advice to aspiring landscapists: "In drawing landscapes, always express a fair horizon, shewing the heavens cloudy or clear. . . . Take great care to augment or lessen every object, proportionably to its distance from the eye; and also to express them strong or weaker. . . . Let your landscapes be always conformable to additional graces, as the farm-house, wind-mill, woods, cattle, travelers, ruins of temples, castles and monuments."[76] These elements appear in the scenes of student notebooks made in Scottish academies from the mid-part of the eighteenth century into the nineteenth century.

Most measuration scenes were included to illustrate a set of mathematical exercises. For each exercise, there was one scene. The exercises usually preceded from easiest to hardest. This means we can be relatively sure about the order in which the exercises and their corresponding scenes were created. The indirect result of this kind of arrangement is that we can be reasonably certain about the order in which the scenes were composed. This is an extremely helpful feature of the notebooks, one sometimes not present in drawings made as marginalia or on loose sheets by other young learners at the time.

Learning to enact principles of landscape composition in realtime with black lead pencils, pens, black ink, and an assortment of watercolors on a piece of paper took training and practice, particularly when it came to strokes used to create different kinds of light and shadow. The scenes in initial exercises of the 1790 *Perth Academy Notebook*, for example, reveal that its keeper had already learned these skills and was ready to learn how to draw ink contour lines and how to wash the different elements of the compositions in watercolors. This explains why the early scenes are devoid of ink contour lines and color, that is to say, they were rendered solely in graphite. As the exercises progress, the notekeeper slowly added increasingly complex combinations

---

76. Robertson (1767, 21–22).

of ink contour lines to highlight important buildings and watercolors to make the scenes look more natural or to fill in the frame surrounding the image (figure 6.14).

More specifically, in the early scenes of the 1790 *Perth Academy Notebook,* the washing was done only in grey watercolors, that is to say, a single color that could be made to achieve different tones by mixing in white paint for a lighter hue. These scenes represent a form of visual training, one in which the student learned to manipulate the tones of colors. Once mastered, the technique was then used to manipulate the tones of other colors in the following scenes as well. Near the middle of the series the student reduced the grey watercoloring in order to introduce blue for the sky and green for foliage, but leaving part of the scene devoid of color. In the remaining scenes that occur at the end of the series of exercises, the student retained similar patterns of coloring but introduced pink and lighter shades of green so that the entire scene was colored. The end result in the final exercises is a series of appropriately contoured and fully colored scenes (figure 6.14). The progression of the scenes indicates that the notekeeper was using them as learning devices that enabled him to practice increasingly difficult tonal and colorization techniques that incrementally improved his ability to mix paint and combine brushstrokes.

The scenes presented in the mathematical exercises depicted in other notebooks show that some students already possessed respectable washing skills when they entered the course. Take for instance the gauging scene depicted in the first practical geometry problem of the anonymous 1804 *Practical Mathematics* notebook featured in figure 6.13.[77] Save a few strokes on the mountains and distant foliage, it displays the semi-advanced washing skills evinced in the final scenes of the 1790 *Perth Notebook* depicted in figure 6.14. But as the scenes in the *Practical Mathematics* notebook progressed, the student incrementally introduced different shades of green, blue, and sometimes pink. The additions were accompanied by a rising level of understanding of the ways in which the paints could be mixed with water or each other to create darker and lighter hues (figure 6.15).

By the time we reach the final scenes in the 1804 *Practical Mathematics* notebook, the drawings demonstrate a variety of landscapes and seascapes depicted vibrantly with an impressive assortment of shades and strokes (figure 6.16). The watercoloring techniques that appear in the scenes of the *Practical Mathematics* notebook, like the scenes featured in other school and academy notebooks, reveal that students were learning how to ink and color

---

77. Anonymous Bound MS (1804), f. 7.

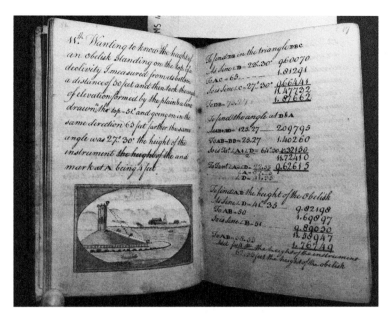

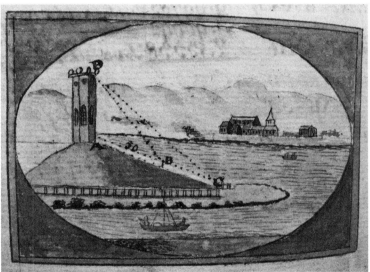

FIGURE 6.14. Pen and wash drawing of a practical geometry exercise involving the height of an obelisk on an ascending slope. Anonymous, *Perth Academy Notebook* (c. 1790) (anonymous notekeeper), Bound MS, NLS MS 14291. Reproduced with permission of the National Library of Scotland. The tableau is painted in blue, green, grey, brown, and pink watercolors. The lines of measurement are stippled in black ink (or possibly watercolor). Top: The open notebook shows open showing folios 16v and 17r. Bottom: Close-up of the tableau featured in folio 16v.

FIGURE 6.15. Pen and wash drawing of a practical geometry exercise involving the height of a tower on a declining slope. Anonymous, *Practical Mathematics* (1804) NLS MS 14285, ff. 18–19. Reproduced with permission of the National Library of Scotland.

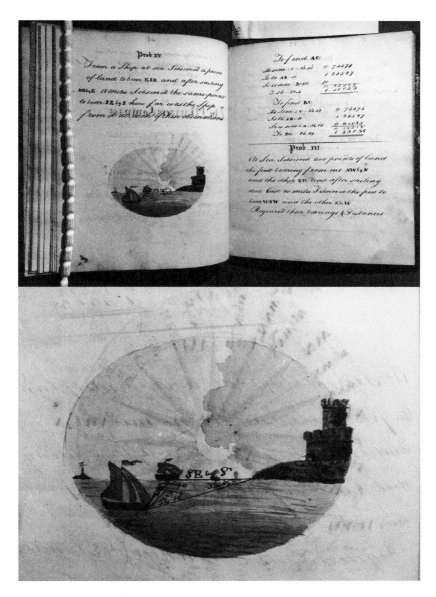

FIGURE 6.16. Pen and wash drawing of a practical geometry exercise involving the distances between three points on the sea. Anonymous, *Practical Mathematics* (1804) NLS MS 14285, ff. 34–35. Reproduced with permission of the National Library of Scotland.

a sketched scene through stroking and washing skills that, like the mathematical exercises, moved from simple to complex combinations.

The scenes in the foregoing notebooks represent the kinds of advanced visualization skills learned in Scottish schools and academies that specialized in practical geometry. Not all students were able to execute such a high degree of precision. Most students created the scenes for the purpose of learning skills that could be used to visualize objects and processes relevant to the improvement of their future prospects or the betterment of society at large. For these students, drawing and watercoloring were in fact modes of learning how to observe and represent useful information on paper.

Observation and the Utility of Perception

Student notebooks repeatedly used the verb *observe* to describe the various skills performed while drawing. After rendering the figure of a concave lens reflecting an arrow (figure 6.17), for instance, the anonymous notekeeper of the 1787 *Perth Academy Notebook* wrote "Observations" that identified its important features, relationships, and meanings:

OBSERVATIONS
I.   The image made by a concave lens is always betwixt the object and lens.
II. The object and image appears under the same angle seen from the center of the lens and is therefore less than the object.
III. The image is always erect.

Here, observation involved the consideration of several optic principles by drawing them *and* through writing out a verbal account of the important features.[78] The inclusion of verbalized observations, particularly the reference made to a "concave lens," could offer helpful guidance when reading the notes in the future or when the student discovered that he had mislabeled the figure as a "convex" lens. Through rendering figures on the pages of their notebooks, students learned to integrate both writing and drawing skills that enabled them, in the words of Omar Nasim, to "observe by hand."[79]

The interactive modes of observing through drawing and writing notebook

---

78. Anonymous Bound MS (1787, vol. 3, ff. 74r–74v).
79. Nasim (2013) emphasizes that drawing served as a mode of learning and perceiving. Scottish university students used such observational practices to create "useful pictures" in their lecture notebooks as well. Eddy (2015).

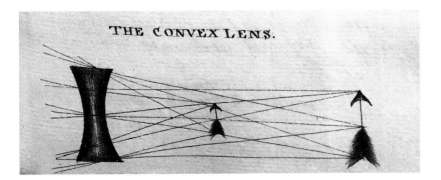

FIGURE 6.17. Pen and ink drawing of the reflection of an arrow in a concave lens (mislabeled as convex lens). Anonymous, *Perth Academy Notebook* (1787), ol. 3, NLS MS 14296, 74r. Reproduced with permission of the National Library of Scotland.

figures and scenes over time slowly shaped a student's perception of how objects and processes might be represented through various kinds of drawn or painted media. As shown in the work of Jonathon Crary and Lorraine Daston, there was a rising awareness throughout Europe and colonial contexts that perception was a mental operation that could be trained through repeated acts of observation.[80] In Scotland, educators influenced by Lockean developmental psychology and the emphasis its adherents placed upon actively shaping the mental operations of students were particularly amenable to this view of observation. This being the case, it is worth considering how the handmade figures and scenes featured in Scottish school notebooks fit into the emergence of observation as a modern mode of perception. Put another way, how might we understand the deeper relationship between drawing and observing learned by students as they kept their notebooks?

One way to answer this question is to distinguish two overlapping kinds of observation that influenced Scottish students. The first was pursued by what could be called an aesthetic observer, an aesthete interested in the visual relationship between mimesis, memory, and beauty. The second kind of observation was practiced by what we might call a technic observer, that is to say, a professional, scholar, or artisan interested in the relationship between mimesis, memory, and *utility*. Both kinds of observers believed that visualizations should be true to nature. The difference, however, was which aspect of the truth one was trained to see and then visualize. Let us explore these two

80. Crary (1992) and Daston (2008, 2011).

distinctions with a view to understanding how they offer insight into the role played by the drawings and watercolors in school notebooks in their capacity as media technologies that aided a student notekeeper's learning experience.

The aesthetic observer employed classical, oftentimes elite, principles of taste with a view to understanding whether a drawing was beautiful or, later in the century, sublime. This kind of observing was underpinned by a nascent tradition of art history, particularly the history of painting, which sought to establish a sense of continuity between modernity and the classical past.[81] Aesthetes who observed the world through this lens sought to create or view images that evoked a sense of beauty or sublimity. More often than not, this kind of observation was oriented toward how viewers should perceive a drawing or a painting that had already been completed, thereby detaching the act of observation from the original processes that an artist used to create an image.[82] Such observing served an important mnemoaesthetic function that was summarized in George Turnbull's aforementioned *A Treatise on Ancient Painting* in the following manner: "The Eye formed by right Instruction in good Pictures, to the accurate and careful Observance of Nature's Beauties, will have, in recalling to mind, upon seeing certain Appearances in Nature, the Landscapes of great Masters he has seen, and their particular Genius and Tastes."[83]

The technic observer operated according to principles of improvement with a view to creating a mnemonic image that was useful to its maker or society at large. This kind of observer was interested in what Stephen A. Marglin has called technic knowledge, that is, a way of knowing that did not insist on universal claims and was oriented more toward "creation and discovery rather than verification."[84] As such observers, student notekeepers treated the act of drawing as a mode of learning that helped them to perceive and remember detailed facts or to understand complex objects or processes.

The mnemotechnic function of notebook images worked best when they were simple or schematic.[85] This ensured that various elements of the drawing's structure or composition did not compete with the information it was

---

81. My thoughts here concerning the general framework that art history provided eighteenth-century aesthetics draws from Didi-Huberman (2005). For the larger classical orientation of aesthetics in eighteenth-century art history and theory, see Potts (1994).

82. Helpful guides to the relationship between art and beauty in Scottish philosophical circles are in the editorial introductions in Friday (2004) and Graham (2015). For the wider relationship between art and beauty in Scottish literate society, see Brown (2016).

83. Turnbull (1740, 146).

84. Marglin (1996, 230).

85. Historians of science and technology such as Tufte (1983), Ferguson (1992), Daston and Galison (2007), and Bredekamp, Dünkel, and Schneider (2015) underscore the schematic simplicity of modern technical figures and diagrams. The role of visual simplicity within the mnemotechnic tradition is mentioned throughout Yates (1966).

meant to represent. In other words, when conducted in an educational setting, observing through drawing was a kinesthetic mode of describing that enabled the drawer to develop powerful mental and manual skills of perception that in turn could be used to create useful visualizations. When the image was finished, its mnemotechnic design served as a cue that helped the drawer better remember the observational skills required to make it or the information that had been attached to it. Most of the Scottish students who made drawings for their school notebooks practiced this kind of technic observing in that they were learning to observe by hand through making useful pictures.[86]

To be sure, I do not wish to suggest that aesthetic and technic modes of observing were mutually exclusive. Nor do I wish to suggest that students did not experience a sense of enjoyment or satisfaction when they drew. The teenage Mary Somerville, for example, experienced much pleasure while learning to employ drawing and painting skills during her younger years in Edinburgh during the late eighteenth century. Reflecting back on this time, she noted, "I must say the idea of making money had never entered my head in any of my pursuits, but I was intensely ambitious to excel in something, for I felt in my own breast that women were capable of taking a higher place in creation."[87] Although student notekeepers did not comment on their efforts, it is safe to assume that some, perhaps many, experienced a similar sense of aesthetic accomplishment when they rendered figures and scenes in their notebooks.

Yet, when it came to the overarching utilitarian and pedagogical purposes that framed how students used drawing to perceive and represent objects and processes in their notebooks, they understood observation in a way that was more mnemotechnic than mnemoaesthetic. As we have seen above, students learning how to sketch, ink, wash, and paint everything from instruments to landscapes needed to make sure that their creations were true to nature in a way that highlighted the properties in objects that were amenable to different kinds of calculation or classification. This explains why, aside from the student sketchbooks of artists such as Ramsay and the advanced brushing and shading techniques evinced in some landscape scenes in Perth Academy notebooks, the mnemotechnic purpose that underpinned students' drawings led them to favor schematic techniques over painterly ones.

The skill of building, or perhaps layering, drawings of natural objects on

---

86. Eddy (2015).
87. Somerville (1873, 60). Somerville continued to paint landscapes and portraits as an adult. Somerville College, Oxford University, houses several, including *An Extensive Landscape of a River Valley and Distant Mountains*, *Castle by a Lake*, *Self-Portrait*, *Thatched Hut by a River*, and *Wooded Landscape with Attending Cattle in the Foreground*.

simple geometric structures such as circles, squares, and triangles was frequently used in advanced school courses on mathematics and natural philosophy as well, especially in highly technical diagrams. The graphite lines of such drawings in student notebooks, evinced so clearly in Fowler's incomplete sundial, were usually erased when they were retraced with pen and ink. The practical purpose that underpinned the schematic nature of figures influenced the kinds of objects and landscapes that students learned to observe and represent when they progressed to working with watercolors. The watercolor scenes in notebooks reveal that instructors concentrated less on freeform painting and more on what could be called schematic painting. The former was a technique in which minimal preliminary drawing was done. Hard contour lines were avoided. The latter was a technique in which objects and scenes were first sketched and then washed. Black or dark grey contour lines then were added to individual, oftentimes diagrammatic, elements to make them stand out.

Figure 6.14 presents a good example of the watercolorings that students learned to paint over preliminary graphite sketches. The boat in the river exhibits the remnants of the original graphite lines that the student left unwashed. Once the washing was finished, the student emphasized the schematic contour lines of the castle tower and its boundary wall with dark grey watercolors. Likewise, he stippled the lines and labels of the superimposed diagram in black ink. The schematic nature of the darkly contoured objects and triangles made it easier for the student to see, to perceive, which part of the scene corresponded to the mathematical problem it was drawn to represent.

As a general rule, the correspondence between schemata and narrative provided a way for students to effectively communicate a wealth of information through the compact scenes that they created for their notebooks. The use of schematic forms of representation in this manner enhanced the work created by contemporary artists operating in other contexts as well. The poet William Blake, for instance, repeatedly used schematic contouring in the watercolors he created to illustrate his poems. He used the lines to highlight elements that corresponded to ideas and imagery that he wanted to emphasize.[88] In the case of Scottish students, and for Blake as well, the modes of visual observation and representation employed to create an image were

---

88. W. J. T. Mitchell uses the term "linear schemata" to describe the schematic elements that appear in Blake's images. He identifies a number of linear patterns that occur repeatedly in Blake's work, particularly circles, orbs, spirals, vortices, arabesques, S-curves, arches, and inverted U's. Blake used these shapes and patterns to represent specific kinds of natural objects and forms of human movement. Mitchell (1978, 58–77). The practice continues today in cartoons. Forceville (2011)

framed by an overarching purpose that corresponded to how it was going to be used when it was finished.

When we step back and consider the larger relationship between representing, perceiving, and observing that took place as students drew in their notebooks, several noteworthy points emerge. Knowing how to perceive, conceive, and execute parallel and rectilinear relationships between objects arranged as figures on paper, for instance, could be applied to a multiplicity of real or imagined objects when students became adults.[89] Once learned, the skills helped students to conceptualize shapes as objects to which various kinds of meanings could be attached. The lines that delineated the space of a square, for example, were significant because they were proportionate, perpendicular, and equal. Only once these qualities were understood they could be used to determine the area, volume, or movement of objects.

The processes and practices students employed to make notebook drawings also reveal that the material and scribal assumptions that underpinned the *tabula rasa* metaphor were enhanced by the abilities required to observe, create, understand, and use a *tabula figura*, that is to say, a manuscript page that presented a figure. When we consider all of the drawings evinced in student notebooks, it becomes clear that exploring the skills through which students composed them enriches our understanding of the *tabula rasa* metaphor. The perceptual and representational centrality of such skills indicate that there was a sophisticated developmental interface between hand-drawn media and the mind, one in which a blank sheet could function as a starting point for an interactive process of observational learning that took place through the spatial relationships of objects on the page and in the world.

---

89. Whitmer (2015, 54–59).

CHAPTER 7

MAPPING

Mapkeepers and Knowledge Systems on Paper

On clear summer days the young Mary Somerville loved to watch the ships sail across the dark blue waters of the Firth of Forth. Gazing from the rocky shore of her home nestled in the quiet shores of Burntisland in Fife, she witnessed vessels sailing through the shimmering waves of the bay to Leith, the main port of Edinburgh. Ship gazing was a popular pastime for children who lived in Scotland's seaside towns. But Somerville's eye was drawn to the ships for a very personal reason. Her father was a naval officer and spent much of his time at sea. His long absences undoubtedly generated a host of questions when Somerville looked at the ships. Where were they going? Where had they been? Did the frigates resemble her father's ship? Did the packet ship that day bring a letter from her father?[1]

Similar questions were asked by other British children over the course of the long eighteenth century, a time when Britain's expanding empire included colonies in every corner of the globe. A helpful tool that schoolchildren like Somerville could use to find answers was a map. Like many students of the time, Somerville had learned how to use geographical maps from the village schoolmaster. She also had sourced a copy of John Robertson's popular *The Elements of Navigation*, which featured a variety of maps and instructions that explained how students might plot the paths of ships.[2] Thus, when she gazed at the ships on the Firth of Forth she possessed an understanding of how to use a map, one that she could use to appreciate her father's many journeys.

---

1. Mary Somerville's memories of her seaside Burntisland childhood appear throughout the many scientific works she published as an adult. The largest number occur throughout Somerville (1873, chaps. 1–6). **2.** Somerville (1873, 47, 49).

Her knowledge of navigational mathematics would later inspire her to write about the larger world of physical geography and natural philosophy.[3]

Yet though we know that students like Somerville were exposed to a variety of maps, and though we have a good idea of how maps were produced by Scotland's printing industry, we know relatively little about the day-to-day forms of training that underpinned the process of becoming a map-minded learner.[4] What skills needed to be mastered? What kinds of materials and instruments were required? How was the experience dependent upon kinesthetic and visual modes of learning? It is here where notebooks become very helpful because they show us how students learned to conceive maps as interactive tools, that is to say, as handmade forms of representation that could be inserted into an integrated informatic device.

In this chapter I explore the interactive relationship between handmade maps and school notebooks. I treat mapmaking as a kinesthetic and mnemonic process that shaped the cognitive capabilities of a learner. Like the figures we examined in previous chapters, when students folded paper, designed layouts, wrote words, and drew figures, they were learning how to observe and represent knowledge. In addition to reconstructing how, precisely, students became map-minded learners, I wish to show how their maps and surveys present a hitherto neglected facet of Locke's *tabula rasa* metaphor.

From a technical perspective, student notekeepers who created maps followed visualizing processes similar to those they used to craft figures for other school subjects such as algebra, geometry, and trigonometry. This was the case in Scotland and in other European countries where students learned to make and use terrestrial and maritime maps.[5] But as we will soon see, some maps from student notebooks were not crafted solely inside the classroom. Survey maps required them to record observations while traveling across fields, around rivers, and over hills. While in the field, students used a variety of notekeeping and mensuration skills to generate, assess, and inscribe data on-site. After they had interfaced with the lay of the land, they then returned to the classroom or study to composite their numeric and visual data into one collective artifact, a notebook, that served as a realtime media technology for years to come.

---

3. Somerville's thoughts on these subjects reached fruition in *The Mechanism of the Heavens* (1831) and *Physical Geography* (1848). Her many scientific works are discussed throughout Neeley (2001). The close relationship between mathematics and navigation in educational settings in Europe and its colonies is addressed in Levy-Eichel (2015) and Schotte (2019).
4. For the place of maps within Scotland's printing industry, see Fleet (2011).
5. Schotte (2019).

Historians and anthropologists working on the epistemology of scientific images have long emphasized that the skills required to make diagrams, figures, and maps are sometimes different from the skills used to view them. They point out that those who possess the ability to make such images tend to treat them more as approximations of the objects they are meant to depict because different kinds of media and training place constraints on how an object is represented.[6] This being the case, school and academic notebooks provide an opportunity to investigate the conditions surrounding how students learned to conceptualize the epistemological advantages and limitations of the maps they were making. Since many of the maps were created with the purpose of being bound into a school notebook, each student notekeeper was also learning how to be a mapkeeper who could see in advance how a map could be designed in a way that allowed it to function within a knowledge system on paper, that is to say, within a codex that operated as a paper machine.

In educational settings, notebook maps were images created through multiple drafts. They were designed by learners wishing to understand the spatial relationships between objects, inside countries, and across landscapes. Many students sought to pursue careers as teachers, surveyors, captains, gunners, architects, levelers, or gaugers. In other words, learning to make a map was a process of setting the mind in motion. It involved sensory skills that were learned on and around multiple forms of graphite, ink, paint, and paper and through different kinds of inscription. Crafting a map that worked as a component in a notebook also extended core categorization and codection skills. The skill of mapping was most frequently learned by students who attended the healthy number of school courses or private tutorials on geography, dialing, navigating, or surveying offered throughout the century in Scotland.[7] In what follows, I will first look at the larger sensory world that influenced how students experienced working with a geography map at school or at home. I then move on to discuss how survey maps in student notebooks were theorized in the classroom, enacted in the field, and then integrated with handwritten notes to form a useful codex.

---

6. For the history of scientific maps, see Nickelsen (2006), Jordanova (2012), Bleichmar (2012), Kusukawa (2012), Nasim (2013), Wittmann (2013), and Eddy (2014; 2015). For the anthropology of mapmaking, see Ingold (2007, 72–103), Ingold and Vergunst (2008), and Roberts (2012).
7. For the steady rise of geography, dialing, surveying, and navigating courses across Scotland during the long eighteenth century, see Withers (1999).

Map-Mindedness and Embodied Experience

Let us first explore how the material and visual experiences of tracing and sewing maps contributed to a potential notekeeper's nascent sense of map-mindedness, the ability to see a map as an artifact that could be used in different ways to experience and interpret the world.[8] Prior to creating maps on their own, students were exposed to those that were included in geography primers. Geography was a popular school subject. Once students left school they continued to interface with maps as adults via their inclusion in travel books or in collectable prints. As noted by historians of geography, making and using a map involved a multiplicity of skills that, when treated collectively, constituted a performance on and across paper.[9] They have also observed that the epistemology surrounding the interpretation or circulation of a printed modern map was significantly influenced by *who* was using it, *where* it was used, and *why* it was being used.[10] This point is especially relevant to student notekeepers because, in addition to using maps at home and in school, they were asked to *make* them as well.

Though interfacing with a printed map did not necessarily teach students how to make one, it taught them how to recognize important landmarks and how to interactively use it as a learning tool. Geography instruction in Europe at this time addressed both ancient and modern topics. For modern geography, British instructors focused on issues relevant to European countries and their global relationship with their colonies. For ancient geography, the main locations mentioned were usually Greece, Rome, and areas of the Middle East relevant to the Bible. There were numerous geography primers published in Scotland, with those of Gawin Drummond, Thomas Salmon, and Alexander Adam being popular during the early, middle, and later parts of the century, respectively.[11]

Globes and printed maps were familiar learning tools in Scotland and, more broadly, within Europe and colonial contexts.[12] Students learned how to use them in different ways, and these efforts laid the foundation for any maps they might draw in the future. Many instructors, for example, did not simply ask students to *look* at maps and globes. They also asked them to *engage* with

---

8. My usage of the term "map-mindedness" is based on Edney (1994). See also Withers (2002).
9. Della Dora (2009) and Lorimer and Lund (2003).
10. Withers (2008; 2001), Livingstone and Withers (1999), and Withers and Ogborn (2010).
11. Drummond (1708), Salmon (1767), and Adam (1794; 1795).
12. For Scottish primers, see Withers (2001) and Eddy (2010b). The relationship between learning and geography in British print culture is examined in Mayhew (1998), Withers (2006), and Dodds (2020).

them as material objects. This pedagogy of material engagement was used to learn a range of other subjects as well.[13] Put more clearly, even before students started to create the maps that would eventually end up in their notebooks, they learned that using one was a process carried out across time and space. It involved different kinds of movement and materials that collectively set the mind in motion across the page and around the classroom.

The skill of mapmaking could be acquired in a number of ways. Within domestic settings, for instance, some girls and boys learned to set maps in motion through sewing and cutting. Sewing was an important skill at the time. In 1795, for instance there were no less than four schools in Edinburgh where students could learn advanced skills relevant to becoming a seamstress or decorative artist. They were run by Mrs. Dods, Miss Henderson, Mrs. Morison, and Mrs. Reid. Such schools enrolled female students, but boys learned to sew as well.[14] As shown in previous chapters, particularly in my discussion of the philosopher Elizabeth Hamilton's positive standpoint on the cognitive benefits of making the binding of a notebook, stitching and embroidery were sometimes treated as a mode of learning by hand that could be used on its own or as an enhancement to other forms of inscription.

In addition to stitching the spine of a notebook or embroidering floral designs onto silk, the presence of prints such as *A New Map of Scotland for Ladies Needlework*, published in 1797 by James Whittle and Robert Laurie, reveals that there was a market for students interested in learning geography through the point of a needle.[15] The detailed contours and colorized features that could be achieved in sewn maps sometimes exceeded that provided by those printed in standard primers. Margaret Montgomery's cross-stitched *A Map of Scotland*, created sometime around 1800 when she was a student, for instance, is framed by lines of longitude and latitude and presents the borders and names of Scottish counties and shires in several shades of blue, red, brown, and green (figure 7.1).[16] Her every silk stitch carried the mnemonic equivalent of a pen stroke and yielded an exquisite, colorful artifact that influenced the ways in which she understood how she might create, manipulate, or interpret a hand-drawn map in future.

The experiential aspect of the needlework used by Scottish girls and women such as Montgomery to create a map on cloth played an important

---

13. Secord (1985) and Tyner (2016).
14. Fothergill (1908, 180) states, "I have had through my hands the samples of several Scotch lads which are as well done as those stitches by the opposite sex."
15. Whittle and Laurie (1797).
16. Montgomery cloth and thread ephemera (1800).

FIGURE 7.1. Hand-stitched map sampler. Margaret Montgomery, *A Map of Scotland* (1800), NLS, EMS.s.701. Creative Commons Attribution (CC-BY) license. Reproduced with permission of the National Library of Scotland.

role in how they learned to contemplate the world around them. Historians of modern material culture have shown that needlepoint and embroidery, as opposed to the more general skill of sewing, was an engaging mode of interface that girls and women could actively use to shape their own minds. As early as the sixteenth century, Mary Queen of Scots famously used her needlework as a subversive reflection on the machinations of the English state.[17] By the eighteenth century, ornamental needlework allowed women to project their femininity and train their industry and modesty through keeping their minds engaged with the images they were creating on a piece of cloth. According to Bridget Long, needlework was a form of entertainment, consolation, and reflection.[18] When seen from this perspective, and when considered in light of Elizabeth Hamilton's standpoint on the benefits of transforming string into a

---

17. Jones and Stallybrass (2000, 154–156).
18. Long (2016). For similar approaches to needlework and other forms of creativity pursued by women, see Vickery (2009), Martin (2011), and Goggin and Tobin (2016).

binding, the stitches on cloth that Montgomery used to craft her needlepoint map were just as cognitively significant as the strokes on paper used by note-keepers to craft a notebook page.

Another domestic tradition of learning geographical knowledge that formed the experiential backdrop of the kinds of mapmaking evinced in student notebooks was the practice of cutting out the countries of a map so that they could be reassembled manually on a table. The process turned the map into a gigantic jigsaw puzzle that functioned as a game for householders or schoolchildren. Some families encouraged children to identify the cut-outs of the countries on such "dissected maps" while blindfolded, a task that required them to feel the edges of the paper with their fingers with a view to using shape as a means of identification. In many respects, the cutting and feeling of borders of countries functioned as a kind of interactive tactility through which learners sharpened their ability to use nonvisual modes of knowledge acquisition.[19]

In contrast to stitching and cutting maps in domestic settings, the male and female students attending burgh schools learned the skill of tracing, a realtime mode of interface that involved following lines across maps with their fingers. The skill was helpful for students who would go on to be professionals, merchants, or educators. The practice was explained in Evan Lloyd's *A Plain System of Geography*, published near the end of the century. Lloyd was a schoolmaster living in Workington near the Scottish borders. He taught his students to use tracing as a cartographic finding aid. In his words: "But, suppose you should want to find any particular place in a map, having its latitude and longitude given, the matter is equally easy; you have nothing to do but to look for the given longitude at the top or bottom of the map, and for the given latitude on either the right or left hand side thereof; then, trace the parallel of that latitude with your finger until you come directly opposite to the given longitude, and there you have the place required."[20]

Lloyd extended the approach later in the book to the borders of countries. After mentioning that "Germany is divided into nine large districts, which are called the nine circles of empire," he suggests: "To trace the form and situation of these circles and towns upon the map, will be to you a pleasing and useful employment."[21] Using the act of tracing to find a location, to understand a car-

---

19. An account of cutting and feeling the contours of countries as a young child living in the English Midlands during the 1780s is given by Mary Anne (née Galton) Schimmelpenninck (1858, 2). Wealthy parents bought pre-made dissected maps that had been glued to wood or cardboard. The larger relationship between pedagogy and paper-based games is addressed in O'Malley (2003).
20. Lloyd (1797, 69–71). Emphasis added.
21. Lloyd (1797, 90).

tographic pattern, or even to plot a journey made it easier for many students to understand the multisensory potential of a map, thereby endowing them with knowledge that they could use if they tried to make one on their own.

In a similar vein, tracing was a kinesthetic learning tool helped school students to remember information better, and it familiarized them with the kinds of lines they might wish to draw in a school notebook map. But it served another important function as well. Learning to read a printed map in a primer would have presented difficulties for some students, particularly those with visual learning conditions or those who had not been exposed to many forms of graphic representation during the early years of their education. Tracing the lines of maps in the manner suggested by Lloyd presented a multisensory teaching strategy based partially on touch that could be used to mitigate the visual confusion caused by the many competing lines running across the face of a map.[22]

There were other techniques through which students learned how to synchronize bodily movement with the multisensory process of experiencing a map. One was for teachers to read out the name of a country and then ask students to physically identify it on a map hung at the front of the classroom. Students rose from their desks, walked across the room, stood next to the map, and then stated relevant historical and political information about the location. Sometimes teachers asked them to clarify further what they had said, an act that allowed them to give a fuller answer to the question. The process was likely used in a variety of educational contexts ranging from homes to a grammar school. This skill of being able to associate bodily movement, such as walking, with the information on a schoolroom map would become very useful for students who went on to create survey maps for their notebooks.

A helpful account of how students interfaced with maps displayed in a classroom is given by the educational reformer Patrick Bannerman in *Letters Containing a Plan of Education for Rural Academies*, a work that sought to use the teaching innovations developed by Scotland's burgh schools and city academies to reform countryside education in contexts normally serviced by parish schools. He summarized the process in the following manner: "For I have known children, before they had reached their ninth year, capable of giving a distinct account of the chronology, the great events and most illustrious characters of ancient history. During the course of their narration, they

---

22. Examples of the various multisensory learning methods used today to help children with specific learning disorders are given in Peer and Reid (2001). Barbara Tversky has published several salient studies on the cognitive relationship between movement and map-based learning. See Tversky (1981; 1993) and Tversky and Schiano (1989).

pointed out upon maps the countries where the scenes of those actions lay, with the progress of the successive conquerors of the world."[23] Bannerman's account reveals how the mnemonics of map-based knowledge was reinforced through tactile and visual demonstration.

In addition to the foregoing examples of multisensory learning that developed a student's ability to value and eventually draw notebook maps, students attending progressive grammar schools learned graphic forms of geographical interface as well. From the mid to late eighteenth century, Scottish grammar school headmasters faced increasing pressure from parents to modernize the curriculum so that students could learn a wider range of subjects and skills. One such subject was geography. Some masters resisted this pressure by attempting to portray the graphic skills required to create a map as being unworthy of inclusion in a liberal education.[24] This view was perhaps most clearly expressed south of the border by the English headmaster and Anglican priest Vicesimus Knox in his book *Liberal Education*, a work that was written in part to revive the dwindling enrolment of his grammar school in Kent. In his view, mapmaking was for the *hoi polloi*, a physical labor unworthy of his elite students: "The drawing of maps, and other minute labours in the pursuit of geography, may be desirable to a person who is designed for some employment connected with surveying or navigation, but are an unnecessary toil to the liberal scholar."[25]

The graphic component of mapmaking seems to have been valued more in Scotland, not least because the economic vitality of so many Scottish families was dependent upon the military and mercantile applications of the surveying and navigating skills that were imperiously dismissed by the likes of Knox. This difference in attitude explains why Scottish grammar schools, including the influential Edinburgh High School, introduced geography as a subject in the second half of the century. It also explains why Scottish headmasters, James Barclay for instance, promoted forms of geographical instruction based on drawing.

Barclay was the master of Dalkeith Grammar School during the middle of the century. In his *Treatise on Education*, he describes a geographical exercise in which he explained "*viva voce* the figure of the earth, her annual and diurnal motion, all the greater and lesser circles, zones, climates, &." Based on his oral description, the students attempted to draw the features in chalk, with a

---

23. Bannerman (1773, 22–23).
24. Mayhew (1998).
25. Knox (1781, 162). Mayhew (1998, 745–746) discusses this quotation and Knox's views as a grammar school headmaster.

view to creating a rough map on a slate board. They were then given a globe and, after examining its features via touching and turning it, they decided who had drawn the most accurate chalk depiction. The multisensory nature of the activity strengthened their skills of listening to a description, drawing a chalk map, and rotating a globe, which explains why Barclay called the exercise a "performance" and why he thought that every student participant was a "little artist."[26]

Desk Maps as Crafted Constructions

The term "map" in geography primers was often used to refer to depictions featuring countries or continents. Simply looking at this kind of map was not a guarantee that a student would be able to construct one by hand or to use it in conjunction with other hand-drawn maps bound together in a notebook. Crafting such a map was a process that involved sketching, inking, and watercoloring, a process that was called *construction* at the time. The exercises used by students to learn the skills of constructing usually involved their moving in various ways around maps as they created them at their desks. It is for this reason that I call them *desk maps* with a view to differentiating them from the *field maps* that we will encounter later in this chapter. Being created at a desk in no way diminished such a map's status as an artifact of mental, material, and kinesthetic interface. The distinction simply draws appropriate attention to the location of the activity.

Constructing a map that was meant to serve as part of a notebook required training. When it came to thinking about what could be included in a hand-drawn map, some primers offered students generic exemplars designed to highlight standard vocabulary and basic topographic features. A good example of this practice occurs in "A Map of a Country exemplified," a figure of a map featured in S. Harrington's 1773 *A New Introduction to the Knowledge and Use of Maps* (figure 7.2). It offers a selection of basic features such as peninsulas, rivers, cities, villages, mountains, and continents. But like many primers, it does not offer detailed instructions on how to *make* one.[27]

A similar situation is presented in the brief chapter on "The Construction and Use of Maps" in John Mair's popular *A Brief Survey of the Terraqueous Globe* that was reprinted several times from the 1760s onward. Mair's narrative describes the basic features of a map: Rivers are "black lines," and "Mountains

---

26. Barclay (1749, 207).
27. The map occurs as the frontispiece of Harrington (1773).

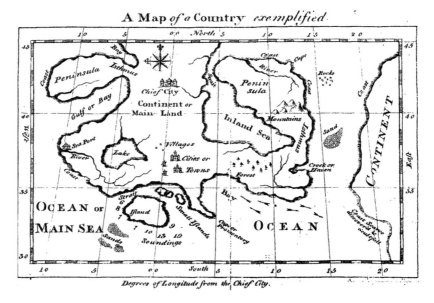

FIGURE 7.2. "A Map of a Country exemplified," frontispiece, S. Harrington, *A New Introduction to the Knowledge and Use of Maps* (London: Crowder, 1773). Author's personal collection.

are represented by a sort of cloud; forests or woods by a kind of shrub; bogs or morasses by shades; sands or shallows by small dots; roads usually double lines; and towns by [an] o, or by the shape of a little house." But Mair's account refrains from giving further instructions on how to create a geographical map from scratch with the aid of specific kinds of media and instruments. The rest of the chapter on map construction, for instance, focuses on the problems and advantages of stereoscopic and orthographic projections of the globe.[28]

This is not to say that more detailed instructions for constructing maps were not available in print. Books for advanced audiences offered brief, dense descriptions of maps that would have been difficult for students to sketch, draw, and label without help from a teacher. A good example of this kind of instruction occurs as the appendix, titled "Shewing the Method and Construction of Maps," of *A New Geographical, Commercial and Historical Grammar*, published by "A Society at Edinburgh" and edited by Alexander Kincaid in 1790. The prolix instructions run for only a few pages and would have been hard for young or graphically inexperienced students to follow. Students seeking to draw their own cartographic projections would have struggled to inter-

[28]. Quotations taken from Mair (1762, 55–58).

pret the appendix's figures without the help of an instructor. Even if students managed to copy the maps depicted in the book's figures, it was unlikely that the process would have led them to a sound understanding of the mathematical principles or drawing skills required to convert the globe and its continents onto a gridded plane.[29]

Overall, beyond simple guidance, geography primers avoided explicit map-making instructions, a situation that strongly suggests that, like penmanship and drawing, many students learned to make maps directly from family members, or from private tutors who did not wish to put their techniques in print. From a practical perspective, the writing and drawing techniques required to construct a map, much like those taught by masters of orthography or draftsmanship, were in many ways trade secrets, tacit forms of knowledge that geography instructors did not want to put in print for fear of losing students.

Like any figural form of representation, the placement of lines on maps carried implicit or multiple meanings. Some primers reminded students of this consideration, explaining that the lines used to represent the equator, poles, and tropics were, in the words of Gawin Drummond's popular *Geography* primer, "imaginary parts" that did not correspond to physical objects.[30] Students had to learn, for example, that even the edges of the paper upon which they drew carried locational meanings. The naval instructor and explorer Sir John Barrow (1764–1848) explained this kind of affordance in the following manner: "In the construction of figures in navigation, let the upper part of the book or paper represent the north, and the lower part the south; then the right-hand side of the meridian will be the east and the left west."[31]

Barrow's principle is clearly evinced in the fine specimens exhibited in the *Maps* notebook created by the schoolgirl Jemima Arrow in 1815 (figure 7.3).[32] Its pages present one of the largest sets of extant hand-drawn geographical maps rendered by a Scottish student in a notebook at the time. Arrow crafted her maps with skills learned from a geography teacher at home or perhaps in a private school. Though we do not know which school she attended, geography was a subject regularly taught to "young ladies" in institutions such as Robert Nicholson's Commercial Academy in Glasgow.[33] Likewise, Edinburgh's

---

29. Kincaid (1790, 543–546).
30. Drummond (1708, 2–3).
31. Barrow (1792, 92). Other primers emphasized the correspondence between the top, bottom, right, and left of the page to the north, south, east, and west compass points. See, for example, Adam (1794, 125–126).
32. Arrow Bound MS (1815).
33. Young ladies were expected to have a sound knowledge of geography. Glover (2011, 159).

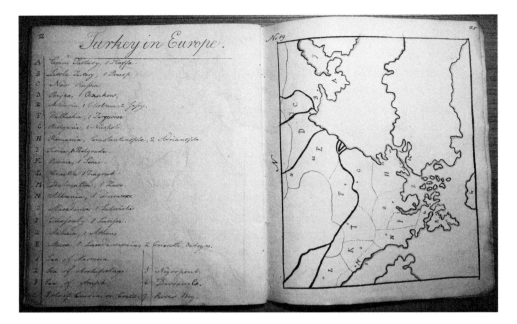

FIGURE 7.3. Map of "Turkey in Europe" (with legend) rendered in ink and black and grey watercolors. Jemima Arrow, *Maps* (1815), Bound MS, NLS MS14100, ff. 32, 35. Reproduced with permission of the National Library of Scotland.

circa 1800 postal directories list geography teachers, some of whom taught drawing as well, a skill that would help students make maps.[34] By 1815, the year in which Arrow finished her maps, Edinburgh's John Fraser had founded a coeducational academy for writing, arithmetic, and geography that was based at 26 James' Square, and William Lennie was teaching geography at Alison Square.[35]

Aside from her notebook, we know very little about Arrow's life.[36] The notebook contains her hand-drawn maps of France, the United Provinces, Germany, Italy, and Turkey (figure 7.3).[37] For each country she inscribed two maps. The first map contained no headings or names. The image was initially drawn

---

34. For Nicholson's Commercial Academy, see Anonymous, "Education," *Caledonian Mercury*, 20 May 1786. Thomas Aitchison's *The Edinburgh and Leith Directory, to July 1800* (1799) indicates that geography was taught by James Leslie of Bailie Fyfe's Close, John Fraser of Carruber's Close, William Walker at the head of Leith Walk, and Mr. Douglas of Nicolson's Street. Walker taught drawing as well, which most likely included geographic mapmaking.
35. Postermaster-General for Scotland (1814, 94, 150). William Scott, the popular teacher of elocution, also taught a "system, of geography" to his students. See Scott's obituary in *The Edinburgh Magazine, and Literary Miscellany, a New Series of The Scots Magazine* (1818, 598).
36. Withers (2001, 253).
37. Arrow Bound MS (1815).

in graphite lines that were then traced over with ink. For the second map, she turned the page and drew (or possibly traced) the same image, but she went further by watercoloring its borders, writing a heading inside each country, and designing a legend in the form of a matrix on the opposite page.

The technoscientific spirit of Arrow's mapmaking exercise afforded two valuable mnemonic devices. The first, a blank map, could be used to test her memory without recourse to the legend. The second, a labeled map, could be used to test her memory against the legend of the facing page. Designing and crafting the map were a mnemonic exercise as well. For instance, she plotted the alphabetical headings of the regions in areas of white space so that they could function as unambiguous labels. Contemplating and then enacting this kind of plotting, even if it was based on an exemplar, was a mode of learning that facilitated the memorization of information and the internalization of the skills required to lay out facts on a map.

It is worth noting the paper-based components that Arrow used to transform her hand-drawn maps into a codex. The notebook is comprised of a quire of bifolia. The tackets and string of the binding reveal that, though the pages of the quire seem to have been kept in order, they were unbound at the time that the maps were made. As discussed in the chapter on codexing, binding a quire made it difficult to lay all the pages flat. The result was a slightly bent page that affected the precision of the lines that a student wanted to draw. Using unbound pages, therefore, allowed Arrow to stretch the paper flat on a desk, table, or drawing board, enabling her to draw, trace, watercolor, and label the map with more precise strokes.

Keeping the paper flat made it possible for Arrow to perform two other tasks that improved the precision of the image. The first, which occurred after she had drawn the graphite lines, was to moisten the surface of the paper by laying a slightly damp cloth over it for a few minutes, an act that prevented the forthcoming ink and paint from bleeding. The second was to write the labels in ink after the sheet had dried so that the edges of the lines would remain sharp. The final result was a remarkably clear map with unfaded grey watercolored borders and crisp lines, the ink of which still appears sharp and glossy.[38]

When we consider the mnemonic function of Arrow's notebook in light of what we learned above about the map-based processes used by Bannerman's students to memorize and demonstrate geographical knowledge, we can see

---

38. Advice concerning the wetting of the paper and the writing of labels on a dry surface is given in Grey (1737, 158). He recommended wetting the backside of the paper with a mixture of roach alum and spring water to "prevent the colours from sinking" and to "give them an additional lustre, and preserve them from fading."

that she was not simply learning to use a preexisting visualization. Rather, she was learning how to make the equivalent of a geographical memory theater in which each country could be associated with information that she or her instructor deemed it necessary to memorize. When she sketched, inked, and washed the maps of her notebook, she was learning how to design, not simply use, her own geographical mnemotechnic device.

The mnemonic purpose of Arrow's map explains why she did not include lines of longitude or latitude. Rather than trace the precise coordinates of an imagined journey, the map's purpose was to evoke information that she had learned to associate with each country. The end result was a cartographic memory theater in which the countries operated like rooms to which Arrow associated information. Her use of mnemotechnics was by no means unique, and we will learn below how similar skills were learned by other students who made and used notebook maps.

Field-Mindedness in the Classroom

Surveying large parcels of land was an essential skill for students seeking to gain employment in the many road-building and construction projects that took place in Britain and beyond over the course of the long eighteenth century. During the middle part of the century the Jacobite Rebellion led the government in Westminster to commission a national survey of Scotland.[39] Artists such as Paul Sandby were asked to capture this massive process through watercolors (figure 7.4).[40] Outside of the military, surveyors were in demand because the courts used them to resolve property disputes. From midcentury forward, individual surveyors could charge high fees when determining the paths of canals in the Lowlands or for situating the many buildings that were part of the urban expansion taking place in Edinburgh and other Scottish cities.[41]

Before aspiring students traveled to a field and attempted to make a map for their notebooks, they needed to acquire a sense of field-mindedness, that is, they needed to know how to adapt a number of notekeeping skills such as cyphering (including various forms of calculating), writing, drawing, and codexing to the context of making a survey. Once they arrived at a given property that needed to be surveyed, they then learned how to orient themselves

---

39. Fleet, Wilkes, and Withers (2011; 2016), Anderson and Fleet (2018), and Roy (2007).
40. Sandby, *View by Kinloch Rannoch* (1746), British Library Maps, K.Top.50.83.2.
41. The pursuit of surveying as a knowledge-making practice in and around Edinburgh is addressed in Rodger (2017).

FIGURE 7.4. Paul Sandby, *View by Kinloch Rannoch*, 1746. Pen, ink, and watercolors. British Library Maps, K.Top.50.83.2. Public Domain, Creative Commons.

within a landscape in a way that allowed them to use their skills to observe and represent the surrounding environment as a survey field map. In this section, I would like to examine the ways in which students laid the foundation for field-mindedness via the experiences afforded through the use of classroom primers and instruments as well as through drafting copies of field maps and field book entries.

Many students started their journey to field-mindedness at home. The tools and skills of surveying and its associated forms of notekeeping were offered in manuals that explained how to plan out family gardens, orchards, and groves. As early as the seventeenth century, books such as John Reid's *The Scots Gardener* encouraged householders to divide their land into triangles, squares, rectangles, and octagons so that they could measure the side of each shape. Knowing the area and the shape of the garden allowed householders to lay out the beds according to symmetric principles (figure 7.5). To avoid the occurrence of what Reid called a "dissatisfied mind," the entire plan, called a "draught," needed to be committed to paper, because, in his words: "All draughts not drawn by the Scale ar[e] but suppositions; the Scale makes them stand directly on Paper as on the Ground."[42]

42. Reid (1683, 4–5). The book was republished in 1766 under the same title.

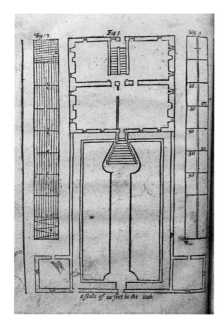 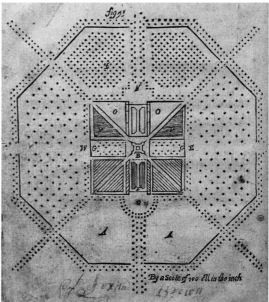

FIGURE 7.5. Rectangular and octagonal garden designs achieved by domestic surveying. In John Reid, *The Scots Gardener in Two Parts, The First of Contriving and Planting Gardens, Orchards, Avenues, Groves: With New and Profitable Ways of Levelling; and How to Measure and Divide Land* (Edinburgh: Lindsay, 1683), figures 2 and 7. Reproduced with permission of the Huntington Library, San Marino, CA.

Reid's manual reveals how the principles of practical mathematics could be applied to domestic surveying. Since the manual was meant to be used in a household, and since we know from previous chapters that both Scottish girls and boys learned mathematics from the late seventeenth century forward, it is worth noting that it was possible for both female and male householders to be exposed to, or familiarized with, a sense of field-mindedness. The presence of this capability is further evinced in *The Ladies Diary*, a magazine that circulated in Scotland from the early eighteenth century forward. Its many editions regularly posed questions for domestic readers interested in calculating the area of a piece of land and were marketed as being, in the words of a 1717 edition, "Delightful and Entertaining" for the "Use and Diversion of the FAIR-SEX" (figure 7.6)[43]

[43]. Questions relevant to land surveying appeared in magazines marketed to female readers across Britain. For a sample, see Anonymous (1717; 1740) and Leybourn (1817). The review of Charles Hutton's *Darian Miscellany* in the natural history, medicine, and mathematics new books section of the *Scots Magazine*, **38** (1776, 262), stated that the miscellany contained mathematical extracts from *Ladies Diaries* dating

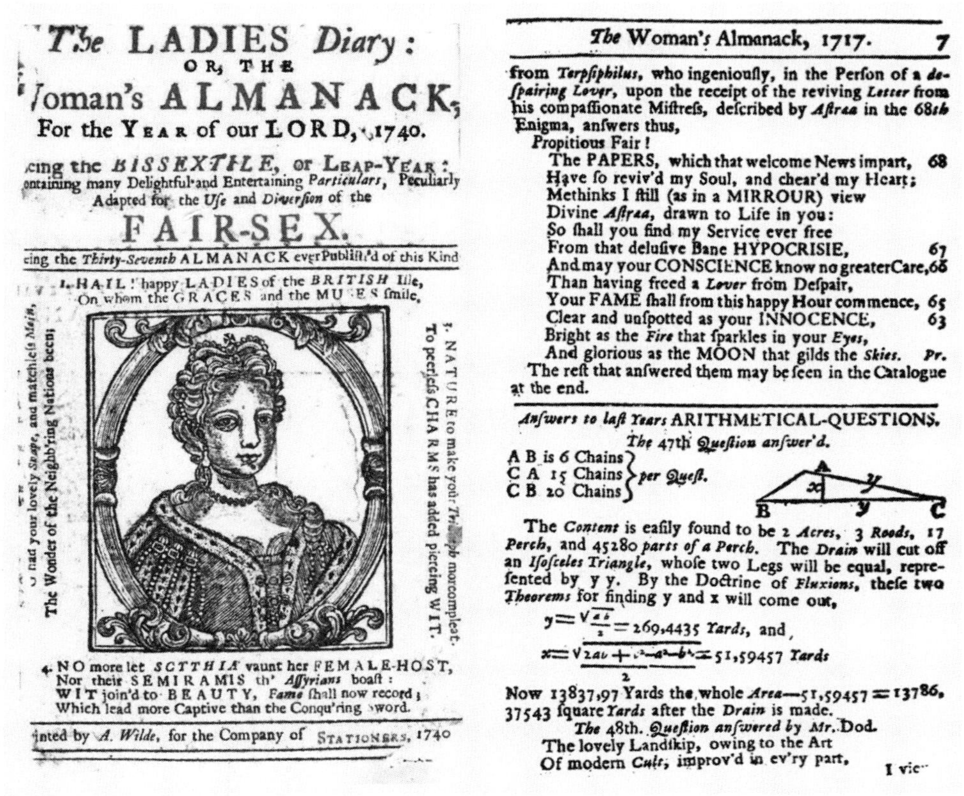
The LADIES Diary:
OR, THE
Woman's ALMANACK,
For the YEAR of our LORD, 1740.
Being the BISSEXTILE, or LEAP-YEAR:
Containing many Delightful and Entertaining Particulars, Peculiarly Adapted for the Use and Diversion of the
FAIR-SEX.
Being the Thirty-Seventh ALMANACK ever Publish'd of this Kind.
1. HAIL! happy LADIES of the BRITISH Isle,
On whom the GRACES and the MUSES smile,
... had your lovely Shape, and matchless Mein,
The Wonder of the Neighb'ring Nations been;
3. NATURE to make your Triumph more compleat,
To peerless CHARMS has added piercing WIT.
4. NO more let SCYTHIA vaunt her FEMALE-HOST,
Nor their SEMIRAMIS th' Assyrians boast:
WIT join'd to BEAUTY, Fame shall now record;
Which lead more Captive than the Conqu'ring Sword.
Printed by A. Wilde, for the Company of STATIONERS, 1740

The Woman's Almanack, 1717. 7

from Terpsiphilus, who ingeniously, in the Person of a despairing Lover, upon the receipt of the reviving Letter from his compassionate Mistress, described by Astraa in the 68th Enigma, answers thus,
Propitious Fair!
The PAPERS, which that welcome News impart, 68
Have so reviv'd my Soul, and chear'd my Heart;
Methinks I still (as in a MIRROUR) view
Divine Astraa, drawn to Life in you:
So shall you find my Service ever free
From that delusive Bane HYPOCRISIE, 67
And may your CONSCIENCE know no greater Care, 66
Than having freed a Lover from Despair,
Your FAME shall from this happy Hour commence, 65
Clear and unspotted as your INNOCENCE, 63
Bright as the Fire that sparkles in your Eyes,
And glorious as the MOON that gilds the Skies. Pr.
The rest that answered them may be seen in the Catalogue at the end.

Answers to last Years ARITHMETICAL-QUESTIONS.
The 47th Question answer'd.
A B is 6 Chains
C A 15 Chains } per Quest.
C B 20 Chains

The Content is easily found to be 2 Acres, 3 Roods, 17 Perch, and 45280 parts of a Perch. The Drain will cut off an Isosceles Triangle, whose two Legs will be equal, represented by y y. By the Doctrine of Fluxions, these two Theorems for finding y and x will come out,

$$y = \sqrt{\frac{ab}{2}} = 269{,}4435 \text{ Yards, and}$$

$$x = \frac{\sqrt{2ab + \ldots - a^2 - b^2}}{2} = 51{,}59457 \text{ Yards}$$

Now 13837,97 Yards the whole Area—51,59457 = 13786,37543 square Yards after the Drain is made.
The 48th. Question answered by Mr. Dod.
The lovely Landskip, owing to the Art
Of modern Cult, improv'd in ev'ry part, I vie-

FIGURE 7.6. Land survey question in a magazine. *The Ladies Diary* sometimes offered practical geometry questions to female readers that addressed surveying topics. Left: Queen Caroline depicted on the title page of *The Ladies Diary: Or, The Woman's Almanack* (London: Wilde, 1740), Google Books, original from Princeton University Library. Right: A surveying question offered in *The Ladies Diary: Or, The Woman's Almanack* (London: Wilde, 1717), part 2, 7, Google Books, original from Bodleian Library, Oxford, Creative Commons License.

Field-mindedness included an awareness of the instruments and texts that could be used to gain the skills required to enter an open piece of property and transform the surrounding space into a survey map. If students wanted to conduct surveys at home or elsewhere, then they had to learn the skills that would enable them to integrate their field maps with their field notebooks. The main goal of a survey map was to determine the area of a parcel

from 1704 to 1779 and that it was "a truly valuable work." The larger context of the readership for *The Ladies Diary* is given in Costa (2002) and Perl (1979).

of land. Aside from notekeeping tools such as pencils, compasses, rulers, and different kinds of paper, the other instruments required to determine the size of a parcel's parameters were chains and theodolites. Such instruments were described in mensuration manuals. The chain (approximately 22.6 meters) was a unit used by surveyors to measure distance.[44] A theodolite was a rotating telescope fixed to the apex of a tripod. It was used to measure angles from fixed points in a landscape. As can be seen by the surveyor stooping toward his theodolite near the center of Sandby's 1746 drawing *View by Kinloch Rannoch* (see figure 7.4), the instrument usually was used to measure large parcels of land.

Acquiring a sense of field-mindedness involved learning how to use manuals or textbooks aimed at teaching methods of field surveying. Textbooks that extended practical geometry into the domain of everyday life or commerce reached at least as far back as the late Middle Ages.[45] In eighteenth-century Scotland, the skills required to survey large stretches of land or coastline were addressed in a variety of textbooks, with Alexander Ewing's *A Synopsis of Practical Mathematics* and Alexander Ingram's *The New Seaman's Guide, and Coaster's Companion* being good examples of the primers written and used in Scotland over the course of the century.[46] Notekeepers consulting these and other affordable primers would have encountered only a few figures that might help them understand how to use surveying instruments, how to range and record the measurements into handwritten tables, how to draw out the parcel of land being surveyed, or how to combine all the data into one notebook.

More expensive books printed in London provided plates that depicted both common and specialized writing and drawing instruments used by surveyors. Examples of this genre are George Adams's *Geometrical and Graphical Essays* (1791) and John Robertson's *A Treatise of Such Mathematical Instruments, as Are Usually Put into a Portable Case* (1775).[47] The figures in these books provide insight into the kinds of graphic skills that *could* be learned from using surveying instruments while making a notebook in school or in the

---

44. Scotland's weights and measures were slightly different from those used in England. For a fuller picture, see Connor and Simpson (2004).

45. Bennett (1998), Camerota (2006), Lilley (1998), and l'Huillier (2003).

46. Ewing's *Synopsis* (1799) was a popular text in Scotland and had numerous reprintings in the 1790s alone. Ingram's *The New Seaman's Guide* (1800) included information similar to that offered by textbooks published earlier in the century.

47. Adams (1791) offers one of the most detailed descriptions of the surveying process, including the many kinds of instruments and notebooks required for experienced surveyors. The first plate of Robertson (1775) gives the everyday instruments used by students learning to make survey maps.

field. In other words, they present a helpful picture of what advanced students could learn in theory and what kinds of visualizations might inspire them to become sophisticated surveyors.[48] But when it comes to understanding how such figures were used in the production of a notebook map, they give historians a limited indication as to how students learned to use them, or whether they even learned to use them at all.

A similar problem existed for scholarly treatises that used geometric parcels of land ostensibly to illustrate the utility of Euclidean geometry. A good example of this kind of book is David Gregory's *Treatise of Practical Geometry*. Gregory began his career as Edinburgh's professor of mathematics in the 1680s and then went on to become the Savilian professor of astronomy at Oxford University. While in Scotland he gave lectures that used surveying examples to exemplify Euclidean geometry. He eventually turned the lectures into a Latin treatise, which was then translated into English and published in 1745. In the introduction of the translation, Colin Maclaurin, Edinburgh's professor of mathematics during the middle of the century, stated that surveying had been taught at the university ever since Gregory's tenure with a view to showing students how to apply Euclid's *Elements* by exercising "the Application of Geometry in Practice."[49] Though an impressive scholarly treatise, it was theoretically oriented in that it did not clearly explain how to move beyond the study or classroom into the field.[50]

As intimated above, though students could glean important conceptual elements of field-mindedness from printed resources available at home or within a school, the experience of using such works was not without its drawbacks. Most offered only a few visualizations that helped students bridge the gap between print and the real world. Add to this fact that using a printed primer as a learning or reference device was complicated by the placement of its maps as figures on plates. For instance, like most surveying primers, Macgregor's *Practical Mathematics* crammed scores of figures onto a handful of plates. Every time students flipped the scores of pages that occurred between a survey's calculations and its corresponding map, they had to devote time

---

48. The embodied relationship between pure and practical geometry, the material world, and metrological instruments in George Adams's *Geometrical and Graphical Essays* (1791) is explored in Damerow and Lefèvre (1985), which appears in their edited volume of George Adams and is discussed further in Lefèvre (2019).
49. Gregory (1761), iii–iv. The *Treatise* went through several editions from the 1750s to the 1790s.
50. Lectures that addressed various aspects of practical geometry continued to be given in Scottish universities well into the nineteenth century, with the mathematics course taught by Professor James Millar (1762–1831) at Glasgow University during the 1810s being a good example of the practice. Millar Bound MS (1815–1818).

to making sure they were looking at the correct figure on the plate, a move that further taxed their working memory.[51] Since there were many such calculations and maps in surveying primers, the cognitive load created by flipping pages and navigating plates was significant and arguably increased the amount of time it took to process, understand, or remember the information.

No matter how much bookish knowledge students might acquire, knowing how to create and keep field maps required a chorus of notekeeping skills that they had to hone on paper. To be sure, some of the interactive aspects of drafting a survey map were mentioned in textbooks in discussions concerning the ways in which a parcel of land might be divided into geometric shapes. This was the approach taken by George Adams in his *Geometrical and Graphical Essays*, a textbook that offered stylized figures that suggested how a student might think about surveying different kinds of landscapes, a property adjoining a river for example (figure 7.7).[52] But the banks of rivers and other kinds of property barriers were variable and there was no replacing the multisensory experience of physically drafting a survey map in situ for oneself.

A similar situation existed for "field books." This was a notebook into which a student entered all on-site measurements.[53] In order to expand both their notekeeping skills and their developing sense of field-mindedness, students had to practice drafting field books and field maps on-site, that is to say, in the words of Edwin Hutchins, "in the wild."[54] Whether a student attended an academy or a university, the ability to enact classroom-based knowledge had to be mobilized through practice. Indeed, some practical geometry students of Professor James Gregory took to the streets and made modest surveys, which they called "plans," of local sites around Edinburgh University.[55] Unlike surveying the streets and courtyards of Edinburgh, learning to navigate the terrain of a rural field while drawing its contours and recording its measurements was a process that provided a wider range of specialized drafting and codexing skills. Put more clearly, to bridge the experiential divide between printed

---

51. Macgregor (1792).

52. Adams (1791, plate 27, figures 25, 30, 31).

53. The term "field book" was used in two ways in eighteenth-century educational and mercantile contexts. The first, discussed in chapter 4, referred to a notebook in which a farmer kept track of the goods produced in the fields of the farm. The second referred to a notebook used to record the measurements taken in a field, a plot of land, that was being surveyed. Surveying field books of this nature are discussed throughout Adams (1791).

54. Hutchins (2002) underscores the importance of paying attention to how maritime navigation is learned through tacit performances enacted in the field on ships at sea.

55. The maps were made prior David Gregory's departure to become the Savilian Chair of Mathematics at Oxford University in the autumn of 1690. Withers (2010, 75–76).

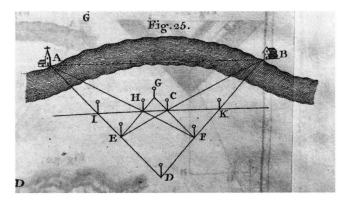
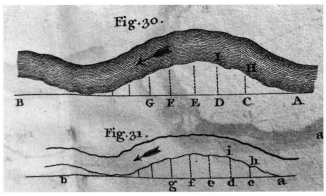

FIGURE 7.7. Stylized riverbank surveys. George Adams, *Geometrical and Geographical Essays* (London: Hindmarsh, 1791), plate 27, figures 25, 30, and 31. Reproduced with permission of the Huntington Library, San Marino, Burndy 702807.

textbook maps and field maps, students traveled to a property so that they could kinesthetically extend their classroom-based knowledge of sketching, tracing, cyphering, watercoloring, and codexing.

This being the case, it is worth asking what graphic skills were being learned by students who planned to transform their on-site observations into a map for their notebook. Students prepared themselves in the classroom for the process of recognizing and re-creating the visual elements of a piece of property by drafting exemplar field maps in graphite on loose-leaf paper or quires. They traced over the lines with ink and, in some cases, washed the map with watercolors. Such skills were taught for an extra fee in the surveying classes offered by Scottish academies. The level of this kind of instruction at the Perth Academy was particularly advanced, especially during the tenure of the drawing master and surveyor Archibald Rutherford.[56] To familiarize themselves with the elements of a field map, students began by drawing a small piece of property with relatively straight borders and no topographical features. Next, they tessellated the interior into triangles and quadrilaterals, that is to say, shapes that could be used to more easily determine the area of a property.[57] Since field maps were based on a field book of measurements, students practiced cyphering the lengths of the sides of the property's polygons into columns and then calculated the area as if they were in the field.

The foregoing process was repeated in the classroom several, sometimes many, times. Each new field map draft introduced more complicated features such as irregular borders, rivers, roads, fences, bridges and foliage, with the corresponding field book measurements increasing proportionally as well. An instructive series of drafted field maps and field notations that move from simple to complex specimens occurs in an anonymous *Mathematics Notebook* from 1816 that is housed in the National Library of Scotland.[58] Like other practical geometry notebooks kept in Scotland for the past century, the movement from simple to complex figures and calculations allowed the student to slowly acquire the writing, drawing, cyphering and codexing skills required to eventually make a land survey. The notebook is a helpful example of cartographical learning because its many figures, maps and cyphers are similar to those presented in other surveying notebooks and primers made over the course of the long eighteenth century. Thus, the purpose of such visualizations and calculations was not to be original, rather, the purpose was to be mnemonically useful to a student learning the conceptual and kinesthetic skills required to make a survey map.

The student keeping the *Mathematics Notebook* first drew basic shapes of planar geometry and then moved on to field map specimens of simple parcels of land. The first parcel was rectangular and the second was triangular. The student then progressed to a series of irregularly shaped quadrilateral fields, tessellating the interiors into triangles. Next came parcels of land that expanded the notekeeper's abilities to plot and draw increasingly irregular borders, the interiors of which were further tessellated into multiple triangles

---

56. Crawford (1906, 61). There is also a brief entry for Rutherford in Jeffares, online edition at http://www.pastellists.com. Accessed 12 April 2020.

57. The verb *tessellate* refers to the process of completely dividing a shape into a mosaic of smaller geometric shapes in a way that leaves no freeform gaps. For further insight into the historical usage of the verb, see the *OED*.

58. Anonymous Bound MS (1816). Another good example of student notebook that follows a similar format and employs watercolors to highlight the borders of the maps is Anonymous Bound MS (1809–1812).

MAPPING | 247

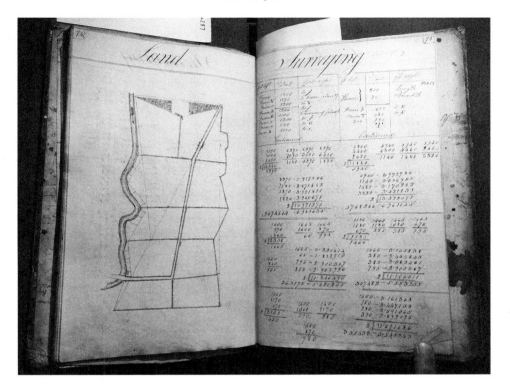

FIGURE 7.8. Pen and ink property map and handwritten field book calculations. Anonymous, *Mathematics Notebook* (1816), Bound MS, NLS, MS 14287, f. 74–75. Reproduced with permission of the National Library of Scotland. Left: Property map. Right: Draft field book data and calculations. The map and the calculations are an adaptation of plate III in John Ainslie's *Comprehensive Treatise on Land Surveying* (1812). See figure 7.9.

and quadrilaterals. Importantly, the figures, maps, and calculations for the exercises appear on the same page, or on a facing page, thereby making it easier for the student to craft and then, when the notebook was finished, access all the information together in the same place.

The *Mathematics Notebook*'s final map, pictured in figure 7.8, is the largest in the notebook, both in terms of the complexity of the tessellations and the number of cyphers that accompany it. The map and its cyphers were both based on plate III of John Ainslie's *Comprehensive Treatise of Land Surveying*, which was published in 1812, only four years prior to the completion of the notebook (figure 7.9)[59]

---

59. Anonymous Bound MS (1816, 74–75); Ainslie (1812, plate III). The plate was used in later editions as well. See Ainslie (1849, 24). Some of the descriptive material in the 1816 notebook was based on

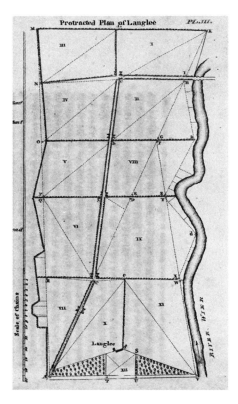

FIGURE 7.9. Engraved property map titled "Protracted Plan of Langlee." John Ainslie, *Comprehensive Treatise on Land Surveying, Comprising the Theory and Practice in All Its Branches; in which the Use of Various Instruments Employed in Surveying, Levelling, &c. is Clearly Elucidated by Practical Examples* (Edinburgh: Silvester, Doig and Stirling, 1812), plate III. Internet Archive, original from the University of Michigan Library.

The student's reproduction of Ainslie's map and cyphers in the pages of his notebook offers further insight into how images were translated between print and manuscript forms of representation in school settings. Notably, the student adapted the map and its accompanying data to suit his needs. His map, depicted in figure 7.8, for instance, does not contain the dense narrative used by Ainslie to describe the map in the *Treatise*. The student's map features other striking alterations. The orientation, for instance, is vertically flipped. Ainslie's original alphabetical labels were omitted. Add to this the fact that over half of the internal tessellations of Ainslie's parcels are absent. The stu-

---

the surveying-related entries in Hutton (1815). Compare, for instance the surveying advice offered on Anonymous Bound MS (1816, 70), to that offered in "The Field Book" section of the surveying article in Hutton (1815, 475).

dent also introduced deviations to the outline of the border and altered several topographical features. These adaptations reveal that the student's goal was not to blindly replicate all of the visual and textual information presented in Ainslie's book.

Instead, the goal was to practice the kinds of notekeeping skills that Ainslie's book assumed but did not explain. Put more clearly, though Ainslie's plate, like so many plates offered in surveying primers, showed what might need to be traced, sketched, ranged, and cyphered, it did not have much to say about how a student might acquire the tracing, sketching, ranging, and cyphering skills that were needed to successfully enact a field map and its associated calculations on paper. What this situation reveals is that the ways in which students learned to draft field maps from surveying instructors was similar to the ways which they had learned orthography from writing masters and mistresses. In other words, like the elaborate plates of alphabets shown to orthography students, the maps presented in surveying textbooks were difficult to realize on paper without the kind of face-to-face graphic instruction offered by a knowledgeable teacher and a fully engaged student.

When seen as graphic devices that were constructed over time by a student who wished to learn or extend mapkeeping skills, the draft field maps in the *Mathematics Notebook* become mnemotechnic platforms that provided the opportunity to learn in realtime how to select and then draw sophisticated tessellations, offsets, and topographic features such rivers, roads, and foliage.[60] This act of strategic selection also occurred while students were ranging and copying sample tables that contained surveying measurements. For instance, the numeric columns written by the student next to his final map (again, featured in figure 7.8), though based on those depicted in Ainslie's book, were altered, presumably to fit his own needs or skills. Rather than using Ainslie's four columns to list the surveying measurements, the student compressed two of the columns together into one, making the structure slimmer and more suitable for the space of his notebook paper.

When viewed from start to finish, the increasing complexity of the map exercises presented over the pages of the 1816 *Mathematics Notebook* reveal that students used their notebooks to learn the core graphic skills required to create a field map and to arrange and calculate the measurements collected in field books. Like Jemima Arrow's geography maps, the exercises in surveying notebooks were desk maps in that they were made at home or in a school

---

60. Anonymous Bound MS (1816, 74).

setting. Though they were effectively drafts, they served to inculcate a sense of field-mindedness that could be applied outside the classroom.

Field Maps and Visualized Data

As shown in Peter Damerow's work on the emergence of mathematical systems in ancient Mesopotamia, surveying was, and has remained, a fundamentally embodied enterprise in which a mapmaker must learn how to coordinate physical activity, manipulate instruments, tessellate land parcels, and keep different kinds of field records.[61] In the case of Scotland, students moved beyond classroom-based forms of interface by leaving the school building and going to properties around the city to extend their measuring, drawing, and calculating skills on and across paper. In pursuing this kind of embodied learning, they participated in the indispensable tradition of making two kinds of field maps. The first was a rough map and the second was a composite map that featured a neater version of the parcel with measurements superimposed over it.

Translating the features of a parcel of land onto a two-dimensional on-site map naturally took a significant amount of writing, drawing, and watercoloring. Once on-site, students familiarized themselves with the property and drew a rough map of its borders. At some point, watercolors were added to differentiate major geographical markers like rivers, fields, property lines, and roads. The rough map often featured only the shape of the parcel and notable landmarks. The interior was usually blank. Having made the rough map, students then conducted the survey, recording the data of the requisite measurements. Extant student notebooks suggest that on-site drawings and measurements were done on loose-leaf pieces of paper that could be easily transported. Then they used the rough map and their measurements to create the composite map. This second map was also watercolored and rendered on loose-leaf paper. It was usually neater and combined the topographic features of the rough map with the measurements collected in a field book. Unlike the rough map, a composite map's depiction of the interior of the property featured triangular and quadrangular tessellations and the respective lengths of each line.

When the steps required to make rough maps and composite maps are considered in tandem, it can be seen that the process involved a number of advanced graphic skills. It is in the pencil, pen, and brush strokes of these

---

61. Damerow (1996, 149–171, esp. 161–163).

maps in motion that we can study how using surveying and notekeeping instruments enabled students to learn, in the words of Wolfgang Lefèvre, "the application of geometrical knowledge in real space" in a way that revealed the "application limitations of geometrical ideas."[62] Lefèvre's larger point is that knowledge of the kinds of space presented by a piece of property involved a good amount of knowhow.

Students acquired the skills associated with mapkeeping knowhow by going to local properties to make field observations. Students of the Perth Academy, for instance, went to the North and South Inch of Perth, a stretch of land near the city of Perth along the River Tay. Perth's surveying masters sent students to the Inch because it was near the school. Instructors elsewhere in Scotland followed a similar practice of using local plots, a tradition that added a distinct geometric feel to the ways in which students learned to experience the local environment.[63] Most of the exercises, as well as most of the maps, contained in an anonymous *Surveying Journal and Accounting Ledger*, a school notebook kept from circa 1809 to 1812, for instance, were based on sites in the Yarrow area of southern Scotland, indicating that surveying was most likely being taught by an academy or master based in the area.[64]

Though the components of many school notebooks hint at the drafting, measuring, and compositing skills required to create a fully functional survey map, the process is illustrated nicely in two maps bound in the mensuration section of an anonymous *Perth Academy Notebook* made around 1790. The draft and composite maps in the 1790 notebook represent the same piece of parkland located on the North Inch of Perth. Like other Perth students, the notekeeper had been sent there to practice his field mapping skills. He first made a rough map, that is, a quarto-sized draft of the property (figure 7.10). It was rendered in graphite, ink, and blue and green watercolors.[65] Its creased corners, soiled smudges, multidirectional folds, and ripped edges indicate that it was made on-site. It represents an early attempt at drawing the perimeter of the piece of land. Notably, its preservation in the notebook is fortuitous. Though most students crafted such rough field maps, they often were not included when surveying notes were bound into leather books.

The rough map in the 1790 *Perth Academy Notebook* functioned as a practice round for the observation, mensuration, and calculation skills the student

62. Lefèvre (2019, 162).
63. The larger relationship between local surveying and eighteenth-century ecology is addressed in Spurr (1951).
64. Anonymous Bound MS (1809–1812).
65. Anonymous Bound MS (1790, 64v).

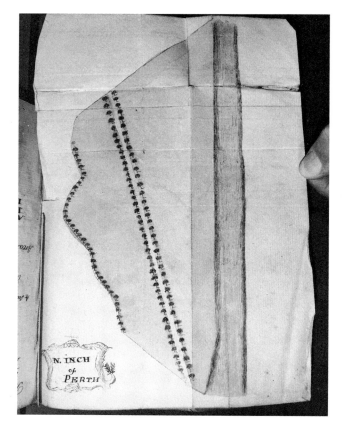

FIGURE 7.10. Pen and wash on-site survey field map draft. Anonymous, *Perth Academy Notebook* (1790), Bound MS, NLS MS 14291, f. 64v. Used by permission of the National Library of Scotland.

notekeeper would need to translate the property onto a more finely drawn composite map crafted on another piece of paper. The composite map, reproduced in figure 7.11, is neater, smaller (octavo-sized), and rendered in the same graphite, ink, and watercolors as the rough map.[66] Rather than being left blank, its interior features all the lines, measurements, and calculations used by the student to determine the total area of the property. More specifically, the property's interior space is tessellated into contiguous geometric shapes, the sides of which represent the lines running across the landscape that the student had measured on-site with chains.

The reduction of the quarto-sized rough map to the octavo-sized compos-

---

66. Anonymous Bound MS (1790, 62v). Although this map of the Inch was made second, it occurs first in the notebook. It is not clear why the binder inserted the rough map after the recopied version.

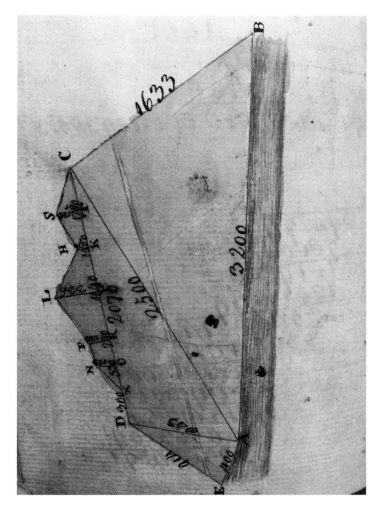

FIGURE 7.11. Redrawn on-site survey map. Anonymous, *Perth Academy Notebook* (1790), Bound MS, NLS MS 14291, f. 62v. Used by permission of the National Library of Scotland.

ite map represents an impressive process through which the student translated visualized data from one medium to another. Here I am using the verb *translate* to capture the movement of the map and its data through media in stages. In this reading, the process of translation taking place in, around, and across a student's notebook maps involved skills and decisions that occurred in stages, each of which was, in the words of the visual historian Ludmilla Jordanova, "a creative act in its own right, rather than a technical exercise."[67] But

---

67. Jordanova (2012, 215).

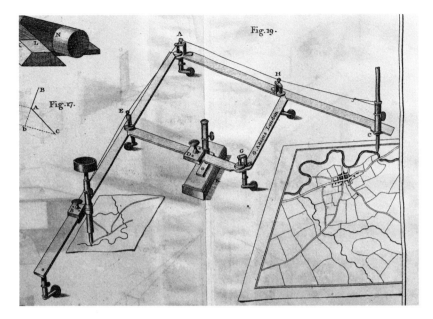

FIGURE 7.12. Pentagraph being used to replicate a map. George Adams, *Geometrical and Geographical Essays* (London: Hindmarsh, 1791), plate 31, figure 19. Reproduced with permission of the Huntington Library, San Marino, Burndy 702807.

what precisely was the overarching process through which a student practiced such acts? The clues offered by the form and fabric of the maps and other components of the notebook suggest three possibilities.

The first option was for the student to trace the rough map. Tracing, however, would have been difficult in this case because the composite map's paper was thick, which meant that its limited transparency worked against tracing another image. Add to this fact that, as already noted, the maps are not the same size. The second option was to replicate the rough map freehand, an act that required considerable skill that could have been learned only through much prior practice. The third option was to use a pantograph, an act that required the skill of assembling the instrument and then using it efficiently.[68] We encountered this instrument in previous chapters. Figure 7.12 shows how it could be assembled to trace one map (on the right) in a way that simultaneously created a scaled down version on another piece of paper (on the left). But a pantograph consisted of many moveable parts and required skills to assemble and use. The precision and materials needed to make one

---

68. For an example of a pentagraph being used to replicate a map, see Adams (1791) plate 31, figure 19).

also ensured that it was an expensive instrument. It is, therefore, likely that the student redrew the composite map freehand.

The precision of the composite map's lines indicate that the student tessellated the land on-site while recording the requisite measurements. The paperwork required for this kind of tessellation was omitted from the *Perth Academy Notebook* when it was bound, which means that many on-site sketches and calculations, like the rough notes and Latin translation exercises of notebooks that other school students kept for literary subjects, were not deemed necessary for inclusion. It was similar for surveying notebooks made by other students as well. But students would not have seen the effort required to copy the omitted paperwork as a loss, especially since Lockean developmental psychology and humanism generally placed a high value on using the acts of re-writing and redrawing as learning tools.

The relatively long and straight borders of the composite map are represented by lines DE, EA, AB, BC, and DC in figure 7.11. The interior of the plot shows in detail how students learned to divide as much of the land as possible into large triangles. Since they had to work with the lay of the land, they sometimes were required to make odd-sized divisions. The far left border of the map, for instance, is variegated, which made it more difficult to tessellate. To solve the problem, the student inserted a straight line, DC, across the variegated border. The student then created offsets at right angles against the straight line for the remaining, irregular parts of the property. After completing all of the plot's tessellations, the student transposed metrological data onto the composite field map by adding the measurement lengths of each line. Finally, he labeled the angles with alphabetical headings.

The *Perth Notebook*'s composite map is followed by several pages of neatly ranged ciphers and calculations that were based on the measurements of the triangles and offsets inside of the plot. As with all geometric calculations based on on-site parcels of land, each kind of shape within the plot—triangle, square, trapezium—required its own formula to determine the area. At the end of the calculations the notekeeper stated the result: "The Area of the North Inch of Perth" was 32 acres, 2 roods, 8 falls and 11 ells. The result reveals that, in addition to having to calculate the area of each shape inside the plot, the student had to convert the answer into the complicated system of squared units used at this time in Scotland.[69]

It is in the student's efforts to translate the rough map and its correspond-

---

69. The calculations occur over Anonymous Bound MS (1790, 65–67). For Scottish weights and measures, see again Connor and Simpson (2004).

ing field book measurements into a neater, composite iteration that the practice of making maps in the classroom, evinced so clearly in the pages of the *Mathematics Notebook* depicted in figure 7.8, became very helpful. The precision of the composite map in the 1790 *Perth Academy Notebook*, though not as neat and tidy as the classroom exercises, represented a significant culmination of many notekeeping skills required to make a survey map from start to finish.

Like the desk maps that surveying students practiced in their school notebooks, the measurements and calculations related to the 1790 composite field map of the *Perth Notebook* occurred on the adjacent pages. Placing the calculations near the composite map made the notebook easier to use when it was finished. Here we can see again an important mnemonic difference between print and script because surveying textbooks usually did not place sample maps next to the text that described them. Since the information did not occur in the same place in textbooks, students had to devote more of their working memory to understanding the relationship between the calculations and the visual elements of the map. A representative example of this mnemonic taxation occurs in the 1792 Edinburgh edition of Macgregor's *Practical Mathematics*. The plate that features the field map occurs 150 pages after Macgregor's discussion of the corresponding field book calculations (figure 7.13).[70] Put more clearly, the placement of maps in student notebooks made them a more user-friendly, and arguably more suitable, learning technology for students.

Maps as Mnemonic Devices

The handcrafted maps of school notebooks functioned as memory theaters, the layout and contents of which served to remind students of the experiences and skills that had enabled them to become makers, users, and keepers of maps. Put more clearly, there was a strong relationship between the arts of memory and the skills that students learned while working with maps during the long eighteenth century. I noted this presence of mnemotechnics in previous chapters, particularly in my discussions relating to how students learned to design their notebook pages so that certain spaces on the page carried different kinds of meanings. The ability to enact architectural mnemonics, especially visual cues, was extended when students learned to create maps. I

---

70. Macgregor (1792). The field book calculations appear on page 277; all of the plates appear after page 431.

MAPPING | 257

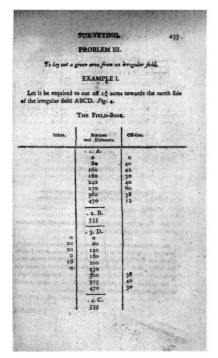 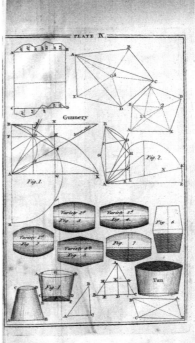

FIGURE 7.13. Printed field book measurements and an engraved field map. John Macgregor, *A Complete Treatise on Practical Mathematics: Including the Nature and Use of Mathematical Instruments* (Edinburgh: Bell and Bradfute, 1792), 277 and plate IX. Left: A field book specimen featuring the plot's measurements. Right: A visualization of the plot's perimeter. It appears on the top left of plate IX at the end of the book on the unnumbered pages following page 431. Internet Archive, original from the University of Michigan Library.

would like to conclude this chapter by exploring this mnemotechnic aspect of notebook maps in more detail.

Ever since classical times there had been longstanding links between geography and humanist mnemotechnics. In Scotland, these links were implicitly made by teachers influenced by the widespread presence of humanism in university courses and through books such as Richard Grey's *Memoria Technica*.[71] In order to understand the connection between mapmaking and the visual mnemonics of humanism, we must note that one of the most successful humanist memory techniques was to use an imaginary house in one's mind to store and organize information. The technique involved associating each

---

71. The longstanding connections between British humanism and geography are addressed throughout Mayhew (1998).

room with a general theme. Objects were then placed in each room that represented specific kinds of information relevant to the theme. To remember something, one needed only to enter the imagined house, go to the room, and then find the object. Sometimes called "technical memory," the technique was advocated by classical orators, especially Cicero and Quintilian, whose works continued to be used in educational settings throughout Europe from the medieval period all the way up to the eighteenth century. It was disseminated in Scotland through the Latin and vernacular editions of Cicero and Quintilian read in grammar schools, academies, and burgh schools.[72]

The mnemonic process that underpinned technical memory skills was summarized by the Scottish educationalist John Adams in the following way: "Some have proposed the imagining of a house or town, and of representing to themselves therein the different parts, in which were placed the things or ideas they designed to remember."[73] Instead of a house, one could also use a theater or a palace, leading some scholars to refer to the mental picture as a "memory theater" or "memory palace."[74] Within this tradition, maps were used as memory theaters well into the nineteenth century. As noted by Veronica della Dora, scholarly Mediterranean travelers treated mythical mountain peaks as "large-scale analogues of the objects contained in the rooms of Quintilian's 'memory house,' or in Renaissance memory theatres."[75]

One of the more elaborate mnemotechnic maps designed for teaching geography to students was created during the late eighteenth century by Gregor von Feinaigle, a Cistercian monk from the area of Lake Constance in Switzerland. It involved replicating a globe within two paper cubes (figure 7.14). He had developed the practice while operating as an itinerant mnemonist, offering tuition in many European cities, including Glasgow and Edinburgh.[76] Each cube appeared as a plate in his book. When cut out and then folded together, the cubes became rooms with interior walls that depicted the Northern or Southern Hemisphere. The walls of the rooms were described in the printed edition of his lectures in the following manner: "The four quarters of the

---

72. Classical and medieval architectural mnemonics such as houses, colonnades, arches, vestibules, roads and towns are discussed in Carruthers (2008, 89–98, 173–174). Cicero summarized his method in *De Oratore*, II.lxxxvi.354. For an interlinear translation, see Cicero (1942, 467). Quintilian advocates the use of a memory house in *Institutio Oratoria* xi.ii.18. For an English translation, see (1876, 336).
73. Adams (1789, 341).
74. For the memory palace metaphor, see Spence (1985). The emergence of modern memory theaters is traced in Yates (1966). Most editions of her book include a helpful foldout plate of the memory theater (as well as rooms and houses) of Giulio Camillo. Helpful insights into the place of memory theaters within scholarly contexts are also given in Blair (1997), West (2006), and Berger (2017).
75. Della Dora (2008, 22).
76. Laver (1979) and Stray (2002).

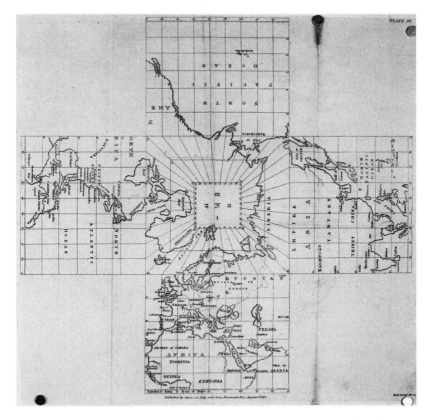

FIGURE 7.14. Gregor von Feinaigle's cartographic memory theater shaped as a paper cube. By the eighteenth century, mnemotechnic skills were firmly integrated into the ways in which students learned to make and use maps. Gregor von Feinaigle, *The New art of Memory, Founded upon the Principles Taught by M. Gregor von Feinaigle*, John Millard (ed.) (London: Sherwood, Neely, and Jones, 1813), plate IV. Internet Archive, original held by the University of Michigan Library.

northern [and southern] hemisphere being arranged on the four walls, when we are in the room, we can, in an instant, see every part of the hemisphere."[77]

For Scottish students, using a map as a memory theater was a technique in which individual countries served a similar purpose to that of rooms of a house. Looking across a map was like remembering a floor plan. Focusing on one country was like entering a specific room containing related thematic information. Once "inside" the country, students could then use visual cues such as the symbols for cities, mountains, rivers, and the like to remember the topographic, economic, and political facts that they had memorized. The

77. Feinaigle (1813, 69).

memory theater technique is clearly illustrated in the way that instructors such as Patrick Bannerman and students such as Jemima Arrow used the countries of geographical maps as visual cues for place-specific information that had been memorized.

Memory theater techniques were also used by students who made surveying notebooks. For the sake of space, I will concentrate on how the straight lines, primarily of borders and tessellations, were cues that could help students remember the skills they had learned while making survey maps. When viewed from the perspective of a student keeping a school notebook, an important visual element of either a draft or composite field map was the straight line. Without straight borders and diagonals, the map's measurements would be inaccurate, which also meant that the calculations of the plot's area would be incorrect. Students were repeatedly reminded of this problem in primers and in classrooms. The problem was mitigated by their ability to see and then represent straight sightlines across a field and a page.

Achieving a straight line in a parcel of land and in a field map was easier said than done, especially since borders and diagonals had to be measured on-site and then rendered on paper. The process involved numerous writing and drawing skills, as well as those that involved calibrating instruments, gauging the lay of the land, and selecting survey points, all of which were based on the student's desire and ability to create a straight sightline across the land. Add to this the fact that surveying by definition could not be done alone. At the very least it required two people; one, sometimes called the foreman, who selected the points and another to carry the chains over the landscape between the points. As evinced in Sandby's depiction of a foreman (with the theodolite) and the two carriers walking in the field, large plots often required several people (see again figure 7.4).

Students learned to create straight survey lines in the field by selecting two fixed points on the plot and then determining the distance with chains. The foreman stood with a pole or theodolite on the first point and the chain carrier walked to a pole planted on a second point. The foreman kept the line straight by constantly gazing past his pole, or through the theodolite's crosshairs, to the second pole in the distance, which then allowed him to see if the carrier was deviating from the straight line that needed to be measured. In the words of the surveying instructor John Love, the skill of creating a "direct line" required the foreman to "take care that they who carry the chain deviate not from a straight line; which you may do by standing at your instrument, and looking through the sights. If you see them between you and the mark observed, they are in a straight line, otherwise not."[78] Here we can see that projecting a sightline across a field was a crucial observational skill that underpinned the process that students learned through making maps for their notebooks.

Though the material engagement with the landscape and surveying instruments would have been a new experience for most students, the skill of sightlining, particularly the ability to imagine and represent straight lines across a field of vision, was already familiar to them because they had developed it via the process of making the headings that appeared in the modules that they used to organize the space of their notebook pages. As we learned in chapter 5, sightlining was the skill of spotting or creating lines of sight between two visual cues. Students learned and then honed this skill by structuring the blank field of each page in their notebooks with headings that operated as visual cues. The headings drew the attention of the eye from one cue to another, oftentimes in a straight line. When they consulted the completed version of the page, the cues and sightlines served as a reminder to them of the ranging skills that they had used to create such an image. In this sense, every page of a student notebook was a modest memory theater of writing, drawing, and categorizing skills.

Like the written page of a notebook, the sketched, inked, and watercolored pages of field maps served as memory theaters that reminded surveying students of the many skills and attendant forms of on-site movement that were learned through the mapmaking process. When seen from this perspective, the plot of land depicted on a student-made composite survey map was like the floor plan of a house. The lines of the map were crafted with specific on-site skills that involved multiple forms of interface that took place between the student's body, instruments, the landscape, and the other helpers who were participating in the survey. This meant that looking at a specific line had the potential of evoking memories of the skills that had been used to make it. Like the rooms of a memory palace, the lines of a hand-drawn composite survey map could be used as memory cues. Put more clearly, though it was possible to use the lines as cues to remember facts about the landscape, they also served to remind students of the *skills* and associated bodily experiences that they had learned while making the map.

Take, for example, the borderline running along the bottom left side of the open meadow depicted in figures 7.10 and 7.11 (line DE in the latter). Looking

---

78. Love (1792, 55).

at the line, which we will remember represented a sightline used by the student in the field, evoked the memories of the skills that the student had employed to create and measure that part of the border. In other words, the line served to evoke memories of how to select points that worked with the lay of the land, how to keep the chains straight by marking the end of each link with a stake, how to successfully operate a theodolite, how to sketch a graphite line in situ while gazing across open land, how to communicate with a chain carrier when trees blocked the line of site, how to use objects such as roads and houses inside the plot as distinctive markers, and how to use the larger features of the landscape such as mountains and rivers to gauge the relative position of the plot to nearby towns or villages.

The lines of the other tessellations and boundaries in figure 7.11 also served to remind the student mapmaker of the skills required to divide, label, and calculate the area inside the plot. Likewise, the borderlines had the capacity to evoke memories of how to tessellate the internal space of the plot so that it consisted only of simple geometric shapes, how to proportionally enlarge or reduce the size of the map so that it would fit better in a notebook, and how to plot handwritten labels and measurements in the open space around the lines in a way that made them easier to spot.

The straight sightlines that students learned to replicate in the borders and geometric divisions in their field maps and composite maps were of course not the only elements that could serve as mnemonic cues. There were other important elements such as angles and curves, and there were a host of other skills performed in the field that enabled students to measure and represent a plot of land. In short, the mnemonic function of the lines, angles, and curves in the maps that students created for their notebooks was effectively the same in that that they collectively created a *tabula memoria*, a picture that served as a memory theater that reminded students of the mapmaking skills they had learned in the field.

The importance of recollection, which was considered an operation of the mind at this time, alerts us to the larger connections between student notekeeping and the skills of reasoning that were identified in previous chapters. Aside from cuing memories of skills and facts that aided ratiocination, how might we further reflect on how creating notebook maps was understood to be a rational exercise? One way to further excavate the positive relationship drawn between reasoning and mapping is to note the high value attributed to practical geometry within the science of mind as it was understood at the time. Perhaps the clearest commentator on the subject was Dugald Stewart,

Edinburgh University's professor of moral philosophy and a leading expert on human cognition.

We encountered Stewart's views on the value of notekeeping in earlier chapters, particularly the formative impact that it had upon memory and other operations of the human mind. In Stewart's discussions of the mental operation of abstraction in his lectures, for instance, he reflected on the importance of gaining "practical or experimental skill" that enabled a person to have "a talent for minute and comprehensive and rapid observation." Such skills were "acquired only by habits of active exertion, and by a familiar acquaintance with real occurrences."[79] His comments on such skills were motivated by his larger skepticism of speculative philosophy and his desire to more firmly connect the operations of the mind to the everyday life of human cognition. But they also reveal how there was scope within the science of mind for the simultaneous modes of "active exertion" and "comprehensive observation" being learned in realtime by students who studied practical geometry.

Stewart's observations are directly relevant to the ways in which surveying students had to learn to abstract the potentially endless stream of data offered by a landscape into the kinds of measurements and descriptions that allowed them to create a notebook map. When seen as a process of abstracting important particulars from the world, the maps become evidence of how realtime skills involving multiple instruments and media technologies were being learned in the predigital era. Through making maps, students became like the tradesmen who could abstract the dimensions of a room by simply looking at it. In Stewart's assessment, such a skill demonstrated the same modes of abstraction employed by the classical geometer Euclid.[80] Such abstractions were admired by Stewart, as it was evidence of the ways in which mental operations could be trained and then applied in the real world. In this reading, on-site surveying was an important training exercise in the skill of abstracting, one in which students also learned how to use the lines of a map as a collective memory theater that reminded them how and why such skills had been learned.

Stewart's lectures on memory further highlighted how practical geometry served to strengthen the operation of attention in a way that allowed the mind to perceive "the relations which exist between different quantities, and the connexions which exist between different relations." Such connections and re-

79. The quotations in the foregoing sentences occur in Stewart (1792, 227–228).
80. For Stewart's discussion of the tradesman and Euclid, see Stewart (1792, 158).

lations could only be made when a person went out into the world and measured the height of landmarks and "the magnitudes or distances of objects." If we extend Stewart's view to the modes of measuring conducted by surveying students, it can be seen that they were pursuing a process of thinking that, like other forms of practical geometry, helped to shape the operations of their minds. Such a process, in Stewart's view, made it possible "to carry our enquiries from facts which are exposed to the examination of our senses, to the most remote parts of the universe."[81] This process, as we have seen, would have been very difficult to learn without the aid of a notebook. Thus, the skills of mapping being learned by notekeepers were crucial to the kinds of thinking processes that helped them to better understand the kinds of facts that could be extended to general principles about the regularities of the world.

Overall, this chapter has advocated a historical stance that seeks to more clearly identify and interpret the materially grounded mapkeeping skills used by notekeepers that helped them think more efficiently and to simultaneously balance a variety of skills that operated in realtime. In this context, crafting a map served as a learning exercise that helped them superimpose a geographical and geometrical sense of space upon objects and landscapes in a way that framed how they experienced the relationship between the mind and body on paper and within the world around them. I have shown that, when we foreground the materials and skills required to make a notebook map, it is possible to catch a glimpse of the dynamic world of predigital media technologies.

---

81. All quotations in this paragraph occur in Stewart (1792, 434–435).

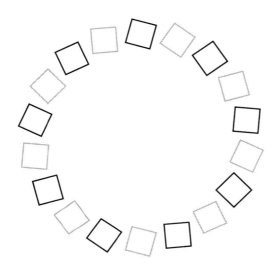

PART III

BEYOND THE *TABULA RASA*

Parts I and II of this book revealed that students used a host of skills to set the *tabula rasa* in motion inside and around the pages of a school notebook in its capacity as a paper machine. In part III, I look beyond the *tabula rasa* as a single entity and examine how it was extended across groups of notebooks designed to be used collectively as a single system of knowledge. More specifically, I explore how university students crafted sets of lecture notebooks in a way that took their notekeeping skills to a new level of artifice, one that enabled them to compress vast knowledge systems into paper machines containing interlinked components that operated across multiple volumes. In order to accomplish this task, they extended what they had learned at school in a way that allowed them to acquire what I call the skills of systemizing, diagramming, and circulating. In the following chapters, I foreground the characteristics of these skills with a view to showing how notebooks were realtime artifacts that played a central role in shaping how students learned to manipulate and access the knowledge economy so characteristically associated with the Scottish Enlightenment.

My exploration of the learning experiences of student notekeepers will be based on the lecture notebooks they kept while attending the universities of Edinburgh, Glasgow, Aberdeen, and St. Andrews from around 1700 to the 1830s. Using a vast and relatively untapped cache of source material, I further enrich the meaning of the *tabula rasa* metaphor by exploring how the components of lecture notebooks shed much light on the relationships that students learned to draw between notekeeping and reasoning. I suggest that such relationships influenced the ways in which students came to understand how they might shape their own minds via the ways in which notekeeping could be used to train the simultaneous combinations of mental operations such as perception, recollection, attention, and abstraction.

One of the points that I wish to underscore is that Scottish universities fostered dynamic communities of student notekeepers. It was a world in which students actively used their notekeeping capabilities to create symbiotic scribal relationships between themselves and their professors. This in turn helped them to see the contingent nature of knowledge and exposed them to the painstaking practices required to preserve and construct knowledge systems. Making and using a lecture notebook in Scottish universities at this time was, therefore, a core part of the learning process. It was fundamentally tied to forms of observation and representation that fused the five senses with the material and kinesthetic acts of setting knowledge in motion on paper. As such, notekeeping played an integral role in the acquisition of knowledge and reinforced abilities and values that remained with students long after they finished their university studies.

CHAPTER 8

SYSTEMIZING

The Syllabus as a System and a Machine

Sometime around 1751 the teenage law student John Millar received an exciting piece of news. Glasgow University had appointed a progressive professor of logic, who, it was rumored, was going to revitalize the syllabus. Having matriculated into the university at the age of eleven, Millar had already taken many of the arts courses and recognized the advantages of being exposed to a new approach to logic. True to his resolve, Millar attended the new professor's lectures on logic and, a bit later, those on moral philosophy.

The concepts and principles that Millar learned in the lectures had a lasting impact on his intellectual development and his larger view of society because the new professor was none other than Adam Smith, the future author of *The Theory of Moral Sentiments* (1759) and *An Inquiry into the Nature and Causes of the Wealth of Nations* (1776), two books that continue to influence political, economic, and philosophical discourse to this day. Inspired by the questions that Smith raised about the relationship between society and the state, Millar eventually became Glasgow's Regius professor of civil law.[1]

Millar's experience as a student in Smith's classroom raises a number of intriguing questions about the relationship between learning and notekeeping within Scottish universities. What kinds of skills, for instance, were needed to make sense of the lectures in realtime? What kinds of techniques were best suited to remembering or recording the lectures? In many respects, these questions are not unique to Millar or Smith and they could be asked of any lecture course given during the modern era.

Like a number of graduates who had attended Scottish universities during

---

1. Miller (2017) and Lehmann (1960).

the long eighteenth century, Millar wrote reflections on his time as a student later in his career. His account explains that the secret to Smith's success in lecturing was a well-ordered syllabus. Smith organized his moral philosophy into four parts, each of which was further ordered into a set of "discourses" that covered a specific topic. Each discourse was in turn broken down into a collection of "propositions." Every time students attended one of his lectures, Smith covered one proposition. At the start of every lecture, he stated the proposition and then, in Millar's words, extemporaneously "endeavored to prove and illustrate" it.[2]

Millar's account of Smith's lectures reveals that Glasgow's moral philosophy students were not simply learning about the relationship between nascent capitalism and political economy. They were also learning how to construct a highly organized knowledge system in realtime over the course of the year that they spent attending Smith's lectures. The fact that Smith's syllabus was organized as a system will come as no surprise to those familiar with his understanding of knowledge formation. He frequently spoke of systems of "political oeconomy," systems of law, and even systems of astronomy. As intimated in Millar's account, Smith's course was organized hierarchically. He divided the syllabus into headings that summarized the parts that were further divided into discourses and then propositions. Students used the headings to follow the lecture as they listened and to organize their notes as they wrote. In following this regime, they were learning to think systematically about the rich range of topics covered by Smith in his lectures.[3]

When considered in light of what we have learned in previous chapters about Scottish school notebooks, Smith's usage of headings to organize his lectures might sound familiar. Like the schoolgirl Margaret Monro, the anonymous Perth Academy students, and a host of other school-aged learners, Scottish university students learned to see headings as categories that organized the knowledge system they were recording in their notebooks. From the standpoint of the students who worked diligently to keep notes, the lectures collectively added up to a gigantic system on paper that they were writing, categorizing, codexing, and sometimes drawing with their own hands.

The process of using a system to structure the learning experiences of university students was set in motion the minute they received the syllabus.

---

2. Millar's reflections are quoted in Stewart (1795). Millar's assessment of Smith's lectures appears in pages xvii–xviii.

3. For an example of how notekeepers used headings to organize the notes they took in Smith's course, see Anderson Bound MS (1750s), 316. The notes are dated, reprinted, and discussed in reference to Adam Smith's ideas in chapter 4, "New light on Adam Smith's Glasgow Lectures on Jurisprudence," in Meek (1977). See also Haakonssen (2016).

It continued as they scrolled quire after quire of rough notes. They then returned to their rented rooms and recopied their notes as a transcript. As they listened, they wrote. As they wrote, they thought about the syllabus. As they were writing, listening, and thinking they were learning the art of systemizing, that is, the ability to conceive labeled packages of information as categories that could be enacted over time and space via the material and visual affordances of what would ultimately become a codex. In short, Smith's students were not simply learning a system. They were also learning to be systemizers.

The systematic element of Smith's course was not unique. It was present in many courses taught in the arts, law, medicine, and divinity faculties of universities in Europe and beyond at this time. In Scotland, elements of it can be seen in the notebooks kept by students who attended schools and academies or were being educated at home. The value assigned to the system as a developmental tool in Scotland and other Reformed educational contexts was in part an extension of the Calvinistic theological tradition that had shaped the learning experiences of students from the sixteenth century onward. Within Scottish universities, following the lead set by Jean Calvin in his *Institutio Christianae Religionis* (1536), regents, and then professors, endeavored to create large theological systems that were formed around a core set of principles and organizational categories based on the Bible.

By the eighteenth century, many Scottish professors took the systematic element of the syllabus for granted. A good example of this viewpoint occurs in the final lecture of the theology course given by Edinburgh's professor of divinity Andrew Hunter during the last decades of the century. He offered the following reflection: "I have now Gentlemen, through the Providence of God finished my Course of Lectures on the System of Theology. I can with pleasure say that the deliberate researches into theological subjects, into which I have been naturally lead, have been serviceable in increasing my acquaintance with the doctrines of natural & revealed Religion, and in establishing me in the belief of their truth & excellence."[4] As we will soon see, long before Hunter had even become a professor, systems thinking had spread to all the faculties in Scotland's universities, where even medical courses used the word "institutes" to describe the ordered syllabus used to organize the lectures.

From the late seventeenth century, Scottish university students used their writing, drawing, and codexing skills to transform the systems offered by professors into material artifacts that acted as informatic devices. Whereas

---

4. Hunter Bound MSS (1779–1807, 6: lecture 121). Hunter used the script in these volumes for the basic outline of the points that he wanted to make in his lectures.

FIGURE 8.1. A transcribed set of student lecture notebooks. Sets were often bound and regularly ran over several handwritten volumes. Joseph Black, *Lectures on Chemistry*, 6 vols. (1778), Paul Panton (notekeeper), Bound MS, CHF QD14.B533 1828. Reproduced with permission of Science History Institute.

schoolchildren usually compiled the notes from all their subjects into a single codex, university students could keep up to ten volumes of notes for one course (figure 8.1). This means that the systematic reach of the data they were organizing operated on a large conceptual and material scale. What made a set of lecture notebooks work as an efficient media technology in this context was the fact that students arranged their notes according to the syllabus in its capacity as an organized system.

The existence of knowledge systems as paper machines at this time was part and parcel of the information structures used by universities and other large institutions.[5] Such structures were underpinned by adaptable notions of order and, as such, were fluid artifacts crafted and recrafted to suit the needs of their designers.[6] Smith recognized this connection, and it contributed to his view that "systems in many respects resemble machines." He observed that "a machine is a little system, created to perform, as well as connect together, in reality, those different movements and effects which the artist has occasion for.

---

5. Higgs (2004), Soll (2009), Cevolini (2016), de Vivo (2007), Friedrich (2018), and Head (2019).
6. The classic cultural history of modern systems is Foucault (1970). Foucault argued that the historical formation of scientific systems was closely tied to the ethos of the institutions in which they were designed. For this reason, systems were fluid structures, which he sometimes called "grids," because they emerged, thrived, and dissipated over time. Subsequent works on the history of scientific systems include Atran (1990), Bowker and Star (1999), Bowker (2005), and Renn (2020).

A system is an imaginary machine invented to connect together in fancy those different movements and effects which are already in reality performed."[7]

Smith's view of a system was shared by many of his contemporaries and resonates with today's notion of systems being, in the words of the historian of science Jürgen Renn, "complex aggregates of knowledge."[8] From a historical perspective, Smith's view that systems were machines makes more sense when viewed in light of the fact that the students who attended his lectures were learning how to use their notebooks as handmade mnemotechnic devices. Based on their experience in the classroom, students were becoming systemizers who were "artists," or more specifically artificers, learning to craft a multivolume paper machine that connected the data and observations presented to them in a university course.

In the larger world of student notekeeping that existed within Scotland's universities, the system presented in Smith's syllabus was modest. Edinburgh's regius professor of natural history John Walker, for instance, offered a system that attempted to comprehensively classify all of the animals, plants, and minerals of the natural world.[9] Many professors enhanced the syllabus with demonstrations, posters, handouts, and slips of paper that contained supplemental material such as diagrams, definitions, tables, bibliographies, and more. Students had to learn how to use these aids alongside the syllabus and the oral component of a lecture when they were scrolling or sketching in the classroom, and when they made their recopied transcript in the evening. At the end of the course, they added paratextual elements as well. Learning to manage all the data spread across such variegated forms of material culture was part of the process through which university students learned to craft a set of notebooks that functioned as a useful paper machine.

The overarching purpose of this chapter is to examine how students learned to transform a set of notebooks from a university course into a paper-based knowledge system. In following this path, I treat student notekeepers as nascent systemizers and notekeeping as a mode of systemizing. My aim is to recast lecture notebooks as artifacts that were crafted through a multisensory process that involved the syllabus, scroll books, transcripts, and other forms of media. To make the process easier to understand, I have visualized it as a flowchart in figure 8.2. The pictograms represent key stages that students used to create a complete set of notebooks. The figure shows that, in addition to being paper machines in their own right, university lecture notebooks were

---

7. Smith (1795, 44).
8. Renn (2020, 65).
9. Walker (1792). Walker's career and system are explained in Eddy (2008).

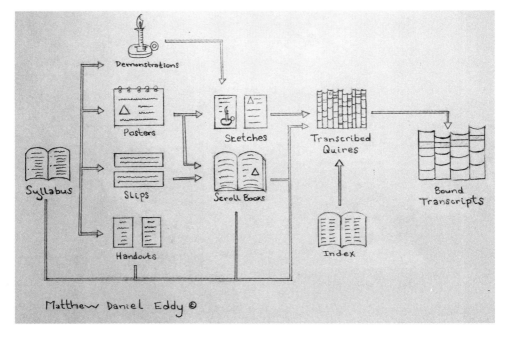

FIGURE 8.2. How students made a bound set of lecture transcripts. © Matthew Daniel Eddy.

created within a much larger paper machine consisting of multiple components. Throughout the chapter, my discussion reveals that, within the manuscript communities surrounding Scotland's universities, notebooks were both knowledge systems and communication systems. Within this context, the syllabus functioned as a multivalent structuring mechanism in the process that students employed to recognize, create, and personalize media technologies.

Lecture Notebooks and Knowledge Formation

Before we launch into how students created the various components of their university notebooks, it would be helpful to mention a few points about the context of lectures in Scottish universities and beyond. Studies on the history of scholarly notekeeping have gained momentum in recent decades.[10] Some of the practices used to assemble university notebooks were elaborate. Students in German universities used a technique called *Schreibechor*, "writ-

---

10. Jackson (2001), Daston (2004), Sherman (2009), Soll (2010), Blair (2010a; 2010b), Grafton (2012), and Yeo (2014).

ing chorus," in which teams attempted to capture every word spoken by a preacher or professor.[11] Students penciled notes in their pockets in Holland, copied Newtonian notations in Cambridge, replicated manuscript notebooks at Harvard, scribbled marginalia in St. Andrews, implemented notetaking procedures in Rome, and employed commonplacing in Paris.[12] Rather than using the same routines, students developed diverse skills in relation to the kinds of information that they needed to learn. Thus, while the notes of some Cambridge students were influenced by what they were taught by elite coaches in small tutorials, the *Schreibechor* technique was developed and honed by students taking notes in sermons.

In Scottish universities, the process surrounding the creation of student notebooks was similarly tailored. They existed within a curriculum in which lectures increasingly constituted the main (and sometimes only) method of formal instruction before students were examined at the end of their degree.[13] Since it was common for adolescents as young as eleven to matriculate in universities, many lecture notebooks were made while students were still learning how to organize knowledge on paper in a manner that extended the notekeeping skills taught in schools. This means that university notebooks, like school notebooks, can be used to unravel the historical emergence of the material and kinesthetic routines that shaped the cognitive development of adolescent notekeepers.

The positive impact of notekeeping was promoted by many parents who sent their sons to Scottish universities. Sometime around 1800, for instance, William Wallace Currie, a student at Edinburgh University, received a useful study tip from his father, the physician and journalist James Currie. Reflecting on his own student years at Edinburgh during the 1770s, James underscored the mnemonic efficacy of keeping lecture notes. He communicated his views on the matter to William with the following comment: "I entirely approve of your taking notes at lectures. . . . He who takes notes is obliged to consider the sense, and to exercise his mind in putting it into new and more concise terms. This is at once to exercise the attention and the judgement, and to impress the memory."[14] In this advice William's father expressed the widely held view that notekeeping helped students learn how to simultaneously balance attention

---

11. The *Schreibechor* process and its application to Kant's lectures is summarized in Blair (2008, 61–62). Blair explains that relay notekeeping most likely emerged from practices used to record sermons.
12. The foregoing examples are discussed, respectively, in Eskildsen (2012), Warwick (2003), Knoles, Kennedy, and Knoles (2003), Nelles (2007), and Blair (2010b; 1989). For further information on the marginalia in the library books of St. Andrews University, see Simpson (2000).
13. Matheson (1979).
14. Currie (1831, 234).

and recollection, two operations of the mind, in a way that made it easier to judge different kinds of information in realtime. I will have more to say about this aspect of notekeeping later in the chapter.

Other parents used their university notekeeping experience to teach their own children. In previous chapters, for instance, we explored the ways in which Amelie Keir worked with her father, the Scottish inventor and industrialist James Keir, to create a codex titled *Dialogues on Chemistry between a Father and His Daughter*. Though the information was given in the form of a conversation, each of the dialogues focused on one core topic and various related subtopics.[15] Keir's father had studied medicine at Edinburgh University and had been exposed to this method of arrangement via the syllabus outlines that professors provided students who were keeping lecture notes.

James Keir found the organization offered by a university outline of headings so effective that he used it to help his own daughter understand the order of the topics in the codex as she was writing them. The narrative of the final sections of *Dialogues* explains how the script was laid out according to a pre-set "outline," that is, according to the core substances and principles of chemistry discussed in each dialogue. Here Keir's understanding of an outline was effectively the same as that used by Edinburgh's professors, and it reveals that some university-educated fathers used the notekeeping skills that they acquired at university to teach their own children.

Historians writing about Scottish universities traditionally have used student lecture notes to investigate the disciplinary content of a course and, as a result, most studies seek to catalogue extant notebooks,[16] or to gain insight into the ideas or personal characteristics of the professors who gave the lectures.[17] This approach sheds light on the facts recorded in lecture notes (or in subsequent publications based on the notes),[18] but it offers limited insight into the thinking skills, particularly those relevant to system building, that students were learning through the acts of writing, drawing, and quiring that were required to craft a lecture notebook, thereby leaving a number of provoking questions about what the process of notekeeping can tell us about knowledge formation.

---

15. Keir Bound MS (1801).

16. Cole (1982), Hatch (1998), Taylor (1978; 1986), Irvine (1981; 1978), Horner (1993). Gregory (1998, 45–46). See also the "second generation" notebooks taken in the medical lectures given by Nathan Smith, an alumnus of the universities of Edinburgh and Glasgow, at Dartmouth College and Yale College. Hayward and Putnam (1998, 339–342).

17. Bevilacqua (1965), Lothian (1963), Mossner (1965), Bryce (1983), Cairns (1988), Lehmann (1970), Meek (1977), Suderman (2001), Wood (2017; 1991), and Macintyre (2012).

18. Abbott (1998; 1989).

Student notebooks are hard to find for many eighteenth-century and early nineteenth-century university contexts. When they do exist, it is often difficult to find more than a few that were made at the same time in the same institution, or which were made by the same student for different subjects. This often forces historians to focus on the notebook of one student. In rare cases historians may have access to the notes of a handful of university students who studied in a single institution with a specific professor or those who studied the same subject. In these cases, it is possible to treat the notes as a collective object of study that can be used to reconstruct and research a community of notekeepers.[19]

Fortunately, the large number of extant notebooks kept by students attending Scottish universities make it possible to reconstruct such a community. In my research, I have discovered hundreds preserved in university collections located across Europe, North America, New Zealand, and Australia. When these specimens are considered in tandem, they amount to possibly the largest extant corpus of pre-nineteenth-century student manuscripts in existence. The Scottish corpus not only has notebooks made by the same student in different courses, it also contains multiple sets of notebooks that were kept at relatively the same time in the same course.

Scottish students kept notes in a university context that, as shown by Rosalind Carr, was immersed in an atmosphere of civic masculinity.[20] There is evidence, however, that indicates female students attended lectures offered by medical professors. The 1783 student list for Professor John Hope's Edinburgh botany lectures, for instance, has an entry for a "Miss Grant of Grant."[21] Likewise, Thomas Young, a leading surgeon and "man-midwife" in Edinburgh in the middle of the century, welcomed women who wished to attend his clinical lectures. They were granted "the Privilege of attending the Labours, paying two Shillings for each Labour, till sufficiently qualified for practice."[22] He also issued certificates to document their attendance.[23] Several decades later, James Hamilton, Edinburgh's professor of midwifery, estimated that over one thousand female midwives had been trained at the university between 1780 and 1815. Outside the university, the extramural natural philosophy lectures of

---

19. For studies that have used student notebooks to reconstruct the skills, routines, and beliefs of scribal communities, see Blair (2008, 59–62); Knoles et al. (2003); and Eddy (2013).

20. Carr (2014, 36–72) and Barclay and Carr (2018).

21. Corrie (2009, 32).

22. Young (1750, 2). Young's career is detailed in Hoolihan (1985).

23. The certificate of lecture attendance that Young wrote for the midwife Margaret Reid is extant and quoted verbatim in Hoolihan (1985, 336, f. 31). Hamilton's estimate and Reid's certificate are discussed in Sanderson (1996, 58–60). For a survey of the university lectures available to female midwives in Scotland, see Cameron (2003).

Glasgow's Professor John Anderson attracted female attendees as well.[24] No lecture notes written by female students attending any of the foregoing lectures seem to have survived.

Within the halls of the universities, in addition to admitting students who attended academies and grammar schools, Scottish professors attracted attendees who had received alternative forms of education at home or as apprentices. The practice existed because it was possible to pay a professor directly for admission or to attend only a few lectures on a topic of interest. The price was usually three guineas per course, but aristocratic students sometimes paid up to five pounds and concessions were offered to sons of the clergy.[25] The flexible matriculation practices of Scotland's universities, combined with the country's general rise of prosperity from the middle of the century onward, ensured that students increasingly came from the rising middle class. Many professors catered to this audience by calibrating the pace and content of their courses in a way that appealed to those coming from different educational backgrounds.[26]

Since most students took several courses per year, keeping notes was often a full-time job. The number of lectures given per week for each course varied across the century. But they occurred often, usually three to five times per week. In Glasgow, for instance, John Anderson gave a prodigious number of lectures in his natural philosophy course. He summed up his format in the published version of his lecture headings. On Mondays, Wednesdays, Fridays, and Saturdays he lectured on the history of physics and on "reasonings" about the "facts of the material world" and their relationship to geometry (plane and solid), arithmetic, and algebra. In these lectures he used "Figures and Machines." His Tuesday and Thursday lectures were "totally different" in that "No Mathematical Reason is employed in them."[27]

When it came to preserving and organizing what they had learned during their university years, students used several kinds of notebooks, so we must

---

24. Sexton (1894, 10). Anderson donated his wealth, books, and instruments to found a pubic institute for education in Glasgow and his will stipulated that it must regularly offer lectures that women could attend. See Anderson (1796). Anderson's commitment to "level the educational playing field" is addressed in Chernock (2009, 48–52, quotation on 50).

25. Professors recorded the names of students in class lists. Some lists state the student's degree and country of origin. For a table of the 650 names that appear in the class lists kept by John Walker during his tenure as professor of natural history in Edinburgh's medical school, see Eddy (2008, 229–250).

26. Mathew (1966). A synthetic study of the curricula offered by the faculties of Scottish universities during the long eighteenth century has yet to be written. Helpful summaries of the curricula are given in Bower (1817; 1822), Wood (1993), Coutts (1909), Rosner (1991), and Eddy (2008).

27. The Anderson quotations in this paragraph are taken from Sexton (1894, 8–9). Sexton states that he found them in the 1786 edition of Anderson's *Institutes of Physics*. The material was not included in the 1786 fourth edition that I consulted. This suggests that the lecture summary appeared in only some of the copies of the 1786 edition, or that Sexton conflated it with the fifth edition (1795).

be clear about which kind we are about to examine. There were at least three notebook genres, each of which required different kinds of notekeeping skills. Here I want to differentiate the purpose and usage of each genre with a view to introducing the kind of notebook that was most prevalent in Scottish university contexts.

Some university students at this time made manuscript textbooks, that is, notebooks written as the professor slowly dictated from notes. The goal was to create a relatively verbatim copy that functioned as the course textbook. In other words, it was a notebook made directly from dictation. It was an ancient practice throughout Europe and it was often a mandatory component of a course. One of the reasons that dictation had developed within Europe was because books were expensive and sometimes scarce in university settings and in places where graduates would eventually find themselves after their studies. With the rise of print culture and the explosion of books from the seventeenth century forward, however, there was less of a material need for students to create manuscript books of this nature.

The practice of dictating notes to students to make manuscript notebooks was called *dyting* in seventeenth-century Scotland.[28] At that time, students were assigned to one professor, a regent, who took them through most of the curriculum over the course of several years. In his history of Glasgow University, James Coutts summed up the process in the following manner: "Teaching was to a large extent done by means of slow lecturing—notes dictated by the regents to students ('dictates')." Ideally, the process was conducted in Latin.[29] Students then used the manuscript to debate a collection of questions, which were called *theses*, set by the professor.[30]

While there is evidence of dictation-style notetaking in Scottish academies and burgh schools and, further afield, in colonial North America, the practice was dying out in Scottish universities by the late seventeenth century.[31] The decline of dictation began in the 1640s, when the General Assembly of the Church of Scotland, which played a regulatory role in educational matters,

28. The University of St. Andrews Special Collections has *dictata* from its own colleges (St. Leonard's College and St. Salvator's College) and from the universities of Edinburgh, Glasgow, and Aberdeen (Marischal College). The manuscripts date from the 1670s to around 1710. See, for example, Vilant Bound MS (1708–1709). Seventeenth-century university *dictata* were largely devoid of figures; however, there are a notable exceptions. See, for example, the likenesses of philosophers drawn in the following notebook taken at St. Salvator's College, St. Andrews University: Glegg Bound MSS (1648–1649).
29. Coutts (1909, 173). Prior to writing his book, Coutts served as the registrar of the university.
30. For a sample of theses and an explanation of the role they played in the curriculum, see Turnbull (2014, ix–xxvi, 43–74).
31. For the Perth Academy, see Anonymous Bound MS (1780s–90s). For North America, see Knoles et al. (2003), Morton (1940), and Norton (1935).

appointed a commission to investigate the state of university teaching. The report of the commission determined that "the *dyting* [dictating] of long notes has, in time past, proved not only a hindrance to the necessary studies, but also to the knowledge of the text itself."[32] Accordingly. the General Assembly recommended that the practice be replaced with a "*viva voce*" summary given as a lecture.

A second kind of student notebook was the commonplace book. It was made while students attended university and it functioned as a storage and organizational device for quotations, bibliographies, and personal observations.[33] Those that feature mainly extracts from books are sometimes called "compilations."[34] Commonplace books were information management tools that preserved facts and ideas relevant to the subjects that students were studying (or wanted to study).[35] They were generally not used, however, by Scottish students to record knowledge gleaned directly from lectures, the main reason being that students simply did not have the time to do so.

A third genre was lecture notebooks. They contained the notes taken by students attending the lectures of a professor or demonstrator. In Scotland, the content of such notebooks was linked to a student's notekeeping abilities and to the manner in which the course was delivered. Scottish lectures often did not focus on one set textbook per se, rather, each course was a commentary on the categories that the professor used to systemize the subject matter under discussion. Scottish student lecture notebooks were based on these commentaries, but they were not an official requirement for a university degree. Nevertheless, students used them when writing essays and to prepare for the oral and written exams that occurred at the end of their studies.[36] Most Scottish university students made this kind of notebook and, consequently, its various forms will be the focus of the following chapters.

Studies on the emergence of modern scholarly culture have emphasized that lecture notes were usually created in a series of stages. Drawing from her extensive work on student manuscript culture, Ann Blair has suggested that

---

32. The commissioner's report is quoted and discussed in Jardine (1825, 17–18). For the full 1642 text, see Universities Commission (1837, 206). The pedagogical problems engendered by dictates are also addressed in Coutts (1909, 179–180).
33. The practices of eighteenth-century commonplacers are addressed in Allan (2010) and the introductory material of Moore (1997) and Byrd (2001).
34. A helpful study of a notebook made by the student compiler Edmund Leigh at Brasenose College, Oxford, is discussed in Serjeantson (2013).
35. For a commonplace book kept by a student attending Edinburgh University during the 1760s, see Blagden Bound MS (1767).
36. Professors set topics and readings for students interested in writing essays. Some professors, Edinburgh's John Stevenson, for instance, kept the best essays and had them bound. McCain (1949, 67).

rough notes and copied lecture notes should be seen as primary and secondary stages of notetaking.[37] Based on his research on Jesuit colleges, Paul Nelles extended these two categories. He offered a six-part editorial process of student notebook production and usage that included stages of "pre-reading" material, lecture notetaking, review exercises, extending lecture notes (with library books), using lecture notes in disputations and compositions, and finally compiling a course summary.[38] The stages offered by Blair and Nelles correspond to the evidence available from the German, French, Spanish, and Italian contexts examined in their work. Using their approach as a guide, it seems that there was a seven-stage process of lecture notebook composition and usage in Scotland, with each stage requiring a different set of skills.

The process of making a lecture notebook in Scotland had evolved into a robust enterprise by the middle of the eighteenth century, rivalling the complexity and organization of similar inscription practices in other European universities where Scottish students sometimes studied.[39] The first stage consisted of reading a professor's syllabus and, sometimes, his publications. The second stage involved taking rough notes in scroll books, a process I called scrolling in my discussion of school notebooks in previous chapters.[40] In the third stage, students copied out their notes, sometimes leaving blank spaces for them to pursue the fourth stage of adding hand-drawn figures.[41]

Once students had created a set of unbound, recopied quires, they were ready to begin the fifth stage of the process, in which they edited their notes through various modes of interleaving, annotating, and paratexting. This stage could sometimes take a bit of time, but it usually led to a sixth stage that involved various modes of binding the notes. Notably, the lecture notebooks created in the third to sixth stages were called *transcripts*. They were circulated in local and global contexts. In this seventh stage of the process, students and professors used and circulated transcripts in ways that, ironically, both elevated and problematized their status as authoritative reference

---

37. Blair (2008, 39–40; 2010a, 303–316) discusses two orders of notekeeping. See also Blair and Stallybrass (2010).

38. Nelles (2007, esp. 85).

39. A frequent destination for Scottish students from the 1690s to the 1730s was Leiden University. Underwood (1977). As the century progressed, Halle, Montpelier, Goettingen, and Uppsala also became popular, especially for medical students.

40. Historians of the scholarly tradition who work on student notebooks sometimes call rough notes "*Mitschriften*" or "primary notes." For the use of "*Mitschriften*" to denote "lecture notes," see Hallett and Majer (2004, xiii–xiv).

41. When referring to recopied student notes, historians of the scholarly tradition sometimes use the collective term "*Reinschriften*" or "secondary notes." More specifically, "*Reinschriften*" is used to denote extended notes or "fair copies." For the former English translation, see Cooper and Mace (2010, 170).

works. A transcript for one course usually contained several handwritten volumes, but some sets ran to more than ten volumes.

The foregoing process enabled students to record as much as they could in lectures. They also consulted with each other outside the lectures to fill any gaps. This means that student notes functioned as a core educational technology that utilized personal and collective forms of observation. The process of writing and rewriting lecture notes involved hours of concentration and served to reinforce the content of the lectures through recursive scribal activities. Such work was demanding, and perhaps this explains why there are remarkably few doodles in the margins.[42]

The Syllabus and Its Organizational Technologies

Professors worked very hard to create a syllabus of lecture headings that was easy to use and ordered the subject hierarchically into a system that students could replicate in their notebooks. As we learned in previous chapters on school notebooks, headings were commonly used by student notekeepers for a wide variety of subjects. I observed that the use of headings as organizational technologies for large knowledge systems reached as far back as ancient orators such as Cicero and Quintilian, who called them *loci*, the Latin term for "places." During the Renaissance, headings became firmly integrated into the efficient notekeeping systems of humanists. By the time that Scottish professors were using them during the long eighteenth century, headings were being employed to organize everything from natural history works such as Carl Linnaeus's textbook *Philosophia Botanica* to manuscript cookbooks.[43]

Within English-speaking countries and territories of the Atlantic world, the headings used to organize a lecture syllabus were collectively called *heads*, *outlines*, or *institutes*. Sometimes professors in North America who had studied in Scotland referred them as *lecture pamphlets*.[44] Scottish booksellers operating in Edinburgh, Glasgow, Aberdeen, and St. Andrews sold printed copies of lecture headings for a few shillings. In 1793, for instance, the Edinburgh bookseller Peter Hill's catalogue listed the lecture headings of Andrew Duncan, professor of medicine at Edinburgh University. Hill offered one copy

---

42. This stands in contrast to other notekeeping contexts where marginalia were more common. See Sherman (2009), Jackson (2002), and Simpson (2000).

43. For the role played by headings in Carl Linnaeus's published books and notebooks, see Eddy (2010a). For the emergence of headings as organizational technologies during the modern period, see Ong (2004), Yeo (2001), Allan (2010), and Dacome (2004).

44. Carlson (1981, x). The place of printed lecture headings within Scottish publishing culture is address by Sher (2008, 105–106).

of Duncan's "heads of lectures on medical jurisprudence" and two copies of Duncan's "heads of lectures on the theory and practice of medicine." The price for each one was 3s 6d.[45] University libraries sometimes kept copies of printed lecture headings for students who could not afford to buy them. Students using Glasgow's divinity library, for instance, could consult "Heads of a course of Lectures on the Study of History" that William Wright, professor of history, used to organize his course.[46]

Not all professors opted to print their syllabus of lecture headings. Some, William Cullen for instance, circulated handwritten headings for students to copy.[47] A number of the outlines that Cullen distributed during his time at Glasgow University still exist. They reveal that, like those of his contemporaries, the outlines consisted of headings that summarized the main topics.[48] Lecture headings represented a monumental amount of writing and rewriting for the professors who created them and for the students that used them to keep notes. The personal papers of professors also reveal that they kept notes on how to improve their headings in ways that enhanced the learning experience of their students.[49] Francis Home, a professor in Edinburgh's medical school, was so proud of his 1770 *materia medica* syllabus that he asked the artist David Allan to include it in his portrait. It appears as "Meth. M. M." alongside all of his other publications on the second shelf of his bookcase (figure 8.3).[50]

To make the syllabus easier to follow, some professors used their lecture headings to create a branching diagram, a *clavis*, that resembled a tree.[51] This form of representation was a powerful ordering device in manuscript culture.[52] Its importance in Scotland as a visual ordering tool stretched as far back as the early seventeenth century when the works of the French logician Petrus Ramus were introduced into the curriculum of Scottish universities.[53] Elsewhere the *clavis* played a central mnemotechnic role in continental universities where Scottish students studied. It was used by influential professors

45. Hill (1793, 14, nos. 5689, 5690, and 5691). To see how a student used the headings, see Duncan Bound MS (1792–1793).
46. Glasgow University (1790, 54). See also Wright (1767).
47. Coutts (1909, 488).
48. Cullen, *"A Chemical Examination of Common Simple Stones & Earths*, Loose-Leaf MS (n.d.).
49. Corrie (2009, 34).
50. The printed syllabus of headings appeared as Home (1770). The headings are organized hierarchically according to the Linnaean system of class, order, and genus. The headings of Home's 1758 syllabus are organized according to the traditional hierarchy of book, part, and section.
51. Historians have given many names to the branching *clavis*, including branching diagram, Ramistic tree diagram, brackets, braces, diagram, and dendrogram.
52. See Ann Blair's informative account of the *clavis* in Blair (2010b), 144–158.
53. Reid (2011, 201–232). The eighteenth-century importance of the *clavis* in Scottish universities is raised in the appendix of Bentham (1816). The spread of Ramism in Scotland and continental Europe is addressed in McOrmish (2018) and Hotson (2007), respectively.

FIGURE 8.3. Professor Francis Home with his personal library. *Francis Home* by David Allan (EU0445) © University of Edinburgh Art Collection, late eighteenth century.

such as Georg Wolfgang Wedel at the University of Jena, Hermann Boerhaave at Leiden University and Carl Linnaeus at the University of Uppsala.[54] Scottish professors sometimes printed their branching diagrams of lecture headings separately or included them at the start of a course's printed lecture headings.[55] Others designed handmade posters or handheld exemplars for students to copy (figure 8.4).[56]

Students used the headings of the syllabus in the classroom to follow the

---

54. Linnaeus learned to make a *clavis* from 1725 to 1727 when he attended the Växjö Gymnasium. Linnaeus Bound MS (1725–1727). The most elaborate dichotomous table in his notebook occurs on f. 160. This visualization was a reordered table extracted from Wedel (1677).

55. For examples of a printed *clavis*, see John Walker (1781), which summarized the classes that he used in his Edinburgh mineralogy lectures from the 1780s to the 1790s. Several midcentury *clavis* examples are housed in the John Hope Collection in the RBGE. Their titles are: *Methodus D. F. B. Sauvages, ex foliis clavis* [n.d.], *Methodus A. Caesalpini* [n. d.], *Methodus A. Q. Rivini* [n.d.], *Methodus Calycina C. Linnaei* [n.d.], *Clavis Classium Raii* [n.d.], *Clavis Classium A. Von Royen* [n.d.], and *Methodus D. I. P. de Tournefort* [n.d.].

56. William Cullen arranged the headings of his early mineralogy as a *clavis*. See Cullen, loose-leaf MS [n.d.]. Students copied branching diagrams of lecture headings into their notebooks as well. See Hope Bound MS (1780, 156r, 157r); Cullen Bound MS (c. 1765, 2: ff. 30, 34, 50, 60); Cullen Bound MS (1765–1766, ff. 138, 140); and Black Bound MS (1776, vol. 2). A helpful account of the presence of the *clavis* in modern classification texts is addressed in Scharf (2007).

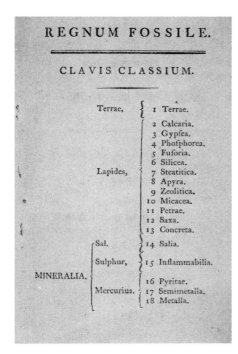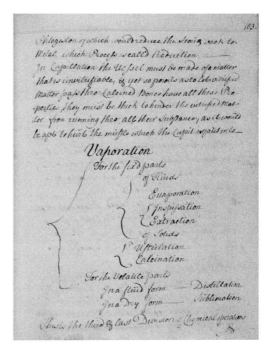

FIGURE 8.4. Branching diagrams of lecture headings. Left: The main classes of Professor John Walker's printed mineralogy syllabus, which doubled as the headings of the mineralogy lectures he gave in his Edinburgh natural history course. John Walker, *Schediasma Fosilium* (Edinburgh: 1781). Personal collection of the author. Right: A student's notebook page featuring a branching diagram with the subheadings of William Cullen's lecture on the chemical operation of "Vaporation." It was most likely based on a poster that Cullen hung in the classroom or a handout that he passed around to students. William Cullen, *Chemical Lectures, vol. 1* (1760) (anonymous notekeeper), Wellcome MS 1918, Bound MS, f. 103. Reproduced with permission of Wellcome Trust Library, London.

topics covered in each lecture and when they recopied their notes in the evenings, thereby strengthening the cognitive processes that united categorizing with notekeeping. Since there was increasing diversity in the kinds of students taking university courses, professors endeavored to capture the attention of their students and to present information in a memorable manner. They also endeavored to include new, original information and to offer students accessible and well-structured lectures that were conducive to keeping notes. Here the syllabus proved to be a central organizational artifact in that it functioned jointly as a systematic arrangement of the subject *and* a notekeeping aid.

The elegance and utility of the syllabus created by Aberdeen's Professor Robert Eden Scott (1769–1811), who taught a variety of courses during his tenure, was noted in the 1811 funeral oration that his students subsequently published: "In his academical lectures he combined original observation with judicious selection, and accurate scientific arrangement with that simplicity and perspicuity and neatness of style which is the best and most captivating vehicle of instruction."[57] With these words his students were saying that the key to a well-designed syllabus was its "scientific arrangement," that is to say, the organization of its headings as a knowledge system.

A professor's syllabus often corresponded to a stack of loose-leaf lecture slips, each of which addressed a specific heading that he used to give his lectures. Professors read out the heading from the syllabus and then used the slips as a cue for the information that they wished to communicate. The result was an environment in which professors could lecture extemporaneously from their slips. The "slipification" of the lecture was a mnemonic practice that professors had developed during the late seventeenth century when the General Assembly banned them from reading their lectures from a fixed manuscript.[58] As the personal lecture notes of professors reveal, the use of headings written on loose pieces of paper allowed them to constantly rework the content and order of their lectures, thereby treating the course as an adaptable, evolving knowledge system.

Perhaps the largest extant collection of professorial lecture slips was written by Adam Ferguson, Edinburgh's professor of moral philosophy. Today they are housed in the Special Collections of Edinburgh University. He wrote his "heads or short notes" on slips and then replaced and reordered them on an annual basis in a way that transformed them into a paper machine of moveable units of knowledge. The combinatorial method allowed him to constantly refine his system based on his evolving interests and the needs of his students. This means that the order and content of the notebooks kept by students attending his lectures were different from year to year, reflecting his changing views of the subject.[59]

An informative illustration of how professors used paper slips as they lectured was captured by the artist James Kay in a 1787 etching of Joseph Black, Edinburgh's professor of chemistry (figure 8.5). Kay depicts Black standing behind a lecture table. To his right and left are the teaching aids that he used to dazzle and inspire his students. Strewn across the middle of the table are the slips that he used during his lectures. The tableau presents Black with a slip

---

57. The oration was republished in Ogilvie (1891, 640–642, quotation on 641).
58. My use of the term "slipification" here is taken from Burke (2012). Burke used the term to refer to the important role played by paper slips as paper technologies during the modern era.
59. The loose-leaf papers and slips used by Adam Ferguson are preserved as Ferguson Bound MSS (1776–1785). A summary of Ferguson's method of lecturing from "heads or short notes" is given by his editor in Ferguson (1792, v–vii). The historical context of the topics covered in Ferguson's lectures is addressed in Smith (2018). For printed examples of Ferguson's lecture headings, see Ferguson (1766; 1769; 1792).

FIGURE 8.5. Professor Joseph Black holding lecture slips. *Joseph Black*, etching by James Kay (1787), reproduced with permission of Wellcome Trust London.

in his hand, gazing across the table into his classroom, perhaps listening patiently to the question of an inquisitive student.[60]

Further insight to the conceptual fluidity illustrated by lecture slips is offered by Edinburgh's Professor Alexander Hamilton in his *Outlines of the Theory and Practice of Midwifery*, which was a revised extended version of his lecture syllabus published near the end of his career. He recalled that the headings for his first midwifery syllabus were "begun and carried on in a hurry, having been actually written and published in little more than two months."[61] Like most such publications, the subsequent editions of his lecture syllabus eliminated and added new headings as required. Some professors intentionally pointed out this changeable nature of the syllabus. Early in his teaching career at Edinburgh's medical school, Professor Andrew Duncan summed up this situation to his students in a lecture with the following observation:

---

60. Joseph Black's use of slips was most likely inspired by the "short notes" used skillfully by William Cullen in his Glasgow medical lectures during the 1740s. Coutts (1909, 488). For an indication of what was on Black's slips during the 1780s, see Black Bound MS (1787), McCartny (notekeeper).
61. Hamilton (1784, 419–420).

"When, however, you attend separate teachers, who take different views of the same subject, they will not only be led to arrange facts in a different manner, but to enlarge on different particulars."[62]

Rather than using slips as lecture prompts, other professors used headings and epitomes written on small bifolia. Such was the case for William Cullen, who shuffled topics from year to year in a way that fit with his research and the interests of students.[63] Cullen's extant prompts reveal that he used full sentences, which he expanded as he read.[64] They show that when treating a difficult subject such as the mind and body problem, Cullen worked very hard to maintain clarity, an attribute that helped his students more easily understand and record his lectures. The theology lecture notes read by Edinburgh professor Andrew Hunter followed a similar format. He was particularly organized in that he recorded the composition and delivery dates on the notes he read for each lecture.[65]

By using a syllabus and its related slips and bifolia as adaptable media technologies that doubled as thinking tools, professors communicated the fact that knowledge systems, represented in this case as the ordered lecture headings, were fluid artifacts that could be ordered, reordered, expanded, and contracted based on new evidence or the needs of the orderer. A lecture based on slips or loose bifolia also ensured that the information given under each heading could vary from year to year, leading some students to enhance their notes by taking the course more than once. When students repeatedly consulted the notebooks kept by their peers and others who had attended a given course in recent years, they became used to the idea that the syllabus as a system changed over time.

The mnemonic utility of lecture slips was powerfully demonstrated to students who attended the courses of professors who had used the slips to memorize most of the information that was included in their lectures. From the early decades of the century professors such as Edinburgh medical school's Alexander Monro Primus knew some of their lecture material so well that they only used a list of words or phrases as headings that functioned as mnemonic cues. This method allowed Monro Primus to deliver his lectures in a person-

---

62. Duncan (1776, 13).
63. Wolf (2015, ix, 168), discusses Cullen's lecture prompts housed at the Royal College of Physicians of Edinburgh.
64. Wolf (2015, 208).
65. Hunter Bound MSS (1779–1807). The notes are now bound, but they were in a loose-leaf format when he read them.

able style that appealed to students and most likely set them at ease when keeping notes.[66]

The mnemonic and logical expediency of notekeeping for both students and faculty led some professors to explicitly underscore the value of inscription as a reason-building exercise. Dugald Stewart used his lectures on moral philosophy to highlight the link between scribal tactility, visuality, and learning. He took care to point out the benefits of the preparatory books made by the ancient Greeks and to underscore the impact that writing and drawing had upon cognitive development.[67] He even went so far as to intimate that lecture "heads or outlines," by which he meant the syllabus, assisted students "in tracing the trains of thought."[68] He and others encouraged notekeeping because they firmly believed in the fundamental mnemonic power imparted through creating an ordered set of notes.

John Anderson, Glasgow's professor of natural philosophy, explicitly flagged the cognitive benefits of using a printed syllabus of lecture headings as a guide to notetaking: "In this manner [of using the syllabus to take notes], it seems proper to guard against the inaccuracies into which young Students are apt to fall, while the publication of more than such Outlines might lead them to lay aside the custom of taking notes, a custom, by which their attention and ingenuity are constantly exercised, and the Lectures and Experiments become, as it were, their own."[69] Even if professors did not offer explicit notekeeping instructions, they worked very hard to provide lecture headings that were designed to help students strategically listen to the oral presentation of an ordered knowledge system in a way that enabled them to successfully filter, scroll, and later remember what they had heard in the classroom. It was this skill of using a syllabus of lecture headings as a strategic listening and writing tool that prepared students for managing the large paper systems that imperial institutions such as the Church, the East India Company, and even Parliament employed to circulate data.

---

66. The memorization of lectures was difficult for some new teachers. The young John Monro, the son of Alexander Monro Secundus, once became so nervous in a lecture that he forgot the words he had memorized. He was doubly unlucky because he had left his notes at home. But luck soon returned when he realized that the subject of the lecture was a set of anatomy specimens that he himself had prepared. Using the specimens as a guide, he was able to give a successful extemporaneous lecture that was well received. Struthers (1867, 24).

67. Stewart (1792, 53–55, 441–444).

68. Stewart (1811, 329). For a sample of the syllabus of lecture headings used by Stewart (1793), see Stewart (1793). George Jardine, Glasgow's professor of logic, also commented on the use of syllabus lecture headings. See Jardine (1825, 278–279).

69. Anderson (1777, 3).

The multisensory facet of using the syllabus to keep notes partially explains why many professors repeatedly emphasized the importance of attending lectures. When Professor John Robison, for instance, published a textbook based on his Edinburgh natural philosophy lectures, he advised his readers to remember that "the student must be mindful that this book will not supersede the necessity of carefully attending the lecture. Many things, illustrative and interesting, will be heard in the class."[70] For Robison, the "many things" included extemporaneous commentaries on the visual aspects of diagrams and discussions of the material properties of water, metals, and mechanical engines.

Former students also noted the importance of the oral component of lectures. When reflecting on his undergraduate days at Glasgow University during the 1780s, the judge and journalist Francis Jeffrey, Lord Jeffrey, offered a revealing observation on this point in reference to regius professor of civil law John Millar. As we learned at the start of this chapter, Millar had been inspired in his student years by Adam Smith's early lectures. Jeffrey reflected that the books published by Millar, many of which were based on his lectures, revealed "nothing of that magical vivacity which made his [Millar's] conversations and lectures still more full of delight than of instruction."[71] Rather than reading solely Millar's books, his students learned to remember and preserve his ideas by keeping notes in his lectures. This was the case for most students attending Scottish universities. By following this approach to notekeeping, they learned to master a set of syllabus-oriented listening and scrolling skills that they could employ once they had finished their studies.

Dugald Stewart's positive view of the interface between orality, systematicity, and notekeeping is helpful here as well. The preface of his 1793 printed moral philosophy syllabus explains that he designed his lecture headings specifically as memory devices for students to use in realtime as he lectured. The goal, in his words, was "to state, under each head, a few fundamental principles, which I was either anxious to impress on the memory of my hearers; or which I thought might be useful to them, by relieving their attention during the discussion of a long or difficult argument."[72] As can be seen in his syllabus, the "principles" were effectively the lecture headings, all of which were

---

70. Robison (1804, v–vi).

71. Young (2016, 122). The lecture style of other professors inspired students as well. When reflecting on the 1790s Greek lectures of Andrew Dalzel, Edinburgh's professor of Greek, the future journalist and jurist Henry Cockburn, Lord Cockburn, noted that the delivery evoked "delicious dreams of virtue and poetry." Dalzel (1861, 114).

72. Stewart (1793, vi). Other professors made brief comments about the utility or purpose of their syllabus of lecture headings. See Bruce (1780, vii).

CONTENTS.

CHAPTER II.

Of the various Branches of our Duty.

CONTENTS.

PART II.

OF THE ACTIVE AND OF THE MORAL POWERS OF MAN.

CHAPTER I.

Classification and Analysis of our Active and Moral Powers.

FIGURE 8.6. Table of contents for the nineteenth-century published edition of Dugald Stewart's moral philosophy lecture headings. Dugald Stewart, *Outlines of Moral Philosophy*, 9th ed., James McCosh (ed.) (London: Low, Low and Searle, 1876). Internet Archive, original from the John M. Kelly Library, University of St. Michael's College, University of Toronto.

arranged hierarchically to form a coherent system of chapters, parts, and sections.

The system of headings worked so well for Stewart's students that he and his subsequent editors used it in many printed editions of the syllabus. Indeed, it was used as late as 1876 in the edition that was edited by James McCosh, the president of Princeton University (figure 8.6).[73] Likewise, as late as the 1830s the students and professors in the divinity faculty of Edinburgh University still found Stewart's syllabus so useful that the theology library took the unusual step of keeping three copies in the stacks for consultation.[74]

Stewart's ordered syllabus is a good example of the positive commitment

73. Stewart (1876).
74. Cunningham (1829, 230).

that most Scottish professors had toward the cognitive utility of encapsulating, organizing, and rewriting headings and their corresponding data in real-time. The structure of the syllabus also explains why students did not see notekeeping as a mindless act of replication. Some students even felt disappointed when professors offered printed copies of the syllabus because it meant that they did not have to copy them out by hand. Alexander Coventrie, who came from the Clyde Valley to study medicine at Edinburgh University during the 1780s, for instance, felt disappointed when a professor printed his lecture headings or "any part of his discourse" because it "deprived" students of hearing "the charm of novelty" and rendered them "less attentive."[75]

Coventrie's view was anchored in the rich variety of notekeeping skills that students practiced in Scottish universities. In the remaining sections of this chapter, I will explore these skills by first explaining how students learned to keep scroll books during a lecture. I then move on to examine the transcribing skills they employed to transform their scroll books into a transcript, by which I mean a recopied, expanded, and bound set of notebooks. I close with a discussion of how the creation of scroll books and transcripts was an exercise that reinforced systemization skills in a way that harmonized with the cognitive model promoted by eighteenth-century developmental psychology.

Scroll Books and the Strategies of Realtime Learning

Once university students acquired their syllabus and writing materials, they were ready to go to a lecture and start keeping notes in their scroll books. Prior to the twenty-first century, historians often treated rough notebooks as a manuscript genre that paled in comparison to complete notebooks or even printed texts. Yet scrolling was an essential skill that helped students manage the large systems of knowledge being presented in lectures.

Some professors offered advice on what to scroll. Professor John Robison not only encouraged his Edinburgh natural philosophy students to keep notes, but he observed that the human mind could not possibly remember all the formulae covered in any one lecture. To overcome this problem, he advised his students to inscribe "formulae, or other symbols of mathematical reasoning, as occur in the lecture. These will frequently give a compendious expression of a process of reasoning which he may otherwise find very difficult to remember with distinctness." Here it can be seen that Robison held

---

75. Coventrie (1904, 108).

that scrolling, even when it involved numbers and symbols, was an essential part of the "process of reasoning."[76]

Though professors provided detailed syllabi and underscored the importance of scrolling, they often left it up to the students to work out their own techniques. Some students extended notekeeping skills that they had gained through extracurricular forms of employment. The young James Finlayson, Edinburgh's future professor of logic, for example, spent some of his time as a student during the 1780s serving as an amanuensis to Glasgow's professor John Anderson.[77] Others, as shown in previous chapters, had been exposed to sophisticated writing and drawing skills at school or at home prior to their matriculation into the university.

Scrolling notes in an eighteenth-century Scottish university was not for the faint hearted. The need for more teaching space from the middle of the century onward ensured that courses were overcrowded.[78] The rooms, halls, and theaters were fitted with rows of tightly aligned benches, each of which faced an angled board that functioned as a desk for students keeping notes. Based on rare sketches housed by the University of Aberdeen Special Collections, we can see that the Marischal College classroom used by Professor Patrick Copland to teach mathematics and natural philosophy followed this arrangement, with an increasing number of benches being added as student numbers rose over the years (figure 8.7).[79] By the early nineteenth century there was barely space left in his classroom for the lectern.[80]

Even the best notekeepers would have found their scrolling and sketching skills challenged in a classroom such as Copland's, especially when the entire bench was full. It was a similar situation for the overcrowded lectures given by Edinburgh's professor of anatomy and surgery Alexander Monro Secundus. Close quarters of this nature undoubtedly fostered distractions such as the accidental elbow jabbing that occurred when left- and right-handed students sat next to each other. Comfort breaks for students sitting in the middle of the bench were of course out of the question. The sounds of nibs scraping and the sight of quills or pens constantly bobbing across the page or into an ink pot would have been a constant challenge for distractible students. But there

---

76. Robison (1804, vi).
77. For Finlayson's role as an amanuensis, see Thomson (1853, 318).
78. Haynes and Fenton (2017, 11–55).
79. Anonymous, *Pencil Sketches of Professor Patrick Copland's Marischal College Classroom*, University of Aberdeen Special Collections, MS M363.
80. Another layout of a circa 1830 Marischal College classroom is reproduced in Wood (1993, 37). For more on Copland's career as a professor, see John S. Reid (1985; 1982; 1990).

FIGURE 8.7. *Pencil Sketches of Professor Patrick Copland's Marischal College Classroom,* University of Aberdeen Special Collections, MS M363. The top sketch depicts Copland's late eighteenth-century classroom. Note the poster on the wall near the window and the portable blackboard at the front of the room. By the first decade of the nineteenth century, as shown in the bottom sketch, more benches had been added. There also are more posters on the wall and the blackboard is larger.

were also advantages to such close quarters, one being the fact that students could see the notes of those around them, which made it possible to gain observations or facts they had missed. Though students could follow the lecture by consulting the syllabus, scrolling was a difficult task because professors barraged them with countless examples, definitions, illustrations, observations, and citations. Add to this other distractions such as running out of paper and ink, or the fact that many professors used a variety of printed and manuscript teaching aids like lecture headings, maps, and handouts.

Students scrolling in medicine, natural history, and natural philosophy lectures had to contend with diagrams, tables, and figures that professors offered on posters and chalkboards.[81] The drawings of Copland's classroom depicted in figure 8.7 offer a glimpse of how professors displayed such visualizations at the end of the eighteenth century. Both drawings feature a lectern as well as a chalkboard on a stand next to the window. Copland used the chalkboards to inscribe diagrams and equations as students watched.

Although many students diligently scrolled all the lectures given in a course, extant copies are fragmentary, usually consisting of a handful of notes or drawings taken in only a few lectures.[82] Scrolled notes were sometimes thrown away after they were recopied. Alternatively, they functioned as scrap paper or kindling. Scrolled notes on occasion were included accidentally in collections of loose personal papers that were bound either when they were donated to a library or after the notekeeper died.[83] They were also preserved as makeshift bookmarkers in sets of bound lecture transcripts. An example of this practice can be seen in a quire of scrolled notes preserved in the recopied notebook kept by the student George Sligo in the law lectures of Edinburgh's Professor David Hume (1757–1838). Though the content of the recopied text is based on the rough notes, Sligo extended or trimmed the original content in the copying process.[84]

Scroll books oftentimes exhibit a number of tell-tale characteristics. The script usually was laid out as one large narrative column that took up the entire page. Some students, such as Sligo, included freestanding headings to help them find information at a later date; however, the speed at which they needed to scroll often prevented this graphic luxury.[85] Scrollers tended not

---

81. For botanical teaching figures, see Noltie (2011a).

82. The different paper sizes and different penmanship styles (e.g., ranging from semi-neat to scrawled) sometimes make it difficult to determine whether they are a rough or recopied set. A good example of a "borderline" set is Tytler Bound MS (1800–1801). The ephemeral nature of rough notes at this time is detailed in Stallybrass et al. (2004). Historians of science have also addressed these issues in reference to Robert Boyle (Yeo [2014]) and Carl Linnaeus (Müller-Wille and Charmantier [2012a]).

83. A good example of different kinds of notes being bound in one volume is Hill Bound MS (1770s). The volume begins with "Lectures on Chymistry by Joseph Black MD Edinr. Novr. 1771" and then moves on to "Lectures on Humanity, Delivered in Edinburgh University." See also the different paper notebooks of John Borthwick bound up as Hill Bound MS (1802–03).

84. Hume Bound MSS (1816–1817). George Sligo's quire of rough notes is tucked into volume 4. The rough set is very hard to read and carries the title "Conveyancing," and corresponds to the subject matter of ff. 307–311.

85. Sligo structured his rough notes with numeric headings set to the left margin, and centred headings that indicated the general topic of the lecture. See his rough notes which are tucked in of Hume Bound MS (1816–1817, vol. 5).

to differentiate key words in the narrative with underscoring, capitalization, italicization, or majusculization. Likewise, scrollers created hardly any paratextual material like a full title page, index, or table of contents. The sentences inside the narrative column usually were written in relatively straight lines and, though pressed close together, the spaces between the sentences varied slightly from line to line, indicating that many students did not draw a graphite grid to guide their writing. Nevertheless, as can be seen in the scrolled notes of an anonymous student who attended Professor James Millar's Glasgow mathematics course during the 1810s, rushed cursive sentences occurred even when scrollers had the foresight to draw ruled lines in graphite prior to the lectures.[86]

An excellent collection of extant scroll books is housed in the Wellcome Trust Library in London. They were kept by the eighteen-year-old Sir Charles Blagden in the Edinburgh chemistry lectures given by Joseph Black during the 1760s (figure 8.8).[87] He scrolled his notes in octavo-sized paper books. Like most notes of this nature, his inscriptions are small and cramped. The handwriting is rushed and consequently hard to read in places (figure 8.9). The pages are crammed with as much information as possible. Black's frequent use of experiments and attention to detail led many of his students to scroll in this manner. Despite scrolling quickly, Blagden still managed to fit in a few freestanding headings that stated the lecture numbers. Other students taking rough notes in Black's lectures managed to fit in centered headings as well.[88] Although rough notes of this nature take some skill to decipher, they are of great value because they shed light on how students learned to judge information quickly, in realtime.

Keeping a scroll book, the result of which was sometimes called "short notes," was a prudent option for students who attended courses run by professors who packed a large amount of information into every lecture. Black's chemistry lectures were, in the words of his former student the American senator Samuel Latham Mitchell, "greatly enhanced by the experimental illustrations with which he accompanied them. Those experiments made so considerable a part of his course." Consequently, Mitchell spent so much time observing the lectures that taking brief notes was the only option for him. He

---

86. Millar Bound MS (1815–1818), anonymous notekeeper. The student's rough notes begin at f. 38.
87. Black Bound MS (1766–1767). A good set of rough and recopied notes also exists for the Edinburgh chemistry lectures of William Cullen, who was Black's mentor. Cullen Bound MS (1757) and Cullen Bound MS (1757/8).
88. See the centred headings throughout chemistry notes in Hill Bound MS (1770s).

FIGURE 8.8. A set of scrolled notebooks for one lecture course. Joseph Black, *Lectures on Chemistry*, vols. 1–9 (1766–1767), Charles Blagden (notetaker), Bound MSS, MSS 1219–MSS 1227. Reproduced with permission of the Wellcome Library, London. Top: View from above. Bottom: View from the side.

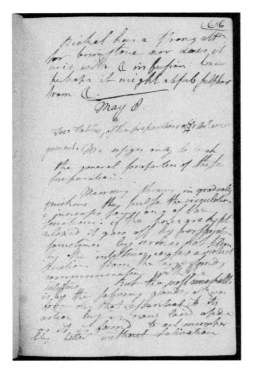

FIGURE 8.9. Scrolled and transcribed lecture notebook pages. Left: A page from Charles Blagden's scroll book. Joseph Black, *Notes of Dr. Black's Lectures*, vol. 9, 1766–1767, Charles Blagden (notetaker), MS 1227, f. 606. Reproduced with permission of the Wellcome Library, London. Right: A page from Charles Blagden's transcript of Joseph Black's 1766–1767 lectures. Sir Charles Blagden Papers, OSB MSS 51, box 8, "Black's Lectures," f. 33. Reproduced with permission from the Beinecke Rare Book and Manuscript Library, Yale University.

summarized the effect that this practice had on his scrolling in the following manner: "I remember very well the amount of my notes, and I attended him two sessions, was frequently very small when reduced to writing."[89]

Short notes were sometimes recommended by parents as well. Drawing from his own experience as a student in Edinburgh's medical school during the 1760s, Benjamin Rush, a leading medical professor at the University of Pennsylvania and signatory of the American Declaration of Independence, offered the following advice when his son, James, set off for his university studies: "Take short notes of all the lectures you hear. . . . Put indexes to all your manuscripts."[90] Likewise when reflecting on the time that his grandfather,

---

89. Mitchell (1976).
90. Rush (1948, 281). James Rush followed his father's advice. See Rush Bound MS (1809).

Levi Myers, spent at Edinburgh medical school, Robert Poole Myers took care to mention that scrolling required his grandfather to practice "habits of close attention."[91] Other erstwhile students and their biographers made similar points about the notekeeping practices used in Scottish universities.

The foregoing reflections of Benjamin Rush and Levi Myers reveal that scrolling was understood to be a dynamic activity that enriched the powers of the mind in ways that strengthened attention, memory, and judgment. These operations were widely seen as attributes of a mind that possessed reason. Attention, along with perception, filtered out superfluous sensations and thoughts. Recollection retrieved facts and observations from the storehouse of memory and was oftentimes dependent upon printed and written devices. Judgment was the ability to place all the operations of the mind in the service of making sound decisions in realtime; it was the mode through which understanding was achieved.[92]

Professors emphasized the mnemonic advantages of scrolling epitomes of lecture topics because it exercised the mental operations of attention and abstraction. A key assumption that often accompanied professorial discussions of these operations was the fact that a person's attention was limited. The more time and attention that one devoted to writing, the more one would become distracted and, correspondingly, the less one would remember from a conversation or lecture.

Aberdeen's Professor Robert Eden Scott, for instance, summed up the fine line between attention and distraction that occurred during the act of writing. In his *Elements of Intellectual Philosophy* (1805), he reasoned that: "To write a great deal, cannot be highly useful to Memory; for the attention is but too apt to be diverted from the matter itself, to the mere manual operation."[93] This view, which was shared by other professors, was made in a context where, as we will soon see, students extended their short notes into a fuller version in the evening.

Scott designed his *Elements* to be a textbook for the students who attended his philosophy course. The inverse relationship that he saw between remembering and writing was directly relevant to the skill of scrolling because he believed the more a student wrote during a lecture, the less he would remember. It was for this reason that short notes worked best for students keep-

---

91. Levi Myers was from South Carolina and was one of the first Jewish students to study medicine in Edinburgh. For the notes he took in Edinburgh, see Myers (1903, 7).
92. The centrality of attention, perception and judgement within the philosophy of mind in Scotland is addressed throughout McCosh (1875).
93. Eden (1805, 299).

ing scroll books. Scott suggested writing "short passages" with a view to using them as mnemonic cues to which one could "afterwards conveniently refer." Scott's comments were directly applicable to the processes of thinking and writing that students sought to improve while keeping notes in a classroom, in their rented rooms, or, later in their lives, at home in their study.

The foregoing cognitive advantages of scrolling short notes partially explains why so few lecture notes were written in shorthand. Even for students who did not learn stenography in school, it was a skill that could be acquired through books published by stenographic masters such as William Williamson and Reverend William Graham. Graham's *Stenography* was even dedicated to university "students of divinity, or law, and of physic."[94] But as we just learned, scrolling involved a student who was thinking and learning while writing and listening. Rather than valuing a verbatim set of notes, students valued the capabilities that could be acquired through the act of scrolling. From their perspective, notekeeping was a formative exercise, and the goal was not simply to replicate precisely what a professor had said. In other words, students were not passive conduits through which the words of a professor automatically flowed onto a page.

The cognitive limitations of shorthand were succinctly summarized in a letter that the aforementioned Edinburgh student William Currie received from his father. Drawing from his own experience as a student at Edinburgh University several decades earlier, William's father observed: "I do not know that I should wish you to take every thing *verbatim*, even if you possessed a shorthand that would admit of it . . . in taking the whole *verbatim*, one exercises the attention on the expression principally, and in recording the individual words the sense probably escapes you, or makes so transient an impression that it is speedily forgotten."[95] Put more simply, William's father was saying that scrolling was an interactive mode of observation, a way of judging and remembering what an expert was saying in realtime. It was for this reason that his father advised him not to use shorthand and went so far as to offer the following recollection based on his own time as an Edinburgh medical student: "We used to observe that those who had the professors' lectures most fully in their short-hand, had least of them in their memory."[96]

Using shorthand not only worked against the goal of becoming a critical listener who was able to filter the lecture for key pieces of information. It also

---

94. Williamson (1775); the title page states the Williamson was "Teacher of that Art in London, Late of Edinburgh." Graham (1787) and Mavor (1792).
95. Currie (1831, 234).
96. Currie (1831, 235).

worked against the role that scroll books played as mnemonic cues when students expanded their notes into a transcript after the lecture. To understand this function of scroll books, we must now turn to the different techniques that students elected to use when writing their short notes. As explained in the diary of Sylas Neville, a medical student who attended the Edinburgh University during the 1770s, there were several scrolling strategies. They were not mutually exclusive, and students combined them as they saw fit.

The first approach, used by Neville, was to "take a good deal of the principal observations, but not near the whole lecture."[97] Such notekeeping required an attuned ear, one that could use the syllabus alongside the oral cues of the professor to determine the key points and then to epitomize them on paper. The second approach, used by Neville's friend and fellow medical student Richard Dennison, was to identify key points in the lecture by using prior knowledge of the subject. This could be gained through reading books from the library or that were lent to them by professors. The library register of Edinburgh University shows that a number of students checked out books on behalf of professors as well. This indicates that students served as research assistants, a situation that no doubt affected how they utilized printed sources to gain background knowledge before or while they took lecture notes.[98]

A third approach, also used by Dennison, was to attend the lectures several times over a period of years and make notes each time.[99] A fourth strategy, unmentioned by Neville, was to take notes directly on the printed copy of the course's syllabus. Using the headings as a guide, students could discern which ideas and terms needed to be noted or ignored.[100] The Huntington Library's 1771 edition of Hugh Blair's syllabus of headings titled *Heads of Rhetoric and Belles Lettres*, features annotations most likely taken by a student in this manner (figure 8.10). As indicated by the rubricated underscoring in UCLA's 1782 edition of Alexander Fraser Tytler's syllabus of lecture heads titled *Plan and Outlines of a Course of Lectures on Universal History, Ancient and Modern*, some students also used underscoring to customize the text in the classroom or perhaps even at home (figure 8.11).[101]

The foregoing scrolling techniques helped students develop the skills of

---

97. Neville (1950, 140). Hereafter I will refer to this source as Neville (1767–88/1950).
98. Edinburgh University, Bound MS (1768–1781).
99. Neville (1767–88/1950, 140).
100. A good example of this kind of rough notetaking occurs in the Huntington Library's copy of Blair (1771).
101. See the marginalia in the "Chronological Table" at the end of William Andrews Clark Memorial Library's copy of Tytler (1782, 228–229). The names of the emperors in this copy were underlined in red ink by either a student or a later reader.

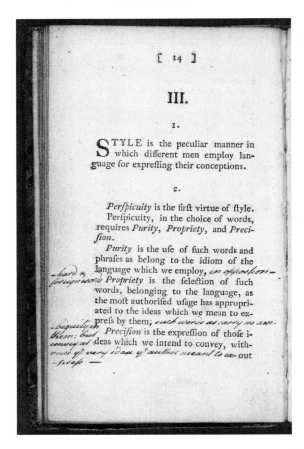

FIGURE 8.10. Notes scrolled on a printed syllabus of lecture headings. Hugh Blair, *Heads of Rhetoric and Belles Lettres* (Edinburgh: Kincaid & Creech, 1771), Call No. 378392, 14. Reproduced with permission of the Huntington Library, San Marino, CA.

knowing when to listen for key ideas and then knowing how to write them down in a way that let them go back to listening. They also illustrate how capturing, in Neville's words, the "principal observations" of the lecture resonated with the positive cognitive utility of the "short notes" praised by parents and the "short passages" recommended by professors such as Robert Eden Scott. Thus, students, professors, and parents understood scrolling to be a skill that had to be actively managed in realtime and that it worked best when notekeepers recorded short passages from a lecture that could be expanded when they "filled in" their transcripts into a fuller account during the evening.

Edinburgh medical students had to develop further strategies to keep organized scroll books in the Royal Infirmary, where they had to attend clinical

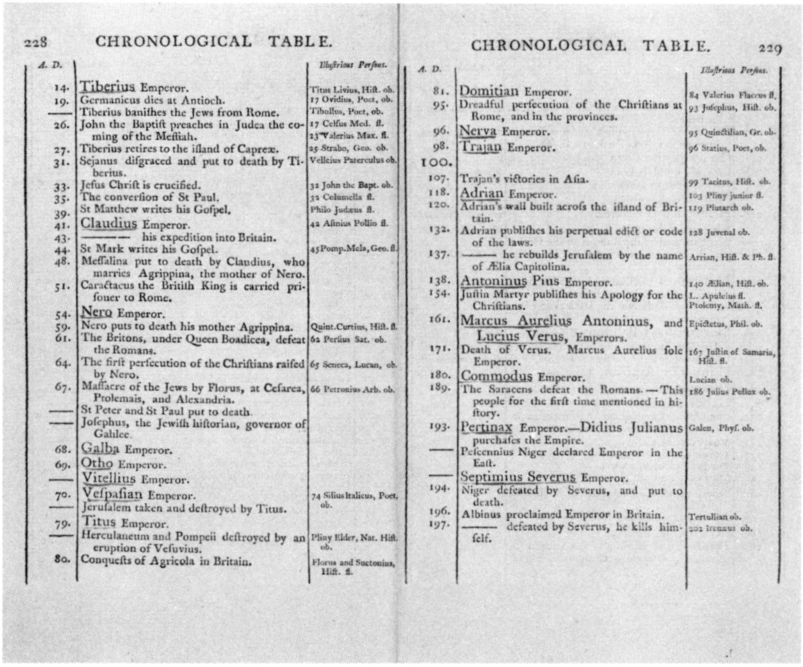

228 CHRONOLOGICAL TABLE.

A. D.		Illuſtrious Perſons.
14.	Tiberius Emperor.	Titus Livius, Hiſt. ob.
19.	Germanicus dies at Antioch.	17 Ovidius, Poet, ob.
——	Tiberius baniſhes the Jews from Rome.	Tibullus, Poet, ob.
26.	John the Baptiſt preaches in Judea the coming of the Meſſiah.	17 Celſus Med. fl. 23 Valerius Max. fl.
27.	Tiberius retires to the iſland of Capreæ.	25 Strabo, Geo. ob.
31.	Sejanus diſgraced and put to death by Tiberius.	Velleius Paterculus ob.
33.	Jeſus Chriſt is crucified.	32 John the Bapt. ob.
35.	The converſion of St Paul.	32 Columella fl.
39.	St Matthew writes his Goſpel.	Philo Judæus fl.
41.	Claudius Emperor.	42 Aſinius Pollio fl.
43.	—— his expedition into Britain.	
44.	St Mark writes his Goſpel.	45 Pomp. Mela, Geo. fl.
48.	Meſſalina put to death by Claudius, who marries Agrippina, the mother of Nero.	
51.	Caractacus the Britiſh King is carried priſoner to Rome.	
54.	Nero Emperor.	
59.	Nero puts to death his mother Agrippina.	Quint. Curtius, Hiſt. fl.
61.	The Britons, under Queen Boadicea, defeat the Romans.	62 Perſius Sat. ob.
64.	The firſt perſecution of the Chriſtians raiſed by Nero.	65 Seneca, Lucan, ob.
67.	Maſſacre of the Jews by Florus, at Ceſarea, Ptolemais, and Alexandria.	66 Petronius Arb. ob.
——	St Peter and St Paul put to death.	
——	Joſephus, the Jewiſh hiſtorian, governor of Galilee.	
68.	Galba Emperor.	
69.	Otho Emperor.	
——	Vitellius Emperor.	
70.	Veſpaſian Emperor.	74 Silius Italicus, Poet, ob.
——	Jeruſalem taken and deſtroyed by Titus.	
79.	Titus Emperor.	
——	Herculaneum and Pompeii deſtroyed by an eruption of Veſuvius.	Pliny Elder, Nat. Hiſt. ob.
80.	Conqueſts of Agricola in Britain.	Florus and Suetonius, Hiſt. fl.

CHRONOLOGICAL TABLE. 229

A. D.		Illuſtrious Perſons.
81.	Domitian Emperor.	84 Valerius Flaccus fl.
95.	Dreadful perſecution of the Chriſtians at Rome, and in the provinces.	93 Joſephus, Hiſt. ob.
96.	Nerva Emperor.	95 Quinctilian, Gr. ob.
98.	Trajan Emperor.	96 Statius, Poet, ob.
100.		
107.	Trajan's victories in Aſia.	99 Tacitus, Hiſt. ob.
118.	Adrian Emperor.	103 Pliny junior fl.
120.	Adrian's wall built acroſs the iſland of Britain.	119 Plutarch ob.
132.	Adrian publiſhes his perpetual edict or code of the laws.	128 Juvenal ob.
137.	—— he rebuilds Jeruſalem by the name of Ælia Capitolina.	Arrian, Hiſt. & Ph. fl.
138.	Antoninus Pius Emperor.	140 Ælian, Hiſt. ob.
154.	Juſtin Martyr publiſhes his Apology for the Chriſtians.	L. Apuleius fl. Ptolemy, Math. fl.
161.	Marcus Aurelius Antoninus, and Lucius Verus, Emperors.	Epictetus, Phil. ob.
171.	Death of Verus. Marcus Aurelius ſole Emperor.	167 Juſtin of Samaria, Hiſt. fl.
180.	Commodus Emperor.	Lucian ob.
189.	The Saracens defeat the Romans. —This people for the firſt time mentioned in hiſtory.	186 Julius Pollux ob.
193.	Pertinax Emperor.—Didius Julianus purchaſes the Empire.	Galen, Phyſ. ob.
——	Peſcennius Niger declared Emperor in the Eaſt.	
——	Septimius Severus Emperor.	
194.	Niger defeated by Severus, and put to death.	
196.	Albinus proclaimed Emperor in Britain.	Tertullian ob.
197.	—— defeated by Severus, he kills himſelf.	202 Irenæus ob.

FIGURE 8.11. Rubricated underscoring on a chronological table in a printed syllabus of lecture headings. Alexander Fraser Tytler, *Plan and Outlines of a Course of Lectures on Universal History, Ancient and Modern* (Edinburgh: Creech, 1782), 228–229. Reproduced with permission from The William Andrews Clark Memorial Library, University of California, Los Angeles.

lectures and walk the wards with professors. These activities were required for students who wished to graduate with a medical doctorate. Each noteworthy case consisted of two components. The first was the case lecture given by the professor, which treated relevant medical theories and practices. If the patient had died, a summary of the autopsy was given as well. The second component was a chronological list of symptoms and treatments that had occurred over the days, weeks, or months that the patient had stayed in the infirmary.[102]

After a professor had given a lecture on a specific case, students had to record the weekly entries for the given patient that the clerk of the infirmary had entered into the ward ledger. But getting information from the ledgers was not a straightforward task. During the middle part of the century, students had to scroll the entries during special sessions when the clerk of the infirmary read the ledgers out loud. This kind of dictation was laid down in the statues of the

102. Risse (1986).

infirmary: "Every Saturday afternoon at three O'clock he [the clerk] shall begin to read aloud the week's articles of his ledger to such privileged students of the infirmary."[103] As enrolment increased from the 1750s onward, the clerk's weekly diction proved inadequate and the rules were relaxed so that students could check out the ledgers.

Since the process of keeping clinical notes was potentially overwhelming, the students were given advice by professors. The Edinburgh college minutes summarized the process in the following manner: "Students have likewise an opportunity of attending the ordinary physicians of the Royal Infirmary, visit the patients every day, examine their cases and prescribe in presence of the students, keeping regular journals of the progress of the disease, or the various prescriptions, or the effect of the medicines, and of the final issue of the case, all of which are open to the inspection of the students, from which they may make extracts."[104] Some professors, William Cullen for instance, offered specific advice on how to take notes in situ. He advised students to first record the patient's age, temperament, and profession, and then to note the disease's remote causes, genesis, and symptoms.[105]

Overall, the cognitive training afforded by scrolling in university contexts was sometimes subtle and not fully realized by students until after they had graduated. This was certainly the case for Benjamin Rush and James Currie, both of whom had put their notekeeping skills to good use in careers that required the management of large amounts of information across multiple genres of paper. Likewise, after Charles Blagden finished his Edinburgh studies, he continued to keep notebooks that acted as *aide-mémoire*. Crucially, after reading the hundreds of manuscript pages that Blagden bequeathed to posterity via his diaries and letters, Hannah Wills concluded that "Blagden's student manuscripts instilled in him a discipline for daily writing, a habit that continued until his death."[106]

Scrolling inculcated skills of listening, writing, and thinking in a manner that strengthened a student's capability to epitomize, abstract, and recollect facts, observations, categories, and arguments. It instilled powerful inscription skills that enabled students to remember and preserve oral and written data that was organized according to a coherent system. But it was also a

---

103. Stedman (1749, 38–39).
104. Risse (1986, 340).
105. The appendices in Risse (1986, 341) reproduce medical cases extracted from clinical lecture notebooks kept by Edinburgh students in the 1780s and 1790s. Some Edinburgh publishers tried to market "pocket" sized medical reference books to students keeping notes in the infirmary. Nisbet (1793).
106. Wills (2019, 67). Wills discusses the relationship between Blagden's notekeeping, association, and memory throughout the article.

capability-building exercise in that it required student notekeepers to practice many interlinked manual and conceptual abilities, especially the skills of writing fast and navigating the lecture headings listed in the syllabus. Over time, students improved their observational skills and their capability to simplify, epitomize, and order complex ideas through the act of writing. Since many endeavored to neatly copy their rough notes after the lecture, scrolling was often a propaedeutic exercise in which the skills of concentration and inscription were implicitly directed toward the creation of a systemized, permanently bound set of notebooks.

Transcripts and the Extension of Memory

The act of extending scroll books into a fuller set of notes, a transcript, was a core scholarly skill. As we learned above, it was an interactive mode of notekeeping because it required students to "fill in" observations and facts from the lectures that were preserved in their memory. Additionally, the material construction of the codex and the visual structure of the rewritten page kinesthetically reinforced the scholarly connections drawn between ordered knowledge and the ordered mind. In this sense the stroke of a pen or pencil was an epistemic event, an action that imbued a student with a related set of beliefs that both justified and underpinned its existence as a meaningful performance.

Students usually used octavo- or quarto-sized paper for their transcripts, but irregular formats and folio editions do exist.[107] To avoid forgetting information, students filled out their notes in the evening after the lecture. William Currie's father reinforced this facet of student notekeeping by advising his son to remember that "nothing will be more useful than to take notes which, if you copy them into a book on the same day, you may extend most to an abridgement."[108] In addition to using their scroll books to make a transcript, students employed the course syllabus, handouts (distributed by professors), and the notes of other students.[109] Combining all of these resources strengthened a student's sorting and ordering capabilities.

A syllabus of lecture headings was a particularly helpful organizational tool and students regularly used it to systematically order their rewritten notes. Some even resorted to copying the headings of the syllabus into their tran-

---

107. For a folio edition of copied notes, see Hume Bound MS (1810) (anonymous notekeeper).
108. Currie (1831, 234).
109. For an example of one of the most thorough syllabi of lecture headings given to Scottish students, see Anderson (1777).

script when they had missed a lecture. An anonymous student attending Alexander Fraser Tytler's universal history lectures around 1800 wrote the following at the start of his notebooks: "For the six preceding heads of this Lecture, see the Outlines."[110] Likewise, an anonymous student attending William Cullen's Edinburgh lectures on chemistry during the 1760s cross-referenced a section of his notes to the lecture headings by writing "Vide Syllabus Page 9."[111]

The movement of large amounts of manuscript material from one notebook to another was of course not a practice unique to university students. It had been an indispensable information management skill used in commonplace books since the Renaissance. By the late eighteenth century even traveling botanists transcribed excerpts from their field notebooks into a "register" notebook in a process that has been called "writing after the fact."[112] Since students keeping their own notes in Scottish universities used observational skills to hear new facts and to see new objects in lectures (especially in medical courses), they were also writing after the fact when they filled out their rough notes.[113] Since transcribing involved the insertion of more information from their own memories and the notes of others, students were learning to weave scribal skills into kinesthetic routines that treated transcripts as expandable files that could be extended to fit their intellectual or educational needs.

Learning to treat transcripts as expandable systems, though eventually rewarding, was expensive and time-consuming. The cost of notekeeping materials was flagged by students in their correspondence to relatives. The American student William Quynn, for instance, wrote a letter back home in which he grumbled about the fact that "there are many other expenses that accrue from Purchasing Book Paper" for notes.[114] The immense labor involved with writing out a lecture transcript was articulated in 1785 by the medical student Alexander Coventrie with the following observation: "I took notes from all the lectures, generally the leading topics, which I filled up at my lodging, which kept me from bed till two in the morning." Here "leading topics" refers to lecture headings and "filled up" refers to the rewriting and remembering process used to transform scroll books into transcripts. Looking back on his studies

---

110. Tytler Bound MS (1800–1801). The syllabus in question was Tytler (1782).
111. Cullen Bound MS (1765–1766), Blagden (transcriber), 51.
112. The acts of "writing after the fact" and using a notebook as an "open file" are discussed in Bourguet (2010).
113. The observational skills engendered by student notekeeping are addressed in Nelles (2010).
114. Quynn (1936, 197). Earlier in the century, during the 1730s, William Sinclair, a Scottish medical student attending Leiden University, recorded a 12s purchase of "paper, pen and ink." This low price suggests that he brought supplies in his large travel chest. Sinclair (1995).

near the end of his first year at university, Coventrie wrote in his diary that "my late hours revising my notes taken at the lectures, wore on my constitution, and I longed for the approach of May and the end of the lectures."[115]

A similar account of painstakingly extending notes is given in the diary of Sylas Neville. After attending a 1771 anatomy lecture of Professor Alexander Monro Secundus, he returned to his rented room, recopied his notes, and then wrote in his diary: "Tues. Nov. 12. Allowed R. Byam, a gent. from Antigua who lodges in our house, my notes from Monro's 4th lecture to copy. Dennison says it is almost a full copy—sincerely or not I do not know. . . . Did not get to bed till 1/2 past 12 o'clock. Extending my notes taken at the Chemical and Anatomical lectures employs my whole time and prevents my doing any thing else. Tired, uneasy & low-spirited."[116] As indicated throughout Neville's diary, students lent each other their scroll books and transcripts, especially in cases where a lecture had been missed or was difficult to understand.

Students occasionally copied the notes of courses that they had not attended. The medical student John Bacon succinctly summed up this routine at the front of his copied version of the notes taken by another student in John Gregory's lectures on medical practice: "N.B. These lectures were written at Edinburgh in the years 1772 and 1773. The Manuscripts from which I copied them, were lent to me by my ingenious and worthy Friend Doctor Remmet of Exeter."[117] Since the weekly effort required by the transcribing regime was demanding, some students lightened their load by creating notekeeping consortia. Evidence for this practice appears in the inscription "Thomas Parke and Co.," which occurs throughout the eleven hundred pages of notes kept by Edinburgh medical student Thomas Parke in one year of study during the mid-1770s. Such consortia facilitated a collective mode of observation that maximized the efforts of several students.[118]

Regardless of the subject matter, all students had to graphically design notebook pages that were easy to access. To make their transcript headings stand out as visual cues, students ranged their pages according to the standard module that was used by school notekeepers. The main elements were freestanding, usually centered, headings followed by a block of text. Some students also included wider margins into which they wrote subheadings. Figure 8.12 is a heuristic representation of the gestalt image created by note-

---

115. The Coventrie quotations come from Coventrie (1904, 109, 110).
116. Neville (1767–88/1950, 140).
117. Gregory Bound MS (1772–1773) features John Bacon's comment about copying the notes. It is quoted in Gregory (1998, 46).
118. Bell (1951, 250).

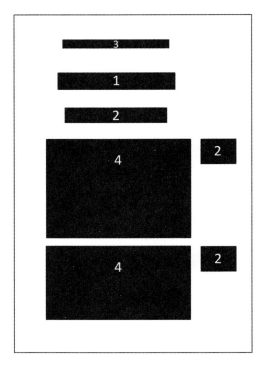

FIGURE 8.12. Stylized graphic layout of a student notebook page. The (1) main heading, (3) running head, and (4) paragraphs are plotted within a large column running down the middle and left side of the page. The (2) subheadings are plotted either in the large column or in a smaller column running down the right side of the page. © Matthew Daniel Eddy.

book headings and margins. Knowing how to range the foregoing elements of a notebook page was something that university students had to learn and value.

Crafting the pages of lecture notebooks required a steady hand, tools of inscription, and, crucially, time.[119] When university students repeatedly selected all or some of the elements, they effectively became realtime compositors because they were ranging a layout pattern that ran across all the pages contained in all the volumes of their transcripts. In most cases, the visual pattern of the layout, particularly the consistent placement of the headings, ensured that the elements would be plotted in approximately the same place on every page. Students selected a combination of the elements that best suited their visual needs and strengths.

A few students attempted to replicate the typography and visual format

119. Eddy (2013).

featured in the layout of a professor's syllabus. An excellent advanced example of this practice can be seen in Sir David Pollock's ten-volume set of the 1790s lectures given by Professor John Walker in his Edinburgh natural history course (figure 8.13).[120] These notes force us to consider the extent to which a student could "replicate" the layout of the headings that appeared in a printed syllabus. More specifically, even in this set of beautifully transcribed notes, the flowing hand and connected letters of cursive writing styles could never produce a mimetic copy of the independent, predominantly disconnected, letters used in printing.

To make their transcripts even neater, many students followed the wider scribal practice of pre-drawing a graphite grid on the paper. The presence of the grid effectively allowed students to inscribe their notes into a rectilinear column of information, thereby creating a visually and conceptually ordered page. Notably, they usually erased the grid after they had written their notes in ink. In some sets, the top, side, and bottom edges of the grid were cut off when the notes were rebound.[121] There are, however, a number of transcripts in which a grid is still present in some form.[122]

Students usually made a title page for each volume. Since it appeared at the front, it was often damaged or lost over time. It usually stated the name of the professor, the date, and the title of the course. The last of these was sometimes an abbreviation, epitome, or emendation of what the student thought the course should be named.[123] The title was normally handwritten, but law student David Johnstone went so far as to have his printed as *Notes on the Law of Scotland, Taken from the Lectures of David Hume, Esq. Advocate*.[124] Aside from the absence of the printer's name, Johnstone conveyed the basic information one would find featured on the title page of a printed book.

In addition to creating title pages, students made tables of contents for their transcripts. Here the syllabus proved to be a helpful tool once again. While some students chose to use it as a premade table of contents by pasting it to a flyleaf,[125] many fully embraced the mnemonic power of copying the

---

120. Compare the graphic layout of David Pollock's lecture notes in Walker Bound MS (1797, 5: 18–19) to Walker (1787, 24–25). The content of Walker's mineralogy lectures is treated throughout Eddy (2008).
121. See, for example, the volumes of Hume Bound MS (1810–1811).
122. Whole or partial graphite grids are still visible in Hume Bound MS (1810, vol. 1) and Finlayson Bound MSS (1795–1796, vol. 1).
123. The dates on the cover pages of lecture notes need to be treated with care. Students and transcribers occasionally wrote the year in which the notes were copied and not the original year in which the professor gave the lectures. Cole (1982, 53–55, ref. 20).
124. Johnstone used printed title pages for all of the volumes of Hume Bound MSS (1810–1812).
125. Compare the anonymous student notes in Millar Bound MS (1771) to the printed syllabus in Millar (1771).

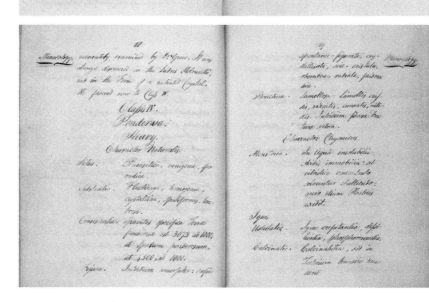

FIGURE 8.13. Lecture headings printed in a professor's syllabus and written in a student notebook. Top: The heading for Ponderosa minerals in John Walker, *Classes Fossilium* (Edinburgi, 1787), 24–25. Reproduced with permission of the Linnean Society of London. Bottom: The same Ponderosa heading in John Walker, *An Epitome of Natural History*, vol. 5 (1797), David Pollock (notetaker), Bound MS, GEN.707D, ff. 18–19. Reproduced with permission of Edinburgh University Library Special Collections.

syllabus of lecture headings into their notebooks. The content and layout of this paratextual apparatus was relatively easy to conceptualize in these cases because students could use the professor's lecture headings as a guide.[126]

But the skills used to read the headings as a useful template were a bit different from the skills required to write them out on the page. Put another way, writing out the headings as a table of contents at the beginning of a notebook required a range of writing skills, particularly those that allowed students to lay out the words as a rectilinear matrix on the page. The presence of different graphic factors explains why students laid out their tables of contents in ways that converged or diverged with the various typographic features of the professor's lecture headings. Some feature numbered headings but do not include page numbers, others do not include numbered headings but list page numbers.[127]

A good number of transcripts had an index. Since most transcripts comprised several volumes, the placement of the index varied. Some students made a cumulative index in the last volume. Others put one at the end of each volume. Most indices listed key terms alphabetically.[128] The absence of savvy indexical systems, such as John Locke's commonplace method based on the vowels of entries, suggests that students found straightforward alphabetical listings more useful.[129] Designing a notebook index involved listing, alphabetizing, numbering, sorting, and repeatedly shuffling through an entire set of transcribed notes. Students had to select the terms that they wanted to order before reading and rereading their notes so that they could collect the page numbers where the terms occurred. They also had to use plotting, indenting, aligning, and other ranging skills to create a personalized and hence useful layout for the index. The customized nature of this practice allowed some students to lay out their index in columns and others to use different arrangements such as tiled boxes (figure 8.14).[130]

The additional time and effort required to design and collate an index explains why many transcript indices made by students and professional tran-

---

126. Printers followed the practice of turning lecture headings into tables of contents when they published a professor's lecture notes as a book. See the tables of contents in Cullen Bound MS (1761, f. 3) and Cullen (1775, 494–495).

127. For lists of unnumbered and unpaginated lecture headings that served as a tables of contents, the beginning of each volume in George Sligo's law notes: Hume Bound MSS (1816–1817). For lists of numbered and paginated lecture headings that served as tables of contents, see the front pages of each volume of David Johnstone's law notes: Hume Bound MSS (1810–1812).

128. Hume Bound MSS (1810–1812).

129. Locke's commonplacing method is explained in Yeo (2001, 110–115; 2014, chap. 7). Yeo explains that Locke's entries were indexed by the first letter and vowel of the first word of the heading.

130. See the tiled index in Cullen Bound MS (1760, v. 1). For an example of one that occurs as a single page, see Leechman Bound MS (c. 1780), (anonymous notekeeper), 181.

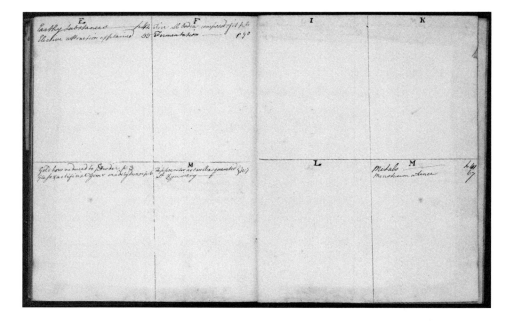

FIGURE 8.14. A tiled index. Like many indices found in student notebooks, much of it is incomplete. William Cullen, *Chemical Lectures*, vol. 1 (1760) (anonymous notetaker), Bound MS, MS 1918. Reproduced with permission of the Wellcome Library, London

scribers are incomplete (see again figure 8.14). An anonymous transcriber of Adam Smith's 1760s *Lectures on Justice* even went so far as to replicate the exact pagination of the original notebook so that the page numbers in the index would not have to be changed when it was copied. To achieve this goal, he wrote in a smaller hand and had to sometimes simply stop writing even if he had not reached the end of the page. The end result was that one page might contain twenty-six written lines while another contained only twenty. It seems that whoever commissioned the copy felt that having an index was more important than filling every line on every page with narrative.[131]

Transcripts contain lists of information that students copied from either printed handouts that professors sold through local bookshops or from inscribed posters that professors hung in their classrooms. The lists included but were not limited to keywords, definitions, book titles, experimental protocols, and noteworthy dates.[132] As with the syllabus, students sometimes used

---

131. Cannan (1896).
132. For a list of definitions written by John Lee in the notes he kept for John Hill's Edinburgh philological lectures, see Hill Bound MS (1797, ff. 149–159). For a list of experiments, see the notes kept by James Cunningham in John Hope's Edinburgh botany course: Hope Bound MS (1781, f. 333).

SYSTEMIZING | 313

FIGURE 8.15. A branching diagram of lecture headings for a law course. "General Scheme of the Arrangements," David Hume, *Notes of lectures on Scots Law*, vol. 1 (1790), William Patricks (notekeeper), EUL Bound MSS Dc.6.122. Reproduced with permission of Edinburgh University Library Special Collections.

printed lists as reference tools as they wrote their transcripts. Evidence of this practice comes from Blagden's scroll books, where he tucked away a loose-leaf copy of a chemistry preparation list that had been printed by Professor Joseph Black.[133]

As evinced in the notes that William Patricks took in the Scots law lectures given by David Hume at Edinburgh University, some students inscribed a branching diagram, sometimes called a *clavis*, of the lecture headings. This practice imparted a host of spatial and kinesthetic categorizing skills that enabled them to visually map out an entire knowledge system on one sheet of paper (figure 8.15).[134] The title of Patricks's *clavis* was "A General Scheme of Arrangements" and its headings corresponded to those that he used to organize his law notebooks. The density and symmetry of the *clavis* suggests that it was copied from an exemplar distributed by Hume. But it is possible that Patricks

133. *The Preparation of Mercury*, found by the author tucked inside Blagden's Black Bound MS (1766–1767, vol. 3).
134. See William Patricks's *clavis* in Hume Bound MS (1790, vol. 1).

designed it himself, which would have required him to first work out its structure on other pieces of paper. Either way, it took much effort to create, and in the end the final product served to function as a mnemotechnic key to the entire system designed by Hume and correspondingly to the order of all the notebooks kept by Patricks.

Some students practiced small but important acts of innovation by supplementing, augmenting, or rearranging the graphic elements of the syllabus. John Lee's 1797 transcripts from John Hill's Edinburgh course on philosophy are a particularly good example of this practice. When compared to the syllabus printed by Hill in 1792, it can be seen that Lee adapted the layout in the pages of his notebook (figure 8.16). Lee also selected and then recombined headings and subheadings that did not occur together in the syllabus. The first page of his transcript, for example, combines centered headings and subheadings that are featured on different pages of Hill's syllabus.[135] These adaptations show that students selected elements from the printed lecture headings that worked best for their personal approach to managing information on paper.

Industrious students continued to amend their notes even after they had been bound. Since most sets of transcripts were inscribed on the recto page only, the verso page was available for future annotations, observations, or corrections. Some students expanded the system offered by the syllabus by adding further notes to the blank endleaves,[136] while others used this space as scrap paper for financial calculations or the occasional doodle.[137]

Many sets contain sidenotes written on the blank verso page. They functioned like footnotes, supplementing the main narrative with further information or bibliographic references gained from sources that were read after the course had been taken. In terms of their form and function, sidenotes hovered somewhere between footnotes and marginalia. They existed as a scribal genre primarily because the text of printed books occurred on both the recto and verso faces of the pages, which eliminated the possibility of writing annotations on the verso page. Students usually wrote superscripts to label their

---

135. Compare the headings and subheadings featured on page of Hill (1792, 1, 8), to that written by Lee in Hill Bound MS (1797, 2). It is possible that Lee might have been using the second edition of Hill's lecture syllabus (1785); however, the order of headings and subheadings on this edition's first page does not match the format used by Lee.
136. For annotations, see the "Catalogue of Books" at the end of Hume Bound MS (1815–1816, vol. 3, and the appendix). One entry refers to a book published in 1830. Many pages are left blank at the end, suggesting that this volume was bought prebound and blank.
137. See the accounting tables scribbled by John Bruce in notes he kept for Hugh Blair's Edinburgh course on rhetoric and belles lettres: Blair Bound MS (1779, vol. 2, front and back flyleaves).

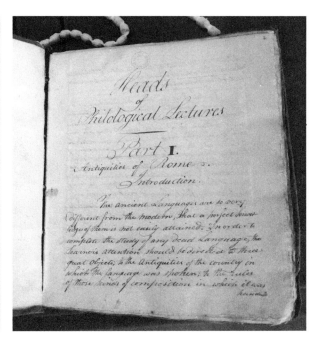

FIGURE 8.16. Lecture headings and narrative printed in a professor's syllabus and written in a student notebook. Left: John Hill, *Heads of Philological Lectures, Intended to Illustrate the Latin Classicks*, 3rd ed. (Edinburgh: Smellie, 1792), 1. D.S.a.1.23 Centre for Research Collections, Edinburgh University. Right: John Hill, *Heads of Philological Lectures. Intended to Illustrate the Latin Classics, in Respect to the Antiquities of Rome; the Rules of General Criticism; and the Principles of Universal Grammar* (1797), John Lee (transcriber), Bound MS, Dc.8.141, f. 2. Both reproduced with permission of Edinburgh University Library Special Collections.

sidenotes. These tended to be Roman letters, Greek letters, crosses, asterisks, or occasionally hashtags, inserted in the narrative of the recto page and then written again on the blank facing verso page alongside the new note.[138]

Annotating transcripts helped students better understand how to correct or improve a text that functioned as part of a paper system. Such additions came in different stylistic and material forms. David Johnstone, for instance, made corrections by pasting new handwritten law notes over old notes.[139] The most common way of annotating notes was by crossing out words or sentences and then writing the correct information, if space permitted, above the line or on the verso side of the facing page. Such corrections ranged from fix-

138. For sidenotes, see Hume Bound MS (1810–1811, vol. 4).
139. Hume Bound MS (1810–1812). For the pasted notes, see vol. 1, 216.

ing misspelled words to adding new material like the dates that the lectures were given. Originally, Johnstone had not recorded the dates and numbers of the lectures, so he later went back and penciled this information into the margins.[140]

Lines and the Media of the Mind

In previous chapters sections, we learned that notekeeping skills carried cognitive significance. Notekeeping was a mode through which students could strengthen mental operations such as perception, attention, association, abstraction, and recollection. But simply exercising such operations was no guarantee that a notekeeper's mind would be shaped in a positive way. Certain kinds of scribal activity could produce irrational associations as well as misperceptions, inattention, and distraction. A goal for university students and professors alike, therefore, was to use notekeeping activities in ways that were conducive to having a positive effect upon mental operations.

Bearing the foregoing factors in mind, the question that I would like to pursue in this final section of the chapter is this: how might we interpret the material evidence offered by student lecture transcripts in a way that connects to the potentially positive cognitive relationship that existed between notekeeping and thinking? Though there are many potential answers, I would like to explore the question by building on the way in which the *tabula rasa* metaphor placed the material culture of transcribing in conversation with the culture of the mind advocated by Scottish professors and pedagogues.

At the time, conscious thought was repeatedly described as a "train" of ideas arranged as a straight line that flowed past the mind's eye. The words "chain" and "string" were used as well. Conversely, irrational trains were glossed as being bent. The acquisition of ideas and their flow as lines past the mind's eye was influenced by mental operations such as perception, attention, association, and recollection.[141] The notion that thought was a train, a line, was a fundamental element of the Lockean cognitive model that enjoyed wide circulation in Scotland.[142]

---

140. Hume Bound MS (1810–1812). Other good examples of pen and ink corrections occur n John Lee's edition of Finlayson Bound MS (1795–1796).
141. The notion that thought was a line of ideas appears across Western history. During the period considered by this book, it was described via the metaphors of a "chain," "train" and "string." In secondary literature, it is sometimes referred to as the "chain of reasoning," "rational chain," "association chains," and "train of thought." For the presence of the metaphor in Scotland, see Eddy (2011) and Mortera (2005). For an example of the metaphor's usage outside of Scotland, see Yeo (2007). For the larger cognitive context that supported the metaphor, see the early chapters of Hacking (1975).
142. Eddy (2016b).

As shown by Tim Ingold's work on the history of the role played by lines as cognitive models and as memory tools, lines are artifacts to which different cultures attach diverse meanings and through which they learn important skills and ideas.[143] Within eighteenth-century Scotland, the straightness of a line was widely interpreted to be a sign of orderliness, a concept that the reading public throughout Britain considered an attribute of rationality and its related performances of self-organization.[144]

For those who subscribed to Locke's cognitive model and its commitment to the *tabula rasa* metaphor, it was frequently assumed that there was a similitude between words written as a line on the page and ideas ordered as a train in the mind. The simultaneous experience of writing in sentences and thinking in trains created what might be called the "line of reason," an extended metaphor that combined the train of ideas and the *tabula rasa* into an enactive understanding of metacognition in which ideas were moveable entities that could be ordered within the mind by manipulating words on paper.[145]

Scottish scholars who wrote about the science of mind often used the extended metaphor when attempting to understand the nature of ordered, conscious thought. Some held that the train was made up mainly of ideas. Others such as Thomas Reid, Glasgow's professor of moral philosophy in the middle of the century, held that the train was composed of ideas and mental operations, leading him to use the term "trains of thinking." His student, Dugald Stewart, whom we have encountered several times in this book, followed a similar interpretation and used the term "train" to describe lines of ideas and thoughts.[146]

The metaphor of reason being a line experienced simultaneously in the mind and on the page transformed the multiple acts of student transcription performed daily in Scottish universities into significant and worthwhile activities. When students wrote lines of words in straight sentences, columns, or rows, they participated in a process that their professors framed as a real-time mode of ordering "straight" lines of ideas as chains in the mind. In other words, rewriting scrolled notes as recopied notes concurrently served as an

---

143. Ingold (2007; 2015).
144. The relationship between order and self-organization is treated throughout Sheehan and Wahrman (2015).
145. Aspects of the simultaneous relationship between words and ideas within Scottish theories of mind, language, and representation are discussed in Eddy (2006; 2011).
146. Thomas Reid (1785, chap. 4, esp. 406) offers a discussion of the "train of ideas." Stewart (1792) offers a more robust terminology by referring to "trains" of inquiry, reason, thoughts, ideas, particulars, and thinking.

ordered mode of thinking, one that implicitly transformed student transcription into a cognitively impactful exercise.

Most university students would have already been familiar with the notion that writing sentences on the hand-ruled lines of their notebooks had a positive effect on the chains of thought in their minds. As mentioned in chapter 2's discussion of the ways in which schoolchildren learned to keep notes, the term *lines* was generally used to refer to handwritten homework that involved composition, calculation, or transcription. Teachers regularly employed this kind of linear language to describe thought. Likewise, grammar school students would have been aware that the Latin word *sententia*, from which we get the English word *sentence*, referred to a thought or way of thinking. After students arrived at university, they were further exposed to more applications of the linear metaphor.

An important, oftentimes subtle, application of the line of reason appeared in discussions of morality. In order to understand the connection between morality and its general relevance to university transcribing, it is worth remembering that one of the putative responsibilities of a Scottish regent or professor as an educator was to encourage behavioral patterns that prevented a student's mind from becoming "bent." James Fordyce, who taught teenage boys as a professor of Marischal College, Aberdeen, succinctly summed up this view in his *Dialogues Concerning Education*: "Too much Pains therefore cannot be taken, in watching over the Mind in its unformed, but most susceptible State; in preventing wrong *Associations*, in teaching it to make such as are allied by Nature, and in counter-working a perverse *original Bent*, by those Associations and Exercises, which are most effectual to battle it."[147]

Though some of Fordyce's progressively minded contemporaries would not have agreed with his assumption that the mind was originally "perverse," the meaning he assigned to the linear metaphor was clear and consistent with others who wrote about the environmental aspects of cognition: rational thought was straight and irrational thought was not. Crucially, the straightness, the order, of the mind could be influenced by using mental operations, association in Fordyce's case, as formative tools. His reference to the linear metaphor makes it easier to see why writing and rewriting sentences as straight lines in lecture transcripts carried regulative undertones. When seen as a capability-building exercise, transcription was an activity through which students learned to use script to organize their thoughts.

The regulative aspects of transcribing become clearer if we turn to Thomas

---

147. Fordyce (1757, 191). Italics in original.

Reid's understanding of the mental operation of imagination for further insight. He held that imagination was "the constant ebullition" of "thought barely speculative." It involved a train of ideas that also included mental operations, "sentiments, passions, [and] affections." The key point to note in terms of our interests in university transcribing is that Reid believed that there were two kinds of trains: "spontaneous" and "regulated." In Reid's words, "the train of ideas of thought in the mind are of two kinds; they either such flow spontaneously, like water from a fountain, without exertion of a governing principle to arrange them; or they are regulated and directed by an active effort of the mind, with some view and intention."[148]

Reid held that learners, or thinkers more generally, could gain control over their spontaneous thoughts by practicing "regular trains of thinking," an activity that he described as "chiefly, copying what they see in the works and discourse of others" who think according to ordered principles.[149] His references to "works" and "discourse" reveal that the principal modes of "copying" that he had in mind were those that involved language. Here he is signaling the regulative function that he and other professors ascribed to language within the line of reason metaphor. His views speak to a pedagogical context in which the lines of the sentence functioned as a workshop where writers could practice placing words, which corresponded to ideas, in order. The act of transcription, therefore, was a form of training that allowed the transcriber to learn how to order the words in the line of a sentence in realtime.[150]

Put more clearly, the regulative link that Reid drew between transcription and the train of ideas is a good example of the pervasive, but oftentimes assumed, reciprocal relationship between the evolving structure of a learning mind and the ways in which students interfaced with lecture notebooks as paper machines, that is, as artifacts with a multiplicity of visual and material components that functioned as well-ordered media systems. When conducted according to an organizing principle, writing ordered headings, sentences, or lists was a metacognitive event in that it concurrently instigated the act of thinking an ordered train of ideas and operations in the mind. Since most lecture transcripts were organized according to headings that professors had selected according to governing principles, transcribing functioned simultaneously as a process through which the skills of systemization were learned.

---

148. Reid (1785, 406–409).
149. Reid (1785, 418).
150. I am using Reid as a representative example of the broader notion within Scottish universities that mental operations could be trained to institute patterns of behavior that regulated the passions. Another influential iteration occurs in Smith (1759). For the historical context of passions as an object of enquiry at this time, see Hirschmann (1977).

The realtime relationship between the linear flow of words on the page and ideas in the mind was underscored by other Scottish professors, most notably at Edinburgh University in the rhetoric lectures of Hugh Blair and the moral philosophy lectures of Dugald Stewart, whose courses were not only popular with arts students, but also with students reading law, divinity, and medicine degrees.[151] In Stewart's lectures on "the power which the mind has over the training of its thoughts," he reflected on the relationship between thinking and writing in the following manner: "But the principal power we possess over the train of our ideas, is founded on the influence which our habits of thinking have on the laws of Association; an influence which is so great, that we may often form a pretty shrewd judgement concerning a man's prevailing turn of thought, from the transitions he makes in conversation or in writing."[152]

It is in the relationship between the linear order of words and ideas that the act of repeatedly writing and ranging thousands of sentences on the grid of a lecture transcript emerges as an especially powerful cognitive activity. James Beattie, Aberdeen's professor of moral philosophy, underscored the importance of this kind of enactive transcription in a chapter on memory in his *Dissertations Moral and Critical*. After praising the perceptual benefits of reading books aloud and "with propriety," he made the following observation: "If we transcribe slowly, in good order, with scrupulous nicety in punctuations and spelling, and with reasonable distance between the lines, we shall have a better chance to remember what we write, than if we were to throw it together confusedly, and in haste."[153]

If we carefully read Beattie's thoughts on transcription, it is possible to discern that, in addition to improving the mental operation of recollection, the act of transcribing slowly improved a writer's attention, and the act of observing correct "punctuation and spelling" along the straight and evenly spaced "lines" of sentences facilitated the ordered succession of ideas in the mind. In this sense, the stroke of a pen on a transcript page was a metacognitive event because it was an action that was believed to have a positive impact on the operations of the mind in a way that justified the act of notekeeping as a rational, meaningful performance.

The linear imagery of the train of ideas fostered another cognitive connection between transcribing and thinking within Scotland's universities. In the early chapters of this book, I mentioned that Dugald Stewart used the word

---

151. Mortera (2005) and Eddy (2011).
152. Stewart (1792, 292).
153. Beattie (1783), 32).

picture to describe the structure of a disorganized commonplace notebook page. He was a firm believer in the power of organized notes because he held that "the advantages of order in treasuring up our ideas in the mind, are perfectly analogous to its effects when they are recorded in writing."[154] In Stewart's view, the origin of the disorganization evinced in commonplace books was the fact that commonplacers often did not use a "method" to select the passages.

Based on what we now know about how university students used lecture headings to systematically order and label their notes, it can be seen that Stewart's combined reference to "method" and "picture" intimates that a disordered commonplace book was one in which the transcriber had not thought in advance about the conceptual and visual arrangement of the data, a move that prevented the creation and insertion of topical headings that were ordered according to a coherent theme or principle.[155]

To help his students learn how to recognize and use headings that were ordered into a system according to a principle, Stewart, as we learned earlier in this chapter, provided students with a syllabus of ordered lecture headings that they could use to trace, in his words, "trains of thought."[156] Here his reference to mental trains alerts us to the fact that he used the linear metaphor to understand the cognitive salience of his lecture headings. If we look closer at Stewart's syllabus, it emerges that its headings are not usually presented as fully formed sentences. As can be seen in figure 8.6, Stewart's headings, like those presented in the syllabi of many university courses, were usually terms or phrases. The fact that the headings were not full sentences means that their arrangement in the syllabus in its capacity as a system represented a linear form of order, a train of thought, that was slightly different from that offered in the line of a sentence. What kind of order was Stewart referencing?

To answer this question, we need to move from looking horizontally at the words in sentences to looking vertically at the words in lists. More specifically, we need to look at Stewart's lecture headings in their capacity as categories that added up to a knowledge system. The headings on his syllabus were vertical lines of word clusters the represented the categories in the system he

---

154. Stewart (1792, 423).

155. The absence of thematic headings was a notable data management concern. As shown throughout Allan (2010), eighteenth-century commonplacers often treated their notes as loosely organized memoranda of passages and thoughts that arose spontaneously according to their interests over time.

156. Stewart (1811, 329). For a sample of the syllabus of lecture headings used by Stewart, see Stewart (1793). George Jardine, Glasgow's professor of logic, also commented on the use of syllabus lecture headings. See Jardine (1825, 278–279).

used to organize his moral philosophy lectures. As shown by the anthropologist Jack Goody many years ago, vertical lists of words are an important form of linear order that, unlike the left-to-right flow of sentences, requires a top-to-bottom scan down the face of a tablet, obelisk, or page.[157] If seen as a vertical line of categories, the list of headings offered in Stewart's printed syllabus of lecture headings becomes a downward flowing linear pattern, or, in his words, a "picture."

The descending flow of the pattern was organized according to the nomenclature of hierarchical categories—parts, chapters, and sections—that Stewart used to order his headings.[158] Within the structure, the headings flowed according to the method of classification Stewart had used to arrange them into a system. Thus, when students used Stewart's syllabus to order the notes that they were transcribing by hand, the vertical order of the headings was based on the classification system that he had invented. The basic architecture of the pattern, which remained relatively the same in reprinted editions (see figure 8.6), was presented to students in the table of contents. There they could cast their eye down the page from heading to heading within the hierarchical order offered by the parts, chapters, and sections.

Far from being a unique occurrence, the vertical train of headings offered by Stewart was arguably the most common layout pattern employed by Scottish professors to organize their printed and handwritten syllabi during the long eighteenth century. Professors at other European universities used similar patterns for their headings as well.[159] Stewart's headings are, however, noteworthy because his lectures clearly explained how the overarching visual organization on the page was directly relevant to the pervasive line of reason metaphor that shaped the developmental context in which students kept their lecture notebooks.

In order to have a recognizable impact, the relationship between transcribing and thinking promoted by Stewart and other professors had to be honed through repeated acts of notekeeping that took place over a considerable period of time. This was precisely what was happening every time a student organized a set of transcripts by writing the lecture headings as freestanding labels that served to organize chunks of information ranged as sentences on the page. When students wrote the headings in this way, they were effectively designing a vertical line of words, a list, that manually extended the system-

---

157. Goody (1977, 74–111).
158. Stewart (1793).
159. Examples of syllabus headings used in other European universities occur throughout Ong (2004), Blair (2010b), and Eddy (2010a).

atic order of the syllabus down the center of every page of every volume of their transcripts.

If we step back and recall the scale of materials, skills, and activities required to create a set of lecture notebooks, it can be seen that the process of student notekeeping was a form of training that was not easily experienced outside the context of a university. When it came to the social reproduction of knowledge systems, the process of keeping lecture notebooks played two significant roles within the student communities based in and around universities. On the one hand, they offered students the opportunity to create the components of an ordered set of notebooks that could be operated as an enormous paper machine, thereby conferring the capability of seeing systems as dynamic and rearrangeable entities. On the other hand, the process implicitly encouraged students to value the active manipulation of specialized forms of manuscript media in ways that shaped the evolving capabilities of their own minds.

Stewart's reflections on the form and meaning of the syllabus and its relationship to transcribing invites scholars to remember the historically important role played by student notebooks as realtime learning devices that made sense within a context that recognized the developmental relationship between media and the mind. When he spelled out the significance of using the syllabus as a template to transform the page of a notebook into a word picture, a *tabula verba*, he was identifying a core transcription skill that was widely believed to positively impact the order of ideas in the mind. At the same time, he was offering a cognitive rationale for the serialized layout that each student extended across an entire set of lecture transcripts in their capacity as a fully interactive system of human interface. The process of crafting such a robust media technology empowered students to incrementally strengthen their skills of systemization, which, over time, enabled them to understand and manipulate the large data systems they encountered after they left university.

CHAPTER 9

DIAGRAMMING

Paths and Diagrammatic Knowledge

During the 1720s, Professor Robert Simson was haunted by a menacing specter. Try as he might, he could not solve a geometric problem that he had encountered as a student. As Glasgow University's professor of mathematics, Simson felt it was his duty to distinguish himself as an academic by solving an ancient and unresolved porism proposed by Euclid. Though he was widely admired for his mathematical prowess and his retentive memory, the solution evaded his normal problem-solving routines. The problem was impervious to the long hours that he dedicated to studying mathematics each day. It resisted his frequent walks of one hundred paces across the university quadrangle and gardens. Then, when he least expected it, the answer came to him in the most curious of ways.[1]

One day Simson decided to walk with friends along the banks of the Clyde River. Somewhere along the path he found himself separated from the group and began to reflect on his old mathematical foe. As he surveyed the landscape and riverbanks, the answer came to him in the form of a diagram. Deprived of pen and paper, he produced a piece of chalk and drew it on the bark of a nearby tree. The diagram proved to be essential for his research and enabled him to make noteworthy contributions to mathematics that attracted students to his lectures and augmented the reputation of the university. Throughout the rest of his career, he reveled in retelling the story of his riverside epiphany to friends, family, and students.[2]

The Simson episode illustrates the vital role played by diagrams within

---

1. Robert Simson's habits of thinking are addressed in Trail (1812, 73–78).
2. The story of Simson's tree diagram is recounted in Trail (1812, 19–20).

modern university settings. They functioned as crucial thinking tools that gained deeper and long-lasting meaning through the act of drawing them out on paper, or, in the case of Simson, on whatever medium was near to hand. In this context, the process of sketching a diagram was fundamentally a mode of interface that united the body with the mind.[3] Put more clearly, Simson's diagram was not simply a shape rendered on a tree. It was the instantiation of a collection of diagramming skills that he had learned as a student and then exercised as a professor. Diagramming was part of the process that he used to manage knowledge in his mind and on paper.

Simson's use of the diagram operated in conjunction with modes of bodily movement that he had regularized into lines across surfaces. His daily walks, for instance, were paths along which he moved in a manner that enabled him to think as he shuffled his feet, swung his arms, counted his steps, and passed notable geographic markers. Creating knowledge in this manner was in many ways similar to how the wayfarers discussed in Tim Ingold's work engaged in thinking routines when they used paths as lines of observation that ran along a landscape.[4] Simson's case shows that he also was thinking along the lines of his diagram as he drew it and that, correspondingly, forming such lines in his mind and with his hand was a crucial element of solving the problem presented by the porism. In short, the act of drawing was intimately connected with creating, understanding, and remembering his diagram. It was a mode of redrawing reason on the bark of a tree.

Many Scottish professors held that, like the words of sentences, lines were signs to which meanings were attached. The origin of this notion reached back to Locke's theory of mind, particularly his discussion of diagrams in his *Essay on Human Understanding*. In the book's section on general knowledge, he points out that mathematicians had to learn to associate geometric "propositions" (theorems) with shapes like circles and squares. In order to extend the knowledge to other similar shapes, they had to repeat the process many times. He states that if mathematicians wanted to extend the mental associations made between a shape such as a triangle and a proposition, then they had to renew the "demonstration in another instance" before they "could know it to be true like another triangle, and so on."[5]

---

3. For the interactive nature of diagrams as thinking tools, see Tversky (2017).
4. The kinaesthetic aspects of wayfaring and pathfinding are central themes that run through Tim Ingold's work. For specific usages of "paths of observation," see Ingold (2007, esp. 166), and for the related notion of "paths of haptic perception," see Ingold (2015, 79–83). Ingold's views on the relationship between moving and knowing draws from the "paths of observation" concept as developed by the ecological psychologist James Jerome Gibson, who summarized his view on the topic in Gibson (1986, 197–198).
5. Locke (1765, 11).

Locke's discussion of the relationship between propositions, represented as words, and shapes, represented as lines, foregrounded the complexity of diagrams as forms of representation. He went so far as to point out that the renewal process required to link propositions to shapes involved an uncountable number of associations in the mind. Indeed, even a mathematician of Sir Isaac Newton's standing would not be able to reconstruct the "admirable chain of intermediate ideas, whereby he at first discovered it [a proposition] to be true."[6] The upshot of Locke's view of diagrams was that they needed to be sketched or consulted many times so that the meaning of the lines could be better preserved in the memory. In addition to strengthening associations, the process also strengthened a diagrammer's attention and recollection.

The connection that Locke drew between recursivity and diagrammatic knowledge was accepted by numerous Scottish professors. Edinburgh's Dugald Stewart summed up the view in the second volume of his *Elements of the Philosophy of the Human Mind*, a work that was, as we learned in previous chapters, based on the moral philosophy lectures that he gave in Edinburgh during the 1790s. He pointed out that theorems had to be "annexed to" a diagram "with which the general truth comes very soon to be associated."[7] He based his assessment on his own experience. As a student at Edinburgh University, he personally had witnessed the diagrams drawn and displayed in the mathematics courses, and in the anatomy and chemistry courses of Alexander Monro Secundus and Joseph Black, respectively.[8] Prior to being appointed Edinburgh's professor of moral philosophy, he had served as the university's professor of mathematics. Rather than expecting students to conjure images based solely on the verbal descriptions they read in books or heard in classrooms, he held that teachers needed to provide exemplars of diagrams so that students could record or memorize them. As I will show in this chapter, many Scottish professors understood that if they were going to employ diagrams, then they needed to use clear descriptive language and offer relatively simple schemata to which students could attach ideas.

Many students had been exposed to diagrams prior to their arrival at university. In previous chapters, we saw that there was a rich tradition of diagramming that existed in Scotland's schools and academies. The diagrams existed in a sophisticated world of material culture in which transforming a notebook page from a *tabula rasa* into a *tabula figura* was intimately tied to

---

6. Locke (1765, 12).
7. Stewart (1814, 114).
8. Eddy (2006, 383).

the skills required to create a *tabula verba*. The tradition of actively engaging with a diagram as a *tabula figura* continued within universities.

In this chapter, I wish to explore this rich material and visual tradition by examining how Scottish university students used their lecture notebooks to learn the kinds of diagramming skills evinced in the Simson episode. To pursue this topic, I treat their diagrams as mnemotechnic artifacts that are objects of historical enquiry in their own right. My aim is to articulate the skills that underpinned the processes of diagramming and to show that students learned to enact diagrams within sets of lecture notebooks that were designed to operate as paper systems organized around the order of the syllabus. I suggest that, within this context, the meaning and utility of diagrams were beholden to the part of the system that they were created to represent.[9]

Schemata as Useful Mnemonic Aids

When reflecting on the diagramming skills learned by Scottish university students, it is helpful to remember that cultural historians have long held that diagrams provided both a material and a manual reference point for the emergence and transmission of concepts that linked objects in the world and ideas in the mind.[10] Within this literature, diagramming is sometimes framed as a process of degeneration or generation.

The degenerative approach tends to focus on a specific diagram that is taken to be a normative object, Darwin's evolutionary tree for instance. The goal is to identify how previous versions were incomplete or how later versions were corrupted through transmission.[11] In contrast, scholars who employ the generative approach are more concerned with how the act of diagramming created a memorable and useful artifact. Their goal is to understand how diagrams were used to generate and reconfigure knowledge.[12] As we will see throughout this chapter, the generative approach is better suited for exploring

---

9. For an overview of the relationship between meaning of diagrams and their historical context, see Eddy (2021).
10. Galileo's diagrammatic annotations are analyzed in Raphael (2017, 98–128). For the diagrammatic thinking in the scientific thought of Robert Hooke, Carl Linnaeus, and Charles Darwin, see, respectively, Yeo (2014, 243–247), Müller-Wille and Charmantier (2012a), and Priest (2018).
11. The mimetic relationship between nature and diagrams from the seventeenth century forward is treated in Daston and Galson (2007). For links that some historians draw between corruption and hand-made diagrams, see Ferguson (1992).
12. The mnemonic relationship between nature and diagramming from the sixteenth to nineteenth centuries is unpacked in Yates (1966) and Nasim (2013). Other studies that treat diagrams as generative artifacts are Kaiser (2009) and Voss (2010).

the kinds of diagramming skills learned by notekeepers who attended Scottish universities during the long eighteenth century.[13]

Students who attended Scottish schools, academies, and domestic tutorials were exposed to diverse sketching, drawing, and watercoloring techniques. After arriving at a university, some used these skills to draw likenesses of their professors in the margins of their notebooks or to execute well-drawn watercolored diagrams that represented the experimental instruments that professors displayed either in situ or on a poster during lectures (figure 9.1).[14] An excellent example of this kind of watercolored diagram was created by an anonymous student who attended William Cullen's Edinburgh chemistry course around 1760. It depicts two experiments involving furnaces and glassware. Whereas the original sketches made with a black lead pencil were erased, the student left a graphite wisp of smoke shooting out of the kettle (figure 9.2).[15] Another striking example is the set of watercolored, three-dimensional renditions of experimental glassware that appear in the notebooks kept by the Welsh student Paul Panton in Joseph Black's 1776 chemistry course (figure 9.3).

The techniques exhibited in the foregoing diagrams, as well as those presented in most of the likenesses, are consistent with the kinds of drawing skills learned by many students before they arrived at university. Nevertheless, on the whole, most of the diagrams in the lecture transcripts do not feature the kinds of watercoloring or three-dimensional shading techniques presented in the notebooks created by academy students, the schoolgirl Jemima Arrow, the juvenile poet John Black, the apprentice George Sandy, or the other young notekeepers we encountered in earlier chapters of this book. This means that even though some students possessed tolerable drawing abilities, they elected not to use them when keeping their university notebooks. Instead, the diagrams rendered in their scroll books and transcripts were largely schematic, sometimes only consisting of a few quickly sketched lines or curves.

---

13. There is no shortage of literature on the history, anthropology, and philosophy of scientific diagrams. Ambrosio (2020) identifies two important interpretations: objectivity without images and objectivity w th images. She attributes the dichotomy more to the viewpoints of those writing about the history of diagrams and less to those who actually made and used them. She offers several plausible ways of overcoming the dichotomy, one of which includes focusing more on the ways that scientists were "doing things with diagrams" and to place "emphasis on the role of diagrams in pedagogical traditions and practices" (371).

14. For examples of professorial likenesses, see Black Bound MS (1767–1768, 3); Monro Secundus Bound MS (1799–1800, 13); Tytler Bound MSS (1793, 1: ff. 38v and 39r, 2: f. 61r); Finlayson Bound MS (1795–1796, flyleaves). Student notebooks contain other kinds of preliminary graphite sketches as well. See for instance the graphite sketch of trees in Hume Bound MSS (1810–1811, 1: f. 135r).

15. Cullen Bound MS (1: 1760, f. 147). Examples of furnaces also appear in Cullen Bound MS (1760, 1: ff. 144, 146–147). For mineshafts, see Cullen Bound MS (1760, 1: f. 126).

DIAGRAMMING | 329

FIGURE 9.1. Likenesses drawn by students of professors. Left: Pen and ink likeness by Thomas Cochrane of Joseph Black, professor of chemistry at the University of Edinburgh. Joseph Black, *Notes on Dr Black's Lectures on Chemistry, 2* (1767–1768), Thomas Cochrane (notekeeper), Bound MS, Old Catalogue Reference No. 34/28, f. 3. Reproduced with permission of University of Strathclyde Special Collections. Right: Pen and ink likeness by George Bruce of Alexander Monro Secundus, professor of anatomy at the University of Edinburgh. Alexander Monro Secundus, *Lectures on Surgery* (1799–1800), George Bruce (notetaker), Bound MS, RAMC/516, f. 13. Reproduced with permission of the Wellcome Library, London.

The schematic element of handmade lecture notebook drawings might at first seem strange, especially since professors used numerous diagrams, figures, and tables in their teaching. Additionally, printed syllabi contained illustrations such as maps.[16] But as I will explain in more detail below, simple diagrams provided a nondistracting platform to which students could attach the various facts, concepts, and processes that were relevant to the knowledge system under discussion in each course of lectures.

Diagrams tended to be used in courses that addressed topics relevant to natural philosophy and medicine, that is to say, mathematics, geometry,

---

16. Tytler (1782). There are six maps placed at the back of the volume.

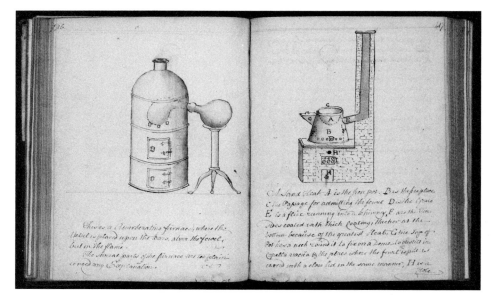

FIGURE 9.2. Pen and wash drawing of chemical furnaces and associated instruments by an anonymous student notetaker. William Cullen, *Chemical Lectures*, vol. 1 (1760), (anonymous notetaker) MS 1918, Bound MS, ff. 146–147. Reproduced with permission of the Wellcome Library, London.

mechanics, astronomy, chemistry, anatomy, physiology, botany, and natural history. Students attending other universities across Europe from the Renaissance forward commonly used modest diagrams to learn these subjects as well. This means that Scottish students were extending a larger tradition in which diagramming served as a mode of observing what they had witnessed within university classrooms, theaters, gardens, museums, or even field trips to local sites of interest.[17]

As explained in previous chapters, prior to arriving at university, many former academy and burgh school students had learned to make notebook diagrams. The practice certainly continued at the university level; however, there were two significant differences that influenced the ways in which students engaged with lecture diagrams. First, schoolroom settings allowed teachers to display visual ephemera for longer periods of time, either on the walls as posters or as handouts that students were given a good amount of time to copy. This was not the case for university courses where diagrams, most of which were handmade and unavailable elsewhere, were considered the intellectual property of the professors and, as a result, were available only at the

---

17. For university field trips, see Cooper (2007). For scholarly diagrams, see Kusukawa (2009).

DIAGRAMMING | 331

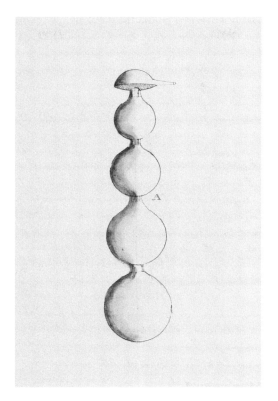

FIGURE 9.3. Pen and wash drawing of a glass instrument by the student notekeeper Paul Panton. Reproduced with permission of Science History Institute. Joseph Black, *Lectures on Chemistry* (1776), Paul Panton Jr. (transcriber), vol. 2, Bound MS, QD14 B533, f. 122.

time students were listening to the lectures. Second, the scope of the data being covered in university courses was much larger, which meant that, as we will soon see in the following sections, students had to develop their diagramming skills with a view to strengthening their ability to perceive, attend to, and recollect visual information that was directly relevant to the knowledge system being presented by any given professor.

Professors designed a variety of lecture diagrams. The simplest were drawn in chalk on blackboards. One of the most advanced offerings occurred in the Edinburgh anatomy lectures of Alexander Monro Primus and his son Alexander Monro Secundus. From the 1730s onward, they gathered a community of anatomical illustrators around the medical school.[18] Likewise, when the anat-

---

[18]. One of the primary locations used for meetings by St. Luke's Academy was Primus's anatomy lecture rooms in the university. In addition to working directly with artists, Secundus likely used rooms around his anatomy lecture theater to provide space for artists and their supplies. Rock (2000).

omy theater of the new college was built in the 1790s, Secundus ensured that it was flanked not only by preparation and dissection rooms, but also "painting rooms" that were "completely fitted up."[19] Secundus also commissioned a series of plates, several of which, when assembled together, presented a life-size engraving of a cadaver.[20]

Edinburgh students were offered an impressive array of diagrams in the botany course convened by Professor John Hope as well. Many of Hope's diagrams can still be consulted at the Royal Botanic Garden Edinburgh.[21] He employed a group of artists to create large colorized diagrams of plants that were related to the morphology, physiology, or cycle of reproduction.[22] It is likely that some of his diagrams were based on prints made by Chinese artists living in southern China.[23] Hope's artists made classroom diagrams with different media, including ink wash, iron gall ink, red chalk, graphite, and watercolors. His personal papers reveal that he was keen to use representations that could be seen easily by students. In 1776, for example, he noted that he needed to ask the artist Andrew Fyfe to create figures that could be "seen at any distance in the room."[24] He hung his posters and circulated his handouts in a large lecture room that spanned the top level of the purpose-built cottage in the university's Leith Walk botanic garden.[25] The cottage now sits in the grounds of the Royal Botanic Garden Edinburgh (figure 9.4).

The visual aids provided by professors such as Hope and the Monros made it easier to understand the emphasis that their syllabi placed upon the morphology and function of anatomical parts. Importantly, the acquisition of this understanding happened simultaneously as the professor explained the diagram alongside specimens or preparations in the classroom. The mnemonic impact of this kind of teaching made it possible for some students to remember the diagrams without drawing them. But even if such students had wanted

19. Duncan (1794, 425).

20. For Monro Secundus's cadaver plates, see M. H. Kaufman (1996; 1999). The efforts of the Monros to produce anatomical plates like those in the Netherlands are recounted in Rock (2000).

21. Hope Bound MS (n.d., ff. 54r, 73r, 102r [figs. 6–12]); Hope Bound MS (1780, ff. 108v, 109v, 110v).

22. See John Hope's teaching posters in the RBGE. Some are reproduced and explained in Noltie (2011a; 2011b).

23. Hope's many teaching diagrams show that, from the 1760s to the 1780s, he used posters drawn and painted by Andrew Fyfe (1754–1824), John Lindsay (d. 1803), Thomas Donaldson (b. circa 1739), William Delacour (d. 1767), John Bell, Jacob More (1740–1793), William Crawfurd, and Agnes Ord Williamson. The identities of Hope's artists are given throughout Noltie (2011b). For Hope's Chinese figures, see page 37. Andrew Fyfe and Jacob More have entries in the *ODNB*. John Lindsay has a brief entry in Desmond (1994, 430). Williamson is one of the few female illustrators known to be involved with university teaching in Britain. She made her drawings in the early 1780s and was the daughter of John Williamson, the chief gardener of the botanic garden. See Noltie (2011b, 9, 29, 30). For more on female illustrators at the time, see Roos (2018).

24. Corrie (2009, 34).

25. Forsyth (2016).

FIGURE 9.4. Professor John Hope's lecture room, Botanic Cottage, Royal Botanic Garden Edinburgh. Photograph by Lynsey Wilson, reproduced with permission of RBGE.

to physically draw all of the diagrams, there were a number of contextual factors that made it difficult to replicate a large number of them.

One mitigating factor was the physical space of the classroom. The interior design of lecture rooms and theaters was often unconducive to creating sketches for fully fledged drawings. As mentioned in chapter 8's discussion of lecture scroll books, many of the classrooms were too small, and this led to overcrowded lectures. The cramped space offered in the classrooms used by professors such as Aberdeen's Patrick Copland and the Monros in Edinburgh made it difficult to sketch the diagrams that they displayed on posters or drew on chalkboards. In the case of Monro Secundus, the maximum capacity of the benches in his octagon-shaped anatomy lecture theater was three hundred students. He had erected the theater during the late 1760s because his classroom in the university's 1617 hall was too small. But Monro's lectures were regularly oversubscribed, sometimes exceeding the original capacity by one hundred students. In 1783, for instance, his lectures attracted 436 students (figure 9.5).[26]

Unlike a normal classroom in which students sat in rows situated on one level, the seats of the theater used by Monro Secundus from the 1760s to 1790s

---

26. Haynes and Fenton (2017, 30–31). A helpful list of primary and secondary sources about the Monros and their teaching is given in the bibliography of Alberti (2016).

FIGURE 9.5. Anatomy and surgery course class ticket of the student John Guest, engraving in blue ink (1785) EUL, EA CA6. Reproduced with permission of Edinburgh University Library Special Collections. Alexander Monro Secundus's octagonal lecture theater is depicted at the bottom left side of the image. The professors of the University of Edinburgh's medical school issued tickets, slightly larger than modern business cards, for their lectures. Students paid their fee directly to the professor and then received the ticket as proof of payment.

rose up around a center stage to afford a better view of the professor and his preparations of anatomical specimens. Sitting cheek to jowl in the tightly stacked rows that rose up around the central lecturing dais, students had to contend with a restricted sketching space as they attempted to peer around the heads of other students in the row below. Watching Monro's preparations, listening to his lecture, *and* recording his observations was challenging. Lecture notebooks indicate that many students spent so much time trying to capture a professor's words in the classroom that they simply did not have the time to sketch the visual material.

The absence of diagrams in some student notebooks was noticed by none other than Monro Secundus. After reading John Thorburn's shorthand notes of his lectures, Monro observed with regret that they lacked the diagrams that he used to teach anatomy. Extant sets of lecture notes taken in Monro's course and in the anatomy or physiology lectures of other professors teaching in Scottish universities reflect a similar situation. Monro explained the omission in the following manner: "Mr. THORBURN, who had no knowledge of Anatomy when he began to write my Lectures . . . had not attempted to copy any of those figures which, in this and in many other parts of the Course, I have been in the custom of drawing with chalk upon the black board, in order to render my lectures more intelligible to students."[27] Monro's assessment reveals that, in addition to the time required to make sketches, re-creating anatomy figures sometimes necessitated a familiarity with the subject matter that some young students did not possess.[28]

Despite the foregoing restrictions presented in Scotland's classrooms, many students managed to sketch or memorize diagrams. They were able to achieve this feat because they strategically reduced professorial visualizations into schemata that were easy to sketch and remember. The nature of such simplified diagrams can be divided into shapes, pictograms, and tables. The skills required to sketch them were central to the learning experience of the students who drew them. In what remains of this chapter, I discuss how students learned to make and use such diagrams as visual thinking tools and as mnemonic cues that corresponded to the system being presented in a given professor's course syllabus.

Shapes as Repurposed Perceptual Devices

Professors intentionally repurposed recognizable shapes such as squares, triangles, circles, and chiasmas so that students could easily attach complicated ideas to them. Students were familiar with such shapes because they were used by schoolchildren to learn geometry and various branches of practical mathematics such as gauging, leveling, and surveying. The structures of simple shapes made them easier to use and explain when professors hung them on posters or drew them on chalkboards.

Within Scottish universities it was generally recognized that shapes could

---

27. Monro Secundus (1794, 40).
28. There is evidence that some Edinburgh anatomy students during the 1790s included anatomical printed plates in their notebooks. See the plates by Alexander Bell included bound with student notes in Monro Secundus and Monro Tertius (1801–1802). The plates complement the handwritten narrative recorded by the student. Monro Tertius began giving many of the course's lectures in 1797. He was widely regarded as an uninspiring speaker whose lectures were not easy to follow. The plates, therefore, might have been included by the student to compensate for the lower quality of the lectures.

serve both as cognitive metaphors and as paper tools that aided thought.[29] As shown in previous chapters, Scottish theories of learning treated diagrams and other kinds of visual patterns as perceptual devices, as objects that assisted the operations of the human mind. The ubiquity and fecundity of Locke's *tabula rasa* metaphor and its extension as a *tabula figura* provided professors with a cognitive rationale for classroom diagrams. This rationale was aided by John Locke's larger influence over the science of mind in Scotland, particularly the priority that it gave to the mental operations of perception and attention.[30]

Locke held that diagrams based on geometric shapes were platforms upon which ideas could be "laid" in ways that aided reason. He called this kind of diagramming "right reasoning." Scottish professors took a similar position. Glasgow's professor of moral philosophy Thomas Reid, for example, referred to the kinds of diagramming performed by geometricians as a "long process of reasoning."[31] Importantly, according to Locke, in order to perceive the shape of a diagram as a visual and kinesthetic nexus of ideas such as proportionality, linearity, equality, or measurability, one needed to literally draw it. In his words, when a person "would shew the certainty of this truth, [that] the three angles of a triangle are equal to two right ones; the first thing, probably, that he does, is to draw a diagram."[32]

Reid and other professors interested in the science of mind, Dugald Stewart for instance, maintained that tracing the lines of a geometric diagram with the eye or drawing them with the hand strengthened the mental operations of perception, attention, recollection, and abstraction in a way that aided ratiocination.[33] Accordingly, when university students drew the shape-based diagrams offered by their professors, no matter how simple, they did so in an educational context infused with cognitive theories that placed a high value on the skills that could be acquired via the act of diagramming.

Professors such as Edinburgh's Monro Secundus, John Robison, and John Playfair used simple geometric shapes over and over again to represent dif-

---

29. As in previous chapters, here I am using the term *paper tool* as it is defined in Klein (2001; 2003; 2019).
30. Eddy (2016b).
31. Reid (1764, 236–237).
32. Though Locke's positive view concerning the proprioceptive relationship between drawing and understanding diagrams is intimated throughout his works, it was most clearly articulated in his widely read letter to the bishop of Worcester, which I have quoted above. The letter appears in Locke (1759, 374). Locke used diagrams, particularly triangles, throughout his work to illustrate various aspects of knowledge acquisition. See Locke (1767, 36, 97, 148, 151–153, 157, 164, 170, 174, 195–196, 272).
33. Reid in particular valued the kinds of mental tracing used when the eye was "fixed" upon the lines of a diagram. Both Reid and Stewart held that diagramming strengthened the mental operations of perception, attention, recollection, and abstraction. Reid (1764, 236–237, 243–244) and Stewart (1792, 157–158, 176–177, 419).

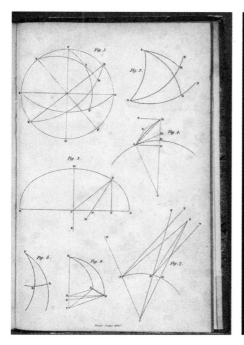 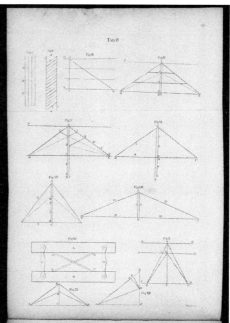

FIGURE 9.6. Lecture diagrams. Left: The first of four unnumbered plates that appear at the end of John Playfair, *Outlines of Natural Philosophy*, vol. 2, 2nd ed. (Edinburgh: Abernethy and Walker, 1816). Reproduced with permission of the Huntington Library, San Marino, CA. Right: Table 2 in Alexander Monro Secundus, *Observations on the Muscles, and Particularly on the Effects of Their Oblique Fibres* (Edinburgh: Dickson and Balfour, 1794), reproduced with permission of the Wellcome Library, London.

ferent forms of movement. Whereas Monro Secundus was particularly fond of using triangles to represent anatomical movement, Robison and Playfair employed combinations of circles, semicircles, quarter-circles, and angles to depict the movement of matter covered in their mathematics and natural philosophy courses (figure 9.6).[34] Though professors offered a variety of shapes to illustrate their lectures, students were selective about which ones they drew in their notebooks.

The shape-based diagrams that students saw fit to draw were those that were simple and could be sketched quickly while they were taking notes in the classroom. This explains why Monro Secundus's simpler diagrams, many of which were different kinds of triangles, seem to have been replicated more often (figure 9.6). Students drew such diagrams in their notebooks to learn

34. Engravings of Monro's diagrams were printed in Monro Secundus (1794). John Robison's appear in Robison (1804). Engravings of John Playfair's diagrams were printed in his lecture headings. Playfair (1812; 1816).

how they worked as thinking tools and to create a copy that they could consult in the future when they became professionals. Such was the case for James Carmichael Smyth, who had studied with Monro around 1760 and went on to become the physician extraordinary to King George III. In 1817 Smyth wrote a letter that described and explained one of the triangular diagrams that Monro used to illustrate the movement of muscles. In his words, "his [Monro's] reasoning on this subject, with a diagram which I copied at the time, I have still in my possession."[35]

Like the shapes drawn by students taking Monro Secundus's anatomy course, those recorded by students attending geometry or mechanical philosophy lectures tended to be those that could be quickly and easily drawn. As evinced in the swiftly drawn lines, numbers, and words inscribed in scroll books, such diagrams were efficient calculating devices that helped students understand the quantitative relationship between time and space in realtime. A similar relationship is presented in the diagrams included in lecture transcripts. Though students developed important diagramming skills during the rewriting and redrawing process, the final product, which we might call a transcript diagram, presents only a partial picture to historians seeking to understand the skills being learned by students, the reason being that transcript diagrams often omit the scribbles, misdrawn lines, and erroneous calculations that were part of the early notekeeping process.

The difference between scrolled and transcribed shape-based diagrams can be seen in the bound notebook titled *Syllabus of Mathematics* that was kept by an anonymous student who attended the Glasgow mathematics course given by Professor James Millar during the early nineteenth century.[36] The content of Millar's lectures was similar to that of the geometry courses offered by academies and universities. The bound manuscript contains both a scroll book and a transcript based on different parts of the lectures. It is likely that the student did not have time to rewrite and redraw the scrolled notes and elected to simply bind them with the transcribed sections. But the presence of these two manuscript genres in one bound volume offers insight into how students used the notekeeping process to learn diagramming skills that involved geometric shapes.

Two sets of scrolled trigonometry problems in the notebook are especially instructive.[37] As can be seen in figure 9.7, the script for the first set runs from the

35. James Carmichael Smyth, "Letter from Dr Carmichael Smyth, November 19, 1817," in Monro Secundus (1840, xiii–xvi).
36. Millar Bound MS (1815–1818).
37. Millar Bound MS (1815–1818, 40).

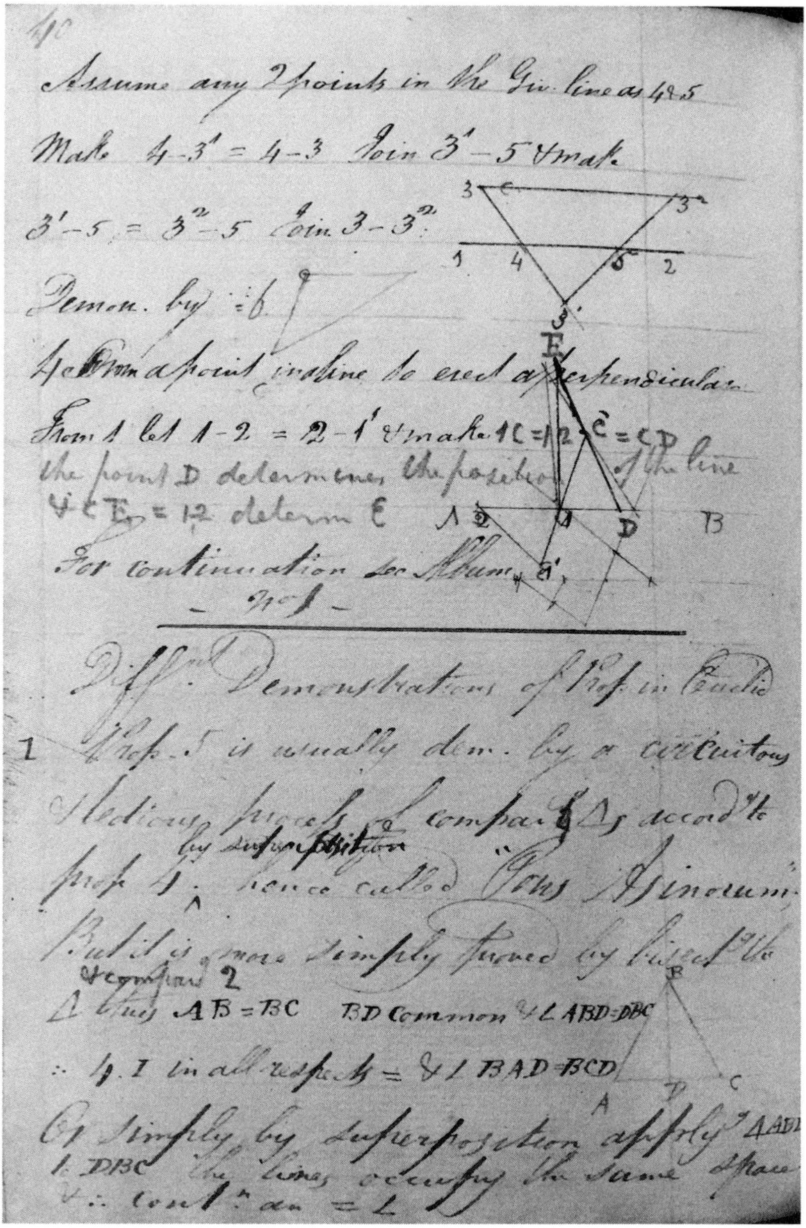

40

Assume any 2 points in the Giv line as 4 & 5

Make 4-3' = 4-3 Join 3'-5 & make

3'-5 = 3²-5 Join 3-3².

Demon. by ∴ 6

4c From a point in a line to erect a perpendicular.

From 1 let 1-2 = 2-1' & make 1C = 12 C = CD
the point D determines the position of the line
& C E = 12 determ E

For continuation see Album

— № 1 —

Diff⁴ Demonstrations of Prop in Euclid

1 Prop. 5 is usually dem. by a circuitous
   tedious process & compar. △s accord⁴ to
   by superposition
   prop 4. hence called "Pons Asinorum".

   But it is more simply proved by bisect 1 &
   & compass
2 thus AB = BC   BD Common & ∠ ABD = DBC

   ∴ △. 5 in all respects = & ∠ BAD = BCD

   Or simply by superposition apply △ ADB
   & DBC the lines occupy the same space
   & ∴ cont⁴ an = ∠

FIGURE 9.7. Lecture diagrams sketched in a scroll book. James Millar, *Syllabus of Mathematics Delivered by Prof. Miller* (1815–1816), Bound MS, HM72759, Burndy Codex 47, f. 40. Reproduced with permission of the Huntington Library, San Marino, CA.

top of the page to the thick break line in the center. It features three diagrams formed from triangles. The second set of problems runs from the break line to the bottom of the page. It features one diagram formed from an isosceles triangle. The student was particularly organized in that he drew a graphite grid of ruled lines for the script prior to the lecture. Close inspection of the page reveals that most of the writing was done in ink during the lecture. The graphite lines and alphanumeric headings of the diagrams were drawn in the classroom as the student listened to Millar's discussion of the problems. At some point after the lecture the student revisited the first set of problems on the top by writing more graphite script and by tracing the two righthand diagrams in ink.

The script and shapes in the *Syllabus of Mathematics* notebook reveal that diagrams in mathematical scroll books were works in progress in that they were constructed over a period of time, both inside and outside the classroom. The notebook also offers some insight into how diagramming played out as a realtime learning activity, that is to say, it offers clues as to how students sketched diagrams while simultaneously listening to a lecture. In figure 9.7, for example, the inverted triangle at the upper right side of the page is positioned within the blank space offered at the end of the third line of writing. The placement of the diagram shows that the student was thinking about how to maximize the space of the page in a way that allowed him to record the greatest amount of information. In other words, the student was practicing a sophisticated form of information management that involved decisions concerning the placement of qualitative and quantitative data. It was in these kinds of decisions that student notekeepers learned to understand the representative potential of manuscript communications media.

Scroll books offered the opportunity to learn realtime skills of diagramming that helped students negotiate the rich and potentially overwhelming relationship between data and media. In the case of the anonymous student who kept the Glasgow *Syllabus of Mathematics* notebook, however, scribal multitasking became more difficult to maintain when the lecture moved to a more complicated problem that involved sketching a diagram comprising several superimposed triangles. The result can be seen in the diagram that occurs just above the breakline in the center of the page presented in figure 9.7. The apex, which is labeled point E, runs through the script written above it. The collision indicates that the student was scrolling in haste and did not have time to fully consider how the lines might be drawn so that they did not intersect with the script he already had written. It is precisely in these kinds of adjustments and overlaps where students learned to negotiate the complex relationship between the abilities that they had learned in school and the diagramming skills that they wished to acquire through keeping notes in a university lecture course.

Shape-based diagrams usually represented ideas, objects, or processes relevant to the knowledge system being presented in the lectures. Nowhere was this more the case than in the chiasma diagrams recorded by students who attended the Glasgow and Edinburgh chemistry lectures of William Cullen and Joseph Black from the 1740s to the 1790s. Formed in the shape of the letter X, the diagram represented the attractions, the affinities, that existed between substances in compounds. In eighteenth-century experimental philosophy, affinity was the main force that chemists used to explain the behavior of micromatter. It was used to understand the majority of the medical, agricultural, and mineralogical experiments demonstrated within the teaching spaces of Scottish universities.

Put more clearly, affinity was to chemistry what gravity was to mechanics and planetary astronomy.[38] Since Cullen and Black did not want to confuse their students, many of whom came from diverse educational backgrounds, their chiasmas were heavily heuristic. Black's rendition of the chiasma is particularly significant to historians of science because, unlike Cullen's, it included numeric ratios and is often taken to be the first modern-day chemical equation (figure 9.8).[39] The purpose of the diagram was to provide a simple, schematic representation to which students could easily assign information.

Black's chiasma was a paper tool in that it consisted of an outer zone of qualitative headings placed on the endpoints of the lines and an inner zone of quantitative headings running around the inside angles. The outer zone of headings were names or symbols of substances involved in double elective chemical reactions. The headings of the chiasma depicted in figure 9.8, which was recorded in the notebooks of Black's student Paul Panton, are volatile alkali (top left), muriatic acid (top right), vitriolic acid (bottom left), and fixed alkali (bottom right). Headings of other substances could be placed on the tips of the chiasma as well. The numeric headings featured in the inner zone of the chiasma angles represented the ratio of attraction—the affinity—between the substances.

Black's chiasma was a chronogram in that it heuristically represented at least three stages of a chemical reaction. The stages can be seen in the styl-

38. For material and conceptual foundations of chemical affinity, see Klein (1994), Kim (2003), Taylor (2006), and Klein and Lefèvre (2007).
39. Cullen Bound MS (1765–1766, ff. 221, 328). Chiasmas occur in most copied notes from Black's lectures. Black Bound MS (1782, 3: lecture 107). For a 1770s version, see Black Bound MS (1776, vol. 3, ff. 107, 493). Several are reproduced in Black (1966/1767, 274–279). The important role played by Black's chiasma within the history of science is explored more fully in Eddy (2014).

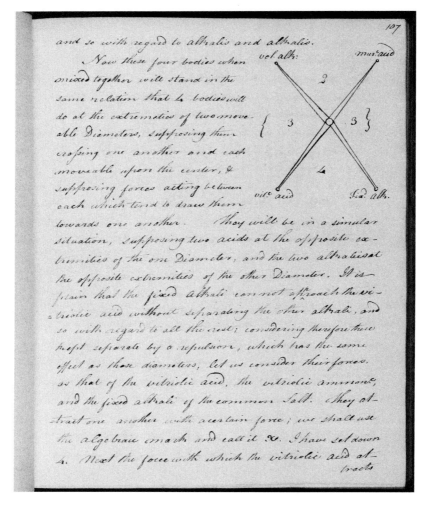

FIGURE 9.8. Double elective affinity chiasma. Joseph Black, *Lectures on Chemistry, vol. 3* (1778), Paul Panton (notekeeper), Bound MS, Chemical Heritage Foundation, QD14.B533 1778, folio 107. Reproduced with permission of Science History Institute. Black used a chiasma to visualize the hypothetical strength of the forces acting in chemical reactions.

ization of the chiasma presented in figure 9.9. In the first stage, there were two compounds, neither of which existed in physical contact. There was compound AB, made up of substances A and B. There was also compound CD, comprising substances C and D. Placing compounds AB and CD in contact with each other created the second stage, in which the affinities between the substances in the two compounds broke. In the third stage, the substances A and C united to form compound AC, and substances B and D united to form

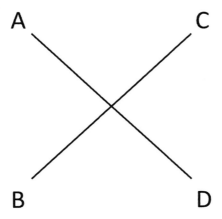

FIGURE 9.9. Stylization of Joseph Black's double elective affinity chemical chiasma. © Matthew Daniel Eddy.

compound BD. To put it algebraically (as Black sometimes did in his lectures), when a compound made up of substance A and substance B was placed in contact with another compound made up of substance C and substance D, the substances broke apart and recombined so that A united with C and B united with D: AB + CD ⇒ AC and BD.

Since the chiasma had four tips, the structure could be used only to represent an experiment involving a maximum of four substances. This meant that Black had to take the form of the diagram into account when he selected compounds for classroom experiments that could be represented as a chiasma in a student notebook.[40] He developed the diagram because his students were faced with a particularly thorny visual problem that plagued chemical knowledge at the time. Whereas the objects of mechanics and anatomy were things like planets and body parts that were readily visible, the objects of chemistry, that is, chemical microparticles, were not entities that had ever been seen nor had the prospect of being seen in the near future. Black's chiasma, therefore, was an attempt to mitigate this problem and was not meant to be a literal representation of material particles or their movements through time or space.

Though an understudied shape, the chiasma served as a learning and thinking tool within many literary and educational contexts.[41] Historians seeking to explain the visual origins of Black's chiasma have traditionally pointed to similar diagrams developed by his teacher William Cullen, and by Jean Be-

40. Eddy (2014) and Crosland (1959).
41. For the metaphorical and spatial role of the chiasma as a shape used for thinking or learning in the modern world, see Engel (2016) and Camp (2008).

guin in his popular seventeenth-century textbook titled *Elemens de Chymie*.[42] Yet, unlike the Cullen and Beguin diagrams, Black positioned numbers on the inner angles of his chiasma. Where did he get this idea and what mnemonic purpose did it serve for student notekeepers? In order to see the origin and utility of Black's inner zone of numbers, we must first remember that many of the students sitting in his lectures previously had attended Scottish schools. These students would have been familiar with a chiasma because it was used by Scottish schoolmasters to teach "the casting of nines" or "casting out the nines," which was a handwritten calculation performed to double-check the answers of lengthy arithmetic problems.[43]

The casting of nines calculation was written out on a chiasma, a practice that probably originated from the shape's longstanding use in the Ramistic tradition.[44] An excellent handmade example of this kind of chiasma can be found in the marginalia written by the children of the Erskines of Torrie in the books of the family library during the middle of the eighteenth century (figure 9.10).[45] Earlier in this book I showed how the Erskine children used the books of the family library during the late eighteenth-century as makeshift notebooks. In the 1704 edition of *Herodiani historiarum libri 8*, one of the children drew a chiasma and, in so doing, revealed the visual origins of arguably the first chemical diagram. The seventeenth-century to nineteenth-century versions of the calculation consisted of several steps and the numeric answer to each one was placed on the inner angles of the chiasma. Thus, the inner zone of numbers that Black used to represent chemical affinity ratios was based on a simple paper tool used by schoolchildren. Like the geometric polygons used to teach university students mathematics and mechanics, Black's chiasma would have been familiar to his students.

But what specifically were student notekeepers learning from Black's chiasma when they drew it? As mentioned above, the numbers in each angle represented the ratio of attraction operating between the two corresponding

---

42. The connection between the chiasmas of Cullen and Black was addressed in print as early as 1803 when John Robison included a reproduction and description of the diagram in Black (1803a, 544–546). The conceptual connections shared by the chiasmas of Black, Cullen, and Beguin are addressed in Crosland (1959). The graphic context and history of Beguin's chiasma is addressed in Smets (2013).
43. Casting of nines was explained in mathematics texts used in Scottish schools. See Panton (1771, 23–24) and Gordon (1770, 25–29). The context of its usage is given in Wilson (1935, 2, 31, 85). Casting of nines is not used very often today, but, when it is employed in twenty-first-century classrooms, the numbers of the computation are lined up in a column. A clear summary of the computation is given in Asimov (1964, 29–34).
44. See the calculation chiasma in Ramus (1569, 124). The connection between the Ramist chiasma in this text and Beguin's chemical diagrams is addressed in Smets (2013).
45. The Erskine chiasma appears on the endpaper at the front of Herodotus (1704). The provenance of the eighteenth-century books from the Erskine family library is addressed in Friends of Duff House (2011).

FIGURE 9.10. Casting of nines chiasma. The diagram was used by Scottish children to double-check multistep addition. The chiasma in this figure was drawn by one of the Erskine children living at Dunimarle Castle during the 1760s. Herodotus, *Herodiani historiarum libri XIII. Recogniti & notis illustrati* (Oxoniæ: G. West & Ant. Peisley, 1704), Dunimarle Library No. 982. Reproduced with permission of the Dunimarle Library, Duff House, Historic Environment Scotland.

substances fixed to the pinnacles of its outer zone. Crucially, the ratios that students inscribed on the diagram in their notebooks did not represent a real unit of measurement. Black had devised the ratios only to conceptualize the relative attractions of the substances visualized in the diagram.[46] The notion of using ratios to represent force in this manner was probably taken from planetary astronomy, where they were employed to compare the unitless planetary distances and perturbations of orbits. It was not until after astronomers had collected data during the 1761 and 1769 transits of Venus that an accurate distance between the sun and the earth was calculated, thereby allowing ratios to be expressed in known units of measurement (like miles).[47] Since those transits occurred during Black's lifetime, he and his students

---

46. Eddy (2014).
47. The exact ratio was not known in Black's lifetime. The precise distance was contested because of the "drop effect" that occurred when Venus first appeared in front of the sun. Schaefer (2001).

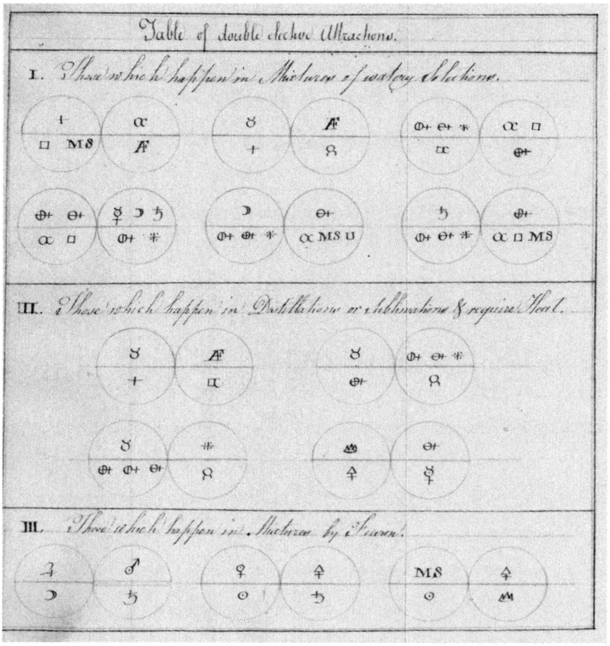

FIGURE 9.11. Circlet chemical affinity. Joseph Black, *A Course of Lectures on the Theory and Practice of Chemistry*, vol. 3 (1782), Bound MS, Royal Society of London, MS/147/3. Used by permission of the Royal Society of London.

probably saw his ratios in a similar way, as formulas simply waiting to be activated with numbers in the event that a viable unit of chemical force was discovered.

The chiasma worked for student notekeepers because it was visually simple, and this perhaps explains why they consistently rendered it in a way that closely resembled Black's version of it. This simplicity, however, was not as straightforward for other shapes that he used. An excellent case in point can be seen in his second affinity diagram, which was made up of two circlets set side-by-side and dissected by a line (figure 9.11). Though the circular structure of the diagrams was familiar to students who recorded them, it took some time and effort to learn how to draw them. This situation is evinced in the scroll books taken by Charles Blagden in Black's 1766 lectures.[48] Blagden would go on to become the secretary of the Royal Society of London. The first time that Blagden encountered Black's circlet diagrams, however, he struggled to reproduce them as sketches in his scroll book. In the end, he eventu-

48. Black Bound MS (1766–67).

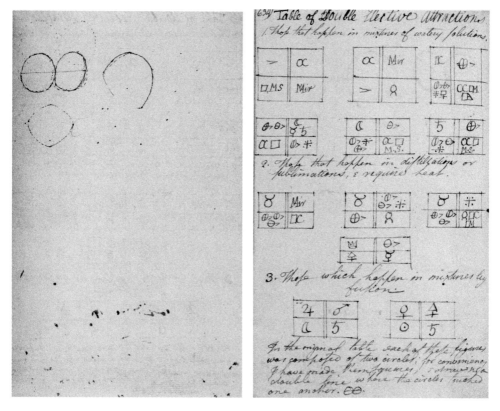

FIGURE 9.12. Left: Charles Blagden's unsuccessful attempt to inscribe Black's affinity circlets. Right: Blagden's reconfiguration of Black's circlets into squares. Joseph Black, *Notes of Dr. Black's Lectures, Notebook 9* (1766–1767), Charles Blagden (notekeeper), Bound MS, MS 1277, f. 634r–634v. Reproduced with permission of the Wellcome Library London.

ally learned to use them by redrawing them as squares, an act that took some time to work out (figure 9.12).[49]

The general point to take from Blagden's unfinished circlets, or even from the many complete chemical circlets and chiasmas that appear in other student notebooks, is that even when the shape of a diagram was familiar, notekeepers still had to learn how to use or make it kinesthetically and how to make sense of it conceptually. In addition to the diagram's geometric structure, students had to learn, for example, the meaning of the chemical symbols and how to follow the flow of information. As Blagden's case illustrates, since

[49]. Blagden's unfinished circles occur in Black Bound MS (1766–67, notebook 9, f. 634r). His square revisualizations occur on f. 634v.

there were no printed versions of the diagrams, some students had to learn how to draw them in their notebooks while they were observing the experiments that Black was conducting in front of them.[50]

Blagden's adaptation of Black's circlets also reminds us that even in the case of simple shapes, students had to learn how to execute their drawing abilities while learning to understand the meaning of a diagram. In the case of Black's chiasma, even though its general structure, components, and chemical relationships remained similar in different sets of student notes, we must not forget that each student experienced them anew for the first time when they attended his lectures. Consequently, students found ways to adapt the diagram according to their personal writing skills and learning needs. Some chiasmas were bigger than others. Some students wrote a verbal description of the diagram and then drew it. Others drew it first and then described it.

When the medical student Thomas Cochrane filled out his transcript of Black's course during the 1760s, for instance, he wrote the verbal description of the experiments first and then delineated the diagrams below them (figure 9.13). The reason Cochrane chose to place the verbal description before the diagram was simple: though the form was familiar, its meaning, like most of the shapes used by Scottish university professors, was not self-explanatory. This was because he was a student and not an experimental expert like Black. In order for him to understand the chiasma in his transcript, he felt it necessary to write the description first so that he could better understand the diagram when he drew it.

Unlike the charts and figures depicted elsewhere in his chemistry transcripts, Cochrane did not attempt to first draw the lines of his transcribed chiasma in graphite. The thickness of the ink strokes at the top of each line of his chiasma in figure 9.13 indicate that Cochrane drew it from top to bottom. The uneven, occasionally squiggled, flow of each line indicates that he drew freehand without the use of a straightedge like a ruler. When compared to the even and fluid lines of the well-practiced letters of Cochrane's handwriting in his transcript, the lines of the chiasma communicate less of an orthographic confidence, suggesting that he was less familiar with rendering it in pen or ink. In other words, drawing the lines of the chiasma took more concentration, more effort, indicating the importance that he attached to making sure it was

---

50. The posthumous printed edition of Black's lectures edited by John Robison featured a representation of the chiasma in the endnotes. Like the some of the chiasmas rendered by students in their notes, the structure was slightly different from that used by Black. The Robison chiasma is depicted inside a square, not as a freestanding chiasma, and the circles representing the substances are labeled with alphabetic heads (A, B, C, D), not with chemical names or symbols. Black (1803a), 544. The Robison chiasma was included in the American edition as well. Black (1807, 390).

FIGURE 9.13. A chemical chiasma drawn by the student Thomas Cochrane in Professor Joseph Black's chemistry course. Top: Enlargement of Cochrane's chiasma. Bottom: The placement of Cochrane's chiasma within the notebook page. Joseph Black, *Notes on Dr Black's Lectures on Chemistry, vol. 1* (1767–1768), Thomas Cochrane (notekeeper), Bound MS, University of Strathclyde Special Collections, Old Catalogue Reference No. 34/28, f. 221. Reproduced with permission of University of Strathclyde Special Collections.

included in his notebook. In exercising this effort, he preserved the diagram, both in his memory and on paper.

Cochrane's chiasma represented simple chemical reactions that took place in the experiments that he had witnessed firsthand in Black's classroom lectures. More specifically, in the year attended by Cochrane, Black used chiasmas to represent only a handful of the hundreds of experiments that he showed to his students. The descriptions offered in Cochrane's notes in these instances follow the normal practice of stating the substances in the various compounds and then describing what happened when they were exposed to heat, water, or each other. Cochrane's inclusion of the chiasma provided a diagrammatic mode of remembering Black's accounts of how the substances in the initial compounds recombined to make new compounds.

In addition to observing what Cochrane's descriptive prose has to say about the meaning of Black's chiasma, it is important to note what it does *not* say about the shape's actual lines. Cochrane did not record, for example, what, precisely, the two lines themselves were meant to represent. Did they represent time, space, or entities? At first glance this might appear to be an absent-minded omission, an instance in which Cochrane simply did not record the part of the lecture in which Black explained their meaning. But Cochrane was a conscientious student and a diligent notekeeper, so the silence is notable.

If we consider the chiasmas drawn in other notebooks kept by Black's students, it can be seen that they too are silent on the meanings of the lines. A comparison to the chiasma sections of Black's own transcribed lecture notes housed at the Royal Society of London also reveals that he did not clearly explain the meaning of the lines either.[51] The overarching picture that emerges from the community of notekeepers that surrounded Cochrane is that the lines were, in fact, not meant to represent anything that existed in the material world at all. This made the chiasmas different from the drawings of scientific instruments that Cochrane and other students drew elsewhere in their notebooks, all of which were schematic representations of objects they had personally seen.

The absence of an explicit meaning for the lines of the chiasma did not indicate that delineating it was a purposeless exercise. Quite the contrary. As we have seen above, the lines were in fact a mnemonic structure, a form of representation that served as a realtime learning tool for Cochrane and the diverse range of students who made notebooks based on Black's lectures. When students drew and redrew the chiasma, it served as a mode of dividing up the space of a page into an outer zone, comprising the tips of the lines, and into an inner zone, comprising the angles created by the axis. Thus, even before students inscribed headings in the respective zones, the act of drawing the chiasma was a way of learning how to create a mnemonic platform that visually sorted different kinds of information. It is in this sense that the mnemonic potential of the lines was multivalent in that, like a grid or a graticule, they could

---

51. Black Bound MS (1782, vol. 2, lectures 61–63 [folios are unnumbered]).

be used to represent other kinds of simple chemical substances that a student might encounter in the future.

Pictograms and Visual Judgment

In addition to using simple shapes as diagrams, university professors exhibited specimens, instruments, apparatus, and posters. One of the most efficient ways that students captured these objects in their notes was through the use of pictograms. The practice of using pictograms as mnemonic cues within a knowledge system was commonly practiced in scholarly culture at the time, especially in medicine and natural history. Perhaps the most influential pictograms were those included on the 1736 plate titled *Methodus Plantarum Sexualis* that was devised by the Swedish naturalist Carl Linnaeus and the German artist Georg Dionysius Ehret.[52] The plate presented twenty-four flower pictograms, each of which served as a visual cue for one of the classes of Linnaeus's botanical taxonomic system. Within the history of the natural sciences, Linnaeus's system was an important development because it laid the foundation for the scientific classification of organisms that we use today.

In addition to the plate's success as visual form of representation within Linnaeus's publications, the schematic nature of the pictograms proved to be an effective learning tool for students who attended the botany lectures that he gave in his capacity as professor of medicine at Uppsala University. There he championed using live and dried specimens in conjunction with the pictograms.[53] His desire to use the pictograms as schemata in this manner originated in his own notebooks, where he employed them as thinking tools when in the field and when ordering information on paper.

The role of pictograms in Scottish student notebooks served a similar function to those developed by Linnaeus. Drawing them was a mode through which students learned to quickly create simple diagrams that represented the features of an object relevant to the system they were writing in their notebooks. Students drew pictograms most often in medical notebooks. Most were simple and schematic. They normally consisted of a few lines.

Most pictograms depicted an image that students had seen on a poster or an object that they had seen in an experiment or demonstration. Scaling

---

52. Ehret (1736).
53. Charmantier (2011). The relationship between the material and visual culture of teaching in Linnaeus is addressed in Eddy (2010a). Charmantier emphasizes that speed and simplicity were important features of Linnaeus's notebook drawings. She groups Linnaeus's pictograms under the rubric of "analytical drawings" that functioned as diagnostic tools. Linnaeus's active engagement with print and manuscript culture is explored in Müller-Wille (2018) and Charmantier and Müller-Wille (2014).

down such images and objects into a pictogram allowed students to sketch or memorize a schema without seriously impeding their scrolling efforts. But this kind of scaling took practice, mainly because designing pictograms required students to select the key contours of specimens or poster images that they wished to draw or omit. Put another way, most of the pictograms in both scroll books and transcripts consist only of a simple set of lines configured to represent the multiplicity of edges presented by the profiles of the objects or images that students chose to draw.[54]

The surface of a recently cut tree trunk displayed in a botanical lecture, for instance, presented many possible edges that could be reduced to pictographic contour lines. If drawn as a two-dimensional cross-section, should there be a line to demarcate the edge between the inner and outer bark? Should the edge of every ring be included? Should the edges of decayed potions be omitted? These questions reveal that achieving fidelity between the lines of pictograms and the experience of the props and posters presented in lectures was by no means a simple affair.

In addition to extending drawing skills that students already possessed, creating a notebook pictogram served as a form of manual, visual, and conceptual training that worked in conversation with the subjects they were studying. A pictogram of plant germination, for instance, needed to include the basic features of the embryo. Other morphological parts were potentially expendable when it came to ensuring that the image remained schematic. Finer details such as root hairs or the texture of the dermis needed to be selected or omitted in regard to whether they were discussed in the lecture and whether they inhibited the pictogram's function as a schema.

Students also had to consider different pictographic styles of representation.[55] Should the image be rendered in a continuous line drawn in an unbroken stroke? Or should it be a series of lines that required multiple stokes? Should the pictograms be built out of geometric shapes like triangles, squares, circles, or ovals? Or should they simply present a freeform silhouette? Should watercolors be used? When students grappled with these and other questions relevant to the form and meaning of pictograms, they were, in the words of Nigel Holmes, learning the skill of "drawing to explain."[56] The explanation was in many respects autodidactic in the sense that the purpose of drawing a

---

54. For a discussion of the multiplicity of decisions required to transform the edges of objects and scenes into contour lines, see Hochberg (2007, 30–59).
55. The ways in which the visual conventions of a group of pictographs can operate as a visual system are discussed in Holmes (2000–2001).
56. Holmes (2001–2000, 143).

notebook pictogram was both to understand the facts, concepts, and objects being presented during the lecture and to learn how to be a pictogrammer, a person who could redeploy the skills of pictogramming in other contexts.

The pictograms created by student notekeepers tended to be side views, side cuts, or silhouettes of objects or images on posters that students had witnessed in the classroom. The aim was to draw a small, simple, and schematic image that helped them remember what they had seen, touched, smelled, heard, and, occasionally, tasted. Sometimes students drew pictograms in midsentence within the narrative of the script. When the student George Bruce attended Monro Secundus's lectures during the 1790s, for instance, he created pictograms of inhalers, scissors, pins, tubes, syringes, bandages, and other instruments associated with surgery and dissection, that is to say, instruments that were displayed or passed around the lecture theater (figure 9.14).[57] An anonymous student attending William Cullen's chemistry lectures during the 1760s devised pictograms of retorts, Florentine flasks, funnels, and other chemical "vessels."[58] Instead of concentrating solely on instruments, students taking John Hope's Edinburgh botany course drew pictograms of images that he displayed on posters.[59]

Student pictograms were miniature diagrams that were meant to accompany the narrative recorded in a lecture notebook. This usage strengthened a student's attention, abstraction, and recollection skills, that is, operations of the mind that, according to the developmental psychology of Scotland, could be shaped through practice. As we learned in the previous section, drawing shapes as diagrams was seen as conducive to learning by professors. The pictograms drawn by students offer the opportunity to explore the practical application of this sentiment.

As shown by Bender and Marrinan, there was scope within Lockean psychology at the time for using diagrams of scientific instruments to understand or remember "array bundles of data."[60] Employing the plates that appear throughout the *Encyclopédie, ou Dictionnaire Raisonné des Sciences, des Arts et des Métiers* (1751–1772) as examples, Bender and Marrinan argue that when used in conjunction with the descriptive narrative of the articles, diagrams of

---

57. Pictograms, inhalers, scissors, pins, tubes, syringes, and bandages were popular. Monro Secundus Bound MS (1799–1800, ff. 90, 96, 156).

58. Representative apparatus pictograms appear in the anonymous chemistry notes of Cullen Bound MS (1762, ff. 57, 88–89, 94–95).

59. A pictographic table of a number of watercolor diagrams used in John Hope's botany lectures was collected by an anonymous student and drawn in Hope Bound MS (n.d., f. 102r). See also the diagrams in Hope Bound MS (1780).

60. Bender and Marrinan (2010, 72–73). See also Ann Banfield (2019).

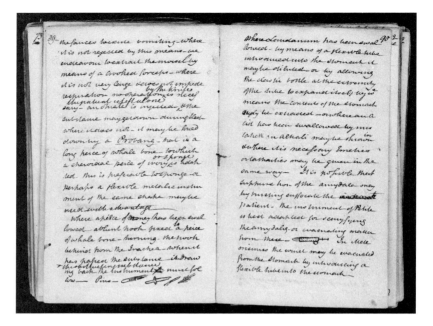

FIGURE 9.14. Pictograms of anatomical instruments drawn by the student George Bruce. Alexander Monro Secundus, *Lectures on Surgery* (1799–1800), George Bruce (notetaker), Bound MS, RAMC/516, f. 89–90. Reproduced with permission of the Wellcome Library, London.

instruments facilitated an important mode of knowledge formation. It was a similar scenario for the apparatus and specimens drawn by Scottish students, especially since these visualizations functioned as learning tools. The only difference was that rather than gaining meaning from the information supplied by a printed narrative (such as the articles addressed by Bender and Marrinan), the pictograms of Scottish university students gained meaning from the narrative that they had reconstructed from what a professor had said and performed simultaneously in realtime in the classroom.

As mentioned above, a noteworthy set of pictograms based on lecture posters appear in the transcripts of students attending Professor John Hope's Edinburgh botany course during the mid- to late eighteenth century.[61] Unlike most university courses, his was held in the summer, when the flowers of many plants could be observed by inquisitive students. But he also needed to lecture on plants that were not in bloom, those that had unexpectedly died, those he did not have in his garden, and those he had included in past physi-

---

61. For examples of pictograms drawn by other students who took Hope's Edinburgh botany course, see Hope Bound MS (1777–1778); Hope Bound MS (1780); Hope Bound MS [n.d.].

ological experiments. To mitigate the absence of such plants, he collaborated with illustrators to present diagrams on posters and on handheld sheets. This enabled him to employ an impressive portfolio of images and specimens while lecturing in the classroom or greenhouse of the university's Leith Walk botanic garden.[62]

When it comes to understanding the process that students used to convert Hope's visual aids into pictograms, the lecture transcripts of his son, Thomas Charles Hope, are particularly helpful.[63] Thomas Hope took his father's botany course in 1783. His transcript features three plant embryo pictograms (figure 9.15).[64] At the time, the course's syllabus was divided into four parts and plant embryos were used in the second part to illustrate the process of germination. Drawing the pictograms served to reinforce what Thomas Hope had heard in the lecture and what he had seen on the posters. The pictograms also helped him remember the realtime manipulation of specimens that he had witnessed in lectures conducted in the botanic garden's greenhouse. There was a haptic element to his experience as well, as specimens were sometimes passed around by students.[65]

Prior to taking the botany course in the summer of 1783, Thomas Hope had learned to use pictograms in the chemistry course that he took with Joseph Black over the 1782–1783 academic year. When he wrote his chemistry transcripts, he drew pictographs between the break lines and even in midsentence to represent experimental apparatus. This experience served as a form of self-training that showed him how to use pictograms to exercise his attention, recollection, and judgment. Once strengthened, these operations of the mind enabled him to see a pictogram as a thinking tool that could be used to remember and further explore the knowledge systems that he was recording in lecture notes taken in other courses.[66]

Thomas Hope's botany transcripts illustrate the ways in which pictograms functioned as artifacts that allowed students to reflect on the implicit theoretical and disciplinary matters raised in lectures. This facet of notekeeping is readily evinced in the plant embryo pictograms. He drew all three of them side by side in the section of his notes that addressed germination. They appear on

---

62. Many of the posters still exist and are housed in the special collections of the Royal Botanic Garden Edinburgh. Noltie has used the word "thumbnail" to describe the pictograms that occur in the notes of students who attended Hope's botany course. Noltie (2011b, 8, 17).
63. Hope Bound MS (1783).
64. Hope Bound MS (1783, lecture 30, f. 18v).
65. The bean plant embryo diagram is discussed throughout Hope Bound MS (1783, lecture 33, ff. 14v–18v; pictogram on f. 18v).
66. Thomas Hope's pictograms occur across Black Bound MS (1782–1783, vol. 2, ff. 11–14, 72v, 25v, 27r, 27v, 28v, and vol. 3, ff. 74v, 88v, 89v).

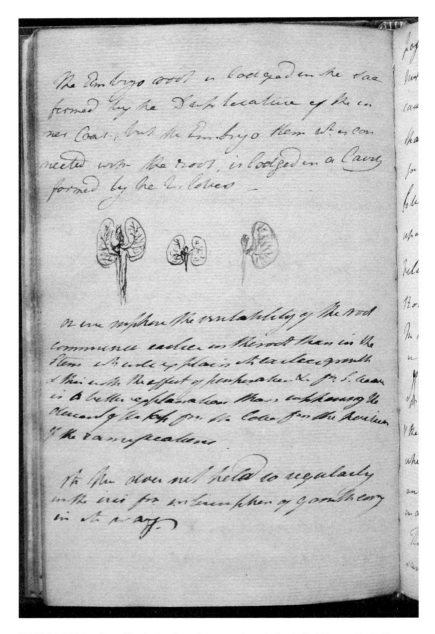

FIGURE 9.15. Ink and graphite plant embryo pictograms drawn by the student Thomas Charles Hope. John Hope, Lecture 32, *Dr Hope's Lectures in Botany* (1783), Thomas Charles Hope (notetaker), Bound MS, f. 18v, from the Archives, Royal Botanic Garden Edinburgh.

a verso page of his transcript, which, like many students, he had left blank for future annotations. Their placement indicates that they were drawn after the lecture and most likely after he had written his transcript, which means the pictograms were based on either a scroll book or his memory of what he had seen in the lectures. Their schematic structure speaks to the ways in which pictograms offered students a mode to observe the posters and objects they witnessed in realtime.

Each pictogram depicts a unique morphological trait displayed in the embryos of different plants. The first offers a double-lobed bean embryo with a long root. It was based on a poster drawn by Andrew Fyfe and exhibited by Professor Hope to students in the greenhouse (figure 9.16). Thomas Hope explained the meaning of the pictogram with the following description: "The Embryo root is lodged in the sac formed by the Duplicature of the inner Coat, but the Embryo stem w^h. is connected with the root, is lodged in a Cavity formed by the 2 lobes."[67] He evidently was unsatisfied with this description, most likely because it did not clearly explain what caused the growth of the root. This led him to add more to the description.

Based on eighteenth-century physiological theory, a possible cause of the embryo's growth was "irritability," a central concept used by physicians to explain the generation of plant, animal, and human tissue.[68] To make the connection between irritability and his pictogram more apparent, Thomas Hope added the following observation, "or we suppose the irritability of the root commence earlier in the root than in the stems w^h. will explain its earlier growth & this with the effect of temperature & f^m. . . . leaves is a better explanation than supposing the descent of the . . . f^m. the position of the ramification." The addition of this supposition reveals that Thomas Hope was using his pictogram as a paper tool, an interactive artifact, that enabled him to enhance his notes with additional information that he deemed relevant to a key physiological theory being debated by leading anatomists throughout Europe.

Thomas Hope's second pictogram features a double lobed plant embryo with a short root. It was likely drawn from memory to illustrate the description of an iris embryo that he had already written in his notes: "In the Iris the growth is singular in this manner, whatever distance the seed be from the surface it always pushes out a [plume?] to the surface & then pushes out the seminal leaf which push out the root which never pushes out fibres like it the equally deeper be on a level with the original seed."[69] The third pictogram

67. The quotations in this paragraph and the next are taken from Hope Bound MS (1783, f. 18v).
68. Steinke (2005).
69. Hope Bound MS (1783, f. 19r).

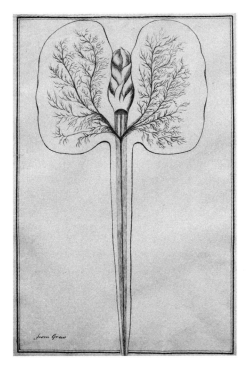

FIGURE 9.16. Pen and ink bean embryo poster on folio-size paper. The poster depicts the vasculature of cotyledons. It was used by John Hope in his Edinburgh botany lectures. Item D.28, John Hope Collection of Teaching Diagrams, Royal Botanic Garden Edinburgh.

presents a single-lobed embryo with a medium-length root. Its description is illegible. Based on the schematic shape of the pictogram, it is relatively clear that its morphology is different from that of the first two. The larger point to note, however, is that Thomas Hope was using the second and third pictograms as thinking tools to recall and process information that he had seen in the lecture.

As mentioned above, Thomas Hope patterned his first pictogram on Andrew Fyfe's poster (see again figure 9.16), which Fyfe had modeled on a plate that he had seen in a book. This means that the image of the bean plant embryo had moved across time and space through processes involving the transformation of a plate, a poster, and a pictogram. Students who had kept school notebooks would have been aware of the potential cognitive enhancement offered in this kind of visual *translation*.[70] As we learned from the school note-

---

[70]. For the transformative and creative aspects of visual translation, see Jordanova (2012, esp. 215–216). A helpful study of the translation of images through different kinds of early modern material culture is Burke (2003), particularly his concise discussion of Trajan's column in Rome on 280.

book figures and maps discussed in previous chapters, the process of translating images and their corresponding data through different kinds of media provided the opportunity for learners to use their notebook pages to combine acts of drawing, thinking, and writing in a way that strengthened their fact-gathering and decision-making skills both inside and outside the classroom. Thus, in addition to revealing how student notekeepers used pictograms as interactive learning artifacts, Thomas Hope's reworking of Fyfe's poster provides a fascinating episode of an image in motion. It offers a way to explore how images were jointly translated and repurposed within a university setting to suit the needs of learners and teachers.

Fyfe's poster was based on a bean embryo diagram taken from a plate in Nehemiah Grew's *The Anatomy of Plants*, which was originally published in 1682 (figure 9.17).[71] A number of the course's lecture posters were based on plates that appeared in the book. Professors teaching other courses in Scottish universities practiced this kind of visual translation as well.[72] Moving images across different media was not a straightforward task, particularly for professors who wanted to offer visualizations that students could remember or record. Accordingly, professors and their illustrators usually adapted images so that they were suitable for students. In the case of Fyfe's bean embryo poster, Grew's original diagram was only a few centimeters high and was featured as one of ten figures on an octavo-sized plate printed at the back of his book. Fyfe extracted it from the plate and enlarged it into a size that worked on a poster that could be seen easily by students in Hope's crowded classroom. To make the image as simple as possible, Fyfe eliminated the cross-section of the bean's root that was depicted in the Grew plate.[73] He adapted many of the poster images based on book plates in this manner.

The foregoing translation speaks to the fluidity in which John Hope's botanical images moved between print and manuscript media. After he had selected them from books, they journeyed through the hands of artists, and then across classrooms and into the notebooks of students. When it comes

---

71. Hope was also fond of using schematic figures from Stephen Hales's *Vegetable Staticks* (1727), a text that played a major role in popularizing what Susannah Gibson has called the "Newtonian vegetable," that is, a mechanical conception of plant growth and physiology. See Gibson (2015, 154–162).

72. The notebooks taken by students attending Robert Eden Scott's Aberdeen lectures on natural philosophy, for instance, feature diagrams most likely adapted from James Ferguson's popular published series of lectures. Compare the figures of telescopes drawn at the end of Scott Bound MS (1810) to those featured in plate XIX in the optics section of James Ferguson's *Lectures on Select Subjects in Mechanics, Hydrostatics, Hydraulics, Pneumatics, and Optics* (1799). Scott also included additional telescopic figures that are not in Ferguson.

73. The bean embryo poster used by John Hope in his Edinburgh botany lectures is housed in the Hope Collection of the RBGE. It was based on a figure featured in Grew (1682, table 2).

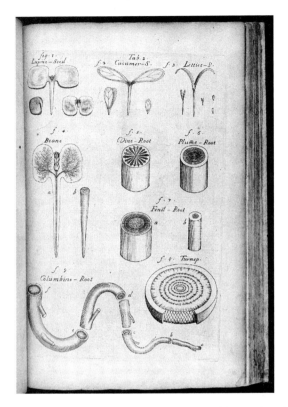

FIGURE 9.17. Bean embryo engraving. Nehemiah Grew, *The Anatomy of Plants with an Idea of a Philosophical History of Plants, and several other Lectures, Read before the Royal Society* (London: Rawlins, 1682), table 2. The bean embryo is depicted under "'f. 4" (figure 4), in the middle of the left column. It served as the basis for a poster that John Hope presented to his students in the classroom. Reproduced with permission of Missouri Botanical Garden's Peter H. Raven Library.

to the role played by such images in Thomas Hope's efforts to record and understand the system of botanical knowledge presented in his father's lectures, one striking aspect of the translations is how they all remained relatively schematic. This was despite the fact that the image was significantly enlarged when Fyfe translated it to a poster. Here we can see that professor, illustrator, and student alike recognized the value of keeping the image simple so that information could be easily attached to it. This preference combined with the schematic nature of most student pictograms reminds us of the pragmatically minded approach to images that permeated Scotland's universities. Like the diagrams based on shapes discussed in the previous section, the meaning of the schema presented in Fyfe's poster and Thomas Hope's pictogram was anchored in the system of knowledge being presented in the lecture syllabus.

I now move on to another set of notebook pictograms with a view to exploring the role that they played in helping students attending Joseph Black's Edinburgh chemistry course to remember spatial and sensory information. Students who took the course, including Thomas Charles Hope and Thomas Cochrane, used pictograms to represent scientific instruments such as thermometers, lamps, glassware, furnaces, apparatus, and more (figure 9.18).[74] In what remains of this section, I use Cochrane's notebooks to explore how chemistry students used pictograms as mnemonic cues that helped them better understand and remember Black's experiments.

When it came to understanding the concept of chemical affinity, drawing pictograms of experimental instruments had a number of advantages. At one level, they connected matter theory with experimental practice because they helped to illustrate what a student could do materially in a lab with the affinity concept. Pictograms were also a form of visual shorthand that worked in combination with the verbal definitions, descriptions, and observations of experiments that students captured in their notebooks. At another level, pictograms functioned as information management tools, serving as mnemonic encapsulations that helped students to intelligently record and understand the step-by-step experimental instructions given during the lectures. In this sense, the pictograms helped them to more readily inscribe what the ethnologist Eric Livingston has called "situated practices of laboratory chemistry."[75]

Like other chemistry students, Cochrane used pictograms in his notebooks to remember how vessels and supporting apparatus were set up during experiments. Practicing this kind of visual shorthand made it easier to sidestep some of the long verbal descriptions that Black gave in his lectures. This aspect of pictogramming was a helpful mnemonic activity because Black's lectures, in the words of his student Samuel Latham Mitchell, "consisted of things *done*, rather than *said*."[76] The act of devising and drawing a pictogram also functioned as a mode of kinesthetic learning that helped to reinforce the meaning of the image and its larger relevance to the chemical knowledge system being presented by Black. Pictogramming further enabled students such as Cochrane to associate the invisible attractions of substances with the visible properties that they observed during the experiments conducted in the classroom.

---

74. Cochrane's pictograms appear in Black Bound MS (1767–1768, vol. 1: thermometer (f. 33), furnace and lamp (f. 74), furnace (f. 77), furnace experiment (f. 95), u-shaped glass tube designed by Black for fixed air experiments (f. 355), jar (f. 356); and in vol. 2: unidentified instrument (3f. 9), glass receiver (f. 66), experimental setup (f. 486)).

75. Livingston (2008, 154).

76. Mitchell (1976, 745–746).

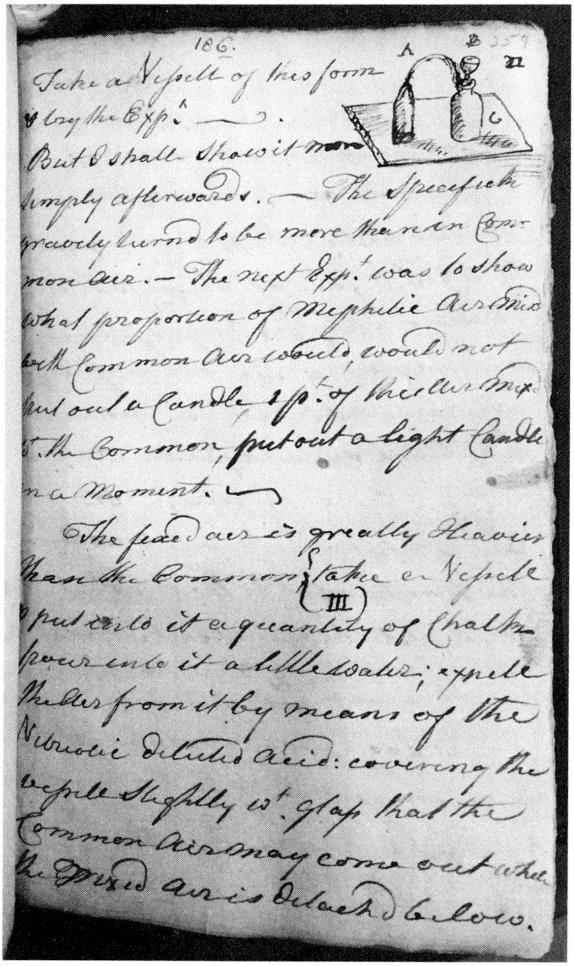
186.
Take a Vessell of this form
& try the Expt. —
But I shall show it more
simply afterwards. — The Specifick
gravity turnd to be more than in Common air. — The next Expt. was to show
what proportion of Mephitic air mix'd
with Common air would, would not
put out a Candle 1 pt. of this air mix.
5 the Common, put out a light Candle
in a Moment.
The fixed air is greatly Heavier
than the Common; take a Vessell
put into it a quantity of Chalk
pour into it a little water; expell
the air from it by means of the
Vitriolic diluted Acid: covering the
vessell slightly wt. glass that the
Common air may come out when
the fixed air is detach'd below.

FIGURE 9.18. Pen and ink pictogram of an experiment drawn by the student Thomas Cochrane. Joseph Black, *Notes on Dr Black's Lectures on Chemistry*, vol. 1 (1767–1768), Thomas Cochrane (notekeeper), Bound MS, University of Strathclyde Special Collections, Old Catalogue Reference No. 34/28, f. 359r.

Cochrane, like other notekeepers attending Black's course, used pictographic association to enhance his memory of the smells, bangs, and colors of the reactions that took place during the lectures.[77] Likewise, pictogramming individual instruments such as flasks, tubes, and furnaces functioned as a mnemonic mode of learning that further inculcated the procedures that Black

77. The importance of avisual forms of evidence in modern chemistry is underscored in Roberts (2005).

used to manipulate substances across time and space, that is, across the instruments and apparatus that he assembled and moved around the demonstration table in realtime.[78] The kinesthetic facet of pictogramming ensured that when students reached the end of the notekeeping process, the pictograms performed the dual function of operating as cues that related to the *facts* given in the lecture and the *skills* demonstrated in situ by the lecturer.

In addition to training students how to remember an experimental procedure, pictogramming showed that it was possible to use images to edit out bits of time, stages in the process, that were less relevant to the element of the system that a professor was attempting to communicate. Drawing pictograms and rewriting their descriptions in notebooks in this way helped students learn how to negotiate various temporal facets of experimentation. When framed as components that existed within a lecture transcript, pictograms were more like composite snapshots, idealized moments of the course that were frozen in time. Students designed them in ways that helped them remember and understand the key points of the experiment.

To keep the pictogram simple, students had to decide what visual elements they wanted to omit in the image they were designing. The art of knowing when to strategically omit information was displayed by Black himself in the realtime demonstration of experiments he performed in his lectures. To keep the demonstrations memorable, he did not dwell on all the failed experiments that had contributed to the isolation of the substances that he was using.[79] Likewise, Black oftentimes streamlined the presentation of experiments by conducting time-consuming operations such as distillations outside of the classroom so that students could witness the final, and more exciting, part of a long experiment.[80] Black understood the heuristic principle that "less was more" and he streamlined the information being presented to his students so that they could efficiently capture the key elements of experiments in their notebooks.[81]

---

78. A number of Black's instruments are still extant. See Anderson (1978). The instrumental context for gravimetric analysis Scotland is given throughout Connor and Simpson (2004).

79. Isaac Newton, whose methods served as a guide for most natural philosophers, omitted calculations and experiments "for brevity's sake" in the later editions of his *Principia*. The mathematical context of these omissions is addressed in Smeenk and Schliesser (2013, esp. 151–152).

80. Aside from the hundreds of experiments recounted every year in Black's course, other good examples of the step-by-step material manipulations that Black gave his students orally can be seen in the printed "handouts" that he made for his students as early as the 1760s. See Black, *The Preparation of Mercury* (n.d.), in Black MS (1766–67, tucked inside vol. 7). Mercury was an important ingredient for drugs.

81. The "less-is-more" effect is explained in Gigerenzer (2008, chap. 2). For the role played by this kind of thinking within human development and the emergence of "homo heuristicus," see Brighton and Gigerenzen (2012)

Learning to streamline the key elements of an experiment into a pictogram took practice. I mean this in the literal sense: students sometimes needed to draw preliminary sketches of the basic contours of their pictograms to ensure that they were portraying the elements of an experiment that served to remind them of the things they had seen, heard, smelled, and felt in the classroom. Preliminary sketches of pictograms are difficult to find, especially since students did not keep them once they had written their transcripts. In the case of Cochrane, however, we can catch a glimpse of the process because there are two preliminary pictograms sketched in graphite on the blank pages that he interleaved into the transcript before it was bound (figure 9.19). Both represent experimental glassware.[82]

It is not clear when Cochrane made the pictograms. He could have sketched them during a lecture and then interleaved them into the transcript. Alternatively, he could have drawn them from memory after he inserted the blank interleaves. Nevertheless, their very presence reveals that Cochrane presketched some of his pictograms with a view to reconstructing the basic set up of the apparatus used by Black in the lectures. When seen from this angle, the sketches show how Cochrane, and arguably other students, designed their pictograms.

The first pictogram consists of a flask and a bottle sitting next to each other on a tray and is depicted at the top of figure 9.19. The mouth of the flask on the left is sealed with a stopple that has a tube running through it. Its base is slightly higher on the tray than that of the bottle on the right, making it appear somewhat further away. The flask's shading makes its body appear cylindrical. The shading on the bottle to the right of the flask was less successful, being so heavy that it makes the surface look flat. The shading and positioning of the flask and the bottle reveal that Cochrane was attempting to introduce three-dimensional elements into the otherwise schematic nature of the glassware in the pictogram. It was most likely this kind of preliminary sketch that helped him practice the shading skills that he used to render other pictograms elsewhere in his notes.

Cochrane's second preliminary pictogram, depicted on the bottom of figure 9.19, reveals different pictogramming skills. It consists of two instruments, a glass bell on the left (A) and a specialized flask on the right (C). The neck of the flask features a spout with a bulbous receiver. A tube connects the top of the bell to the mouth of the flask. The alphabetical labels A and C

---

82. For the extent and variety of Black's glassware, see Kennedy et al. (2018), Cameron, Addyman, and Macfadyen (2013), and Anderson (1978).

FIGURE 9.19. Graphite pictogram predrawings sketched by the student Thomas Cochrane. Joseph Black, *Notes on Dr Black's Lectures on Chemistry*, vol. 1 (1767–1768), Thomas Cochrane (notekeeper), Bound MS, University of Strathclyde Special Collections, Old Catalogue Reference No. 34/28. Top: Close-up of the first pictogram on f. 357r. Bottom: A pictogram on the blank f. 364r.

written in the vessels correspond to labels in the narrative of the experiment being explained on the facing page. This indicates that Cochrane was using the sketch to work out the placement of labels within the pictogram. Additionally, the sketch does not feature the standard stopples that would have been placed at the top of the bell and in the mouth of the flask, revealing that he was practicing the skill of schematizing, that is, the ability to selectively omit, often in realtime, parts of the instruments that were less important to the aspects of the experiment that he wanted to communicate.

When considered in reference to the experiments they were meant to represent, the schematic element of the preliminary pictograms drawn by Cochrane reveals that he was not attempting to create self-explanatory forms of representation. Instead, each pictogram was meant to function as a mnemonic nexus for the facts and skills preserved in his notebooks and in his mind.[83] It was similar for other students who made pictograms in their chemistry notebooks. In other words, the pictogramming techniques they used in their lecture transcripts reveal that it would have been difficult for them to understand their images without originally having seen the experiment in the classroom. It was mainly through observing professors conduct and explain experiments in realtime that students acquired the theoretical and disciplinary background knowledge that enabled them to attach temporal and spatial coherence to the pictograms. This multistable aspect in particular made pictograms versatile in a way that would have been difficult to recognize by those who had not attended the lectures or did not have a good working knowledge of chemistry.

Tables as Kinesthetic Diagrams

A good number of professors offered students tables of words and numbers to help them order and remember information covered in lectures. They were matrices comprised of rows and columns that students wrote and read diagrammatically as patterns to access paths of data. The tables ranged from Alexander Fraser Tytler's printed multipage chronology of world history to Joseph Black's inscribed thermometric scales (figure 9.20).[84]

---

83. Experimental or observational procedures played a central role in the graphic management of scientific information across media in modern settings. Nasim (2013) has shown that it is difficult to grasp the centrality of these routines without paying close attention to scribal iterations that take place over time in scientific notebooks. See especially his comments in the introduction on procedures.

84. The chronological tables in Tytler (1782) extended the chronography tradition used elsewhere in British dissenting academies and German universities. For the former, see Priestley (1769). For the latter, see Gierl (2012). For the larger chronograph tradition, see Rosenberg and Grafton (2013).

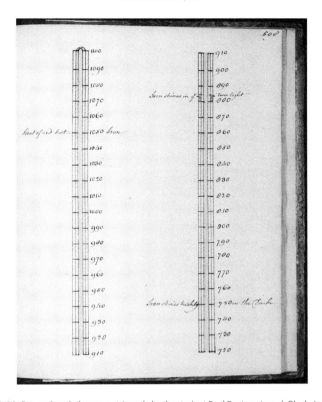

FIGURE 9.20. Pen and wash thermometric scale by the student Paul Panton. Joseph Black, *Lectures on Chemistry* (1776), Paul Panton Jr. (transcriber), vol. 1, Bound MS, Chemical Heritage Foundation, QD14 B533, f. 508r. Reproduced with permission of Science History Institute. The scale depicts the changes in iron that corresponded with different temperatures.

Yet for all their sophistication, student notekeepers tended not to copy the large tables offered by professors in posters or on handouts. The exception to this practice seems to be the table-like "charts" featured in the notes taken by students attending medical lectures.[85] The tables that students saw fit to record seem to have been those that compressed a great deal of information into a format that did not exceed a page. Such tables were efficient media technologies because they offered a great deal of information at a glance.

One of the more sophisticated synoptic tables copied by students was the affinity chemical "chart" designed by Joseph Black (figure 9.21). He used it to represent a single elective attraction, the simplest form of chemical affinity.

85. Several of Black's chemistry students attempted to replicate his tabularized thermometric scales. One of the most detailed iterations appears in Black Bound MS (1776, vol. 1, Paul Panton [notekeeper], ff. 508–511). For a printed version of the scales that is based on Black's lectures, see Martine (1792).

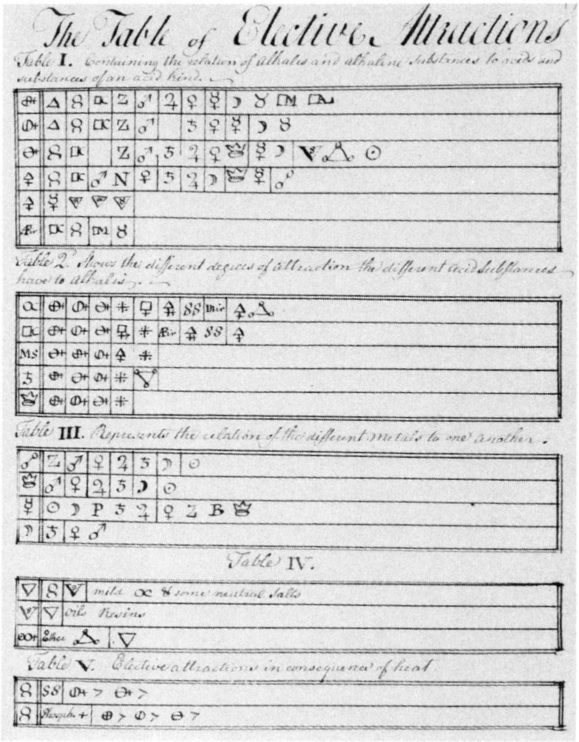

FIGURE 9.21. Paul Panton's chemical affinity table. Joseph Black, *Lectures on Chemistry*, vol. 6 (1776), Paul Panton Jr. (notekeeper), Bound MS, Chemical Heritage Foundation, QD14 B533, f. 17r. Reproduced with permission of the Science History Institute.

It explained how each substance listed in the far left-hand column "elected" to unite with the other substances given in the row next to it. The substances in the rows were listed in accordance to their strength of attraction, with the strongest affinities being on the left and the weakest being on the right.[86]

Most students replicated the affinity table in their notebooks or on a loose-leaf sheet, indicating that they thought it was a paper tool worth reproducing. The order of the symbols in all affinity tables, whether listed in rows or columns, operated on a simple principle of visual proximity. The principle associated nearness with strong attractions and farness with weak ones.[87] In writing

---

86. For more information on how the content of Black's lectures related to other chemistry courses being taught in Europe, see Donovan (1975), Eddy (2014; 2015), and Crosland (1959).
87. Black used a similar principle of proximity in the temperature charts that he developed for his students as well. Nearness to one pole of the column represented hotness, and nearness to the other pole represented coldness.

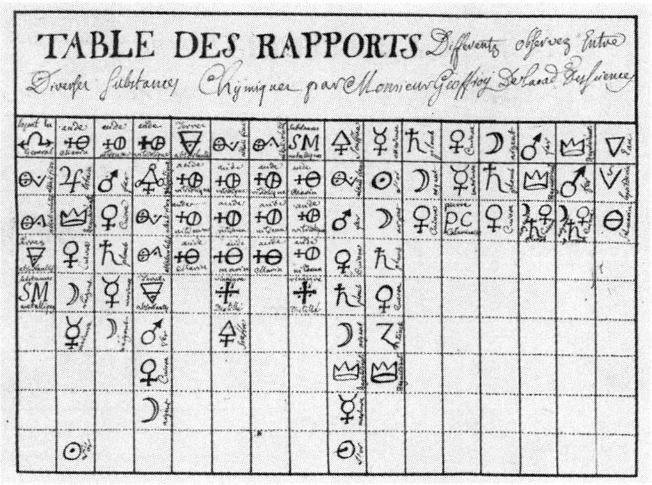

FIGURE 9.22. "Table des Rapports." Gabriel François Venel, *Cours de Chymie Fait chez Monsieur Montet Apoticaire: par Monsieur Venel Docteur et Proffesseur en L'université de Médecine à Montpellier* (1761), Claude Delonne Balme (transcriber), Bound MS, Wellcome MS 4914. Venel based his table on the visual structure originally developed by Étienne François Geoffroy. Reproduced with permission of the Wellcome Library, London.

the table, students manually reinforced their understanding of chemical relationships in a way that helped them to remember the comments Black had made about strong or weak single attractions during the lectures.[88]

Copying affinity tables played a central role in chemical teaching during the eighteenth century. They were used frequently by Black's own teacher William Cullen and in the popular chemistry courses of French teachers like Étienne François Geoffroy (1672–1731) and Gabriel François Venel (1723–1775) (figure 9.22).[89] Cullen found the tables so pedagogically effective that he hung his at the front of his classroom.[90] Following the lead of Cullen, it is likely that Black hung a copy of his table at the front of the room for his students while he was lecturing as well. There were no printed copies of it, which means that its replication and circulation were entirely a manuscript affair.

88. Attractions are discussed in the language of strong and weak in Black (1966/1767, ff. 23, 33, 34, 35, 38, 39, 40, 41, 43, 59, 60, 61, 63, 79, 89, 118, 133, 158).
89. Experimentally based affinity tables had provided a way to think with paper ever since Geoffroy had popularized them at the beginning of the century. For a list of the published affinity tables that appeared from the 1720s to the 1790s, see Duncan (1996, table 4.1). The conceptual importance of affinity tables is treated in Klein and Lefèvre (2007). For the use of affinity tables in Scottish teaching, see Taylor (2008) and Eddy (2015). For France, see Lehman (2010). The pedagogical reaction to the affinity concept in Holland is addressed in Powers (2012, 163–168).
90. Taylor (2014, 262).

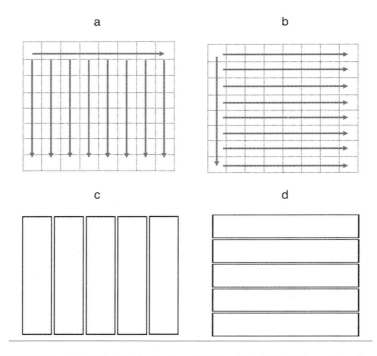

FIGURE 9.23. Chemical affinity table visual patterns. © Matthew Daniel Eddy. (a) The reading pattern of tables based on Étienne François Geoffroy's "Table des Rapports." (b) Reading pattern of Joseph Black's affinity table. (c) The vertical segmentation pattern offered by the columns of Geoffroy's affinity table. (d) The horizontal segmentation pattern offered by the rows of Black's affinity table.

Tables based on Geoffroy's model listed the lead substances across the top row and all the reactions in the downward flowing columns (see again figure 9.22). Many chemists followed his lead. This meant that their tables were designed to be read in two glances. The first started at the left of the top row of the table and moved rightward to the column of the desired lead substance. The second glance then moved down the column across all the other substances that combined with it. A representation of the scanning pattern created within the table can be seen in figure 9.23a. Black broke with this tradition. As noted above, he listed the lead substances down the left-hand column and the corresponding reactions in the rows (see again figure 9.21). The different scanning pattern presented by Black's table can be seen in figure 9.23b.

Students inscribed Black's table in two ways. The first was to simply copy the entire table, treating it as a single structure on a quarto- or folio-sized page. The affinity table in the six-volume set of notes made by Paul Panton while taking Black's course during the 1770s is a good example of this practice (see again figure 9.21).[91] A second option was to copy the table's five parts onto separate pages as five freestanding microtables, each of which was called a "divisio," a "division," or simply a "table." Each of the five divisions offered a succinct overview of how affinity operated in relation to an analogous set of substances reacting in a similar way to create different compounds.[92] The microtables worked better for students seeking to write the affinity table in a landscape orientation on the smaller pages of an octavo-sized notebook. This was the option chosen by Thomas Cochrane in the notebooks he kept during the 1760s (figure 9.24).

Black was able to split his table into convenient divisions because, as noted above, his structure was different from others used at the time. The descending lists of chemical attraction in most affinity tables, including Geoffroy's model, the most influential at the time, were arranged in columns. Tables based on it could only be split vertically into segments of columns, a layout that was conducive to a portrait-orientation on a large-sized notebook page or classroom poster (figure 9.23c). In contrast, the lists of chemical attraction offered in Black's table were arranged in rows that could be split horizontally into five microtables that corresponded to the five divisions (figure 9.23d).

The arrangement of Black's table encouraged his students to see the structure as an adaptable paper tool that could be modified. More specifically, Black's decision to structure his table horizontally and then to divide it into five divisions made it easier for students to see it as a modularizable entity, that is, as something that could be split into smaller parts. This was a noteworthy innovation because the overarching impression of a table based on Geoffroy's model was that of a singular entity. In other words, when seen from the perspective of Black's students, Geoffroy's table was not clearly divided into sections and its visual structure did not easily suggest that it could be modularized. Had Black based his table on Geoffroy's model, it would have been more difficult for his students to see beyond the apparent visual unity it presented.

Let us have a closer look at Cochrane's horizontally modularized microtables with a view to understanding how he made them (see again figure 9.24). They occur together as a group in a quire that also includes a table of affinity symbols and their meanings, and a collection of circlet diagrams used to

---

91. Black Bound MS (1776, 6: f. 17).
92. More specifically, divisions 1 and 2 were saline reactions, division 3 featured metallic reactions, division 4 involved what Black called "mild substances," and division 5 addressed key attractions involving heat.

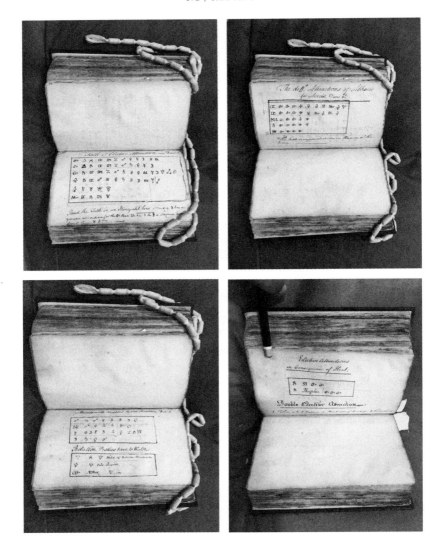

FIGURE 9.24. Joseph Black's affinity table broken by rows into five parts by the student Thomas Cochrane. Joseph Black, *Notes on Dr Black's Lectures on Chemistry*, vol. 2 (1767–1768), Bound MS, University of Strathclyde Special Collections, Old Catalogue Reference No. 34/28, occurring across ff. 447–454. Reproduced with permission of the University of Strathclyde Special Collections.

represent double elective affinity reactions.[93] When Cochrane bound his transcripts, he placed this quire in the second volume. He most likely had made all

---

93. Black Bound MS (1767–1768, 2: ff. 444–460). Cochrane's microtables are reproduced in Black (1966/1767, 161–165). The meaning and visual usage of the circlets is addressed in Eddy (2014) and Crosland (1959).

of the microtables before he transcribed his scroll books, electing to use them as ready-to-hand reference tools while he was rewriting his notes. The microtables are neat and contain only a few mistakes, which is a common characteristic of most transcribed affinity tables. To make the microtables, he began by stacking several sheets of blank quarto paper. He then folded the sheets in half to create a quire with octavo-sized pages. Next, he drew the microtables on the recto and verso sides of the pages, with ink from one microtable occasionally running through to the other side of the paper.

To create the framework of each microtable, Cochrane used the same step-by-step process. Using a black lead pencil, he first turned the page sideways in a landscape orientation and drew a rectangle in graphite. He then drew a grid inside the rectangle. The grid offered boxes that could be used as columns or rows. Following Black's original table, Cochrane used the boxes to make rows. In the leftmost box of each row, he wrote the symbol of a lead substance in ink. In the boxes of the row that followed each lead substance, he wrote the symbols, also in ink, of other substances that were known to combine with it. At some point in the process, most likely near the end, he traced over the graphite lines of the border with ink to make it easier to see.

Students were willing to expend time and energy on crafting Black's table because it helped them better understand chemical affinity, that is, the material force that underpinned most theories of micromatter employed by eighteenth-century chemists. The full table was a central media technology for Black's course because it allowed students to *see* an overview of the affinity concept at a glance in one place.[94] When students kinesthetically engaged with its material and visual elements by inscribing or reading its columns and rows, it became more familiar and hence easier to understand, use, and replicate.[95] Though explicit definitions of various aspects of affinity were sprinkled throughout Black's lectures, it was the table that gave his students a constant visual reference point when they scrolled and transcribed lecture notes, or when they used their notes after they finished their studies. It allowed them to see easily a single elective reaction as one entry in a larger system of knowledge that was based on the theory of affinity. In this sense, the table operated as an essential informatic companion to Black's syllabus.

If we step back and look at the ranging, drawing, and writing skills required to re-create Cochrane's neat, symmetric, and proportionate microtables, it

---

94. The same was true for students studying chemistry in eighteenth-century Paris. See Lehman (2010).
95. The important relationship between material engagement and "bodily learning" is underscored by many anthropologists. Greg Downey's (2011) discussion of dance is especially relevant to the relationship that I am drawing between student notekeepers (as learners) and the material culture of manuscripts.

can be seen that he had probably made a similar gridded structure before, either by practicing it just before he made his notes or, more likely, in a notebook that he made when he was a younger student in school or at home. By making the affinity microtables, he was extending his ranging skills in a way that augmented his perception of how an informatic key could be represented visually in an ordered manner on the page. Inserting the symbols into the rows inculcated a stronger sense of how to transform the space of a page into a mnemonic scale that was relevant to the chemical knowledge system that Black presented in his lectures. Writing out the symbols in this way not only helped students such as Cochrane remember the scale, it enabled them to understand how it was possible to see a row as a scalable entity, as a visualization of probable relationships, the veracity of which was based upon the experiments they had witnessed in the classroom, and on evidence that Black had culled from letters and publications of other experimentalists operating in Scotland and throughout Europe and its colonies.[96]

Traces and Realtime Observation

One of the interesting effects of the diagrams used by Scottish professors is that students remembered their form and meaning long after they had finished university. Some students could even recall the diagram without the use of their notebooks. To gain greater insight into this phenomenon and its relationship to the media of notekeeping, I conclude this chapter by exploring a discrete process that some students used to memorize the form and meaning of diagrams. The process was guided by a skill that I call *tracing*. It involved acts of memorization that required students to attentively observe, either in the classroom or later in the mind's eye, the lines of a diagram as cues to which meaning could be attached.

As mentioned above, Scottish professors drew diagrams in front of students on classroom blackboards. Thomas Webster, who attended Patrick Copland's Aberdeen natural philosophy course during the 1780s, found blackboards so helpful that he used them throughout his career as a geologist and professor. Following Copland's example, Webster drew diagrams in realtime

---

96. Here I am drawing from Tsing (2012) and the notion that "scalability" is a historically traceable entity related to, but not always part of, the history of metrology (numeric quantification). The changing relationships between lines and scales are treated throughout Ingold (2007; 2015). For a helpful discussion on the relation between materiality and scalability that is relevant to the creation of scales within Scotland's student manuscript culture, see Knappett (2011).

for the courses that he gave at the Royal Institution of London around 1800. In a letter dated 1799 to the Royal Institution's patron Count Rumford, Webster outlined the diagramming technique that he used in his lectures: "The figures should be drawn to a large scale by the teacher upon a black board with chalk and the demonstrations gone through slowly and distinctly. Each person should at the same time draw the diagram with ink upon paper and if possible, demonstrate the proposition each in his turn."[97]

The act of watching a diagram being drawn in realtime on a blackboard helped students learn the skill of diagrammatic tracing. Though many students learned the skill, there is little extant evidence that speaks to *how* they acquired it. Fortunately, some helpful exists for the kinds of tracing learned by the students who attended the anatomy lectures of Alexander Monro Secundus. When he discussed his diagrams, he used two visual techniques of description, both of which helped his students mentally trace the lines. In the first technique, he employed well-selected, image-evoking words that, when recorded, helped students trace the diagram in their mind after the lecture. The tracing occurred when students filled out their transcripts in the evening or when they read their notes at a later date. This mode of remembering the diagram moved from words to lines. In the second technique, Monro drew the diagram on a chalkboard as students watched. In this scenario, the students traced the lines of the diagram in realtime as they watched him draw and explain each one. This second mode of tracing moved from lines to words in that it was the form of the diagram that helped students remember the meaning. Both of Monro's techniques were based on observations made in realtime by students during a lecture. Notably, Monro sometimes spoke the words and drew the lines at the same time to describe his diagrams in the classroom.

Let us take a closer look at word-to-line tracing. Monro Secundus actively facilitated the movement from words to lines by selecting nouns, adjectives, and verbs that lent themselves to rectilinear shapes or forms of motion. He sometimes supplemented this terminology with image-evoking language relevant to the striking colors that he artfully injected into anatomical preparations to enhance their visual impact upon his students.[98] His clear and pre-

---

97. Edwards (1972, 47).

98. The use of colorful injections in anatomical preparations was known as the Ruyschian art, after the Dutch anatomist Frederik Ruysch (1638–1731). Monro Primus's use of the technique is addressed in Monro Secundus (1840, xiv–xv) and Lawrence (1988). Monro Secundus used similar techniques. For his use of wax colored with vermillion (red), quicksilver (silver), and spirit of turpentine colored with vermillion (a reddish tint), see Monro Secundus (1783, 18–19). For milk, tincture of madder, tinctures of indigo, tinctures of stone-blue, and yellow wax (most likely colored with saffron), see Monro Secundus (1770, 13, 19–20).

cise language made it easier for students to understand how his diagrams, which usually were constructed from triangles or squares, represented various parts of the body.

Student notes based on Monro Secundus's lectures offer many examples of rectilinear language that lent itself to figural imagery involving straight lines. An example of this practice is presented in a set of notes titled *Anatomical Lectures* kept by John Thorburn while he attended Monro Secundus's course three times during the mid-1770s. In the fifteenth lecture of Thorburn's notes, Monro describes an experiment that he devised to ascertain the growth of blood vessels around scar tissue (*cicatrice*). He first explained the reason behind the experiment in the following manner:

> I shall next show you a new Skin formed, or the Structure of the Cicatrice of the Skin, after having cut the Skin and allowed the part to heal up again; I find a vast number of Vessels in the Cicatrice, which appears more vascular than the Skin does in any other part; we see Vessels coming from all sides to it, and that these are intermixed, but it remains a question whether these new Vessels communicate with each other, as is the case with the rest of the Blood Vessels.

He then describes the experiment, using rectilinear language that likens the incisions around the scar to four straight lines that added up to a simple square:

> In order to determine this, I durst not trust to the Eye, or to the microscope, but I cut into the Belly of an Animal by a longitudinal incision, till the Cavity of the Abdomen was fully opened, I then allowed the part to heal by sewing the opposite sides together, I after that cut transversely above and below, and then made another cut parallel to the former; by which means I had cut quite round a certain portion of the Skin of the Abdomen; I then injected the Vessels from the heart, and I found that there were numerous Vessels filled in that part I had now cut round.[99]

Monro's use of the adjective "longitudinal" to describe the first incision helped students visualize a simple line set in an open field of vision. The incision eventually became a scar, a permanent line on the body. Using the scar line

---

Further insight into the coloring techniques used by the Monros can be gleaned from Pole (1790). A helpful study of the material culture surrounding eighteenth-century anatomical preparations is Hendriksen (2015).

99. Monro Secundus Bound MS (1773–1774, vol. 2, ff. 296–297). Many Monro scholars consider this set of notes, kept by the medical student John Thorburn, the most authoritative record of his lectures as they were given during the middle part of his career.

as a visual reference point, Monro then proceeded to describe the square incision that he made around the scar to examine the flow of blood. He begins by describing the cuts that formed the left and right of the square as lines made "transversely above and below." The "above and below" refers to the fact that the incision of both the left and right sides of the square began above the scar and ended below it. He then describes the lines that formed the top and bottom sides of the square as "parallel to the former," that is, parallel to the horizontal scar. For students who had witnessed the procedure and heard the oral description in the classroom, the geometric language made it possible to evoke the memory of the event, that is to say, to trace the lines of the cuts as a diagram in their minds.

Monro Secundus used a similar geometric language in his lectures to describe many of his diagrams. His straightforward style had a profound effect upon the memories of his students and helped them recall his diagrams many years after they attended his course. In addition to this practical outcome, Monro's style fit well with the larger belief held in Britain since the late seventeenth century that precise and unadorned descriptions of objects served to make a stronger impression on the memory of a writer, reader, or listener.[100] By the eighteenth century, this axiom was firmly integrated into the Lockean theories of mind used by Scottish students and professors to understand the cognitive framework of verbal descriptions.

Monro Secundus's colleague Hugh Blair, Edinburgh's professor of rhetoric and belles lettres, for instance, voiced this view in his lectures, where he emphasized the value of paying attention to the strength, the mnemonic impact, that a clear description could have upon the mind. In his words, "The strength of a description arises, in a great measure, from a simple conciseness."[101] The use of such language, moreover, preserved the order of ideas and operations in the mind. Like many professors, Blair understood this act of preservation as a positive one because "as in all subjects, which regard the operations of the mind, the inaccurate use of words is to be carefully avoided."[102]

Let us now turn to line-to-word tracing. Monro Secundus facilitated this mode of memorization by drawing diagrams on the chalkboard. In the case of the *cicatrice* procedure, the diagram was a simple square. Thorburn recorded this diagram as a pictogram in his notebook. It is present in a copy of his transcript that was made by a professional copyist and then sold to Monro.

---

100. The scope of early experimental philosophy's commitment to clear style is explored in Wragge-Morely (2020, 106–134).
101. Blair (1783, 70).
102. Blair (1783, 16).

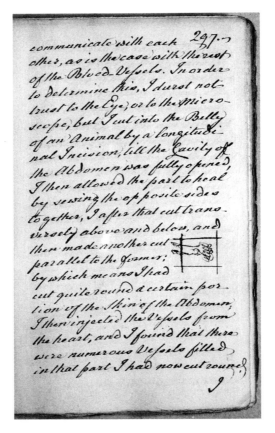

FIGURE 9.25. Description and diagram of the *cicatrice* procedure. Alexander Monro Secundus, *Anatomical Lectures Delivered by Doctor Alexander Monro*, vol. 2 (1773–1774), Bound MS, University of Otago Special Collections, ff. 296–297.

As can be seen in figure 9.25, the scar tissue is depicted as a squiggly line and the incisions form a square around it. The overarching point to note is that it would have been difficult for a person who had not attended the lectures to understand the precise meaning of the diagram without having heard or read its description, thereby further illustrating the fact that its schematic nature worked in reference to what Monro had said and done in the classroom, and, more generally, the larger anatomical system that he was using to organize the course. When Monro drew it on the board, he was showing students in realtime how to associate his geometric description with a rectilinear shape that was relatively easy to remember.

Monro Secundus's technique of drawing his diagrams on a chalkboard helped his students remember how to design them from the bottom up.

Watching the process encouraged students to see his diagrams as temporal artifacts to which meaning could be attached incrementally as he added each new line. This art of performing diagrammatic reasoning in front of students reached as far back as ancient Mesopotamia and Greece, where instructors drew geometric shapes in clay or sand, respectively. In the case of modern Britain, the mnemonic technique of drawing pedagogically oriented shapes in front of students was popularized by John Locke in *Some Thoughts Concerning Education*. After outlining the efficacy of asking students to repeatedly "observe" the various lines on globes, he advised teachers to draw a "Draught" of the solar system, and therein explain "the Situation of the Planets, their respective Distances from the Sun, the Centre of their Revolutions." The combined act of drawing the orbits of the planets as circles in front of students prepared them "to understand the Motion and Theory of the Planets, [in] the most easy and natural Way."[103]

When considered from the perspective of Locke's views on the value of using diagrams as realtime teaching aids, it can be seen that Monro Secundus was using a similar technique. The same could be said for other Scottish professors such as Edinburgh's Joseph Black and Aberdeen's Patrick Copland, who drew their diagrams in front of students in the classroom. When professors used rectilinear diagrams, it made it easier for notekeepers to remember their form and meaning by tracing the paths of the schema's lines with the mind's eye during the lecture. Once memorized, students could recall the diagram by drawing it in their transcript later in the day, or by retracing it in their memory whenever they wished. The ability to trace and retrace a simple schema in this manner helps to partially explain how, despite the conservative number of diagrams recorded in scroll books and transcripts, former students were able to recall them for decades after they left the university. In an age when figural images were much less prevalent than they are today, it is likely that the diagrams made a stronger impression.

Sometimes Monro Secundus's diagrams sprung into students' minds after they left a lecture. In 1764, for example, the medical student John Haygarth asked Monro in a letter to further explain how one of the anatomy course's diagrams represented the movement of intercostal muscles. The letter contained Haygarth's redrawn version of the diagram for reference. Monro duly replied to the great satisfaction of his student.[104] Monro's use of his lecture diagrams alongside preparations also inspired some students to make mechani-

---

103. Locke (1752, 271–272).
104. Monro Secundus (1794, 33–39).

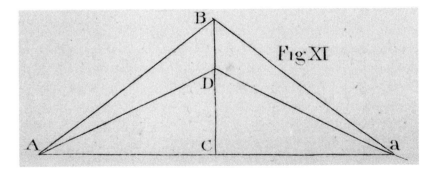

FIGURE 9.26. Anatomical teaching diagram illustrating muscle movement. Alexander Monro Secundus, *Observations on the Muscles, and Particularly on the Effects of Their Oblique Fibres* (Edinburgh: Dickson and Balfour, 1794), table II, fig. XI. Reproduced with permission of the Wellcome Library, London.

cal models out of wood and string that mimicked the movement of the bones and muscles. An account of this kind of creativity is given by Daniel Rutherford, who recounted his activities in the following way: "I was so delighted with the beautiful mechanism of the parts I had seen, and with the ingenious explanation you had given of their action, that I will recollect I soon therefore made a small model to imitate their motion."[105]

Perhaps Monro Secundus's most famous diagrams were the triangles that he used to represent the movement of muscles (figure 9.26). They offer another helpful case study that illustrates the mnemonic efficacy that his two modes of description had upon his students after they left his classroom.[106] The words and lines that he used to describe the diagrams in his lectures made an impression on a number of students who were able to trace them in their minds without the aid of notes years after they had attended Monro's course. William Greenfield, for example, attended Monro's lectures in 1790 and was able to explain an isosceles triangle diagram in a 1793 letter that he sent to Monro.[107]

Reflecting on his student years, Thomas Charles Hope not only remembered Monro Secundus's intercostal muscle diagram, he also could recollect

---

105. Monro Secundus (1794, 62). Daniel Rutherford succeeded John Hope as Edinburgh's professor of botany. Machines made from string, pulleys, and rods were used as learning devices in courses outside of anatomy as well. They enabled mechanical philosophy students, for instance, to trace the movement of force along the lines of the strings in a way that helped them remember the concepts communicated in lectures. For an example of a machine that was likely used by Professor Patrick Copland in his Aberdeen mechanical philosophy lectures circa 1800. See "Pulleys," University of Aberdeen Collection of Historical Scientific Instruments, Object Number: ABDNP:200018a. For the topics covered in Copland's lectures, see Copland Bound MS (1801).
106. The most popular anatomical diagrams are discussed in Monro Secundus (1794).
107. Monro Secundus (1794, 72).

how Monro had "figured" it on the blackboard in realtime. He had attended Monro's lectures during the 1780s and, though he did not copy the diagram into his notes, he was able to retrace it from memory in 1793, that is, years after he had attended the course: "To the best of my recollection, the diagram you [Monro] figured was an isosceles triangle, having a line drawn from the apex perpendicular to the base. In this figure two sides of the triangle represented oblique muscles, and the perpendicular drawn from the angle formed by them represented a straight one."[108] The form and meaning of the diagram was also recalled by other students of Monro well into the nineteenth century.[109] Thus, even when Monro's students did not draw the diagram in their notebooks they were still able to remember it decades after they studied with him.

The mnemonic power of Monro's diagrams reminds us of the sophisticated relationship between representation and observation that existed within Scottish university classrooms. In this chapter we have seen how student notekeepers used various forms of media to acquire diagraming skills that strengthened their ability to understand and manipulate visual information. We learned that the most memorable diagrams were schemata constructed from relatively simple lines to which information from the lectures could be easily associated. This commitment to simplicity was influenced by a Scottish interpretation of Lockean psychology that drew realtime correspondences between lines on the page and ideas in the mind. In this context, transforming a *tabula rasa* into the *tabula figura*, both literally and metaphorically, was a metacognitive process that endowed student notekeepers with diagraming skills that they remembered long after they finished their university studies.

---

108. Monro Secundus (1794, 70).
109. See J. Carmichael Smyth's comments about his Monrovian diagrams in his letter reprinted in Monro Secundus (1840, xiii–xvi).

CHAPTER 10

CIRCULATING

Local and Global Networks

In the summer of 1783, Francis Hamilton was becoming slightly nervous. He was near the end of his medical studies and would soon need to find a job. Like many graduates of Edinburgh's medical school, he had heard stories of those who had made their fortune abroad operating as agents of empire for the military, or for organizations such as the South Sea Company, the Levant Company, the East and West Florida Companies, and the East India Company.[1] Though Hamilton wanted to become financially independent, his heart was set on being a naturalist. To this goal, he joined the East India Company as a medical officer in 1784 and was assigned to a British man-of-war.[2]

Most navy and merchant vessels at this time required passengers to travel light. Nevertheless, in addition to bringing a small number of personal items in his travel chest, Hamilton's scientific interests led him to include the unbound lecture transcripts that he had taken while attending the botany course of Edinburgh's Professor John Hope in 1780. The decision to include his transcripts illustrates the high value, both personal and professional, that Hamilton attributed to them. In an age in which information economics was still partially dependent upon manuscript forms of communication, Hamilton's decision also serves to highlight the role played by lecture transcripts as dy-

---

1. An informative example of Scots who used their medical education to make careers in the Levant Company and the East India Company is the case of the brothers Alexander and Patrick Russell. For more on their careers, see Starkey (2018), Boogert (2010), and Laidlaw (2010). Further examples of Scottish-educated naturalists working with the East India Company are mentioned in Noltie (2017).
2. Hamilton eventually became an expert on the natural history of Asia and published several books on the topic. He was born Francis Buchanan, but changed his surname when he inherited land and a title from his mother's side of the family. See Katherine Prior's *ODNB* entry, "Hamilton [formerly Buchanan], Francis, of Buchanan (1762–1829)," and Higginbotham (1874, 39).

namic artifacts that commonly circulated through networks of students, professors, librarians, and other kinds of collectors that radiated through Britain and across the globe. After finishing their studies, many students transported their notebooks to places where they could be put to good use. Professors participated in the circulation process as well by buying lecture transcripts for their personal consultation. Some transcripts even went to the furthest reaches of the British Empire to be used by graduates such as Hamilton who pursued careers as physicians, lawyers, clergy, and naturalists.

Sometimes the paths of circulation took on lives of their own. Such was the case for Hamilton's transcripts, which, instead of joining him in his duties as a ship's surgeon, ended up in the trunk of another person, a Mr. Boswell, who somehow managed to forfeit them in a battle to Tipu Sultan, the ruler of the Kingdom of Mysore in southern India. Renowned as a scholar and reformer, Tipu had the notes beautifully bound in red Indian leather. They remained in his library until he was defeated in the Battle of Seringapatam, at which time they were recognized by a Major Ogg, a British military officer, and returned to Hamilton. This remarkable story of circulation is told in the 1806 inscription that Hamilton wrote on the front flyleaf of his lecture transcript: "These notes were taken by me at the Botanical Garden Edinburgh in summer 1780. In a voyage to India in 1785 Mr Boswell, then my mate and who remained in the country, had by mistake put them up in his trunk and lost them at the affair near Satimangulum, where they were taken by Tippoo, and by him bound up in their present form. At the taking of Seringapatam they fell into the hands of Major Ogg who has restored them to me."[3]

The circulation of lecture notebooks was by no means unique to British students. As shown by Ann Blair, student notebooks were media technologies that played an important role in circulating knowledge during the modern period. In her words, "Note taking constitutes a central but often hidden phase in the transmission of knowledge."[4] This being the case, it is worth exploring the ways in which the lecture notebooks under discussion in this book circulated across Scotland and beyond. Though an inventory of the notebooks does not exist, there were hundreds, perhaps even thousands, that circulated during the long eighteenth century. Indeed, during the 1790s alone, Edinburgh's Alexander Monro Secundus estimated that there were over four hundred lecture transcripts of his anatomy course in circulation.[5] In this chapter, I explore

---

3. Hope Bound MS (1780, 1).
4. Blair (2004, 85).
5. Monro Secundus (1794). Struthers (1867, 29) estimated that 3,500 students attended Monro Secundus's anatomy course between 1780 and 1790 alone.

how these and other notebooks were spread and adapted within a surprising variety of local and global knowledge networks.

My exploration of notebook circulation is guided by two principal aims. The first is to better establish the kinds of places and networks through which lecture notes circulated between students, graduates, professors, households, and larger institutions such as libraries, markets, and even law courts. My second aim is to show that, in addition to the scholarly skills that students gained while keeping a notebook, there were more socially oriented skills that they learned through circulating notebooks themselves, or through witnessing the ways in which professors, publishers, and judges debated the legal or epistemic status of lecture transcripts as commodities in Britain's information economy.

Personal and Institutional Libraries

Once lecture transcripts were bound, they became manuscript books that circulated between institutions during and after the lifetime of their original keepers. Notably, transcripts were incorporated into the collections of many different libraries, and I would like to touch on two kinds, personal and institutional, with a view to showing that notebooks enjoyed a healthy circulation. The personal libraries that included lecture transcripts were usually curated by university students, professors, and graduates. The institutional libraries tended to be those managed by universities and their associated faculties, or those associated with professional medical or legal societies. Let us explore the place of lecture transcripts within these different locations.

The personal libraries of students were not fixed entities in that they frequently borrowed books from each other, from professors, and from various libraries associated with universities or professional societies. It has often been assumed by historians that students were borrowing printed books; however, as noted in previous chapters, students borrowed manuscript books such as lecture transcripts from each other. It was also possible to borrow lecture transcripts from university libraries and from "class libraries" that provided books and other sources about specialty topics.

A good example of a class library is the "physiological library" managed by Robert Steuart, professor of natural philosophy at Edinburgh University during the 1720s. It was maintained by Steuart, "six curators chosen yearly by the students," and gentlemanly benefactors based in the Lowlands.[6] As noted

6. Steuart (1725, 3).

by Michael Barfoot, one of the students who took Steuart's course at this time and had access to the library was the future philosopher, David Hume.[7] In addition to a variety of printed books, Hume and other students interested in learning more about optics could consult a quarto-sized lecture transcript of James Gregory's mathematics course. Gregory had served as professor at the universities of St. Andrews and Edinburgh from the 1660s to 1670s. He was widely known for his work in optics and the invention of what is now called the eponymous Gregorian telescope. Titled *Optica, Catoptrica, & Dioptrica: Propositions Astronomicae*, the transcript was probably based on lectures given at St. Andrews, as Gregory only served as professor at Edinburgh for a year.[8] Later in the century, the class library managed and owned by John Walker, Edinburgh's professor of natural history, offered students the opportunity to consult lecture transcripts of the botany course given by Charles Alston, professor of medicine at Edinburgh in the middle of the century.[9]

Depending on the wishes of a professor or his spouse, class libraries were either sold or donated to the university after a professor passed away. William Leechman, professor of divinity and principal of Glasgow University, for instance, gifted his library to the divinity faculty's library in the 1780s.[10] Such donations sometimes included lecture transcripts or books that featured the marginalia of former professors. Students consulting Glasgow's divinity library during the 1790s, for instance, were able to consult a "Bible interleaved, with M. S. Notes by Dr. Leechman."[11]

Students usually intended to put their learning to good use once they finished university. To this goal, they assembled small personal libraries that served to preserve or extend the knowledge systems that they encountered during their studies. Since most students lived on modest budgets, and since traveling with a cartload of books was expensive and fraught with risk, they needed to select their books wisely. In addition to making their own manuscript books via the process of notekeeping, they usually chose to buy canonical books recommended by professors. Scottish students traveling to

---

7. Barfoot (1990) argues that it was likely that the books of Steuart's library played a role in shaping Hume's early understanding of natural philosophy.

8. Steuart (1725, 30, entry 153). The entry is listed under "Jacobi Gregorii," but it is possible that it referring to the optics lectures at Edinburgh given by James Gregory's nephew, David Gregory (1661–1708), the university chair in mathematics from 1683, who also lectured on optics. For the content of the younger Gregory's lectures, see Gregory Bound MS (1683–1686) and the Papers of James Gregory, EUL, Coll-33, which also contains lectures on optics.

9. John Walker's library was catalogued by Elliot (1804) after he died. See the entry for Charles Alston's botanical lecture notes.

10. A list of the books from Leechman's class library is appended to Glasgow University (1790).

11. The Bible is listed in Glasgow University (1790, 69) within the section of the catalogue that lists "Books given by Principal Leechman."

mainland Europe took care to return with their manuscript notebooks as well. Following his medical studies in London, Paris, and Leiden during the 1710s, for example, Alexander Monro Primus brought back "numerous manuscript Excerpts from Books, the full Notes of the Lectures he heard and the Description of whatever he saw relating to Medicine that was curious, with Copies of Notes taken by other Students of the Lectures they had from Teachers whom he never attended."[12] Likewise, students who traveled from Europe and the British colonies to study in Scottish universities endeavored to transport their lecture transcripts back home.

Students used their notebooks to study for examinations or as reference works long after they had left university. The book catalogues compiled by Edinburgh's auctioneers to advertise the collections of recently deceased professors or professionals also shows that lecture transcripts were bought by others seeking to augment their personal libraries. The evidence for the presence of the Alston lectures in Walker's professorial library, for instance, comes from the catalogue compiled by the Edinburgh book dealer Cornelius Elliot when the collection was auctioned in 1804.[13]

Peter Hill, the Edinburgh bookseller and friend of the poet Robert Burns, sold lecture transcripts during the later decades of the century. His 1793 catalogue, for instance, offered a 1781 edition of "Dunbar's natural philosophy, MSS, neatly wrote." James Dunbar was the professor of natural philosophy at King's College, Aberdeen. The first part of his course treated natural philosophy and the second part addressed moral philosophy.[14] The fact that the transcripts were sold in Edinburgh, as opposed to Aberdeen where Dunbar taught, shows how student notebooks circulated between university cities.[15] Hill's catalogue also offered a 1788 set of "Notes from Dr Walker's lectures, MSS." This was the very same Professor John Walker who, as we learned above, was himself a collector of lecture transcripts. The 2s 6d price of the Dunbar and Walker transcripts suggests they were not bound. This was common at the time, as those with personal libraries liked to use the same or similar bindings for all of their printed and manuscript books.

Professionals kept copies of lecture transcripts in their libraries as well. The flyleaves of notebooks sometimes offer hints as to how and why they were

---

12. Monro Primus (1954, 82). Sir John Clerk of Penicuik, for example, brought his lecture notes on metaphysics and ethics when he went to study in the United Provinces during the late seventeenth century. Mijers (2012, 129).
13. Elliot (1804).
14. Hill (1793, 14 [lot 5699], 35 [lot 6195]).
15. For the content of the lectures, see Dunbar Bound MS (1791–1793).

purchased. Dr. John Boswell, for example, wrote the following on the flyleaf of a set of midwifery lecture notes that he bought for his own library: "I bought this book att Dr. Whytt's sale for it contains a variety of things amongst others the Dr.'s own ms. notes when a student of old Dr. Young's lectures on his own private practice." He goes on to say that "Dr. Whytt and I were contemporary fellow students of Dr. Young's an. 1730–31."[16] The "Dr. Whytt" named by Boswell was in fact Robert Whytt, president of Edinburgh's Royal College of Physicians and professor of medical theory at the university. Boswell's inscription sheds light on why a medical professional at the time would be interested in owning lecture transcripts. It reveals that they contained unique facts and observations unavailable in the other books that Boswell owned. It also shows that former students sometimes collected transcripts for sentimental reasons.

Further evidence of lecture transcripts in personal collections comes from the library of Dr. David Spence, a prominent Edinburgh physician and midwife who was a licentiate of the Royal College of Physicians and a fellow of the Society of Scottish Antiquaries. The 1786 auction catalogue of his library compiled shortly after his death by the bookseller William Wright features several sets of octavo medical lecture transcripts, including "Monro's anatomical lectures, 2 vol. MS," "Young's lectures, MS," "MS notes on Young's lectures, 3 vol.," and "Clinical cases, 1766,–67, MS."[17] Spence attended Edinburgh University's medical school in the 1760s, which means that the manuscripts are probably the lecture transcripts that he kept while attending the courses given by professors Alexander Monro Secundus on anatomy and Thomas Young on midwifery and various professors offering clinical lectures in the Edinburgh Infirmary.[18]

As intimated above, students frequently took their notes home after their studies to be used as reference books that served their personal or professional interests. John Thorburn, for example, brought his shorthand set of 1769–70 anatomy notes home to Cambridgeshire when he finished his medical studies at Edinburgh. But their use as reference aids in his personal library did not end there. Thorburn died in the 1780s and his son, John Thorburn Jr., inherited them. Following in his father's footsteps, Thorburn Jr. studied medicine at Edinburgh University during the early 1790s and he brought his father's notes with him. Then, like his father, he took Monro Secundus's anatomy

---

16. Quoted in Hoolihan (1985, 239, 7). The notes in question are housed in the Royal College of Physicians of Edinburgh, Ms. 1 (M9.19).

17. Martin (1786, 12 [lots 384–387]).

18. Spence submitted his doctoral dissertation to the Edinburgh medical school in 1767. Spence (1767). He was the author of *A System of Midwifery, Theoretical and Practical* (1784).

course. Monro, who was touched by this episode because he too had used his own father's lecture notes when he studied medicine,[19] recounted this form of intergenerational manuscript circulation in 1794: "Thorburn, who is studying Physic, and has attended my Lectures this and last winter, is in possession of his father's original manuscript, written in short hand in 1770, which he has extended as accurately as he could."[20]

Monro Secundus's observation reveals that the lecture transcripts housed in personal and institutional collections were not fixed media technologies. They were, in Monro's terminology, extendable. To see evidence of this kind of adaptation, we need to look no further than Monro's personal library. His interest in Thorburn's notebooks alerts us to the fact that professors acquired student lecture transcripts for their own use. In an age before audio recording devices, student notebooks offered professors a way to evaluate or remember what they had said in their own lectures. After using only headings and personal lecture notes to prepare the lectures that Monro Secundus gave in his career from the 1770s onward, he purchased the notes of Thorburn and several other students.[21] As Monro's cancellation marks and inserted corrections in Thorburn's notes indicate, he read his former students' notebooks to gain insight into what he had said in his lectures, to see what his students had found noteworthy, and to reclaim striking examples or illustrations that he said extemporaneously.[22]

Far from being a rare occurrence, editing manuscript books in this manner was a common practice for those who owned personal libraries. As we learned in previous chapters in my discussion of Margaret Monro, who was Monro Secundus's sister, boys and girls living in literate households learned manuscript editorial skills at an early age. In addition to interfacing with the manuscript books of his family's library as a boy, Monro Secundus, like most Scottish professors, had been a member of a student notekeeping community during his own university studies. In many ways, his use of his students' notebooks shows that he still benefited from the editorial skills he learned at home and as a medical student even after he had become a professor.

Another reason professors bought lecture transcripts for their personal libraries was to gain insight into the topics being taught by their peers. The

---

19. Monro Primus gave a number of his lectures and commentaries to Monro Secundus. The manuscripts are discussed throughout Taylor (1979). For insight into how Monro Secundus used and augmented Monro Primus Bound MS (1750), see Taylor (1979, 83–85).
20. Monro Secundus (1794, 43).
21. Struthers (1867, 28–29).
22. Monro's annotations occur in Monro Secundus Bound MSS (1773–1774). The annotations are addressed in Taylor (1979, 97–98).

future colonial administrator Hugh Cleghorn (1752–1827), who served as professor of civil history at St. Andrews University during the 1770s and 1780s, annotated lecture notes taken by an anonymous student in the 1786 government course offered by John Millar at Glasgow University.[23] Cleghorn was a young professor at the time and later remarked, "It was my fortune, at perhaps too early a period of life, to be appointed to an office whose object is to examine the Civil History of Men." Perhaps this explains why parts of his lectures were notably similar to Millar's course.[24] The transcript bears Cleghorn's annotations and offers a snapshot of the symbiotic relationship that existed between students and professors in notebooks in their capacity as media technologies that circulated useful knowledge.

In earlier chapters we learned that students practiced several kinds of communal inscription when they rewrote their rough notes. Collective forms of writing in lecture notebooks continued long after they left the setting of a university, particularly when they became part of a household or professional library. The bound lecture notes in such collections bear many traditional marks of book ownership and donation, most commonly signatures (sometimes of multiple owners), donor statements such as "Presented by John Grant, Esq.,"[25] and library stamps on the flyleaves. Occasionally, but not infrequently, there are bookplates or embossed insignias that bear the names of the students or institutions that kept them as library reference works. Bookplates such as those featured in the lecture transcripts of John Borthwick of Crookston (1787–1845) and John Waldie of Hendersyde (b. 1781) bear family crests that shed light on the identity of students. Marginalia in such notebooks also offered hints as to how they were amended or annotated after they had been deposited in a collection.[26]

As a member of a wealthy family of Scottish industrialists based in Newcastle-upon-Tyne, John Waldie assembled a household library after his studies in Edinburgh. The library was organized via a system of call numbers written on bookplates affixed to the inside cover of the books. The bookplates of his universal history lecture transcript, which was based on the courses of

---

23. Millar Unbound MS (1786). Cleghorn's inserted notes were written on slips cut from larger sheets of paper, or on the backs of letters addressed to him.

24. Quotation taken from Clark (1992, 18). For the similarities between the lectures of Cleghorn and Millar, see 27–28. When compared to Millar's lectures, Clark maintains that Cleghorn's lectures owed "much more than he acknowledges."

25. For John Grant's provenance note, see the title page of Tytler Bound MS (1800–1801). Most sets of lecture notes held by research institutions today bear a library stamp (or two) at the front.

26. See John Borthwick's library plate in Hill Bound MS (1802–03) and John Waldie's bookplate in Tytler Bound MS (1800–1801). John Borthwick appears as 13th Lord Borthwick in most Scottish peerages and the life and education of John Waldie is discussed in Livsey (n.d.).

Edinburgh's Alexander Fraser Tytler, read "History. No. 95." The words "History. No." are printed and the number "95" is handwritten. Waldie's bookplate offers a glimpse of how former university notekeepers might have classified their own notes in a household library after they finished their studies. When the bookplate is considered alongside other examples of provenance featured in the Waldie and Borthwick notebooks, it can be seen that aside from the immediate annotations and corrections added in the days and months following a lecture course, there were longstanding opportunities for students to use or amend notebooks in institutional or familial communities that existed outside the corridors of universities.

Since the law and medicine courses given by some of Scotland's professors were highly respected within Britain and abroad, institutions often kept copies for consultation. The members of the library of the Medical and Chirurgical Society of London, for instance, could consult the lecture transcripts taken in several courses given by William Cullen. The catalogue lists quarto-sized copies of "Cullen's Clinical Lectures (MS.), A.D. 1772, 1773," "Lectures on the Practice of Physic (MS.) 1769, 1770, Vols. I, II. IV, V.," and "Lectures on the Institutions of Medicine (MS.) 1772, 1773, 4 vols."[27] Cullen's manuscript influence over the theory and practice of medicine as a discipline were extended across the Atlantic as well. The 1806 catalogue of the medical library of the Pennsylvania Hospital lists two collections of "Manuscript notes taken from the lectures of the late Dr. Cullen," one containing seventeen volumes and the other containing twenty-eight volumes. Like the other books in the library, the transcripts could be checked out by the managers, treasurer, physicians, students, apprentices, and apothecaries associated with the hospital.[28]

The contents of lecture transcripts housed in professional libraries were sometimes updated by members. As shown in the work of Mark Towsey, communal inscription was practiced in libraries across Britain during the eighteenth and nineteenth centuries.[29] The annotations made in a seven-volume bound set of lecture notebooks housed in the old library of the Faculty of Procurators in Glasgow shows how this process worked in a professional context. The faculty was the legal body of lawyers who practiced in the city and

---

27. Medical and Chirurgical Society of London (1819, 85).

28. Pennsylvania Hospital (1806, 31). The members who were allowed to borrow books are stated in the rules of the library on page vi. It is possible that a three-volume 1799 transcript of Joseph Black's lectures on chemistry was still in circulation as late as the 1840s in the library of the New York Hospital, an institution that was cofounded by Samual Bard, a graduate of Edinburgh University. The lectures are listed as "Mass," which is possibly a conflation with the abbreviation "MSS" commonly used to denote "manuscripts." New York Hospital (1845, 20).

29. In addition to investigating the writing practices conducted in libraries, Towsey (2010) excavates the larger importance of local libraries across Scotland. On this point, see also Lochhead (1948, 348–371).

its environs.[30] The notebooks were made by an anonymous student who attended the Edinburgh law lectures of David Hume during the 1810–1811 academic year.[31] The provenance is clearly indicated by the "Faculty of Procurators in Glasgow" lexigram embossed on the front of every volume. The pages contain graphite and ink annotations made in different hands, some of which bear dates several years after the notes were written, suggesting that they became part of the faculty's library when it opened in 1817. The inscriptions indicate that each volume was a living document that was changed and emended to fit the needs of the library's users.

Lecture transcripts also circulated back to the libraries of their university of origin during the nineteenth and twentieth centuries. Some were donated my members of a deceased professor's family. Notes that explain this kind of provenance are sometimes written on the flyleaf or they are given in a letter tucked inside the notebooks. The son-in-law of John Hill, Edinburgh's professor of humanity and philology, for example, used the former practice when he donated his father-in-law's papers to the university. A note on the flyleaf of one of the bound manuscript notebooks reads: "M.S.S. of my learned father-in-law—Saved by me from being cancelled—circa 1808." The note is followed by a scribbled set of initials and then "Bound in 1840."[32]

Other sets of notes found their way into university special collections in North America, Europe, Australia, and New Zealand, most remarkably in the case of the Monro family.[33] From the early eighteenth century to the middle of the nineteenth century, three generations of Monros—Alexander Monro Primus, Secundus, and Tertius—held the Edinburgh chair of anatomy.[34] As each retired, he gave his student notebooks to his son. When Monro Tertius died in 1859, he gave all the notebooks to his son, David Monro, who then emigrated to New Zealand. In time, David Monro gave them to the University of Otago, where today they constitute one of the largest intergenerational collections of modern anatomy lecture notebooks.[35] When considered alongside the

---

30. Denholm (1798, 170–171).

31. Hume Bound MS (1810–1811).

32. Hill Bound MS (1770s). Sometimes the notes of provenance written by the presenter are as short as the inscription "Presented by John Grant, Esq." penciled in the front of Tytler Bound MS (1800–1801). Alternatively, librarians sometimes included further information as well. For example, a note dated 1 October 1966 in Hume Bound MSS (1808–1809) (anonymous notekeeper) states that the notebooks were given by "Mr Justice Larskin."

33. For a North American example, see the notes taken by William Logan Jr. in the Edinburgh lectures of Joseph Black (chemistry), Alexander Monro Secundus (anatomy), Thomas Young (midwifery), and Hugh Blair (rhetoric) that are held by the Library Company of Philadelphia, Logan Family Papers, Series VI, William Logan, Jr. Papers.

34. There were several branches of the Monro medical dynasty. See Macintyre and Munro (2015).

35. The story of the Monro manuscripts is told in Taylor (1979, 9–20). Otago's Monro Collection also contains notebooks taken in the lectures of Joseph Black. Oldroyd (1972).

other transcripts mentioned in this section, the Monro family manuscripts reveal that the diverse circulation of lecture notebooks between libraries is part of the larger story of how useful knowledge spread across the globe from the seventeenth century onward.

Commodities within Knowledge Economies

The presence of lecture transcripts in the catalogues of auctioneers and booksellers alerts us to the role they played as commodities within the knowledge economy of modern Britain. As noted by Samuel Latham Mitchell, the American physician and senator who studied medicine at Edinburgh in the 1780s, there was money to be made in keeping and selling lecture transcripts. In the case of the chemistry course given by Edinburgh's Joseph Black, for instance, Mitchell reflected that "Manuscript copies of his lectures were then commonly enough sold by those who made a profitable Business of taking and transcribing them."[36] The prices of lecture transcripts varied based on the subject matter, the number of volumes, and whether they were bound. The bookseller Peter Hill, as noted above, listed a set of Professor John Walker's Edinburgh natural history lectures and a set of Professor James Dunbar's Aberdeen natural philosophy lectures for 2s 6d each.[37]

As observed by Edinburgh's Professor John Robison in the preface to his 1803 published edition of Joseph Black's chemistry lectures, one lecture transcript taken in the course of a well-known professor could cost up to five guineas, that is to say, well over £5.[38] Robison used several student transcripts to interpret Black's thought, each of which cost him between four and five guineas. Alexander Monro Secundus paid five guineas as well for the aforementioned notes kept in his lectures by the medical student John Thorburn.[39] Like Monro, many professors purchased a set of their own lectures, especially if they were thinking about publishing them as a book.[40] Consequently, in addition to learning facts and practicing mnemonic routines, some students kept notebooks with a view to making money.

The demand for lecture transcripts dovetailed with what might be called a manuscript industry that consisted of a diverse world of professionals and merchants who specialized in making or selling written or drawn media tech-

---

36. Mitchell (1976, 745–746).
37. Hill (1793, 14, 35).
38. Black (1803b, xi).
39. Struthers (1867, 28–29).
40. Thorburn sold his notes from the lectures of William Cullen and Joseph Black for the same impressive price. Monro Secundus (1840, vii).

nologies. The making side of the equation included writing and drawing masters, artists, bookkeepers, writers to the signet, secretaries, professional copyists, and, indeed, students, all of whom created manuscripts that they could sell for profit. The sellers included book dealers and auctioneers, as well as the many stationers and shopkeepers who sold the media required to make manuscripts. This means that for those with resources, it was possible to acquire a lecture transcript without having to liaise directly with a student. As intimated in the observations made by Mitchell and Robison, it was possible to hire a professional transcriber to copy all or part of a student lecture transcript.

Though it is unlikely that many students had the means to pay a transcriber during their time at university, it is worth asking whether any saw fit to do so. Little research has been done on how, if at all, students interfaced with professional copyists, but Silas Neville's diary offers several clues. It suggests that, in addition to occasionally transcribing student dissertations, essays, and research papers, transcribers were employed as clerks in the various societies and law courts of Edinburgh.[41] This meant that students, particularly those who waited until the last minute, sometimes struggled to get a transcriber at short notice. Neville was one such student. In his diary he lamented the difficulty of finding one. When he finally did, he still encountered difficulty: "When he came he said he was clerk to a certain society which meets for business on Tuesdays and that he should be fined if he was absent."[42] Neville offered to pay the fine and the transcriber then agreed to do the work for him.

One of the hallmarks of professionally transcribed lecture notebooks is neat and sometimes decorative handwriting. But even this superior penmanship could have been carried out by a student who had studied orthography with a writing master before entering university. A case in point is the neatly written 1788 diary of George Sandy of Edinburgh that was penned after he had finished school and was waiting to become a legal apprentice. The neatness of his notebook, which was written when he was fifteen years old, easily could be taken as the work of a professional transcriber.[43]

Except for the rare but telling glimpses offered in Neville's diary, there is hardly any further evidence regarding the relationship between students and transcribers. Neville employed a servant while he was in Edinburgh but only mentions a transcriber on a few occasions, suggesting that the latter was more

---

41. During the 1760s and 1770s there were around twenty medical dissertations submitted to the Edinburgh medical faculty every year. This number continued to rise, reaching over two hundred per year in 1820. The names of graduates and the titles of their dissertations are given in Edinburgh University (1867). Page iv states that the total number of dissertations submitted between 1726 and 1866 was 6,622.
42. Neville (1767–88/1950, 222).
43. Sandy Bound MS (1788).

a luxury than a commonly used service.[44] This was despite the fact that he spent a great deal of his time transcribing his rough notes. He also transcribed another student's notes of Professor John Gregory's clinical reports "line by line," and wrote (and rewrote) papers, commentaries, and aphorisms for the student medical society. Even when he used a transcriber, he was dissatisfied with the final result and ended up recopying the material himself.[45]

For notebooks that were most likely copied by professional transcribers, it is difficult to determine with any certainty the identities of those who carried out the task. Neville's diary does not even mention the name of the one that he hired. The only solid identity that I have found is that of the copyist who duplicated the set of anatomy notes taken by John Thorburn. These notes were bought by Monro Secundus during the 1770s. Reflecting on the purchase, Monro noted that, "On the 8th of November 1774, I purchased a copy of Mr. THORBURN'S manuscript, written in ten volumes, by Mr. JOHN WILSON, who, being lame, had the conceit of calling himself *Claudero*."[46] Even though Monro gave the name "John Wilson," he was probably referring to the discredited poet *James* Wilson.[47]

As evinced in the previous chapters of this book, the transformation of a lecture transcript into a media technology that could be bought or sold required an immense amount of student labor. One of the things that impacted a transcript's value as a commodity was the role played by student notekeepers as editors. In addition to countless edits they made to their own transcripts, they also existed in an editorial relationship with their professors. In other words, when professors bought and used copies of notebooks based on their lectures, they effectively were treating students as editors who had collated their ideas into an accessible narrative. This of course saved professors a great deal of time and, ultimately, money.

The value of lecture transcripts was recognized by publishers as well. The 1771 pirated edition of William Cullen's *materia media* course, for instance, was based on student notebooks. The corrections of the authorized 1772 edition of Cullen's *materia medica* lectures were based on other sets of student notebooks, as well. As the subtitle states, the book was republished "with many CORRECTIONS from the Collation of different MANUSCRIPTS by the EDITOR." The editor in question was Cullen himself. Ironically, even though he claimed

---

44. Neville (1767–88/1950, 205, 222–223).
45. Neville (1767–88/1950, 205, 208, 213–214) addresses acts of transcribing, copying, writing, and rewriting.
46. Monro Secundus (1794, 39–40). The capitalization is Monro's.
47. James Wilson's life as a poet is summarized in J. O. (1853, 68). Wilson described himself as "Claudero, Son of Nimrod the Mighty Hunter," in the subtitle of his *Poems on Several Occasions* (1765).

that the pirated edition was filled with "blunders & inaccuracies," his personal notes were not enough to complete the editorial task and he had to use student notes to create a more accurate printed edition.[48] Other professors, Alexander Monro Secundus and John Robison for instance, relied on student notebooks when editing lectures of other professors for publication.[49] These examples from the world of publishing further speak to the symbiotic editorial relationship between professors and their students.

In many ways the chorus of student scrollers, transcribers, and draftsmen operating in Scottish universities resembled the group of editors, compositors, and pressmen who worked collectively to produce books on the shop floor of printing houses. The symbiotic editorial dimension of lecture notebooks further reveals that professors were oftentimes dependent upon student notekeeping and consequently it explains why they spent a notable amount of effort keeping track of the notes that had been taken in their lectures over the duration of their careers.

Cullen's dependence on student lecture notes is particularly striking when one considers that there were other manuscript media technologies on the market that he could have used to preserve what he had said in his lectures. For example, later in his career, particularly from the 1780s onward, he dictated medical consultation letters to an amanuensis and used James Watt's copying machine to replicate them. But scribal alternatives required investment, both in terms of paying the amanuensis and acquiring and servicing a new copying machine. When viewed from a practical perspective, Watt's machine was tricky to assemble and produced an inferior copy. After receiving one as a gift, Joseph Black informed Watt that he was "not satisfied" with the quality of the ink required by its automated stylus.[50] It seems that since students were recording and editing notes anyway, it was easier and probably more cost efficient to simply buy or borrow copies as needed from those living in the Edinburgh area.[51]

When it came to considering the financial impact that student lecture transcripts had on the scholarly marketplace, the circulation of manuscript lecture notes was sometimes a double-edged sword. On the one hand, they commu-

---

48. The use of other student notebooks is intimated in the preface of Cullen (1772). Cullen's assessment of the first edition's accuracy is recorded in Neville (1767–88/1950, 236).

49. Alexander Monro Secundus edited his father's anatomy lectures and John Robison edited the chemistry lectures of Joseph Black. For insight into their use of student notebooks, see the introductory comments in Monro Primus (1783) and Black (1803a).

50. After he assembled his own copying machine, Black helped the duke and duchess of Buccleugh assemble theirs as well. "Joseph Black to James Watt, Edinburgh, 18 October 1780," in Black (2012, 433–434).

51. Hills (1996).

nicated a professor's ideas to the public and served to spread his reputation at home and abroad. This in turn attracted students. Lecture transcripts also served to keep former students updated on a professor's new ideas. For instance, when Dr. Gilbert Steuart, based in Wolverhampton, England, found out that William Cullen was lecturing on new subjects, he asked Cullen to acquire and send a set of student lecture notes to him: "I should deem it as the highest Favour if you could consistently with your own Honour & your Student's advantage order Transcripts to be made for me from the most perfect Copys."[52]

But on the other hand, the circulation of student lecture notes could create financial problems for professors. The presence of too many easily accessible copies could tempt some students to transcribe the notes and skip the course despite the strong links that professors and other students drew between notekeeping and thinking. The same might happen if an entire set of lecture notes was published. In both cases, the professor potentially stood to lose a significant amount of money via the loss of students, especially in university courses where each student paid a three-guinea tuition fee directly to him at the beginning of the year. This explains why professors waited to publish their lectures until after they retired, or in the case of some, after they moved permanently to the colonies.[53]

Some professors worried that the circulation of too many lecture transcripts might discourage the public from buying the published version of the course. In these cases, professors attempted to discredit lecture transcripts in the hope that it would encourage potential readers to buy the book. Hugh Blair, Edinburgh's popular professor of rhetoric and belles lettres, employed this tactic in the preface of the published version of his lectures. He suggested that: "When the Author [Blair] saw them [manuscript notes] circulate so currently, as even to be quoted in print, and found himself often threatened with surreptitious publications from them, he judged it to be high time that they should proceed from his own hand, rather than come into public view under some very defective and erroneous form."[54] Similar statements about the inaccuracy or inadequacy of student notes were made in print by other professors such as Cullen, John Gregory, and Monro Secundus when, or after, they published books based on their lectures.[55]

---

52. Gilbert Steuart to William Cullen, 2 January 1770, RCPE, CUL/1/2/75.
53. After failing to secure a university professorship, William West, a mathematics tutor at St. Andrews University, moved to Jamaica. Prior to his departure he published his lectures as *Elements of Mathematics: Comprehending Geometry, Mensuration, Conic sections and Spherics* (1784).
54. Blair (1783, 1: iii).
55. Cullen (1772). See also Monro's comments throughout Monro Secundus (1794). A pirated printed version of John Gregory's lectures was published as *Observations of the Duties and Offices of a Physician*

While it is clear that even the most gifted notekeeper could not capture every word spoken by a professor, the negative statements about student notes need to be handled with care, especially since, as pointed out above, numerous professors eagerly bought transcripts for their own personal libraries. Indeed, Monro Secundus read many transcripts based on his course and judged them, on the whole, to be accurate. He even went so far as to state that the many transcriptions made from Thorburn's notes over two decades had been "handed down to this time, with fewer corrections and additions than might have been expected."[56] It was a similar case for lecture transcripts taken by students in other courses as well. For instance, when the Delaware physician John Vaughan asked Black's former student Samuel Latham Mitchell whether it was possible for a student lecture transcript to accurately represent the content of Black's entire chemistry course, Mitchell answered in the affirmative. Mitchell explained that "from Dr. Black's very slow and deliberate manner of speaking, I can easily believe a good copy might be taken of all that part which could be expressed in words."[57]

The numerous positive appraisals of the accuracy of lecture transcripts further suggest that protestations of inaccuracy were often motivated by concerns over the negative financial repercussions that might occur if too many notebooks were in circulation. Blair's preface, for instance, states that student notebooks based on his course "were first privately handed about; and afterwards frequently exposed to public sale."[58] Here the reference to "frequently" signals that professorial fears over the circulation of lecture notes were more than a simple intellectual matter. Indeed, Blair was clearly worried that transcripts of his lectures could potentially damage the sales of his book.

Whereas an unauthorized printing of an entire set of lecture notes was problematic for some professors, there were some cases where the publication of a selection of lectures might have worked to a professor's advantage. It is highly likely that an unauthorized 1770 edition of Joseph Black's lectures on heat effectively acted as an advertisement that enticed students to travel to Edinburgh and take his entire course.[59] Unlike courses offered in the arts, divinity, and law faculties, Black's lectures included a large number of experiments, many of which had to be seen, heard, and smelled in person to be fully

---

(1770). He then published his own authoritative edition as *Lectures on the Duties and Qualifications of a Physician* (1772). His comments about the "many transcripts" of student lecture notes occur at the beginning in the "advertisement."

56. Monro Secundus (1794, 39).

57. Mitchell (1976, 745–746).

58. Blair (1783, iii).

59. Black (1770).

understood. This meant that in order to see the experiments referenced in the printed version of Black's lectures, one had to travel to Edinburgh and attend the course.[60]

The teaching aids developed by professors for student notekeepers could also be placed in circulation through publication. A good example of this practice can be seen in William Creech's 1792 reprint of George Martine's *Essays on the Construction and Graduation of Thermometers*. Creech was a successful Edinburgh-based publisher and bookseller. Martine was a Scottish physician who studied medicine in St. Andrews, Edinburgh, and Leiden. He had written the first edition of his book in the late 1730s and it remained in circulation after his untimely death in Columbia in 1741. Joseph Black found the book so useful that he recommended it to students who attended his lectures. Black's fondness for the book caught the attention of Creech. Ever ready to spot a good business opportunity, Creech was keen to increase the salability of the 1792 republication of Martine's book by inserting an appendix that included the thermometric scales Black used in his lectures. Creech already published and sold the lecture headings and books of several professors, which meant that he was intimately familiar with the bibliographic needs and desires of students. Based on this knowledge, he capitalized on the advantages of explicitly flagging Black's tables in the reprint's subtitle: "TABLES of the Different Scales of HEAT, exhibited by DR. BLACK, in his Annual Course of Chemistry"[61] (figure 10.1).

Creech's shop was close to the university, near St. Giles Cathedral. The subtitle of Martine's book was sure to catch the attention of the many students and academics who frequented it. Black's scales would have also interested the influential literati and savants who attended the weekly breakfasts, called "Creech's levees," that he hosted in an upstairs room. For those enticed by the subtitle, he added the following advertisement on the next page: "As this book is recommended by DR BLACK, to the Students attending his class, the Editor has endeavored to render this edition more useful to them . . . by adding, in the appendix, some tables of the scales of heat, which the doctor usually exhibits and explains in his course." It is unclear, however, whether Black authorized the publication of the tables, which, despite the fanfare in the front of the book, only ran for four pages.[62]

---

60. A similar case could be made for the official and unofficial printed editions of the chemistry lectures given by Herman Boerhaave at Leiden University during the early eighteenth century. As pointed out by Blair (2004, 93), the Leiden student William Logan took notes on blank pages interleaved into Boerhaave's *Institutiones Medicæ* (1708).

61. Martine (1792).

62. Martine (1792, 183–186). William Creech's levees are mentioned by Barbara M. Benedict in her *ODNB* entry titled "Creech, William [pseud. Theophrastus]." As shown in Sher (2008), Scottish publishers, includ-

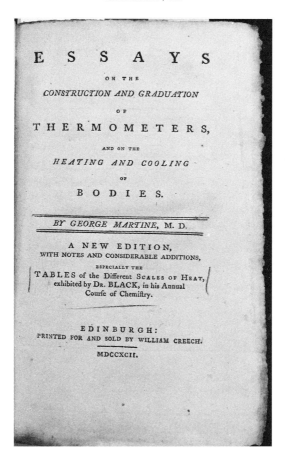

FIGURE 10.1. Title page using the thermometric scales of Joseph Black's chemistry lectures as an advertisement. George Martine, *Essays on the Construction and Graduation of Thermometers and on the Heating and Cooling of Bodies. New Edition* (Edinburgh: Creech, 1792). Reproduced with permission of the Huntington Library, San Marino, No. 473796.

Courts of Law and Public Opinion

Though copyright laws were weak, professors tended to actively oppose the unauthorized publication of material that originated in their lectures.[63] The legal status of transcripts was put to the test in 1771 when the London printer

---

ing Creech, regularly were accused of printing material without permission, or in a format that clashed with the expectations of authors. Creech had published a number of other publications for Black, and this suggests that they had a good working relationship.

[63]. The commercial and legal nuances of British publishing during eighteenth century and nineteenth century are addressed in Sher (2008), St. Clair (2004), and Raven (2014).

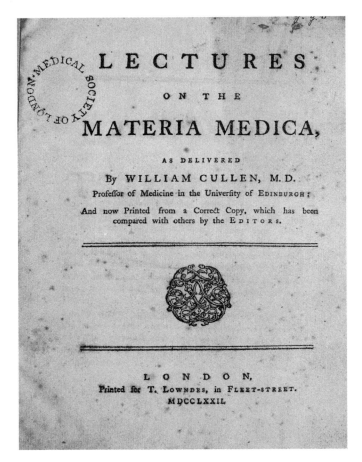

FIGURE 10.2. Title page of the authorized edition of *Lectures on the Materia Medica as Delivered by William Cullen, M.D.* (London: Lowndes, 1772). Reproduced with permission of the Wellcome Library, London.

Thomas Lowndes published the unauthorized edition of William Cullen's *materia medica* course. As recounted in the preface of the 1772 authorized edition, Cullen was so concerned about the circulation of his lecture material in print that, "as soon as he was informed of the Publication, he applied for, and obtained from the Lord Chancellor, an Injunction, prohibiting the sale of the book." (figures 10.2 and 10.3).[64] This brief statement concerning the pirated edition was meant primarily to direct potential buyers to the authorized edi-

---

64. Cullen (1772, vii). For an example of a student notebook based on Cullen's *materia medica* course, see Cullen Bound MSS (1761), Low (notekeeper). A copy of the 1772 authorized printed edition of the course is held by the Wellcome Library, London.

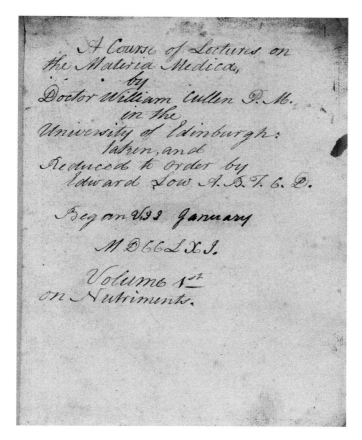

FIGURE 10.3. Title page of William Cullen, *A Course of Lectures on the Materia Medica . . . Volume 1st on Nutriments* (1761), Edward Low (notekeeper), Bound MS, Wellcome MS MSL 22ª. Reproduced with permission of the Wellcome Library, London.

tion. The full story of the case, however, is more interesting and offers insight into how lecture transcripts emerged as intellectual property in the courts of law and within the equally important court of public opinion.

Based in Fleet Street, Thomas Lowndes was known at the time as the proprietor of the eponymous circulating library that boasted over ten thousand books.[65] When he printed Cullen's lectures in 1771, he was expanding his publishing interests to meet the growing demand for books about domestic medicine, popular science, gardening, and animal husbandry. He saw Cullen's lectures as part of this expansion and as a high-profile publication that would entice fashionable readers to visit his shop and read his catalogues. By the

---

65. Lowndes ([1766]), Jacobs and Forster (1995), and Schürer (2007).

1770s, Cullen's medical prowess was well-known in London because many of his students had become successful physicians there.

Theoretically, the book had the potential to increase Cullen's reputation and to entice more students to his Edinburgh lectures. But Cullen ultimately did not see it that way for several reasons. The first was that he no longer lectured on *materia medica*. The pirated edition could not act, therefore, as an advertisement for what he was teaching. But more importantly, Cullen received none of the book's profits. This led him to seek an injunction that, if successful, would prevent Lowndes from printing further copies, thereby putting Cullen in a stronger position to negotiate a publication contract that worked to his benefit.

Cullen's application to seek an injunction constitutes a noteworthy episode in British legal history. His application was one of a series of eighteenth-century legal challenges that laid the groundwork for intellectual property law as it is understood today in the United Kingdom. At the time that the lectures were published, there was precedent within English common law that prevented printers from publishing the manuscripts of authors without permission. Though the precedent existed, Cullen faced two hurdles, one local and the other national. As to the former, the Scottish courts operated independently from the English courts on many matters. This meant that an injunction from a Scottish judge was unlikely to prevent the book from being printed in London, one of the largest publication markets in the world. The second hurdle was that the previous decisions of the English courts to treat book manuscripts as intellectual property were made in cases that involved letters, treatises, and plays, that is, genres which mainly involved one person as the author. Cullen's case involved student lecture transcripts, which meant that it was a different kind of manuscript from those involved in earlier legal cases. Indeed, it was technically students who were responsible for writing the manuscript because Cullen, like many Scottish professors, lectured extemporaneously or from brief handwritten headings, for which there was no clear precedent for legal protection.[66]

The foregoing factors led Cullen to make a daring move. Since the Scottish courts and the English common law courts presented a number of potential problems, he decided to bypass them altogether. Instead, he took his case to the Court of Chancery, an alternative court based in London that heard complicated cases. It was convened by the Lord Chancellor and it was not bound

---

66. The eighteenth-century legal context of the intellectual property rights afforded to authors is explained in Masiyakurima (2020, 25–39).

by the rules of the common law courts.[67] The cases were usually decided on written pleadings and evidence. His choice to use the court was informed by a number of legal advisors. At the time of the case, he was personal friends with several Scottish judges, and his son, Robert Cullen, was a prominent lawyer in Edinburgh. In the end, Cullen's decision to use the Court of Chancery was prudent, as it avoided the local politics that sometimes surrounded Scottish court cases and the decision would apply to all territories that had in principle agreed to honor English court rulings made after the Act of Union in 1707.

Previous to Cullen's case, publishers had attempted to exploit the developing relationship between the book industry and intellectual property law by printing transcripts of plays that had been written during the performance. Though unscrupulous, this kind of publication did not officially break the law, as the publishers were not using a manuscript physically written by the playwright. The practice was put to the test in 1770 when the playwright Charles Macklin successfully obtained an injunction to prevent the printing of a pirated transcript of his play *Love à la Mode*. Macklin's success hinged on the fact that the court had decided to treat the performance a of play as a form of "public disclosure."[68] Thus, when Cullen submitted his bill of complaint to the Court of Chancery the following year, he sought to broaden the meaning of public disclosure to include the content he had delivered publicly to students in his university lectures. In his bill of complaint, he argued that, like the printer who had used the unauthorized transcript of Macklin's play, Lowndes had obtained the lecture transcripts of his *materia medica* lectures through "fraudulent and surreptitious means."[69]

Cullen's bill also argued that, again like the publication of a pirated play, the unauthorized publication of his lectures prevented him from benefiting financially from a form of property that he had created.[70] He established his claim to intellectual labor by emphasizing that he had created the content of the lectures "with considerable trouble and pains."[71] As we learned in previous chapters, such pains included the time and effort he, like so many other professors, spent creating a system that doubled as the syllabus of his course. It also included the efforts that he made to ensure that he offered an original evaluation of up-to-date facts and observations that came from his own research or were taken from the latest scientific literature. Had he been simply

---

67. Gilbert (1758). The court was not without its critics. See Parkes (1828).
68. Masiyakurima (2020, 34–35).
69. Masiyakurima (2020, 36).
70. A copy of Cullen's bill of complaint, dated 12 December 1771, is housed in National Archives of Scotland, NAS, C 12/1033/2.
71. Aplinet al. (2012, chap. 2, sec. 2.45).

reading a fixed set of lectures from a textbook or had the lectures been based predominantly on extracted information, it would have been much harder for Cullen to argue for the originality or uniqueness of the lectures. Since the lecture transcripts of the students varied in terms of their length or specificity, it was the systematic aspects of the syllabus, the organization of the headings that he used to present the information, that implicitly offered continuity within any lecture transcript that the judge may wish to consult.

The bill was received by the Lord Chancellor, Lord Apsley, who agreed to hear the case. As the proceedings could take some time, Apsley granted an injunction until he reached a decision. The fact that Apsley had agreed to both hear the case and grant the temporary injunction suggested to Lowndes that the court could potentially rule against him and that Cullen probably would seek damages, an act that would generate negative publicity at a time when Lowndes was attempting to expand his business. In the end, Lowndes withdrew his claim to publish the book. He and Cullen then negotiated a contract, one from which Cullen benefited financially and acknowledged Cullen's rights as the author of the lectures. Though the court did not rule on the case, the circumstances surrounding it strongly suggested that judges would most likely treat lectures as a form of public disclosure, thereby elevating the risk associated with publishing books based on lecture transcripts. This risk helps to explain why most publishers sought to print authorized lectures from that point forward.[72]

In addition to offering insight into how the intellectual uniqueness of lecture transcripts was defined within a court of law, the episode provides a way to compare and contrast the different reactions exhibited by students and professors within the equally important court of public opinion. For example, though Cullen clearly was annoyed about the publication of the *materia medica* lecture notes, other members of the Scottish medical community and its diaspora throughout the British Empire were happy to see them in print. This sentiment was agreeably communicated in a letter written to Cullen from Dr. Jasper Tough in 1776 from Kilmarnock, a town just over twenty miles southwest of Glasgow. Tough was writing to update Cullen on a patient they were both treating. After outlining the patient's symptoms, he wrote: "I hope soon to hear of the republication of Your lectures on the Materia Medica, for however incorrectly Your Students had Done it, Yet it was the Most useful thing of

---

72. Here it is worth noting that in 1774 the House of Lords decided to limit the period in which an author could claim exclusive rights to a publication. Historians of the book, readership, and authorship are divided as to the impact of the decision. Richard Sher holds that, when one considers the many different markets in which books were sold, the 1774 decision was less impactful. Sher (2008, 27–29).

the kind for Young Men." Here we can see that, notwithstanding his protestations over the pirated edition, Cullen's correspondents and former students were still keen to get a copy.[73]

The popularity of the pirated edition motivated Cullen to find out which students had given their lecture transcripts to Lowndes. As he explained in a 1775 dinner conversation with the ever-present and sociable diarist Silas Neville, the identity of the original notekeeper was revealed without so much as Cullen lifting a finger. Soon after the pirated edition appeared, Cullen received a letter from Dr. Alexander Monro Drummond, Edinburgh University's professor-elect of medical institutes. This post had been vacated recently by Cullen.[74] Upon seeing the pirated version, Drummond realized that it was based upon the notes he had taken in the first year of his studies and, to clear his name and conscience, he proceeded to track down the two people who had copied his transcripts. The result is recounted in Neville's diary: "One was Dr Falconer of Bath, the other a Dr Blair of Virginia, now dead. The former wrote a letter to the Dr [Drummond] acknowledging his having published it."[75]

Drummond's investigation further highlights the central role that students played in disseminating lecture transcripts within the British Isles and across the colonies. It reveals that the provenance of a lecture transcript mattered to many students. The fact that Drummond was able to realize that the pirated edition was based on his notes and the fact that he knew exactly who had copied them shows the intimate familiarity that some students had with their notebooks and with their subsequent lineage of transcription. Cullen was one of the most famous professors in Scotland during the middle part of the century, which explains why some of his students were especially motivated to remember the lineage of the notebooks that they kept in his courses.[76]

The scribal lineage of Cullen's printed *materia medica* lectures adds further detail to the ways in which professors and students operated in a mutually beneficial knowledge economy that was underpinned by the intertwined worlds of script and print. It was a fertile environment in which knowledge was written, erased, rewritten, printed, and reprinted. As illustrated in figure 10.4, the published edition of Cullen's *materia medica* lectures was based on different kinds of script and print that were created and curated by a host of

---

73. Dr. Jasper Tough to Dr. William Cullen, 15 July 1776, RCPE, CUL/1/2/412.

74. Alexander Monro Drummond was offered the Edinburgh chair of medical institutes in 1773. He was living in Naples at the time of the appointment and he eventually declined the offer in 1776. Dalzel (1862, 443–448).

75. Neville (1767–88/1950, 336).

76. For Cullen's popularity with Glasgow and Edinburgh medical students and the conceptual importance of the notes kept in his lectures, see Mulcahy (2019).

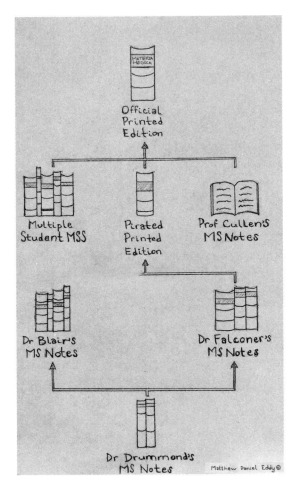

FIGURE 10.4. The lineage of the lecture transcripts used by William Cullen and the printer Thomas Lowndes to create the authorized edition of *Cullen's Lectures on the Materia Medica* (London: Lowndes, 1772). © Matthew Daniel Eddy.

notekeepers. The fact that Cullen's students kept track of who had used and copied their notes was by no means a unique occurrence. There were influential professors in all of Scotland's universities and the flyleaves of notebooks kept in their courses bear witness to the fact that provenance was important for notekeepers.

The first flyleaf of the lecture notebook kept most likely by Patrick or Alexander Walker of Bowland in William Leechman's 1780s Glasgow University course, for instance, recorded the provenance of the notes with the following summary: "This Copy of Dr Leechman's Lecture on Composition Mr. James

Grant Episcopal Minister in Edinburgh wrote off Mr. Walker's Copy and then gave Mr. Walker his in place of his own."[77] Leechman was a respected, but sometimes controversial, minister who had served as moderator of the Church of Scotland in the 1750s. As we learned earlier in the chapter, he was also the professor of divinity and principal at Glasgow University. The note was most likely made when the bound set of the lectures was catalogued in Alexander Walker's expansive personal library during the early nineteenth century.[78]

The line between piracy and plagiarism, however, presented a stronger challenge to the interactive scribal relationship between professors and students. There is evidence that some students published extracts of their lecture notes in magazines, presumably without permission of their professors. Such was the case for J. Rennie, a student who attended the pneumatology lectures given by James Beattie, chair of moral philosophy at Marischal College, Aberdeen. Rennie graduated in 1767, and in 1771 he published eight articles under his name or the pseudonym "Philomathes" in the *London Magazine, or, Gentleman's Monthly Intelligencer*. Based on his notes preserved in the special collections of Glasgow University, it is clear that seven of the extracts were copied directly, virtually verbatim, from the notes that he took in Beattie's lectures.[79]

Rennie told the editors that four of the articles were essays. When the "Essay on Dreams" appeared under Rennie's name in March 1771, for example, it would have looked like he was the author. The same would have been the case for the "Essay on Laughter," "An Essay on the Education of Children," and "An Essay on the Chief Good." Rennie's three other articles, though untitled, would have appeared to be essays to readers as well. It is unclear whether Beattie attempted to stop the publication of the essays. By the time they had appeared, his status as a poet and professor was known to literati throughout Britain. It is unlikely, therefore, that he remained unaware of Rennie's plagiarism. But as we learned above, the copyright laws of the time would have made it difficult for him to take Rennie to court. The fact that the essays were not collected together and published as a pirated edition suggests that Beattie used personal means to dissuade any further publication of his lecture material.

Whereas Cullen and Drummond had to trace the circulation of notebooks to stamp out piracy, Alexander Monro Secundus had to find lecture transcripts to prove that the content of his course had been plagiarized. A particularly

---

77. Leechman Bound MS (c. 1780, front flyleaf). See also *Notebook containing Notes from Lectures on Rhetoric and Logic* Bound MS (n.d.), which the National Library of Scotland attributes to Patrick Walker.
78. The bookplate on Leechman Bound MS (c. 1780) reads "Alexander Walker of Bowland, Esqr."
79. Rennie's extracts from the *London Magazine* are listed and compared to the notes he took in Beattie's lectures, Beattie Bound MS (1767), in Fabian and Kloth (1969) and Tave (1952). See also Mossner (1948a, esp. 112–113).

acrimonious example of this scribal sleuthing can be seen in a priority dispute between him and Gilbert Blane that occurred during the 1790s. Blane had taken Monro's course in the 1769–70 session and had witnessed how his teacher had used geometric diagrams in his lectures. Blane moved to London after his studies and pursued a successful career as an anatomist and expert in military medicine. In 1788 he gave the prestigious Croonian Lecture to the Royal Society of London. He spoke on muscle motion, a topic upon which, as we learned in chapter 9, Monro Secundus was an expert. In the lecture Blane employed one of Monro's geometric diagrams to illustrate muscle movement without mentioning its origin. He then published his paper as a monograph.[80]

The lecture and the monograph enraged Monro Secundus. His reaction stemmed from the fact that Edinburgh's professors often treated the syllabus, ideas, and visual material of their lectures as intellectual property. They let others cite their ideas, but appropriate credit had to be given. It was common, for example, for professors to cite the lectures of their colleagues, a gesture that both showed respect and acted as a form of advertising. When one professor forgot to give appropriate credit, the other professor went out of his way to correct the situation.

To prove his claims against Blane, Monro tracked down the lecture transcripts of former students such as John Haygarth, Benjamin Bell, and James Russell with a view to using their content to expose Blane's alleged plagiarism. All of them had taken notes during the 1760s in Monro's lectures. Additionally, though Monro owned a transcription of John Thorburn's anatomy notes from the 1769–1770 season, he even went so far as to consult the original rough notes held by Thorburn's son. After all these notes were consulted, Monro took his case to the court of public opinion and published a pamphlet which quoted extensively from the Thorburn, Bell, and Russell notebooks to show that Blane's geometric approach to muscle movement was indeed based upon lectures given two decades earlier by his teacher in Edinburgh.[81]

To bolster his case, Monro asked an engraver to reproduce part of the page from his own 1759–1760 notebook, in which he had originally drawn the diagram in the margin (figure 10.5).[82] To underscore the validity of the image presented on the plate, Monro's pamphlet stated that the engraved version was a "literal copy." This claim was underscored in the text next to the printed

---

80. Blane (1790).
81. Excerpts from student lecture notes, as well as from Monro Secundus's own lecture notes, are quoted in the last half of Monro Secundus (1794). Monro included testimonials from former students as well, a tactic that was used by other anatomists in priority disputes at the time. Chaplin (2009, 113–116).
82. Monro Secundus (1794, 31).

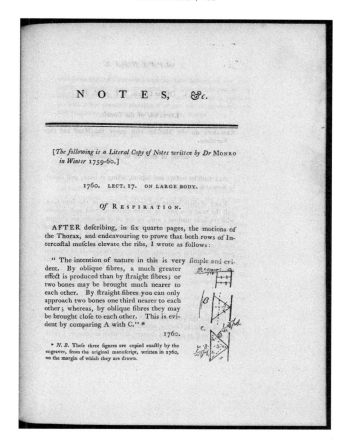

FIGURE 10.5. A "Literal Copy" of Alexander Monro Primus's teaching diagram. Alexander Monro Secundus, *Observations on the Muscles, and Particularly on the Effects of Their Oblique Fibres* (Edinburgh: Dickson and Balfour, 1794), 31. Reproduced with permission of the Wellcome Library, London.

version of the diagram: "*N.B.* These three figures are copied exactly by the engraver, from the original manuscript, written in 1760, on the margin of which they are drawn." In making this move, he transformed a mnemonic device, his notebook page, into a memetic artifact, the printed plate, that functioned as a piece of evidence in what was essentially an early example of a modern copyright dispute.

It is worth emphasizing, as was briefly noted above, that there was scope within the gentlemanly world of eighteenth-century physicians that would have allowed Blane to use the diagram. If appropriate credit was given, then it was possible to use the ideas or images that professors presented to students in lectures. There was even scope for forgiveness if someone unwittingly forgot to credit a professor. Such was the case for Edinburgh's Professor Alexander Hamilton when he wrote up his lectures into a book on midwifery. In the first edition of the book, he described a surgical operation developed by Monro Secundus for obstetrics. Monro had discussed the operation in his lectures, which is where Hamilton had heard about it. It is not clear when Hamilton realized that he had forgotten to credit his colleague, but his subsequent reaction reveals that he was very keen to correct the situation.

Near the end of the revised 1784 edition of his midwifery book, which was based on his lectures, Hamilton stated that "in the former imperfect edition of this work, from the inaccuracy of the language this opinion appears to have been given as my own. I readily make this acknowledgment of Dr MONRO's claim, as I should otherwise detract from his deserved praises." Importantly, Hamilton used a set of student lecture notes, most likely his own, to back up his apology: "The particular method of performing the operation is described so satisfactorily by Dr MONRO, our learned and accurate Professor, in his Lecture, that we shall take the liberty to insert his own words."[83] Hamilton then quotes at length from the notes, revealing the important role they played in the court of public opinion and further illustrating the overarching importance of student notebooks in the knowledge economy of Scotland's universities. Hamilton's response also makes it easier to see why Monro Secundus was so upset with Gilbert Blane. Though Monro was annoyed that his former student had plagiarized the diagram, the true grievance was the fact that Blane continued to claim he had invented the diagram even after Monro asked him to properly acknowledge its origin.

The dispute between Blane and Monro Secundus highlights this chapter's larger point about the legal and epistemological priority enjoyed by student notebooks as media technologies within the knowledge economies of Europe, the Atlantic world, and Asia. The notebooks were artifacts valued by several communities, not least by the students who spent time and effort creating them, but also by the professors who spent a significant amount of labor crafting the courses upon which they were based. In short, the notebooks were a form of economic and cultural capital, the circulation of which reveals the importance of recognizing the flexible boundary that existed between script and print within the scholarly, domestic, commercial, and military information networks of Britain and beyond.

---

83. Both quotations in this paragraph were taken from Hamilton (1784, 347).

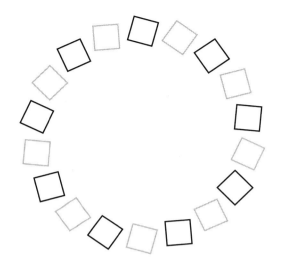

Conclusion

CHAPTER 11

RETHINKING MANUSCRIPTS

The *Tabula Rasa* and Manuscripts

John Locke's *tabula rasa* metaphor, as we have learned, possessed dimensionality. Within homes, schools, academies, and universities, the material components of notebooks made it easier for students to understand the mind as a kinesthetically constructed entity and reason as a skill that was learned over time. Throughout the book I have considered the metaphor in reference to the considerable efforts that students made to transform blank sheets of paper into a media technology that was a purposeful and user-friendly artifact. These efforts, I suggested, shed considerable light upon the sociomaterial preconditions that enabled them to conceive and create a notebook page on which they crafted script as a *tabula verba* and an image as a *tabula figura*. In many cases, the form and function of the page operated as a *tabula memoria*, a memory theater of both facts and skills. The affordances inherent in these forms revealed a developmental richness of the *tabula rasa* metaphor that far exceeded the notion of blankness that scholars have frequently attributed to it.

The metaphor instead reflected a much wider meaning, one that drew from the considerable amount of time, effort, and materials inherent in the skills required to make a notebook's different components. Within this richer meaning, the *tabula rasa* became a longitudinal entity, something constructed in stages over time and dependent upon the resources presented by the educational context of a given student notekeeper. Through expanding the *tabula rasa* metaphor, I have shown that notekeeping was a significant mode of social reproduction that went far beyond the rote transfer of words or figures because it required notekeepers to engage with the notebook as a knowledge-creating material artifact. Notekeepers constantly worked with the form and meaning of their notebook components with a view to creating a powerful media technology. In following this path, they built on skills that they had learned in the past only to learn more skills as they continued to craft their notebooks as artifacts that would be accessible to them in the future. This created a context in which every notebook was a technology that influenced its keeper's perception of the past, present, and future.

Throughout the book I developed a triangular approach that treated notebooks as artifacts, notekeepers as artificers, and notekeeping as artificing. Using evidence created by the students themselves, this book was an exercise in applying this tripartite form of historical analysis in a way that afforded insight into how learners from diverse backgrounds living in Scotland interfaced with everyday objects of learning. I showed that, when viewed as a form of material culture, a notebook can be used to ask questions about the notekeeper as an artificer and about the skills of artifice evinced in the components. In what follows, I would like to offer some concluding observations on how my tripartite approach is relevant to the realtime modes of interface presented by the manuscript culture of predigital knowledge economies.

Manuscripts as Dynamic Artifacts

Though we might be tempted to see notebooks as monolithic, or perhaps fixed, objects today, we learned that they were in fact diverse forms of material culture comprising adaptable components such as bifolia, ink, and bindings that were made and used over time and space. There are of course many kinds of notebooks that we could have examined in greater detail. During the long eighteenth century, humanists created commonplace books and merchants ranged ciphers in ledgers. These and other kinds of notebooks call out for more attention. In many respects, my focus on student notebooks is merely a taster, a sample of the richly diverse world of notebook-based knowledge that was navigated daily by literate society.

Notebook components made through inscription came in many varieties. Simple scribal components occurred as words, numbers, and figures. These served as the foundation for more complex components such as annotations, indices, formulae, diagrams, and tableaux. Likewise, components made through codection occurred in simple forms such as bifolia, modules, matrices, folds, and stitches. These were combined into more complex components such as quires, layouts, and spines or other forms of packaging related to collocation. The larger picture of notebooks as multifaceted forms of communication media was one of richly complex artifacts in motion in which there was fluidity between different kinds of inscription, foliation, and codection. At what point, for instance, did the fold of a bifolium cease to be the edge of a page and become part of a codex's bindings? The inspection of notebook spines revealed that answering this kind of question required a longitudinal account of notebooks as adaptable manuscript technologies kept by specific kinds of agents who operated within the context of a given knowledge community, which, in the case of student notekeepers living in eighteenth-century Scotland, existed in and around a range of related institutions such as homes, schools, academies, and universities.

Each component of a student's notebook as a manuscript provided an entry point into the fascinating variety of preliminary questions concerning the decisions that were made while it was being created. Why was one component combined with the other? Why wasn't a certain kind of component used? Why were the components assembled in a certain order? Why were some materials used, or not used, to make the components?

Answering these questions afforded great insight into how the components of a notebook provide a rich collection of fresh evidence for researchers seeking to understand manuscripts made by actors whose identities are unknown or whose lives have remained undervalued and, hence, underresearched. When approached as evidence of different kinds of material engagement, familiar humanist and commercial notekeeping components such as the annotation, the table, and the diagram become inextricably tied to how notekeepers as inscribers and codexers learned to manipulate knowledge on, around, and through paper.

Other kinds of components blurred the traditional categories used to discuss manuscript culture. Take for instance the multivalent usage of an *observation* or *definition*. Both terms were used in notebooks to denote a way of collecting and representing information about a given topic. But they were also functional labels for a series of pages that formed a section of a notebook in its capacity as a material artifact. These pages were material objects in that they were discrete components of a codex that operated as an informatic machine. Their material attributes affected a notekeeper through the acts of writing, ranging, and quiring that were used while they were being crafted. Consequently, the material activities surrounding the creation of a notebook entry that served as an observation or definition contributed to the skills that underpinned the processes of observing or defining that were experienced by notekeepers.

Teachers and family members asked students to observe and define a great many things when they kept notes. When viewed as a component of a notebook, or perhaps even as a series of pages within a notebook, the observations and definitions themselves become materially mediated forms of representation, the design and preservation of which were inextricably linked to their existence within a communications artifact that comprised an accessible layout, ordered within a codex as a quire.

The material activity of manipulating knowledge in a paper-based artifact that operated as a hand-held machine provided students, householders, and educators with enactive cognitive metaphors that helped them conceive a coherent model of the learning mind. The *tabula rasa*, for instance, supplied a way to use writing and codexing to understand and model the ideas and operations of the processes of thinking. This created a situation in which the materials and forms of a notebook as a manuscript artifact facilitated a helpful metacognitive model that provided a way to think about the kinds of realtime media activities that strengthened mental operations such as perception, recollection, and judgment. Thus, approaching a notebook as a paper machine that was a collection of interchangeable components presents a way to examine the material foundations of key processes that were valued for the role that they could play in strengthening human capabilities.

Manuscript Skills as Artifice

Since the components of notebooks had to be made and combined so that they could function as useful learning technologies, I repeatedly drew attention to the fact that notekeeping was a kind of artificing, an art that consisted of a vast repertoire of materially and kinesthetically grounded skills that underpinned the exercise of reason as it was being learned by students both on their own and in the social or domestic contexts surrounding educational institutions. Transforming graphite, ink, paper, and string into a paper machine, therefore, was hardly an insignificant activity for the many learners and educators who were influenced by the strong emphasis that Lockean educational psychology placed on the cognitive power of literacy and numeracy.

The manuscript skills evinced in the pages of Scottish student notebooks came in many varieties. Creating a figure through the process of inscription involved the skills of impressing, tracing, sketching, circling, dotting, shading, painting, scratching, and powdering—to name but a few. Such skills required different kinds of instruments such as pencils, pens, quills, brushes, nibs, compasses, rulers, penknives, and sandrick powder. When viewed collectively, something as seemingly simple as a figure of a three-dimensional polygon turned out to be not so simple. Setting aside the skills required to render the page on which it was made, crafting such a relatively simple figure involved aligning, sketching, tracing, labeling, shading, and in some cases watercoloring. All these skills collectively amounted to the superskill of drawing. Likewise, more specific diagramming skills such as pictogramming required students to make a variety of realtime decisions regarding the inclusion or omission of visual information.

A similar point could be made about the codection skills of quiring, compositing, and binding that notekeepers used to create the pages on which they rendered script. Each of these skills necessitated the knowledge of different, further specialized skills that had to be learned over months and years and then employed to make small and large manuscript components. Designing a quire, for instance, involved bending, folding (bifoliating), cutting, stacking, interleaving, and stitching. Thus, like the superskills of inscription, there were superskills of foliation and codection.

The larger point to note is that every component of a notebook as a manuscript artifact provides a material focal point that can be used to investigate the kinds of skills required to make it a functioning paper machine. Identifying such manuscript skills reveals they collectively added up to a kind of artifice that was acquired over time and honed through much labor. When seen from this perspective, even something as seemingly straightforward as a crafting a dotted graphite line into a circle turns out to be a skill-based process that could not have been executed without prior practice and familiarity with the thickness of paper, the shape of a nib, the form of a compass (including its interchangeable points), and the movement of the hand. All of these skills were being mastered while students endeavored to learn the subject-specific knowledge of the curriculum.

Put another way, notekeeping skills are themselves important historical objects of enquiry that played a significant role in shaping how students learned to design paper machines that operated as predigital knowledge systems. They lead us to consider the limitations or advantages of foregrounding the kinesthetic and material factors that shaped the emergence of information economies that were dependent upon manuscript forms of representation.

Exploring the preconditions of student notekeeping skills made it possible for me to highlight various modes of artifice that might otherwise seem irrelevant or mundane to us today on account of the hyper-digitized forms of representation that are now so common. Recall, for instance, the skill of ranging, which, allowed students to graphically lay out the page in a rectilinear format that was visually accessible. Like other notekeeping skills, it was acquired through hands-on instruction that worked in conversation with a student's other abilities. It depended upon the physical presence of a tutor who could interface with the student's efforts in realtime, or in some cases through diligent correspondence. Knowing how to range a page was essential for anyone who wanted to keep accounts or who wanted to visually and conceptually order a scientific knowledge system.

The notekeeping skills that facilitated inscription, foliation, and codection required students to materially engage with manuscript tools and supplies in a manner that offered multiple opportunities for them to remember the ideas, facts, rules, and equations that they were replicating and in some cases modifying. Put more clearly, notekeeping, or even writing more generally, even when it involved copying, was a materially grounded way of thinking. It provided students with artificing skills that enabled them to structure knowledge in the mind and on the page. As such an endeavor, it was a foundational kinesthetic activity that influenced how students working with predigital data learned to recognize, create, and evaluate knowledge systems. When notekeeping skills are viewed through this developmental lens, it becomes easier to see that the *tabula rasa* metaphor made sense to students and their teachers because it was based upon an active engagement with the forms of communications media that were used on a daily basis. Such an engagement suggests that manuscripting skills provide a fertile starting point for examining the underlying decision-making capabilities presented in student notebooks as artifacts that were both personal and social technologies of learning.

Manuscript Keepers as Artificers

Throughout the book I used notebooks in their capacity as paper machines to identify sets of skills that spoke to how notekeepers functioned as artificers, as agents who knew how to craft specific kinds of things that were useful. Such an approach allowed me to identify and then explore what the components of notebooks as artifacts had to say about their artificers. The codices that girls designed and kept at home to organize various subjects, for example, were in fact knowledge systems that flowed over hundreds of pages. Based on Lockean developmental psychology, such activities had the capacity to influence how both girls and boys understood the positive impact of their own acts of writing and thinking. Notably, I showed that such activities were pursued by adolescent girls who were often around the same age as male students who attended academies and universities, thereby suggesting a fascinating relationship between gender and the material culture of the mind as it was understood in domestic and educational settings during the long eighteenth century in Britain. Though more work remains to be done on the ways in which female students learned to craft systems on paper, the examples of Margaret Monro, Amelie Keir, and Williamina Belsches reveal that it was possible for them to learn how to use the writing, ranging, and codexing skills of notekeeping to create and manipulate sophisticated knowledge systems on paper.

For scholars seeking to understand the lives of hitherto underresearched historical actors, notebooks are invaluable artifacts because they allow us to build a profile of neglected agents who were diligently learning skills that the modern period associated with reason. Take for instance the "experience" subheading written by the young Margaret Monro that we encountered in our discussion of the annotations made by young writers. It presented us with the opportunity to use her annotating skills to examine her status as an annotator. More specifically, her annotation revealed that she was more than familiar with the sophisticated classification system that her father had proposed for the manuscript. Indeed, the annotation reveals that she was an engaged artificer who fully understood the design principles that underpinned it.

Monro's activities as an annotator provided an attribute, a cognitive clue, that sheds light on her larger abilities as a writer who was capable of coherently ordering a sizeable system of knowledge. My path of investigation started with conceptualizing her annotation as a notebook component, as a kind of a smallish artifact. I moved on to characterize her annotating as a kind of artificing, and then proceeded to explore how Monro, as an annotator, operated as a kind of artificer. When applied to the other annotations in her manuscript, the overarching result of the process was a better picture of Monro, who, as evinced so clearly in her magnificent codex, is a noteworthy personality who has received hardly any attention by historians.

Of course, we could be a bit more creative in selecting the kind of artificers that we would like to investigate. An annotator, for example, is a relatively familiar kind of writer for those working on the history of literacy, literature, or scholarship. But what if we decided to think a bit more outside of the box? What if we chose to look at something like a sightline, that is, something that was certainly there, but that scholars traditionally have not seen as an important historical object of enquiry? What kind of manuscript artificer would this kind of component bring to our attention?

We will recall that a sightline is a path of information that runs across a page or a field of vision more generally. In the case of student notebooks, they often ran between headings and keywords. During the eighteenth century they were sometimes called technical lines. In notebooks, they enabled their keepers and users to more easily find information. The presence of sightlines in notebooks indicates that many students possessed the skill of "sightlining," that is, the ability to position words on the page in a way that created a useful mnemonic pattern. Like a notekeeper who became a sketcher through sketching a figure on a page, students became "sightliners" through sightlining the space of the page.

At first glance the word *sightliner* might sound a bit awkward. But thinking of a notekeeper in this way makes it easier to see students as artificers who used their skills to make decisions about the kind of media technologies that they wanted to design. Put more clearly, if we want to see learning as an interactive and historically traceable process, then treating notekeepers as sightliners offers an example of how we can create new ways to frame manuscripts as artifacts that offer deep insight into how other kinds of manuscript keepers operated as realtime designers and users of personalized technologies that existed within sophisticated knowledge economies.

Time and again we discovered that for every kind of skill there was a specific kind of notekeeper. Tablers made tables. Indexers made indices. Diagrammers made diagrams. Here the processes of artificing signal an important advantage that manuscripts have as forms of historical evidence. Since their users often made them, they were self-made technologies that can be used to better understand how material objects enabled artificers who were creating or manipulating knowledge in realtime. Take, for instance, the notebooks of the medical student Thomas Cochrane, whom we encountered in the third part of this book. In addition to transforming his rough notes into a fuller lecture transcript, he diagrammed equations, drew pictograms, interleaved sheets, and inserted annotations. His many acts revealed him to be a diagrammer, a transcriber, an interleaver, a pictogrammer, and an annotator. These were identities of sorts, which presented various attributes that spoke to his prior education and ambitions in a way that made it easier for me to understand him as a notekeeper and as a medical student.

The status of a notekeeper as an artificer capable of fashioning a self-made technology also helps us identify the limitations and difficulties that students faced. The schoolboy James Dunbar, whom we encountered several times in this book, was a tabler who genuinely struggled to draw the kind of table that he wanted to make. He was a tabler whose difficulties led him to see himself as a mere scribbler. But being a scribbler was an important part of becoming a proficient notekeeper, that is to say, it was essential to becoming an inscriber and a codexer capable of observing and representing knowledge by hand on a page. The example of Dunbar and so many other students reveals that becoming a skilled notekeeper and hence a capable reasoner was often dependent upon trials and errors that occurred on paper every day along the way.

Paying attention to the notekeeper as an artificer takes us beyond simply investigating the designers of manuscripts as agents who dealt solely with the preservation or transmission of ideas. It presents another perspective that allows us to see them as learners or creators who exercised skills that were gained and honed through prior experience that took place within a knowledge community. In sum, it helps us get a better idea of the skills that made it possible to conceive, craft, and use manuscripts as powerful media technologies.

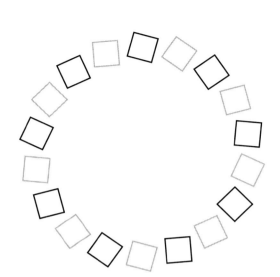

Acknowledgments

There are many people who joined me on the journey that led to this book. Perhaps the most important were those who encouraged me at crucial stages. First on the list is Karen Darling at the University of Chicago Press, who believed in this book through all three of its iterations. No matter what I seemed to research, David M. Knight, Tony Grafton, Seymour Mauskopf, Robert Fox, Ursula Klein, Jed Buchwald, and Moti Feingold found intrinsic value in what I was doing. My ideas over the years have also been challenged and enhanced by the scholars associated at various points in their careers with the Max Planck Institute for the History of Science in Berlin. I offer special thanks to Wolfgang Lefèvre, Staffan Müller-Wille, José Ramón Bertomeu-Sánchez, Daniela Bleichmar, Omar W. Nasim, Christoph Hoffmann, Barbara Wittmann, Elizabeth Yale, Elena Serrano, Christine von Oertzen, Elaine Leong, and Carla Bittel.

Over the course of writing the manuscript I had stimulating conversations or correspondence with Ann Blair, Bill Sherman, Brian Ogilvie, Isabelle Charmantier, Georgette Taylor, Robert R. G. Anderson, Mark Towsey, Charlie Withers, Jim Secord, Anne Secord, Carin Berkowitz, Jacob Soll, Ron Brashear, Richard Yeo, Ludmilla Jordanova, Margaret Schotte, Sachiko Kusukawa, Fredrik Albritton-Jonsson, Anja-Silvia Göing, John Brewer, and Jonathan Rose. Notably, Ann, Ludmilla, Fredrik, and Charlie offered substantial comments on several parts of the manuscript.

My understanding of the history of early childhood education was significantly aided by my wife, Athanasia Chatzifotiou. Sometimes the manuscript versions of this book were curious guests who continued to make cameos at the dinner table or during homework discussions that I had with my two children. I am immensely grateful to my family for suffering through these intrusions and for their unwavering support.

My knowledge of all things Scottish was substantially aided by the expertise of Richard Sher, John Christie, Warren McDougal, Paul Wood, James Harris, Roger Emerson, Donald Kerr, Mark Towsey, Ralph Mclean, Olive Geddes, Thomas Ahnert, David Allan, Henry Noltie, and Jane Corrie. Jane was a gracious hostess, allowing me to stay with her family in the erstwhile house of Robert Chambers, the nineteenth-century publisher and evolutionary theorist. Hopefully some of his erudition rubbed off on me and made it into the pages of this book.

In Durham I benefited from the support and feedback of many colleagues. Robin Hendry, Nancy Cartwright, Andreas Holger Maehle, Emily Thomas, Jeremy Dunham, and Peter Vickers helped me puzzle through the historical relationship between science and philosophy. The history of science, medicine, and technology community provided a supportive environment, particularly the seminars organized by Joe Martin and the conversations I had with Sare Aricanli and Dario Tesscini. Benedict Smith, Simon Paul James, Andreas Pantazatos, Matthew Ratcliffe, and the members of the department of philosophy enthusiastically joined me in exploring various historical topics as well. Parts of this book were presented at workshops hosted by Durham's Centre for Visual Arts and Culture, the Institute for Medieval and Early Modern Studies, the Centre for Nineteenth Century Studies, the Centre for the History of Medicine and Disease, the Centre for Humanities Engaging Science and Society, and the Institute of Advanced Study. My doctoral students Daniel Becker, Jay Bosanquet, Rachel Dunn, Meng Li, Ursula Mulcahy, Tom Rossetter, Lenka Schmalisch, and John Shepherd provided stimulating conversations on various aspects of print and manuscript culture. The sterling research assistance provided by Ian James Kidd, Guy Bennett-Hunter, Naomi Carle, and Laura Dearlove was much appreciated.

Much of this book is based on ephemera housed in uncatalogued boxes or files. I am indebted to the many librarians who helped me find exciting sources hitherto unseen by a modern historian. I am particularly indebted to the staff working in the special collections of the Wellcome Trust Library in London and the universities of Edinburgh, Glasgow, Aberdeen, and St. Andrews. I also wish to thank the staff of Dunn House in Banff, Scotland, the Manuscript Reading Room of the National Library of Scotland, and those working in the libraries of the Royal Society of London, the Royal Botanic Garden Edinburgh, the Linnean Society of London, the Science History Institute, Yale University, and the University of Otago, New Zealand. Additionally, Norman H. Reid generously provided me with further information on the student lecture notes housed by St. Andrews University's Department of Special Collections. Karla Fackler Grafton, archivist and rare books librarian of Westminster Theological Seminary, and Edward Copenhagen, librarian of Harvard University's Gutman Library, graciously provided documents from their respective collections.

Various parts of this book were presented as papers given at events hosted by the Max Planck Institute for Human Development in Berlin, Cambridge University (CRASSH), Uppsala University Department for the History of Science and Ideas, the British Society for the History of Science, the History of Science Society, the Linnean Society of London, the Wellcome Trust Centre for the History of Medicine at University College London, the Royal Society of London, the Royal Society of Edinburgh, the Centre for the History and Philosophy of Science at the University of Leeds, the Max Planck Institute for the History of Science in Berlin, the Science History Institute in Philadelphia, the Huntington Library, the UCLA History Department, Caltech's Humanities and Social Sciences Division, the Centre for the Study of Scottish Philosophy in Princeton, and the Department of the History and Sociology of Science at the University of Pennsylvania.

Early funding for this book was provided by Durham University, an Early Career Research fellowship awarded by the British Arts and Humanities Research Council (AH/H039589/1), and a visiting professorship awarded by Caltech and the Mellon Foundation. I offer heartfelt thanks to Matthew Ratcliffe for reading draft after draft of my AHRC grant application, and to Jed Buchwald and Moti Feingold for offering me the position at Caltech. Further research for this book was made possible by residential fellowships awarded by the Max Planck Institute for the History of Science in Berlin, Durham University's Institute for Advanced Study, and the Huntington Library, San Marino, California.

Bibliography

Abbreviations

AUL	Aberdeen University Library
BAO	Birmingham Assay Office
CL	Clark Library, UCLA
DLDH	Dunimarle Library, Duff House
ODNB	Oxford Dictionary of National Biography
ECA	Edinburgh City Archive
EUL	Edinburgh University Library
GUL	Glasgow University Library
HLSM	Huntington Library, San Marino, CA
NLS	National Library of Scotland
RBGE	Royal Botanic Garden Edinburgh
RCPE	Royal College of Physicians of Edinburgh
SUL	St. Andrews University Library
UOSC	University of Otago Special Collections
USSC	University of Strathclyde Special Collections
Wellcome	Wellcome Trust Library, London
YCBA	Yale Center for British Art, Paul Mellon Collection

430 | BIBLIOGRAPHY

Primary Sources

MANUSCRIPTS AND EPHEMERA[1]

Allan, David. c. 1753. *Exhibition of the Foulis Academy's Paintings in the Inner Court of the University of Glasgow*, engraving on paper, National Galleries Scotland, accession no. P 3122 A.

Allan, David. c. 1788. *Sir William Erskine of Torrie and his family*, National Galleries of Scotland, accession no. PGL 333.

Anderson, John. 1750s. *Common Place Books*, vol. 1, Bound MS, USSC, GB 249 OA/7/1, f. 316.

Anonymous. 1780s–1790s. *Perth Academy Notebook*, Bound MS, NLS MS 14291.

Anonymous. 1742–1746. *Notebook of Juvenile Latin Verses and Exercises*, Bound MS, NLS Newhailes, MS. 25413.

Anonymous. 1744–1787. *Notes Mainly on Classical Authors, Also Latin and Greek Verse and Prose Composition, Including Some Juvenile Exercises*, NLS Newhailes, MS. 24516.

Anonymous. 1787. *Perth Academy Notebooks*, 3 vols., I.M. (notekeeper), Bound MSS, NLS, MS 14294-6.

Anonymous. 1790. *Perth Academy Notebook*, Bound MS, NLS, MS 14291.

Anonymous. 17—. *The Principles of Latin and English Grammar*, DLDH no. 247.

Anonymous. 1804. *Practical Mathematics*, Bound MS, NLS 14285.

Anonymous. 1809–1812. *Surveying Journal and Accounting Ledger*, Bound MS, NLS MS 14283.

Anonymous. 1816. *Mathematics Notebook*, Bound MS, NLS, MS 14287.

Anonymous. n.d. *Napier's Box of Tables*, National Museum of Scotland, H.NL.68.

Anonymous. n.d. *Pencil Sketches of Professor Patrick Copland's Marischal College Classroom*, AUL, MS M363.

Arrow, Jemima. 1815. *Maps*, Bound MS, NLS MS14100.

Beattie, James. 1767. *A Compendious System of Pneumatology, Comprehending, Psychology, Moral Philosophy, and Logic*, J. Rennie (notekeeper), Bound MS, GUL, MS Hamilton 55.

Belsches, Williamina. 1795. *Exercise Book*, Bound MS, SUL, Msdep7, Personal papers, Box 22, no. VII/9.

Black, John. 1797–1798. *Juvenile Poetic works of John Black*, vol. 2, Bound MS, NLS, MS 14233.

---

1. The printed books in this section contain marginalia written by children.

BIBLIOGRAPHY | 431

Black, Joseph. 1766–1767. *Notes of Dr Black's Lectures*, Charles Blagden (notekeeper), Bound MS, Wellcome MS 1219–1227.

Black, Joseph. 1767–1768. *Notes on Dr Black's Lectures on Chemistry*, 2 vols., Thomas Cochrane (notekeeper), Bound MS, USSC, Old Catalogue reference no. 34/28.

Black, Joseph. 1776. *Lectures on Chemistry*, 6 vols., Paul Panton Jr. (notekeeper), Bound MS, Chemical Heritage Foundation, QD14 B533.

Black, Joseph. 1782. *A Course of Lectures on the Theory and Practice of Chemistry. By J B Black, Professor of Chemistry in the University of Edinburgh*, vols. 1–3, Bound MS, Royal Society of London, MS/147/1–3.

Black, Joseph. 1782–1783. *Dr Black's Lectures*, vols. 1–8, Thomas Charles Hope (notekeeper), Bound MSS, EUL Dc.10.91–98.

Black, Joseph. 1787. *Heads of Professor Black's Lectures*, John McCartny (notekeeper), Bound MS, EUL Dc.10.98.

Blacklock, Thomas. n.d. *Kalokagathia* (anonymous notekeeper), Bound MS, EUL Dc.3.45.

Blacklock, Thomas. n.d. *Kalokagathia* (anonymous notekeeper), Bound MS, EUL La.III.80.

Blagden, Charles. 1767. *Memoranda variis e codicibus excerpta*, Bound MS, Wellcome MS. 1234.

Blair, Hugh. 1771. *Heads of the Lectures on Rhetorick, and Belles Lettres, in the University of Edinburgh* (Edinburgh: Kincaid and Creech), HLSM, no. 378392.

Blair, Hugh. 1779. *Lectures on Rhetoric and Belles Lettres*, vol. 2, John Bruce (notekeeper), Bound MSS, EUL Gen. 1990.

Bogle, George. *Schoolboy Diary*, Bound MS, Steggall Collection, Mitchell Library, Glasgow City Archives, TD1681/6/1.

Boyer, Abel. 1720. *The Royal Dictionary Abridged. In two parts. I. French and English. II. English and French* (London: Brown), DLDH 334.

Boyer, Abel. 1782. *The Complete French Master for Ladies and Gentlemen* (Edinburgh: Bell), DLDH no. 986.

Brougham, Henry. 1790. *Translation of English Phrases into Latin* (8 October), loose leaf sheet, ECA, SL137/9/3.

Campbell, Lady Charlotte Susan Maria. 1789. *Memoir and Journal of a Tour to Italy*. NLS Bound MS, Acc.8110.

Copland, Patrick. 1801. *Lecture Notes on Natural Philosophy*, vols. 1–2, Bound MSS, AUL 3995–3996.

Cullen, Archibald. c. 1800. *Translation of English into Latin*, Loose leaf, ECA SL137/9/8.

Cullen, William. 1757. *Rough Notes Taken by David Carmichael from Chemistry Lectures*, Bound MS, RCPE, Cullen Papers, MS 12.

Cullen, William. 1757/8. *Fair Copy Notes Taken by David Carmichael from Chemistry Lectures*, Bound MS, RCPE, Cullen Papers, MS 2.

Cullen, William. 1760. *Chemical Lectures by William Cullen*, vol. 1 (anonymous notekeeper), Bound MS, Wellcome MS 1918.

Cullen, William. 1761. *A Course of Lectures on the Materia Medica*, vols. 1–3 Edward Low (notekeeper), Bound MS, Wellcome MSL 22ª, 22, 22c.

Cullen, William. 1762. *Adversaria Chemia ex prolictionibus Dr Guliemi Cullen* (anonymous notekeeper), Bound MS, Wellcome MS MSL 49. 1771.

Cullen, William. c. 1765. *Chymical Notes*, vol. 2 (anonymous notekeeper), Bound Notebook, Wellcome MS 1923.

Cullen, William. 1765–1766. *Chemistry Notes*, vol. 1, *Taken in the Lectures of William Cullen*, Charles Blagden (transcriber), Bound MS, Wellcome MS 1922.

Cullen, William. 1771. *Bill of Complaint* (12 December), NAS, C 12/1033/2.

Cullen, William. n.d. *A Chemical Examination of Common Simple Stones & Earths . . . by William Cullen with Notes [Incomplete] on Alkali Earths and the Earth's Structure*, Loose-leaf MS, MS Cullen 264. Glasgow University Library.

Deletanville, Thomas. 1794. *A New French Dictionary, in two parts: the first, French and English; the second, English and French* (London: Wingrave), DLDH 328.

Dunbar, James. 1710. *A Volume Completed by James Dunbar in 1710 containing Arithmetic, Introduction to Algebra, and A More Compendious Way of Writeing than Ordinar Called the Short Hand, Making Use of Farthing's Alphabet*, Bound MS, NLS Acc 5706/11.

Dunbar, James. 1791–1793. *Introduction to Natural Philosophy: Institutes of Moral Philosophy*, vols. 1–2, Bound MS, CL, MS.2007.003.

Duncan, Andrew. 1792–1793. *Notes from Dr. Duncan's Lectures on Medical Jurisprudence*, Charles Anderson (notekeeper), Bound MS, EUL Coll-1587.

Edinburgh City Council. c. 1682–1875. *Final Council Minutes*, Bound MSS, ECA.

Edinburgh City Council. c. 1682–1875. *Scroll Council Minutes*, Bound MSS, ECA.

Edinburgh University. 1768–1781. *Library Receipt Book*, EUL Da.2.5.

Elliot, Charles. 1771–1777. *Ledger I*, Bound MS, NLS, John Murray Archive MS 43098.

Erskine, William. 1784. *Latin Exercise Book*, Bound MS, ECA SL137/9/38.

Eutropius. 1779. *Eutropii historiae romanae breviarium, ab urbe condita usque ad Valentinianum et Valentum Augustos . . . In usum scholarum* (Edinburgh: G. Gordon, G. Gray, J. Dickson et C. Creech), DLDH 362.

Ferguson, Adam. 1776–1785. *Lectures on Pneumatics and Moral Philosophy*, 3 vols., EUL Bound MSS DC.1.84–86.

Finlayson, James. 1795–1796. *Notes from the Lectures of Professor Finlayson on Logic*, 2 vols., John Lee (notekeeper), Bound MSS, EUL Dc.8.1421–1422.

Finlayson, James. 1796–1797. *An Epitome of Logic*, 5 vols., David Pollock (notekeeper), Bound MSS, EUL Gen.774–778.

Fleming, Marjory. *Papers of Marjorie Fleming*, loose leaf, NLS Acc. MSS. 1096–1100.

Forbes, James. 1781. *A New System of Practical Geometry Presented to Admiral Forbes by His Most Obedient Humble Servant James Forbes, Ensign in the Coldstream Regiment of Foot Guards, & Member of Mr. Lochee's Royal Military Academy*, Bound MS, NLS MS 9155.

Fordyce, George. 1788. *Chemistry by George Fordyce*, vols. 1–2, Richard Whitfield (notekeeper), Bound MS, St. Thomas's Hospital Collection, King's College London, MS M58-M60.

Fowler, James. 1780. *Schoolbook of James Fowler, Strathpeffer*, Bound MS, NLS MS 14284.

Glegg, Thomas. 1648–1649. *Cursus Logicus*, Thomas Ogilvie (notekeeper), Bound MSS, AUL GB 0231, MS 112.

Gordon, William. 1770. *Universal Accountant and Complete Merchant*, vol. 2 (Edinburgh: Donaldson), NLS no. ABS.2.89.3.

Gregory, David. 1683–1686. *Lectures in Mathematics and Optics given at the University of Edinburgh* (anonymous notekeeper), Bound MS, SUL, msQA33.G8D1 (ms6).

Gregory, James. *Papers of James Gregory*, EUL, Coll-33.

Gregory, John. 1768–1769. *Notes from Dr Gregory's Lectures on the Practice of Physick*, vol. 2, Thomas Cochrane (notekeeper), Bound MS, USSC, 33/34.

Gregory, John. 1772–1773. *Lectures on the Practice of Medicine by John Gregory Professor of Physick in the University of Edinburgh*, John Bacon (transcriber), Bound MS, EUL D.C.6.125.

Greig, Jean. 1763. *Copy Book Belonging to Jean Greig*, Bound MS, NLS Dep 190, Greig Papers, Box 3.

Greig, John. 1762–1764. *Pocket-Book Belonging to John Greig*, Bound MS, NLS Dep 190, Greig Papers, Box 3.

Herodotus. 1704. *Herodiani historiarum libri VII. Recogniti & Notis Illustrati* (Oxoniæ: G. West & Ant. Peisley), Dunimarle Library, no. 982.

Hill, John. 1770s. *Manuscripts of Dr John Hill*, Bound MSS, EUL Dc.8.74.

Hill, John. 1797. *Heads of Philological Lectures, Intended to Illustrate the Latin Classics, in Respect to the Antiquities of Rome; the Rules of General Criticism; and the Principles of Universal Grammar*, John Lee (notekeeper), Bound MS, EUL Dc.8.141.

Hill, John. 1802–1803. *Lectures on Humanity*, John Borthwick (notekeeper), Bound MS, EUL Gen. 841.

Hope, John. 1777–1778. *Lectures on Botany by John Hope M.D. in the Royal Botanick Garden Edinburgh* (anonymous notekeeper), Bound MS, RBGE.

Hope, John. 1780. *Dr Hope's Lectures on Botany*, Francis Buchanan Hamilton (note-keeper), Bound MS, RBGE.

Hope, John. 1781. *Lectures on Botany by Dr Hope*, James Cunningham (notekeeper), Bound MS, AUL MS 564.

Hope, John. 1783. *Dr Hope's Lectures in Botany*, Thomas Charles Hope (notekeeper), Bound MS, RBGE.

Hope, John. n.d. *Clavis Classium A. Von Royen*, John Hope collection of the RBGE.

Hope, John. n.d. *Clavis Classium Raii*, John Hope collection of the RBGE.

Hope, John. n.d. *Dr Hope's Lectures on Botany* (anonymous transcriber), Bound MS, RBGE.

Hope, John. n.d. *Methodus A. Caesalpini*, John Hope collection of the RBGE.

Hope, John. n.d. *Methodus A. Q. Rivini*, John Hope collection of the RBGE.

Hope, John. n.d. *Methodus Calycina C. Linnaei*, John Hope collection of the RBGE.

Hope, John. n.d. *Methodus D. F. B. Sauvages, ex foliis clavis*, John Hope collection of the RBGE.

Hope, John. n.d. *Methodus D. I. P. de Tournefort*, John Hope collection of the RBGE.

Hope, John. n.d. *Notes Taken from Dr Hopes Lectures on Botany* (anonymous note-keeper), RBGE.

Hume, David. 1729–1740. *Collection of Memoranda*, Loose MSS, NLS, MS.23159.

Hume, David. 1790. *Notes of Lectures on Scots Law*, vol. 1, William Patricks (notekeeper), Bound MS, EUL Dc.6.122.

Hume, David. 1808–1809. *Lectures on Scots Law*, 2 vols. (anonymous notekeeper), Bound MSS, EUL Gen. 860–861.

Hume, David. 1810. *Notes of Lectures on the Law of Scotland*, vol. 1 (anonymous note-keeper), Bound MS, EUL Gen. 862.

Hume, David. 1810–1812. *Notes on the Law of Scotland*, vols. 1–3, David Johnstone (note-keeper), Bound MSS, EUL Dc.10.42$^{1-3}$.

Hume, David. 1810–1811. *Notes on the Scotch Laws*, vols. 1–7 (anonymous notekeeper), Bound MSS, EUL Gen. 1391–1397.

Hume, David. 1815–1816. *Notes from Lectures on the Law of Scotland*, vols. 1–3, James Hark (notekeeper), Bound MSS, EUL Dc.3.8-10.

Hume, David. 1816–1817. *Notes from Lectures on Scots Law*, 5 vols., George Sligo (note-keeper), Bound MSS, EUL Bound MSS 2673–2677.

Hunter, Andrew. 1779–1807. *Theological Lectures*, vols. 1–6, Bound MSS, EUL Dc.3.22-Dc.3.27.

Jackson, Robert. 1788. *Geometry Notebook of Robert Jackson, A Schoolboy*, Bound MS, NLS MS 9156.

Johnstone, Elizabeth Caroline. 1816–1826. *Commonplace Books and Notebooks*, Bound MS, NLS, GB233/Ac.8100 120–123.

Keir, James. 1801. *Dialogues on Chemistry between a Father and His Daughter*, Amelie (née Keir) Moilliet (notekeeper), BAO, James Keir Archive, Box 33.

Kincaid, Alexander. 1764. *Latin Exercise Book*, Bound MS, ECA, SL137/9/37.

Leechman, William. c. 1780. *Notebook containing Notes from Lectures on Composition* (anonymous notekeeper), Bound MS, NLS MS.14096.

Linnaeus, Carl. 1725–1727. *Örtabok*, Bound MS, Växjö Library.

Locke, John. 1772. *Some Thoughts Concerning Education*, 14th ed. (London: Whiston, &tc.), NLS C.Fras. 67.

Logan Family Papers, Series VI, William Logan, Jr. Papers, Library Company of Philadelphia.

Lowth, Robert. 1783. *A Short Introduction to English Grammar: With Critical Notes, A New Edition, Corrected* (London: Dodsley and Cadell), DLDH 205.

Mair, John. 1779. *A Radical Vocabulary, Latin and English*, 5th ed. (Edinburgh: Murray), DLDH no. 1791.

Malcolm, Stephana. 1790. *Recipe Book*, Bound MS, NLS Acc.10708/3, ff. 1–8.

Merchant Maiden Hospital Administrational Papers. Edinburgh City Council: NRA 11909 Merchant Co, NRA 32022 Erskine.

Millar, James. 1815–1818. *Syllabus of Mathematics Delivered by Prof. Miller* [sic] (anonymous notekeeper), Bound MS, HLSM mss HM 72759, Burndy Codex 47.

Millar, John. 1771. *Lectures on Government by Mr. Millar P. L. Delivered in Glasgow* (anonymous notekeeper), Bound MS, SUL MS 53/3/9.

Millar, John. 1786. *A Course of Lectures on Government* (anonymous notekeeper, with annotations by Hugh Cleghorn), Unbound MS, SUL ms53/3/10b.

Miller, Hugh. 1815–1826. *Juvenile Poems*, Bound MSS, NLS 7520.

Moir, James. 1775. *The Scholar's Vade Mecum* (Edinburgh: Donaldson), DLDH no. 2130.

Monro Primus, Alexander. 1738. *Essay on Female Conduct in Letters from a Father to his Daughter*, Bound MS, NLS MS 6658.

Monro Primus, Alexander, and Margaret Monro (transcriber and editor). 1739. *An Essay on Female Conduct*, Bound MS, NLS MS 6659.

Monro Primus, Alexander (ed.). 1740–1741. *Proceedings of Latin Reading Club*, Bound MS, UOSC MS 100.

Monro Primus, Alexander. 1750. *Commentary on Monro's Anatomy of the Bones by A.M.P.A. Wrote for the Use of His Son A.M.*, Bound MS, UOSC, M163.

Monro Secundus, Alexander. 1773–1774. *Lectures Delivered by Doctor Alexander Monro Professor of Anatomy etc. in the College of Edinburgh*, 7 vols., John Thorburn (notekeeper), Bound MSS, UOSC, MS 175, MS 176–179.

Monro Secundus, Alexander. 1799–1800. *Lectures on Surgery*, George Bruce (notekeeper), Bound MS, Wellcome, MS RAMC/516.

436 | BIBLIOGRAPHY

Monro Secundus, Alexander, and Alexander Monro Tertius. 1801–1802. *Lectures on Anatomy and Surgery*, RCPE.

Montgomery, Barbara. 1787. *Account Book Begun by Barbara Montgomery*, Bound MS, NLS MS.14098A.

Montgomery, Barbara. 1809. *Embroidered Patterns*, Bound MS, NLS MS.14098B.

Montgomery, Margaret. 1800. *A Map of Scotland*, cloth and thread, NLS, EMS.s.701.

Moore, Mary. 1786–1790. *Letters*, loose leaf, EUL GB 237 Coll-715.

Murray, Amelie. 1744–1745. *Journal*, Blair Castle Archives (Perthshire), loose leaf, NRAS234, Papers of the Murray family, Dukes of Atholl, Box 45a Jac C I (11) items 1–13.

Pape, Joseph. 1799. *Illustrated Journal Kept by Midshipman Joseph Pape*, Bound MS, Royal Museums Greenwich, JOD/205.

Purdie, Peter. 1823. *Mathematics*, Bound MS, NLS MS 14288.

Ramsay, Allan. 1730–1731. *Sketchbook Containing Copies of Old Master Prints and Drawings*, Bound MS, National Galleries Scotland, accession no. D 5109.

Richardson of Pitfour, Robert. 1776. *School Exercise Ledger*, Bound MS, NLS MS 20985.

Richardson of Pitfour, Robert. 1777. *School Exercise Ledger*, Bound MS, NLS MS 20986.

Richardson of Pitfour, Robert. 1778. *Copybook*, Bound MS, NLS MS 20987.

Rush, James. 1809. *Lecture Notes*, Bound MS, Rush Family Papers (1746–1813), Library Company of Philadelphia, vol. 370.

Sandby, Paul. 1746. *View by Kinloch Rannoch*, British Library Maps, K.Top.50.83.2.

Sandby, Paul. c. 1765. *A Lady Copying at a Drawing Table*, YCBA, B1975.4.1881.

Sandby, Paul. c. 1770. *Lady Francis Scott and Lady Elliot*, YCBA, B1977.14.4410.

Sandby, Paul. n.d. *Portrait of a Lady at a Drawing*, YCBA, B1975.4.717.

Sandy, George. *Legal Diary, March–July 1788*, Bound MS, Signet Library, Edinburgh.

Scot, Alexandre. 1784. *Nouveau recueil: ou, melange litteraire, historique, dramatique et poetique; contenant le poeme celebre des jardins de Mons. L'Abbe de Lille* (Edinbourg: Elliot), DLDH no. 293.

Scott, Robert Eden. 1810. *Of Natural Philosophy*, William Watt (notekeeper), Bound MS, HLSM, HM 82554.

Scott, Walter. c. 1800. *Translations of Latin into English and English into Latin*, loose leaf sheet, ECA SL137/9/29.

Sinclair, Alexander. 1809. *Journal of a Tour from Edinburgh to Caithness*, NLS MS.3090.

Steuart, Gilbert. 1770. "Dr Steuart Gilbert to William Cullen," 2 January, RCPE, CUL/1/2/75.

Sym, Katherine. 1788–1838. *Diary and Commonplace Book of Miss Kath Sym*, Bound MS, NLS GB233/Acc.13033.

Tough, Jasper. 1776. "Dr Jasper Tough to Dr William Cullen," 15 July, RCPE, CUL/1/2/412.

Tytler, Alexander Fraser, Lord Woodhouselee. 1782. *Plan and Outlines of a Course of*

Lectures on Universal History, Ancient and Modern (Edinburgh: Creech), William Andrews Clark Memorial Library, University of California, Los Angeles.

Tytler, Alexander Fraser. 1793. *Notes on Universal History*, vols. 1–3 (anonymous notekeeper), Bound MSS, EUL Dc.8.144–146.

Tytler, Alexander Fraser. 1800–1801. *Universal History* (anonymous notekeeper), Bound MS, EUL Dc.6.115.

Venel, Gabriel François. 1761. *Cours de Chymie Fait chez Monsieur Montet Apoticaire: par Monsieur Venel Docteur et Proffesseur en L'université de Médecine à Montpellier*, Claude Delonne Balme (transcriber), Bound MS, Wellcome MS. 4914.

Vilant, Colin. 1708–1709. *Tractatus Metaphysicus*, George Scott (notekeeper), Bound MS, SUL, msBC59.V8.

Walker, John. 1797. *An Epitome of Natural History*, 10 vols., David Pollock (notekeeper), Bound MSS, EUL Gen.703 D–Gen.712 D.

Walker, Patrick. n.d. *Notebook Containing Notes from Lectures on Rhetoric and Logic*, Bound MS, NLS MS.14097.

Wilson, William. 1773. *Elements of Navigation: Or the Practical Rules of the Art, Plainly Laid Down, and Clearly Demonstrated from their Principles; with Suitable Examples of These Rules. To Which Are Annexed All the Necessary* (Edinburgh: Wilson and Donaldson), HLSM, call no. 492548.

Wood, James. 1789. "Dr James Wood of Keithick to William Cullen," 4 September, RCPE, MS CUL/1/2/2469.

PRINTED PRIMARY SOURCES

Adam, Alexander. 1772. *The Principles of Latin and English Grammar* (Edinburgh: Kincaid and Creech).

Adam, Alexander. 1786. *The Rudiments of Latin and English Grammar*, 3rd ed. (Edinburgh: Gordon, Bell, Dickson, Creech, Elliot and Balfour).

Adam, Alexander. 1794. *A Summary of Geography and History, Both Ancient and Modern* (Edinburgh: Bell, Bradfute and Creech).

Adam, Alexander. 1795. *A Geographical Index, Containing the Latin names of the Principal Countries, Cities, Rivers, and Mountains Mentioned in the Greek and Roman Classics* (Edinburgh: Bell, Bradfute and Creech).

Adams, George. 1791. *Geometrical and Graphical Essays, Containing a Description of the Mathematical Instruments Used in Geometry, Civil and Military Surveying, Levelling and Perspective* (London: Hindmarsh).

Adams, John. 1789. *Anecdotes, Bons-Mots, and Characteristic Traits of the Greatest Princes, Politicians, Philosophers, Orators, and Wits of Modern Times* (Dublin: Chamberlaine).

Ainslie, John. 1812. *Comprehensive Treatise on Land Surveying, Comprising the Theory and Practice in All Its Branches; in which the Use of Various Instruments Employed in Surveying, Levelling, &c. is Clearly Elucidated by Practical Examples* (Edinburgh: Silvester, Doig and Stirling).

Ainslie, John. 1849. *A Treatise on Land Surveying by John Ainslie. A New and Enlarged Edition Embracing Railway, Military, Marine, & Geodetical Surveying*, William Galbreath (ed.) (Edinburgh: Blackwood).

Ainsworth, Robert. 1746. *Thesaurus linguæ latinæ compendiarius* (London: Mount).

Aitchison, Thomas. 1795. *A Directory for Edinburgh Leith Mussleburgh and Dalkeith* (Edinburgh: Wilson).

Aitchison, Thomas. 1799. *The Edinburgh and Leith Directory, to July 1800* (Edinburgh).

Anderson, John. 1777. *Institutes of Physics*, 1st vol. (Glasgow: Foulis).

Anderson, John. 1786. *Institutes of Physics*, 4th ed. (Glasgow: Chapman and Duncan).

Anderson, John. 1796. *Extracts from the Latter Will and Codicil of Professor John Anderson* (Glasgow: Reid).

Anonymous. 1717. *The Ladies Diary: Or, The Woman's Almanack* (London: Wilde).

Anonymous. 1732. *The Art of Drawing, and Painting in Water-Colours. Whereby a Stranger to Those Arts May Be Immediately Render'd Capable of Delineating Any View* (London: Peele).

Anonymous. 1740. *The Ladies Diary: Or, The Woman's Almanack* (London: Wilde).

Anonymous. 1772. *Proposals for the Amendment of School Instruction* (London: Wilkie).

Anonymous. 1776. "Review of Charles Hutton's *Darian Miscellany*," *Scots Magazine*, **38**, 262.

Anonymous. 1780. *The Ladies Complete Pocket-Book* (Newcastle-upon-Tyne: Printed by T. Saint, for Whitfield and Co., and W. Creech, Edinburgh).

Anonymous. 1789. *List of Scholars Educated by the Late Mr. James Mundell* (Edinburgh: Mundell and Wilson).

Anonymous. 1811. *The Female Instructor, or Young Woman's Companion: Being a Guide to All the Accomplishments Which Adorn the Female Character, Either as a Useful Member of Society* (Liverpool: Nuttall, Fisher and Dixon).

Anonymous. 1818. *Obituary of William Scott, Edinburgh Magazine, and Literary Miscellany, a New Series of The Scots Magazine, 598*.

Arnot, Hugo. 1779. *The History of Edinburgh* (Edinburgh: Creech).

Baillie, Lady Grizel. 1911. *The Household Book of Lady Grisell Baillie, 1692–1733*. vol. 1 (Edinburgh: Constable).

Bannerman, Patrick. 1773. *Letters Containing a Plan of Education for Rural Academies* (Edinburgh: Ruddimans).

Barclay, James. 1743. *A Treatise on Education* (Edinburgh: Cochrane).

Barclay, James. 1758. *The Rudiments of the Latin Tongue* (Edinburgh: Barclay).

BIBLIOGRAPHY | 439

Barron, William. 1790. *Synopsis of Lectures on Belles Lettres and Logic Read in the University of St Andrews*, 2nd ed. (Edinburgh: Balfour).

Barrow, J. 1792. *A Description of Pocket and Magazine Cases of Mathematical Drawing Instruments* (London: Watkins).

Beattie, James. 1783. *Dissertations Moral and Critical in Two Volumes*, vol. 1 (Dublin: Exshaw).

Bell, A., and C. Macfarquhar (eds.). 1797. *Encyclopædia Britannica; or, a Dictionary of Arts, Sciences, and Miscellaneous Literature*, 3rd ed., vol. 14 (Edinburgh: Bell and Macfarquhar).

Bell, George Joseph. 1802. *Additional Memorial: The Compositors against the Master Printers of Edinburgh* (Edinburgh: Mundell).

Bentham, Jeremy. 1816. *Chrestomathia: Being a Collection of Papers, Explanatory of the Design of an Institution* (London: Payne and Foss).

Bickham, George. 1750. *The Surrey and Southwark Writing-master. A New County Copy-book* (London: Bickham).

Black, John. 1806. *The Falls of the Clyde or, the Fairies; A Scottish Dramatic Pastoral, in Five Acts* (Edinburgh: Creech).

Black, Joseph. 1770. *An Enquiry into the General Effects of Heat with Observations on the Theories of Mixture* (London: Nourse).

Black, Joseph. 1803a. *Lectures on Chemistry, Delivered in the University of Edinburgh by the late Joseph Black, M. D.*, vol. 1, John Robison (ed.) (Edinburgh: Creech).

Black, Joseph. 1803b. *Lectures on the Elements of Chemistry, Delivered in the University of Edinburgh*, vol. 1, John Robison (ed.) (Edinburgh: Mundell).

Black, Joseph. 1807. *Lectures on the Elements of Chemistry*, vol. 1, John Robison (ed.) (Philadelphia: Carey).

Black, Joseph. 1966. *Notes from Doctor Black's Lectures on Chemistry 1767–68*, Thomas Cochrane (transcriber), Douglas McKie (ed.) (Cheshire: ICI).

Black, Joseph. 2012. *The Correspondence of Joseph Black*, vols. 1-2, Robert G. W. Anderson and Jean Jones (eds.) (Farnham: Ashgate).

Blair, Hugh. 1771. *Heads of Rhetoric and Belles Lettres* (Edinburgh: Kincaid & Creech).

Blair, Hugh. 1783. *Lectures on Rhetoric and Belles Lettres*, vol. 1 (London: Strahan; Cadell, Creech).

Blane, Gilbert. 1790. *The Croonian Lecture on Muscular Motion: Read at the Royal Society, November 13th and 20th, 1788. Corrected and Enlarged* (London).

Boerhaave, Herman. 1708. *Institutiones medicæ* (Leiden).

Boutet, Claude. 1739. *The Art of Painting in Miniature* (London: Hodges).

Bower, Alexander. 1817. *The History of the University of Edinburgh: Chiefly Compiled from Original Papers and Records*, vols. 1–2 (Edinburgh: Smellie).

Bower, Alexander. 1822. *The Edinburgh Student's Guide: Or, An Account of the Classes of the University, Arranged Under the Four Faculties* (Edinburgh: Waugh and Innes).

Bowles, Carington. 1796. *The Art of Painting in Water Colours: Exemplified in Landscapes, Flowers, &c.* (London).

Broadie, Alexander. 1696. *The Penn's Practice, A New Copy Book* (Edinburgh).

Bruce, John. 1780. *First Principles of Philosophy* (Edinburgh: Creech).

Bruce, Thomas. 1724. *Arithmetick Vulgar and Decimal; Fully Explained and Directed, after a Plain and Easie Method to the Meanest Capacity* (Edinburgh: Dallas).

Buchanan, Colin. 1797. *Writing Made Easy Containing Specimens and Instructions Sufficient to Enable the Scholar to Write a Fine Business Hand* (Edinburgh).

Buchanan, Colin. 1798. *The Writing-Master and Accountant's Assistant* (Glasgow: Chapman).

Butterworth, Edmund. 1778. *New Sets of Copies in Alphabetical Order* (Dumfries).

Butterworth, Edmund. 1785. *Universal Penman, or, The Beauties of Writing Delineated* (Edinburgh).

Campbell, Alexander. 1802. *A Journey from Edinburgh through Parts of North Britain*, vol. 2 (London: Longman).

Champion, Joseph. 1750. *New and Complete Alphabets; In All the Various Hands of Great Britain* (London: Overton).

Champion, Joseph. 1759. *Penmanship Illustrated in a New Work, Containing a Great Variety of Excellent Copies in All the Useful Hands* (London: Sayer).

Chapman, George. 1773. *A Treatise on Education: With a Sketch of the Author's Method* (Edinburgh: Kincaid & W. Creech).

Chapman, George. 1784. *A Treatise on Education: With a Sketch of the Author's Method* (London: Cadell).

Christie, Hugh. 1758. *A Grammar of the Latin Tongue. After a New and Easy Method, Adapted to the Capacities of Children* (Edinburgh: Sands).

Cicero, Marcus Tullius. 1942. *De Oratore Books I and II*, E. W. Sutton (trans.) (Cambridge, Mass.: Harvard University Press).

Cicero, Marcus Tullius. 1989. *On Old Age. On Friendship. On Divination*, W. A. Falconer (trans.) (Cambridge, Mass.: Harvard University Press).

Cocker, Edward. 1678. *Cockers Arithmetick* (London: T. Passinger and T. Lacy).

Colinson, Robert. 1683. *Idea Rationaria or The Perfect Acomptant* (Edinburgh: Lindsay).

Coventrie, Alexander. 1904. "Extracts From the Journal of a Scotch Medical Student of the Eighteenth Century," L. M. A. Liggett (ed.), *Medical Library and Historical Journal*, **2**, 103–112.

Croker, Temple Henry. 1765. *The Complete Dictionary of Arts and Sciences*, vol. 2 (London: Croker).

Cullen, William. 1772. *Lectures on the Materia Medica, as Delivered by William Culllen, M.D.* (London: Lowndes).

Cullen, William. 1775. *Lectures on the Materia Medica* (Philadelphia: Bell).

Currie, James. 1831. *Memoir of the Life, Writings, and Correspondence of James Currie, M.D. F.R.S,* William Wallace Currie (ed.) (London: Longman).

Dalzel, Andrew. 1861. *Memoir of Andrew Dalzel, Professor of Greek in the University of Edinburgh,* Cosmo Innes (ed.) (Edinburgh: Constable).

Dalzel, Andrew. 1862. *History of the University of Edinburgh from Its Foundation,* vol. 2 (Edinburgh: Edmonston and Douglas).

Darling, W. 1768. *The Complete Letter-Writer* (Edinburgh: Reid).

Denholm, James. 1798. *The History of the City of Glasgow and Suburbs,* 2nd ed. (Glasgow: Chapman, Stewart and Meikle).

Douglas, George. 1776. *The Elements of Euclid* (Edinburgh: Elliot).

Drummond, Gawin. 1708. *A Short Treatise of Geography General & Special to which Are Added Tables of the Principal Coins in Europe and Asia* (Edinburgh: Symson).

Duncan, Andrew. 1776. *An Address to the Students of Medicine at Edinburgh* (Edinburgh).

Duncan, Andrew (ed.). 1794. *Medical Commentaries for the Year 1793* (Edinburgh: Hill).

Duncan, William. 1752. *Elements of Logick* (London: Dodsley).

Eden, Robert Scott. 1805. *Elements of Intellectual Philosophy; or, An Analysis of the Powers of the Human Understanding, Tending to Ascertain the Principles of a Rational Logic* (Edinburgh: Constable).

Ehret, Georg Dionysius. 1736. *Methodus Plantarum Sexualis in sistemate naturae descripta* (Leiden).

Elliot, Cornelius. 1804. *A Catalogue of the Books in Natural History with a Few Others, which Belonged to the Late Rev. Dr. Walker* (Edinburgh: Stewart).

Ewing, Alexander. 1771. *A Synopsis of Practical Mathematics* (Edinburgh: Smellie).

Ewing, Alexander. 1773. *Institutes of Arithmetic* (Edinburgh: Caddel).

Ewing, Alexander. 1799. *A Synopsis of Practical Mathematics, 4th ed.* (Edinburgh: Fairbairn).

Feinaigle, Gregor von. 1813. *The New art of Memory, Founded upon the Principles Taught by M. Gregor von Feinaigle,* John Millard (ed.) (London: Sherwood, Neely, and Jones).

Ferguson, Adam. 1766. *Analysis of Pneumatics and Moral Philosophy* (Edinburgh: Kinkaid and Bell).

Ferguson, Adam. 1769. *Institutes of Moral Philosophy* (Edinburgh: Kinkaid and Bell).

Ferguson, Adam. 1792. *Principles of Moral and Political Science; Being Chiefly a Retrospect of Lectures delivered in the College of Edinburgh,* vol. 1 (Edinburgh: Creech).

Ferguson, James. 1760. *Lectures on Select Subjects in Mechanics, Hydrostatics, Pneu-*

matics, and Optics: with The Use of the Globes, The Art of Dialing, and The Calculation of the Mean Times of New and Full Moons and Eclipses (London: Millar).

Ferguson, James. 1799. *Lectures on Select Subjects in Mechanics, Hydrostatics, Hydraulics, Pneumatics, and Optics* (London: Johnson).

Ferguson, James. 1803. *Lectures on Select Subjects in Mechanics, Hydrostatics, Pneumatics, and Optics: with The Use of the Globes, The Art of Dialing, and The Calculation of the Mean Times of New and Full Moons and Eclipses*, 10th ed. (London: Strahan and Johnson).

Feutrie, Thomas Levacher de la. 1772. *Traité du Rakitis, ou L'art de Redresser les Enfants Contrefaits* (Paris: Lacombe).

Fisher, George. 1771. *The Instructor; or, Young Man's Best Companion*, 23rd ed. (Edinburgh: Alston).

Fisher, George. 1763. *The Instructor; or Young Man's Best Companion* (Edinburgh: Alston).

Fisher, George. 1767. *Arithmetic in the Plainest and Most Concise Methods Hitherto Extant. With New Improvements for Dispatch of Business in All the Several Rules, as Also Fraction Vulgar and Decimal* (Glasgow: Marshall, Knox and Duncan).

Fisher, George. 1786. *The Instructor; or, Young Man's Best Companion. Containing Spelling, Reading, Writing, and Arithmetic, in an Easier Way Than Any Yet Published* (Glasgow).

Fisher, George. 1799. *The Instructor; or Young Man's Best Companion* (Edinburgh: Ruthven).

Fisher, George. 1806. *The Instructor; or Young Man's Best Companion*, 29th ed. (London: Johnson).

Fordyce, David. 1745. *Dialogues Concerning Education* (London: printed privately).

Fordyce, David. 1757. *Dialogues Concerning Education*, vol. 1, 3rd ed. (London: Dilly).

Frazer, Mrs. 1791. *The Practice of Cookery, Pastry, Pickling, Preserving, &c.* (Edinburgh: Hill).

Fulhame, Elizabeth. 1794. *An Essay on Combustion, with a View to a New Art of Dying and Painting* (London: Fulhame and Cooper).

Garnett, Thomas. 1811. *Observations on a Tour through the Highlands and Part of the Western Isles of Scotland*, vol. 2 (London: Stockdale).

General Assembly of the Church of Scotland. 1736. *The Confession of Faith* (Edinburgh: Lumisden and Robertson).

General Assembly of the Church of Scotland. 1737. *The A, B, C, with the Shorter Catechism* (Aberdeen: Chalmers).

General Assembly of the Church of Scotland. 1757. *The Confession of Faith, and the Larger and Shorter Catechism* (Edinburgh: E. Robertson).

George Watson's Hospital. 1724. *The Rules and Statutes of George Watson's Hospital* (Edinburgh: Brown and Mosman).

George Watson's Hospital. 1755. *The Statutes and Rules of George Watson's Hospital, Revised, Amended and Improven* (Edinburgh: Hamilton).

Gerdil, Giacinto Sigismondo. 1765. *Reflections on Education; Relative Both to Theory and Practice* (London: Davis and C. Reymers).

Gilbert, Geoffrey. 1758. *The History and Practice of the High Court of Chancery* (London: Lintot).

Glasgow University. 1790. *A Catalogue of the Books, in the Private Collection, Belonging to the Students of Divinity in the University of Glasgow* (Glasgow: Niven).

Gordon, William. 1763. *The Universal Accountant and Complete Merchant*, vol. 1 (Edinburgh: Gordon and Donaldson).

Gordon, William. 1765. *The Universal Accountant and Complete Merchant*, 2nd ed. (Edinburgh: Donaldson and Reid).

Gordon, William. 1770. *Universal Accountant and Complete Merchant*, vol. 1, rev. ed. (Edinburgh: Donaldson).

Gough, James. 1760. *A Practical Grammar of the English Tongue. Containing the Most Material Rules and Observations for Understanding the English Language Well* (Dublin: Jackson).

Graham, William. 1787. *Stenography; or, An Easy System of Short-Hand Writing* (Edinburgh: Bell).

Gregory, David. 1761. *A Treatise of Practical Geometry*, 4th ed. (Edinburgh: Hamilton, Balfour and Neill).

Gregory, John. 1770. *Observations of the Duties and Offices of a Physician* (London: Strahan and Cadell).

Gregory, John. 1772. *Lectures on the Duties and Qualifications of a Physician* (London: Strahan and Cadell).

Grew, Nehemiah. 1682. *The Anatomy of Plants with an Idea of a Philosophical History of Plants, and several other Lectures, Read before the Royal Society* (London: Rawlins).

Grey, Richard. 1737. *Memoria Technica, or A New Method of Artificial Memory, Applied to and Exemplified in Chronology, History, Geography, Astronomy, 3rd ed.* (London: Stagg).

Grey, Richard. 1756. *Memoria Technica: Or, a New Method of Artificial Memory, 4th ed.* (London: Lintot).

Grey, Richard. 1799. *Memoria Technica, or, A New Method of Artificial Memory* (London: Lowndes).

Hamilton, Alexander. 1784. *Outlines of the Theory and Practice of Midwifery* (Edinburgh: Elliot).

Hamilton, Elizabeth. 1804. *Memoirs of Modern Philosophers*, vol. 3, 4th ed. (Bath: Cruttwell and Robinson).

Hamilton, Elizabeth. 1810. *Letters on the Elementary Principles of Education*, vol. 2, 5th ed. (London: Wilkie and Robinson).

Hamilton, Robert. 1788. *An Introduction to Merchandise: Containing a Complete System of Arithmetic. A System of Algebra. Forms and Manner of Transacting Bills of Exchange. Book-keeping in Various Forms.* (Edinburgh: Elliot).

Harrington, S. 1773. *A New Introduction to the Knowledge and Use of Maps* (London: Crowder).

Hill, John. 1785. *Heads of Philological Lectures, Intended to Illustrate the Latin Classicks*, 2nd ed. (Edinburgh: Smellie).

Hill, John. 1792. *Heads of Philological Lectures, Intended to Illustrate the Latin Classicks*, 3rd ed. (Edinburgh: Smellie).

Hill, Peter. *Supplement to Peter Hill's Sale Catalogue for 1793. A Catalogue of Books, Comprehending an Useful Selection of the Works of the Best Authors* (Edinburgh: 1793).

Home, Francis. 1758. *Principia medicinae* (Edinburgh: Hamilton, Balfour, Neill).

Home, Francis. 1770. *Methodus materiae medicae* (Edinburgh: Kincaid and Bell).

Home, Henry (Lord Kames). 1781. *Loose Hints upon Education, Chiefly Concerning the Culture of the Heart*, 2nd ed., enlarged (Edinburgh: Bell).

Hope, Thomas. 1726. *Minor Practicks, or, A Treatise of the Scottish Law* (Edinburgh).

Hoppus, E. 1799. *Practical Measuring Made Easy to the Meanest Capacity*, 15th ed. (Edinburgh: Hunter).

Hume, David. 1740. *A Treatise of Human Nature: Being an Attempt to Introduce the Experimental Method of Reasoning into Moral Subjects*, vol. 3 (London: Longman).

Hutton, Charles. 1815. *Philosophical and Mathematical Dictionary: Containing an Explanation of the Terms, and an Account of the Several Subjects Comprised under the Heads Mathematics, Astronomy, and Philosophy both Natural and Experimental*, vol. 2 (London: Hutton).

Ingram, Alexander. 1799. *The Elements of Euclid* (Edinburgh: Pillans).

Ingram, Alexander. 1800. *The New Seaman's Guide, and Coaster's Companion* (Edinburgh: Sehaw).

Jardine, George. 1825. *Outlines of Philosophical Education, Illustrated by the Method of Teaching the Logic Class in the University of Glasgow*, 2nd ed. (Glasgow: Oliver & Boyd).

Kames, Lord. *See* Home, Henry.

Kay, James. 1837–1838. *A Series of Original Portraits and Caricature Etchings, with Biographical Sketches and Illustrative Anecdotes*, vol. 1, James Maidment (ed.) (Edinburgh: Paton, Carver and Gilder).

Kincaid, Alexander (ed.). 1790. *A New Geographical, Commercial and Historical Grammar; and Present State of Several Empires and Kingdoms of the World*, vol. 2 (Edinburgh: A Society at Edinburgh).

Knox, Vicesimus. 1781. *Liberal Education: Or, a Practical Treatise on the Methods of Acquiring Useful and Polite Learning* (London: Dilly).

Langley, Batty, with Thomas Langley. 1768. *The Builder's Jewel*, 12th ed. (Edinburgh: Clark).

Lawrie, Andrew. 1779. *The Merchant Maiden Hospital Magazine* (Edinburgh: Darling).

Lewis, James Henry. 1825. *The Art of Making a Good Pen, with Directions for Holding Properly. To which are Added an Analysis of Writing* (Bristol: Lewis).

Lewis, William. 1763. *Commercium Philosophico Technicum: Or, The Philosophical Commerce of Arts* (London: R. Willock).

Leybourn, Thomas. 1817. *The Mathematical Questions Proposed in the Ladies' Diary: And Their Original Answers, Together with Some New Solutions, from Its Commencement in the Year 1704 to 1816*, vol. 4 (London: J. Mawman).

Linnaeus, Carl. 1735. *Systema naturae sive regna tria naturae systematice proposita per classes, ordines, genera, & species* (Lugduni Batavorum: Haak et De Groot).

Linnaeus, Carl. 1751. *Philosophia botanica in qua explicantur fundamenta botanica* (Stockholmiae: Kiesewetter).

Lloyd, Evan. 1797. *A Plain System of Geography; Connected with a Variety of Astronomical Observations* (Edinburgh: Mundell).

Lochee, Lewis. 1776. *An Essay on Military Education* (London: Cadell).

Locke, John. 1693. *Some Thoughts Concerning Education* (London: Churchill).

Locke, John. 1751. *Some Thoughts Concerning Education, The Works of John Locke*, vol. 3, 5th ed., Pierre Desmaizeaux (ed.) (London: Longman)

Locke, John. 1752. *Some Thoughts Concerning Education*, 12th ed. (Edinburgh: Brown).

Locke, John. 1759. *The Works of John Locke*, vol. 1, 6th ed., Pierre Desmaizeaux (ed.) (London: Browne).

Locke, John. 1765. *An Essay Concerning Human Understanding*, vol. 3 (Edinburgh: Donaldson and Reid).

Locke, John. 1767. *An Abridgement of Mr. Locke's Essay Concerning Human Understanding* (Edinburgh: Donaldson).

Locke, John. 1779. *Some Thoughts Concerning Education* (London: Tonson).

Locke, John. 1781. *The Conduct of Understanding* (Cambridge: Archdeacon).

Love, John. 1786. *Geodaesia: Or, The Art of Surveying Made Easy*, 10th ed. (London: Rivington).

Love, John. 1792. *Geodaesia: Or, The Art of Surveying and Measuring Land Made Easy. Shewing by Plain and Practical Rules, to Survey, Protract, Cast Up, Reduce Or Divide Any Piece of Land Whatsoever, Corrected and Improved* (London: Robinson).

Lowndes, Thomas. [1766]. *A New Catalogue of Lowndes's Circulating Library, Consisting of Above Ten Thousand Volumes* (London: Lowndes).

Luckombe, Philip. 1771. *The History of the Art of Printing* (London: Adlard and Browne).

Lundin, Robert. 1718. *The Reason of Accompting by Debitor and Creditor* (Edinburgh).

Macghie, Alexander. 1718. *Principles of Book Keeping Explain'd* (Edinburgh: Watson).

MacGregor, John. 1792. *A Complete Treatise on Practical Mathematics* (Edinburgh).

Maciver, Susanna. 1787. *Cookery and Pastry as Taught and Practiced by Mrs. Maciver, Teacher of Those Arts in Edinburgh, A New Edition* (Edinburgh: Elliot).

Mackenzie, Donald. 1793. "Parish of Fodderty," in John Sinclair (ed.) *The Statistical Account of Scotland, Fodderty, Ross and Cromarty*, vol. 7 (Edinburgh: William Creech), 410–415.

Mair, John. 1736. *Book-Keeping Methodized* (Edinburgh).

Mair, John. 1760. *The Tyro's Dictionary, Latin and English* (Edinburgh: Sands).

Mair, John. 1762. *A Brief Survey of the Terraqueous Globe* (Edinburgh: Kincaid, Bell and Gray).

Mair, John. 1775. *A Brief Survey of the Terraqueous Globe* (Edinburgh: Creech).

Mair, John. 1790. *An Introduction to Syntax: or Exemplification of the Rules of Construction*, 11th ed. (Edinburgh: Bell, Bradfute and Creech).

Mair, John. 1797. *Book-Keeping Moderniz'd:, Or, Merchant-Accounts by Double Entry, According to the Italian Form*, 7th ed. (Edinburgh: Bell & Bradfute).

Malcolm, Alexander. 1718. *A New Treatise of Arithmetick and Book-Keeping* (Edinburgh: Mosman).

Martin, Benjamin. 1759. *Philosophia Britannica: Or, a New and Comprehensive System of the Newtonian Philosophy*, vol. 3 (London).

Martin, Benjamin. 1760. *The Description and Use of a Case of Mathematical Instruments; Particularly of All the Lines Contained on the Plain Scale, the Sector, the Gunter, and the Proportional Compasses* (London: Dollond).

Martin, Benjamin. 1797. *Description and Use of the Pocket Case of Mathematical instruments*, William Jones (ed.) (London: Jones).

Martin, William. 1786. *A Catalogue of a Curious, Uncommon and Valuable Collection of Books, being the Library of the Deceased Dr David Spence* (Edinburgh).

Martine, George. 1792. *Essays on the Construction of Thermometers, and on the Heating and Cooling of Bodies* (Edinburgh: William Creech).

Massey, William. 1763. *The Origin and Progress of Letters* (London: Johnson).

Mavor, William. 1792. *Universal Stenography; or, A New Compleat System of Short Writing* (London: Cadell).

Medical and Chirurgical Society of London. 1819. *A Catalogue of the Library of the Medical and Chirurgical Society of London, Second Part with an Appendix* (London: Taylor).

Merchant Maiden Hospital. 1776. *Rules for the Government and Order of the Merchant Maiden Hospital, Edinburgh* (Edinburgh: Balfour and Smellie).

BIBLIOGRAPHY | 447

Merchant Maiden Hospital. 1783. *Statutes of the Maiden Hospital, Founded by the Company of Merchants of Edinburgh, and Mary Erskine* (Edinburgh: Smellie).

Millar, John. 1771. *A Course on Government* (Glasgow).

Monro Primus, Alexander. 1783. *A Treatise on Comparative Anatomy*, Alexander Monro Secundus (ed.) (Edinburgh: Elliot and Creech).

Monro Primus, Alexander. 1954. "Life of Dr Mono Sr in his own handwriting," H. D. Erlaum (ed.), *University of Edinburgh Journal*, **17**, 77–105.

Monro Primus, Alexander. 1996. *The Professor's Daughter: Essay on Female Conduct*, P. A. G. Monro (ed.) (Edinburgh: Royal College of Physicians).

Monro Secundus, Alexander. 1770. *A State of Facts Concerning the First Proposal of Performing Paracentesis of the Thorax* (Edinburgh: Balfour, Auld and Smellie).

Monro Secundus, Alexander. 1783. *Observations on the Structure and Functions of the Nervous System Illustrated with Tables* (Edinburgh: Creech).

Monro Secundus, Alexander. 1794. *Observations on the Muscles, and Particularly on the Effects of Their Oblique Fibres* (Edinburgh: Dickson and Balfour).

Monro Secundus, Alexander. 1840. *Essays and Heads of Lectures on Anatomy, Physiology, Pathology and Surgery by the Late Alexander Monro Secundus, M.D.*, Alexander Monro Tertius (ed.) (Edinburgh: Maclachlan, Stewart and Company).

Morison, William (ed.). 1806. *Memorabilia of the City of Perth* (Perth: Morison).

Morton, Ralph. 1720. *Recta Scribendi Ratio: or, a Method of Writing Well; Being an Introduction to the Best Forms of Letters, with Copies of the Round Hand* (Edinburgh: Morton).

Moxon, Joseph. 1963. *Mechanick Exercises on the Whole Art of Printing* (1683–1684) (New York: Dover).

Murray, Charlotte. 1808. *The British Garden: A Descriptive Catalogue of Hardy Plants, Indigenous Or Cultivated in the Climate of Great Britain*, 3rd. ed. (London: Wilson).

Neville, Sylas. 1950. *The Diary of Sylas Neville 1767–1788*, Geoffrey Cumberledge (ed.) (London: Oxford University Press).

Nisbet, William. 1793. *The Clinical Guide; Or, A Concise View of the Leading Facts on the History, Nature, and Cure of Diseases* (Edinburgh: Creech).

Panton, William. 1771. *The Tyro's Guide to Arithmetic and Mensuration* (Edinburgh: Reid).

Paterson, James. 1685. *Scots Arithmetician* (Edinburgh: Joshua van Solingen and John Colmar).

Pennsylvania Hospital. 1806. *A Catalogue of the Medical Library, Belonging to the Pennsylvania Hospital* (Philadelphia: Philadelphia Hospital).

Perry, William. 1774. *The Man of Business, and Gentleman's Assistant* (Edinburgh).

Perry, William. 1775. *The Royal Standard Dictionary* (Edinburgh: Willison).

Perry, William. 1777. *The Man of Business, and Gentleman's Assistant* (Edinburgh: Perry).

Phillips, R. 1798. *The Monthly Magazine and British Register*, vol. 4, pt. 2 (London: Phillips).

Playfair, John. 1812. *Outlines of Natural Philosophy, Being Heads of Lectures Delivered in the University of Edinburgh*, vol. 1 (Edinburgh: Neill).

Playfair, John. 1816. *Outlines of Natural Philosophy*, vol. 2, 2nd ed. (Edinburgh: Abernethy and Walker).

Playfair, William. 1786. *The Commercial and Political Atlas* (London: Debrett, Robinson, and Sewell).

Pole, Thomas. 1790. *The Anatomical Instructor: Or an Illustration of the Modern and Most Approved Methods of Preparing and Preserving the Different Parts of the Human Body and of Quadrupeds by Injection, Corrosion, Maceration, Distention, Articulation, Modelling, Etc.* (London: Couchman & Fry).

Postmaster-General for Scotland. 1814. *The Post-Office Annual Directory, from Whitsunday 1815 to Whitsunday 1815 Edinburgh* (Abernethy & Walker).

Preston & Son of London. 1795. *The Multiplication Table, Adapted for Juvenile Improvement as a Lesson for the Piano Forte* (London: Preston & Son).

Priestley, Joseph. 1768. *An Essay on a Course of Liberal Education for Civil and Active Life* (London: Johnson).

Priestley, Joseph. 1769. *A New Chart of History* (London).

Priestley, Joseph. 1770. *A Description of a New Chart of History* (London).

Priestley, Joseph. 1786. *A Description of a New Chart of History*, 6th ed. (London: Johnson).

Quintilian, Marcus Fabius. 1876. *Quintilian's Institutes of Oratory: Or, Education of an Orator*, vol. 2, John Selby Watson (trans. and ed.) (London: Bell).

Quynn, William. 1936. "Letters of a Maryland Medical Student in Philadelphia and Edinburgh (1782–1784)," *Maryland Historical Magazine*, **31**, 181–215.

Ramsey, David. 1750. *The Weaver and Housewife's Pocket-Book* (Edinburgh: Ramsay).

Ramus, Petrus. 1569. *Scholarum Mathematicarum Libri Unus et triginta* (Basle: Eusebius Episcopus).

Reid, John. 1683. *The Scots Gardener in Two Parts, The First of Contriving and Planting Gardens, Orchards, Avenues, Groves: With New and Profitable Ways of Levelling; and how to Measure and Divide Land* (Edinburgh: Lindsay; republished by Edinburgh: Reid, 1766).

Reid, Thomas. 1764. *An Inquiry into the Human Mind, on the Principles of Common Sense* (Edinburgh: Millar).

Reid, Thomas. 1785. *Essays on the Intellectual Powers of Man* (Edinburgh: Bell).

Reid, Thomas. 2004. *On Logic, Rhetoric and the Fine Arts*, Alexander Broadie (ed.) (Edinburgh: Edinburgh University Press).

Reid, Thomas. 2017. *On Mathematics and Natural Philosophy*, Paul Wood (ed.) (Edinburgh: Edinburgh University Press).

Robertson, Hannah. 1767. *The Young Ladies School of Arts*, 2nd ed. (Edinburgh: Ruddiman).

Robertson, Hannah. 1792. *The Life of Mrs. Robertson, Grand-daughter of Charles II. Written by Herself* (Edinburgh: Robertson).

Robertson, John. 1775. *A Treatise of Such Mathematical Instruments, as Are Usually Put into a Portable Case* (London: Nourse).

Robison, John. 1804. *Elements of Mechanical Philosophy, being the Substance of a Course of Lectures on that Science*, vol. 1, *Including Dynamics and Astronomy* (Edinburgh: Constable).

Ruddiman, Thomas. 1779. *Mr. Ruddiman's Rudiments of the Latin Tongue* (Edinburgh: Murray and Cochran).

Ruddiman, Thomas. 1790. *Grammaticae latinae institutiones*, 12th ed. (Edinburgh: Dickson & Creech).

Russell, John. 1772. *Elements of Painting with Crayons* (London: Wilkie).

Salmon, Thomas. 1767. *A New Geographical and Historical Grammar: Wherein the Geographical Part Is Truly Modern* (Edinburgh: Sands, Murray and Cochran).

Sandy, George. 1942. "Legal Diary, March–July 1788," *Book of the Old Edinburgh Club*, **24**, 1–69.

Savigny, John. 1786. *Treatise on the Use and Management of a Razor, with Practical Directions Relative to Its Appendages; Also a Description of the Advantages Attending the form of the Convex Penknives* (London: Bensley).

Scott, William. 1782. *Elements of Geometry* (Edinburgh: Elliot).

Scruton, James. 1777. *The Practical Counting House; Or, Calculation and Accountantship Illustrated* (Glasgow: Duncan).

Sheldrake, Timothy. 1783. *An Essay on the Various Causes and Effects of the Distorted Spine* (London: Dilly).

Sheraton, Thomas. 1802. *Appendix to the Cabinet-Maker and Upholsterer's Drawing-Book* (London: Bensley).

Simson, Robert. 1781. *The Elements of Euclid* (Edinburgh).

Sinclair, George. 1688. *The Principles of Astronomy and Navigation* (Glasgow: Anderson).

Smellie, William (ed.). 1771. *Encyclopædia Britannica, or, A Dictionary of Arts and Sciences*, vols. 1–3 (Edinburgh: Bell and Macfarquhar).

Smith, Adam. 1759. *The Theory of Moral Sentiments* (London: Millar).

Smith, Adam. 1795. "The History of Astronomy," in *Adam Smith Essays on Philosophical Subjects*, Dugald Stewart (ed.) (London: Cadell and Davies), 29–93.

Smith, John. 1755. *The Printer's Grammar* (London: Smith).

Somerville, Mary. 1873. *Personal Recollections, from Early Life to Old Age*, Martha Somerville (ed.) (London: Murray).

Somerville, Mary. 2001. *Queen of Science: Personal Reflection of Mary Somerville*, Dorothy McMillan (ed.) (Edinburgh: Canongate).

Spence, David. 1767. *Dissertatio medica inauguralis de sanguinis ex utero gravidarum et puerperarum praeter naturam profluviis* (Edinburgi: Balfour, Auld et Smellie).

Spence, David. 1784. *A System of Midwifery, Theoretical and Practical* (Edinburgh: Creech).

Stark, J. 1821. *Picture of Edinburgh: Containing a Description of the City and Its Environs* (Edinburgh Fairbairn and Anderson).

Stedman, John. 1749. *The History and Statutes of the Royal Infirmary of Edinburgh* (Edinburgh: Ruddimans).

Steuart, Robert. 1725. *The Physiological Library of Mr. Steuart, and Some of the Students of Natural Philosophy in the University of Edinburgh, April 2. 1724* (Edinburgh).

Stewart, Dugald. 1792. *Elements of the Philosophy of the Human Mind* (Edinburgh: Creech).

Stewart, Dugald. 1793. *Outlines of Moral Philosophy* (Edinburgh: Creech).

Stewart, Dugald. 1795. "Account of the Life and Writings of Adam Smith," in Adam Smith, *Essays on Philosophical Subjects. By the Late Adam Smith*, Dugald Stewart (ed.) (London: Cadell and Davies), ix–xcv.

Stewart, Dugald. 1811. *Biographical Memoirs, of Adam Smith, LL.D. of William Robertson, D.D. and of Thomas Reid, D.D.* (Edinburgh: Ramsay).

Stewart, Dugald. 1814. *Elements of the Philosophy of the Human Mind*, vol. 2 (Edinburgh: Ramsay).

Stewart, Dugald. 1876. *Outlines of Moral Philosophy*, 9th ed., James McCosh (ed.) (London: Low, Low and Searle).

Strange, Robert. 1855. *Memoirs of Sir Robert Strange, Knt., Engraver*, vol. 1, Dennistoun, James (ed.) (London: Longman).

T.H. ["A Lover of Children"]. 1796. *The Child's Guide* (Aberdeen: Chalmers).

Todd, James. 1748. *The School-Boy and Young Gentleman's Assistant, Being a Plan of Education* (Edinburgh).

Tomkins, Peltro William. 1814. "Process for Preparing, with a Description of Some of the Properties of, the Refined Ox Gall, Invented and Prepared," *Philosophical Magazine*, **43**, 350–352.

Topham, Edward. 1780. *Letters From Edinburgh, Written in the Years 1774 and 1775: Containing Some Observations On The Diversions, Customs, Manners, and Laws, Of The Scotch Nation, During A Six Months Residence In Edinburgh: In Two Volumes*, vol. 1 (Dublin: Watson).

Trail, William. 1812. *Account of the Life and Writings of Robert Simpson, M. D. Late Professor of Mathematics in the University of Glasgow* (Bath: Cruttwell).

Turnbull, George. 1740. *A Treatise on Ancient Painting, Containing Observations on the Rise, Progress, and Decline of that Art amongst the Greeks and Romans* (London: Millar).

Turnbull, George. 1742. *Observations upon Liberal Education* (London: Millar).

Tytler, Alexander Fraser. 1782. *Plan and Outlines of a Course of Lectures on Universal History, Ancient and Modern* (Edinburgh: Creech).

Universities Commission. 1837. *Evidence, Oral and Documentary, Taken and Received by the Commissioners . . . for the Visiting of the Universities of Scotland*, vol. 3: *St. Andrews* (London: Clowes).

Vilant, Nicolas. 1798. *The Elements of Mathematical Analysis, Abridged* (Edinburgh: Bell and Bradfute).

Walker, George. 1816. "Directions for Painting Landscape in Crayons," *Scots Magazine*, 104–105.

Walker, John. 1781. *Schediasma fosilium* (Edinburgh).

Walker, John. 1787. *Classes fossilium: sive characteres naturales et chymici classium et ordinum in systemate mineralicum nominibus genericis adscriptis* (Edinburgh).

Walker, John. 1792. *Institutes of Natural History* (Edinburgh: Stewart, Ruthven and Co.).

Watts, Isaac. 1741. *The Improvement of the Mind: Or, A Supplement to the Art of Logick: Containing a Variety of Remarks and Rules for the Attainment and Communication of Useful Knowledge, in Religion, in the Science, and in Common Life* (London: Brackstone).

Watts, Isaac. 1786. *A Discourse on the Way of Instruction by Catechisms and the Best Manner of Composing Them* (London: Buckland).

Webster, William. 1719. *An Essay on Book-Keeping According to the Italian Method* (London).

Wedel, Georg Wolffgang. 1677. *Theoremata medica; seu Introductio ad medicinam, certis theorematibus, juxta ductum institutionum medicarum, absoluta, ad legendum & disputandum proposita* (Jenae: Johannis Bielckii).

West, John. 1784. *Elements of Mathematics: Comprehending Geometry, Mensuration, Conic sections and Spherics* (Edinburgh: Creech).

Whitchurch, James Wadham. 1772. *Essay upon Education* (London: Becket and de Hondt).

Whittle, James, and Robert Laurie. 1797. *A New Map of Scotland for Ladies Needlework* (London: Laurie & Whittle).

Wilkes, John. 1799. *The Art of Making Pens Scientifically* (London: Vigavena).

Williams, David. 1774. *A Treatise on Education* (London: Payne).

Williamson, William. 1775. *Stenography: Or, A Concise and Practical System of Short-Hand Writing* (London: Williamson).

Wilson, William. 1773. *Elements of Navigation* (Edinburgh: Robinson).

Wood, John. 1799. *Elements of Perspective; Containing the Nature of Light and Colours, and the Theory and Practice of Perspective* (London: Cawthorn; Edinburgh: Hill).

Wright, William. 1767. *Heads of a Course of Lectures on the Study of History* (Glasgow).

Young, Thomas. 1750. *A Course of Lectures upon Midwifery: Wherein Will Be Contained the History of the Art, with All Its Improvements Both Ancient and Modern* (Edinburgh).

Secondary Sources

Abbott, Don. 1989. "The Influence of Blair's Lectures in Spain," *Rhetorica*, **7**, 275–289.

Abbott, Don. 1998. "Blair 'Abroad': The European Reception of Lectures on Rhetoric and Belles Letters," in Lynée Lewis Gaillet (ed.), *Scottish Rhetoric and Its Influences* (Mahwah: Erlbaum), 67–77.

Ahnert, Thomas. 2015. *The Moral Culture of the Scottish Enlightenment: 1690–1805* (New Haven: Yale University Press).

Aho, James. 2005. *Confession and Bookkeeping: The Religious, Moral, and Rhetorical Roots of Modern Accounting* (Albany: State University of New York Press).

Alberti, Samuel J. M. M. 2016. "A History of Edinburgh's Medical Museums," *Journal of the Royal College of Physicians of Edinburgh*, **46**, 187–197.

Allan, David. 1993. *Virtue, Learning and the Scottish Enlightenment: Ideas of Scholarship in Early Modern History* (Edinburgh: Edinburgh University Press).

Allan, David. 2008. *Making British Culture: English Readers and the Scottish Enlightenment, 1740–1830* (London: Routledge).

Allan, David. 2010. *Commonplace Books and Reading in Georgian England* (Cambridge: Cambridge University Press).

Alpers, Svetlana. 1983. *The Art of Describing: Dutch Art in the Seventeenth Century* (Chicago: University of Chicago Press).

Alpers, Svetlana, and Michael Baxandall. 1996. *Tiepolo and the Pictorial Intelligence* (New Haven: Yale University Press).

Ambrose, Gavin, and Paul Harris. 2007. *The Layout Book* (Lausanne: AVA Publishing).

Ambrose, Gavin, and Paul Harris. 2010. *The Visual Dictionary of Typography* (Lausanne: AVA Publishing).

Ambrosio, Chiara. 2020. "Toward an Integrated History and Philosophy of Diagrammatic Practices," *East Asian Science, Technology and Society*, **14**, 347–376.

Anderson, Carolyn, and Christopher Fleet. 2018. *Scotland: Defending the Nation, Mapping the Military Landscape* (Edinburgh: Birlinn).

BIBLIOGRAPHY | 453

Anderson, Miranda, George Rousseau, and Michael Wheeler (eds.). 2019. *Distributed Cognition in Enlightenment and Romantic Culture* (Edinburgh: Edinburgh University Press).

Anderson, R. G. W. 1978. *The Playfair Collection and the Teaching of Chemistry at the University of Edinburgh 1713–1858* (Edinburgh: Royal Scottish Museum).

Anonymous. 1838a. "The History of a Black-Lead Pencil," *Saturday Magazine*, 24 March, 109–111.

Anonymous. 1838b. "History of Steel Pens," *Saturday Magazine*, 17 February, 63–64.

Anonymous. 1838c. "History of the Quill Pen," *Saturday Magazine*, 13 January, 14–16.

Aplin, Tanya, Lionel Bently, Phillip Johnson, and Simon Malynicz (eds.). 2012. *Gurry on Breach of Confidence: The Protection of Confidential Information* (Oxford: Oxford University Press).

Arnheim, Rudolf. 2004. *Art and Visual Perception: A Psychology of the Creative Eye* (Berkeley: University of California Press).

Asimov, Isaac. 1964. *Quick and Easy Math* (Boston: Houghton Mifflin).

Atran, Scott. 1990. *Cognitive Foundations of Natural History: Towards an Anthropology of Science* (Cambridge: Cambridge University Press).

Baggerman, Arianne. 2008. "'The Moral of the Story': Children's Reading and the Book of Nature around 1800," in Benjamin Schmidt and Pamela Smith (eds.), *Making Knowledge in Early Modern Europe: Practices, Objects, and Texts, 1400–1800* (Chicago: University of Chicago Press), 143–162.

Baggerman Arianne, and Rudolf Dekker. 2008. "The Social World of a Dutch Boy: The Diary of Otto van Eck (1791–1796)," in Susan Broomhall (ed.), *Emotions in the Households, 1200–1900* (Basingstoke: Palgrave Macmillan), 252–268.

Baggerman, Arianne, and Rudolf Dekker. 2009. *Child of the Enlightenment: Revolutionary Europe Reflected in a Boyhood Diary*, Diane Webb (trans.) (Leiden: Brill).

Bain, Andrew. 1965. *Education in Stirlingshire from the Reformation to the Act of 1872* (London: University of London Press).

Banfield, Ann. 2019. *Describing the Unobserved and Other Essays: Unspeakable Sentences after Unspeakable Sentences*, Sylvie Patron (ed.) (Newcastle-upon-Tyne: Cambridge Scholars).

Barchas, Janine. 2003. *Graphic Design, Print Culture, and the Eighteenth-Century Novel* (Cambridge: Cambridge University Press).

Barclay, Katie, and Rosalind Carr. 2018. "Women, Love and Power in Enlightenment Scotland," *Women's History Review*, **27**, 176–198.

Barfoot, Michael. 1983. "James Gregory (1753–1821) and Scottish Scientific Metaphysics, 1750–1800" (unpublished PhD thesis, Edinburgh University).

Barfoot, Michael. 1990. "Hume and the Culture of Science in the Early Eighteenth Century," in M. A. Stewart (ed.), *Studies in the Philosophy of the Scottish Enlightenment* (Oxford: Clarendon Press), 151–190.

Barfoot, Michael. 1993. "Philosophy and Method in Cullen's Medical Teaching," in Andrew Doig, J. P. S. Ferguson, I. A. Milne, and R. Passmore (eds.), *William Cullen and the Eighteenth Century Medical World* (Edinburgh: Edinburgh University Press), 110–132.

Barrell, John. 1986. *The Political Theory of Painting from Reynolds to Hazlitt* (New Haven: Yale University Press).

Bartine, David. 1989. *Early English Reading Theory: Origins of the Current Debates* (Columbia: University of South Carolina Press).

Bartlett, Roger. 2006. "German Popular Enlightenment in the Russian Empire: Peter Ernst Wilde and Catherine II," *Slavonic and East European Review*, **84**, 256–278.

Bausi, Alessandro, Michael Friedrich, and Marilena Maniaci (eds.). 2019. *The Emergence of Multiple-Text Manuscripts* (Berlin: Walter de Gruyter).

Baxandall, Michael. 1972. *Painting and Experience in Fifteenth Century Italy* (Oxford: Oxford University Press).

Baxandall, Michael. 1985. *Patterns of Intention: On the Historical Explanation of Pictures* (New Haven: Yale University Press).

Baxandall, Michael. 1997. *Shadows and Enlightenment* (New Haven: Yale University Press).

Beal, Peter. 2008. *A Dictionary of English Manuscript Terminology 1450–2000* (Oxford: Oxford University Press).

Beale, James M. 1983. *A History of the Burgh and Parochial Schools of Fife*, Donald J. Withrington (ed.) (Edinburgh: Scottish Council for Research in Education).

Becker, David P. 1997. *The Practice of Letters: The Hofer Collection of Writing Manuals 1514–1800* (Cambridge, Mass.: Harvard College Library).

Bell, Whitfield J. 1951. "Thomas Parke's Student Life in England and Scotland, 1771–1773," *Pennsylvania Magazine of History and Biography*, **75**, 237–259.

Bender, John B., and Michael Marrinan. 2005. *Regimes of Description: In the Archive of the Eighteenth Century* (Stanford: Stanford University Press).

Bender, John, and Michael Marrinan. 2010. *The Culture of Diagram* (Stanford: Stanford University Press).

Benedict, Barbara M. 2004. "Creech, William [pseud. Theophrastus], Bookseller and Magistrate," ODNB.

Benjamin, Walter. 1968. *The Work of Art in the Age of Mechanical Reproduction* (London: Fontana).

Bennett, James A. 1987. *Divided Circle: A History of Instruments for Astronomy, Navigation, and Surveying* (Oxford: Phaidon).

Bennett, Jim. 1998. "Projection and the Ubiquitous Virtue of Geometry in the Renaissance," in Jon Agar and Crosbie Smith (eds.), *Making Space for Science: Territorial Themes in the Shaping of Knowledge* (London: Palgrave Macmillan), 27–38.

Bensaude-Vincent, Bernadette. 2009. *Les Vertiges de la Technoscience: Façonner le Monde Atome par Atome* (Paris: Éditions La Découverte).

Bensaude-Vincent, Bernadette. 2013. "Chemistry as a Technoscience," in Jean-Pierre Llored (ed.), *The Philosophy of Chemistry: Practices, Methodologies, and Concepts* (Newcastle: Cambridge Scholars), 330–341.

Benzaquén, Adriana S. 2004. "Childhood, Identity and Human Science in the Enlightenment," *History Workshop Journal*, **57**, 34–57.

Berger, Susanna. 2017. *The Art of Philosophy: Visual Thinking in Europe from the Late Renaissance to the Early Enlightenment* (Princeton: Princeton University Press).

Bergren, Ann. 2008. *Weaving Truth: Essays on Language and the Female in Greek Thought* (Washington, D.C.: Center for Hellenic Studies).

Berry, Christopher J. 1997. *Social Theory of the Scottish Enlightenment* (Edinburgh: Edinburgh University Press).

Berry, Christopher J. 2013. *The Idea of Commercial Society in the Scottish Enlightenment* (Edinburgh: Edinburgh University Press).

Bevilacqua, Vincent M. 1965. "Adam Smith's Lectures on Rhetoric and Belles Lettres," *Studies in Scottish Literature*, **3**, 41–60.

Bittel, Carla, Elaine Leong, and Christine von Oertzen (eds.). 2019. *Working with Paper: Gendered Practices in the History of Knowledge* (Pittsburgh: University of Pittsburgh Press).

Blair, Ann. 1989. "On Ovid's Metamorphosis: The Class Notes of a 16th-Century Paris Schoolboy," *Princeton University Library Chronicle*, **50**, 117–144.

Blair, Ann. 1997. *The Theater of Nature: Jean Bodin and Renaissance Science* (Princeton: Princeton University Press).

Blair, Ann. 2003. "Reading Strategies for Coping with Information Overload ca. 1550–1700." *Journal of the History of Ideas*, **64**, 11–28.

Blair, Ann. 2004. "Note Taking as an Art of Transmission," *Critical Inquiry*, **31**, 85–107.

Blair, Ann. 2008. "Student Manuscripts and the Textbook," in Emidio Campi, Simone de Angelis, Anja-Silvia Goeing, and Anthony Grafton (eds.), *Scholarly Knowledge: Textbooks in Early Modern Europe* (Geneva: Droz), 39–73.

Blair, Ann. 2010a. "The Rise of Note-Taking in Early Modern Europe," *Intellectual History Review*, **20**, 303–316.

Blair, Ann. 2010b. *Too Much to Know: Managing Scholarly Information Before the Modern Age* (New Haven: Yale University Press).

Blair, Ann, and Peter Stallybrass. 2010. "Mediating Information, 1450–1800," in Clifford Siskin and William B. Warner (eds.), *This Is Enlightenment* (Chicago: University of Chicago Press), 139–163.

Bleichmar, Daniela. 2012. *Visible Empire: Botanical Expeditions and Visual Culture in the Hispanic Enlightenment* (Chicago: University of Chicago Press).

Boivin, Nicole. 2010. *Material Cultures, Material Minds: The Impact of Things on Human Thought, Society, and Evolution* (Cambridge: Cambridge University Press).

Boogert, Maurits van den. 2010. *Aleppo Observed: Ottoman Syria Through the Eyes of Two Scottish Doctors, Alexander and Patrick Russell* (Oxford: Oxford University Press and the Arcadian Library).

Bore, Henry. 1890. *The Story of the Invention of Steel Pens with a Description of the Manufacturing Processes by which They Are Produced* (New York: Ivison, Blakeman & company).

Borsuk, Amaranth. 2018. *The Book* (Cambridge, Mass.: MIT Press).

Bourguet, Marie Noëlle. 2010. "A Portable World: The Notebooks of European Travellers," *Intellectual History Review*, **20**, 377–400.

Bowker, Geoffrey C. 2005. *Memory Practices in the Sciences* (Cambridge, Mass.: MIT Press).

Bowker, Geoffrey C., and Susan Leigh Star. 1999. *Sorting Things Out: Classification and Its Consequences* (Cambridge, Mass.: MIT Press).

Bradley, Steven. 2018. *Design Fundamentals: Elements, Attributes and Principles* (Boulder: Vanseo).

Bredekamp, Horst, Vera Dünkel, and Birgit Schneider (eds.). 2015a. *The Technical Image: A History of Styles in Scientific Imagery* (Chicago: University of Chicago Press).

Bredekamp, Horst, Vera Dünkel, and Birgit Schneider. 2015b. "The Image—A Cultural Technology: A Research Program of a Critical Analysis of Images," in Bredekamp, Dünkel, and Schneider (eds.), *The Technical Image: A History of Styles in Scientific Imagery* (Chicago: University of Chicago Press), 1–5.

Brighton, H., and Gigerenzer, G. 2012. "Homo Heuristicus: Less-Is-More Effects in Adaptive Cognition," *Malaysian Journal of Medical Sciences*, **19**, 6–16.

Brindle, Kim. 2019. "Faded Ink: The Material Trace of Handwriting in Neo-Victorian Fiction," *Victoriographies*, **9**, 242–258.

Broadie, Alexander. 2001. *The Scottish Enlightenment: The Historical Age of the Historical Nation* (Edinburgh: Birlinn).

Brown, Callum G. 1981. "The Sunday-School Movement in Scotland, 1780–1914," *Records of the Scottish Church History Society*, **21**, 3–26.

Brown, Gillian. 2006. "The Metamorphic Book: Children's Print Culture in the Eighteenth Century," *Eighteenth-Century Studies*, **39**, 351–362.

Brown, Ian Gordon. 1984a. "Allan Ramsay's Rise and Reputation," *Volume of the Walpole Society*, **50**, 209–247.

Brown, Ian Gordon. 1984b. "Young Allan Ramsay in Edinburgh," *Burlington Magazine*, **26**, 778–781.

Brown, Leslie Ellen. 2016. *Artful Virtue: The Interplay of the Beautiful and the Good in the Scottish Enlightenment* (London: Routledge).

Brown, Stephen W. 2011. "Paper Manufacture," in Stephen Brown and Warren McDougal (eds.), *The Edinburgh History of the Book in Scotland*, vol. 2 (Edinburgh: Edinburgh University Press), 61–64.

Bryce, J. C. "Introduction," in Adam Smith, 1983. *Lectures on Rhetoric and Belles Lettres*, J. C. Bryce (ed.) (Oxford: Clarendon Press), 1–37.

Burke, Peter. 2003. "Images as Evidence in Seventh-Century Europe," *Journal of the History of Ideas*, **64**, 273–296.

Burke, Peter. 2012. "Past and Future of Note Taking." Lecture at the Take Note conference, Radcliffe Institute for Advanced Study, Harvard University, 2 November.

Byrd, William. 2001. *The Commonplace Book of William Byrd II of Westover*, Kevin Berland, Jan Kirsten Gilliam, and Kenneth A. Lockridge (eds.) (Chapel Hill: University of North Carolina Press).

Cairns, John W. 1988. "John Millar's Lectures on Scots Criminal Law," *Oxford Journal of Legal Studies*, **8**, 364–400.

Calhoun, Joshua. 2020. *The Nature of the Page: Poetry, Papermaking, and the Ecology of Texts in Renaissance England* (Philadelphia: University of Pennsylvania Press).

Cameron, Anne Marie. 2003. "From Ritual to Regulation? The Development of Midwifery in Glasgow and the West of Scotland, c. 1740–1840" (unpublished PhD thesis, University of Glasgow).

Cameron, Ross, Tom Addyman, and Kenneth Macfadyen. 2013. *Old College Quadrangle, University of Edinburgh, Archaeological Excavations: June–October 2010. Data Structure Report* (Edinburgh: Addyman Archaeology).

Camerota, Filippo. 2006. "Teaching Euclid in a Practical Context: Linear Perspective and Practical Geometry," *Science and Education*, **15**, 323–334.

Camille, Michael. 1992. *Image on the Edge: The Margins of Medieval Art* (London: Reaktion).

Camille, Michael. 1998. *Mirror In Parchment: The Luttrell Psalter and the Making of Medieval England* (Chicago: University of Chicago Press).

Camp, Pannill. 2008. "Ocular Anatomy, Chiasm, and Theatre Architecture as a Material Phenomenology in Early Modern Europe," in Maaike Bleeker (ed.), *Anatomy Live: Performance and the Operating Theatre* (Amsterdam: Amsterdam University Press), 229–146.

Canales, Jimena. 2010. *A Tenth of a Second: A History* (Chicago: University of Chicago Press).

Cannan, Edwin. 1896. "Introduction," in Adam Smith, *Lectures on Justice, Police, Revenue and Arms*, Edwin Cannan (ed) (Oxford: Clarendon Press), xviii–xvix.

Capkova, Dagmar. 1970. "J. A. Comenius's Orbis Pictus in Its Conception as a Textbook for the Universal Education of Children," *Pedagogica Historica*, **10**, 5–27.

Carlson, Eric T. 1981. "General Introduction," in Benjamin Rush, *Benjamin Rush's Lec-*

tures on the Mind, Eric T. Carlson, Jeffrey L. Wollock, and Patricia Noel (eds.) (Philadelphia: American Philosophical Society), 1–43.

Carr, Rosalind. 2014. *Gender and Enlightenment Culture in Eighteenth-Century Scotland* (Edinburgh: Edinburgh University Press).

Carruthers, Mary J. 1992. *The Book of Memory: A Study of Memory in Medieval Culture* (Cambridge: Cambridge University Press).

Carruthers, Mary J. 2002. "The Art of Memory and the Art of Page Layout in the Middle Ages," *Diogenes*, **49**, 20–30.

Carruthers, Mary J. 2008. *The Craft of Thought: Meditation, Rhetoric, and the Making of Images, 400–1200* (Cambridge: Cambridge University Press).

Cartwright, Nancy. 1983. *How the Laws of Physics Lie* (Oxford: Clarendon Press).

Cartwright, Nancy. 1989. *Nature's Capacities and Their Measurement* (Oxford: Clarendon Press).

Cavallo, Guglielmo, and Roger Chartier (eds.). 1999. *A History of Reading in the West* (Cambridge: Polity Press).

Cavallo, Guglielmo, and Roger Chartier (eds.). 2003. *A History of Reading in the West* (Amherst: University of Massachusetts Press).

Cavell, Samantha. 2010. "A Social History of Midshipmen and Quarterdeck Boys in the Royal Navy, 1761–1831" (unpublished PhD thesis, University of Exeter).

Certeau, Michel de. 2011. *The Practice of Everyday Life* (Berkeley: University of California Press).

Cevolini, Alberto (ed.). 2016. *Forgetting Machines: Knowledge Management Evolution in Early Modern Europe* (Leiden: Brill).

Chang, Dempsey, Laurence Dooley, and Juhani E. Tuovinen. 2002. "Gestalt Theory in Visual Screen Design: A New Look at an Old Subject," *Proceedings of the Seventh World Conference on Computers in Education: Australian Topics*, **8**, 5–12.

Chang, Ku-ming (Kevin), Anthony Grafton, and Glenn W. Most (eds.). 2021. *Impagination—Layout and Materiality of Writing and Publication: Interdisciplinary Approaches from East and West* (Berlin: Walter de Gruyter).

Chaplin, Simon David John. 2009. "John Hunter and the 'Museum Oeconomy', 1750–1800" (unpublished PhD thesis, Kings College London).

Charmantier, Isabelle. 2011. "Carl Linnaeus and the Visual Representation of Nature," *Historical Studies in the Natural Sciences*, **41**, 365–404.

Charmantier, Isabelle, and Staffan Müller-Wille. 2014. "Carl Linnaeus's Botanical Paper Slips (1767–1773)," *Intellectual History Review*, **24**, 215–238.

Chartier, Roger. 1994. *The Order of Books: Readers, Authors, and Libraries in Europe between the Fourteenth and Eighteenth Centuries* (Stanford: Stanford University Press).

Chartier, Roger, and J. A. González. 1992. "Laborers and Voyagers: From the Text to the Reader," *Diacritics*, **22**, 49–61.

Chernock, Adrianne. 2009. *Men and the Making of Modern British Feminism* (Stanford: Stanford University Press).

Cho, Hwisang. 2020. *The Power of the Brush: Epistolary Practices in Chosŏn Korea* (Seattle: University of Washington Press).

Clark, Andy, and David J. Chalmers. 1998. "The Extended Mind," *Analysis*, **58**, 7–19.

Clark, Aylwin. 1992. *An Enlightened Scot: Hugh Cleghorn, 1752–1837* (Duns: Black Ace Books).

Clements, M. A. Ken, and Nerida F. Ellerton. 2014. *Thomas Jefferson and His Decimals 1775–1810: Neglected Years in the History of U.S. School Mathematics* (Berlin: Springer).

Cohen, Murray. 1977. *Sensible Words: Linguistic Practice in England, 1640–1785* (Baltimore: Johns Hopkins University Press).

Cole, William A. 1982. "Manuscripts of Joseph Black's Lectures on Chemistry," in A. D. C. Simpson (ed.), *Joseph Black 1728–1799: A Commemorative Symposium* (Edinburgh: Royal Scottish Museum), 53–69.

Colley, Linda. 2005. *Britons: Forging the Nation, 1707–1837* (New Haven: Yale University Press).

Connor, R. D., and A. D. C. Simpson. 2004. *Weights and Measures in Scotland: A European Perspective, A*. D. Morrison-Low (ed.) (Edinburgh: National Museums of Scotland).

Cooper, Alix. 2007. *Inventing the Indigenous: Local Knowledge and Natural History in Early Modern Europe* (Cambridge: Cambridge University Press).

Cooper, John Michael, and Angela R. Mace. 2010. *Felix Mendelssohn Barthold: A Research and Information Guide* (Abington: Routledge).

Corey, S. M. 1935. "The Efficacy of Instruction in Note Making," *Journal of Educational Psychology*, **26**, 188–194.

Corrie, Jane. 2009. *Stories from the Historical Archives about Botanic Cottage, the Leith Walk Garden and John Hope's "Other" Life as a Physician* (Edinburgh: Friends of Hopetoun Crescent Garden).

Cosh, Mary. 2003. *Edinburgh: The Golden Age* (Edinburgh: John Donald Publishers).

Costa, Shelley. 2002. "The Ladies' Diary: Gender, Mathematics, and Civil Society in Early-Eighteenth-Century England," *Osiris*, **17**, 49–73.

Costigan-Eaves, Patricia, and Michael Macdonald-Ross. 1990. "William Playfair (1759–1823)," *Statistical Science*, **5**, 318–326.

Coutts, James. 1909. *A History of the University of Glasgow, from Its Foundation in 1451 to 1909* (Glasgow: Maclehose).

Crackel, Theodore J., V. Frederick Rickey, and Joel Silverberg. 2014. "George Washington's Use of Trigonometry and Logarithms," *Proceedings of the Canadian Society for History and Philosophy of Mathematics*, **26**, 98–115.

Crackel, Theodore J., V. Frederick Rickey, and Joel S. Silverberg. 2015. "Reassembling Humpty Dumpty: Putting George Washington's Cyphering Manuscript Back Together Again," in Maria Zack and Elaine Landry (eds.), *Research in History and Philosophy of Mathematics: The CSHPM 2014 Annual Meeting in St. Catharines, Ontario* (Cham: Birkhäuser), 79–96.

Craik, Alex D. D., and Alonso Roberts. 2009. "Mathematics Teaching, Teachers and Students at St. Andrews University, 1765–1858," *History of Universities*, **24**, 206–279.

Cramsie, Patrick. 2010. *The Story of Graphic Design* (London: British Library).

Crary, Jonathan. 1992. *Techniques of the Observer: On Vision and Modernity in the Nineteenth Century* (Cambridge, Mass.: MIT Press).

Crawford, David. 1906. *The Historians of Perth, and Other Local and Topographical Writers* (Perth: Christie).

Crosland, Maurice P. 1959. "The Use of Diagrams as Chemical 'Equations' in the Lecture Notes of William Cullen and Joseph Black," *Annals of Science*, **15**, 75–90.

Cumming, Ian. 1962. "The Scottish Education of James Mill," *History of Education Quarterly*, **2**, 152–167.

Cumming, Sandra. n.d. "The Erskine Family through Their Books" (unpublished MS on deposit at Duff House, Banff, Scotland).

Cumming, Sandra. 2017. "Book Evidence as a Possible Tool in History of Education Research: The Children's Collection in the Dunimarle Library," *History of Education Researcher*, **100**, 92–99.

Cunningham, William (ed.). 1829. *Catalogue of the Theological Library in the University of Edinburgh* (Edinburgh: Balfour and Co.).

Curran, Louise. 2016. *Samuel Richardson and the Art of Letter-Writing* (Cambridge: Cambridge University Press).

Dacome, Lucia. 2004. "Noting the Mind: Commonplace Books and the Pursuit of the Self in Eighteenth-Century Britain," *Journal of the History of Ideas*, **65**, 603–625.

Damerow, Peter. 1996. *Abstraction and Representation: Essays on the Cultural Evolution of Thinking*, Wolfgang Edelstein and Wolfgang Lefèvre (eds.) (Dordrecht: Kluwer Publishers).

Damerow, Peter, and Wolfgang Lefèvre. 1985. "Die alltägliche Seite der Geometrie: Zum Kapitel über die Zeicheninstrumente," in George Adams, *Geometrische und graphische Versuche oder Beschreibung der mathematischen Instrumente, deren man sich in der Geometrie, der Zivil- und Mlitär-Vermessung, beim Nivellieren und in der Per-*

spektive bedient. Nach der deutschen Ausgabe von 1795 (Darmstadt: Wissenschaftliche Buchgesellschaft), 283–300.

Damerow, Peter, and Wolfgang Lefèvre. 1996. "Tools of Science," in Peter Damerow, *Abstraction and Representation: Essays on the Cultural Evolution of Thinking*, Wolfgang Edelstein and Wolfgang Lefèvre (eds.) (Dordrecht: Kluwer Publishers), 395- 404.

Darnton, Robert. 1982a. *The Literary Underground of the Old Regime* (Cambridge, Mass.: Harvard University Press).

Darnton, Robert. 1982b. "What Is the History of Books?," *Daedalus*, **111**, 65–83.

Daston, Lorraine. 1988. *Classical Probability in the Enlightenment* (Princeton, N.J.: Princeton University Press).

Daston, Lorraine. 2004. "Taking Note(s)," *Isis*, **95**, 443–448.

Daston, Lorraine. 2008. "On Scientific Observation," *Isis*, **99**, 97–110.

Daston, Lorraine. 2011. "The Empire of Observation, 1600–1800," in Lorraine Daston and Elizabeth Lunbeck (eds.), *Histories of Scientific Observation* (Chicago: University of Chicago Press), 81–113.

Daston, Lorraine. 2015. "Super-vision: Weather Watching and Table Reading in the Early Modern Royal Society and Academie Royale des Sciences," *Huntington Library Quarterly,* **78**, 187- 215.

Daston, Lorraine, and Peter Galison. 2007. *Objectivity* (New York: Zone).

Daston, Lorraine, and Elizabeth Lunbeck (eds). 2011. *Histories of Scientific Observation* (Chicago: University of Chicago Press).

Dauben, W., and K. H. Parshall. 2014. "Mathematics Education in North America to 1800," in Alexander Karp and Gert Schubring (eds.), *Handbook on the History of Mathematics Education* (New York: Springer), 175–186.

Davis, Natalie Zemon. 1984. *The Return of Martin Guerre* (Cambridge, Mass.: Harvard University Press).

Davis, Natalie Zemon. 1995. *Women on the Margins: Three Seventeenth-Century Lives* (Cambridge, Mass.: Harvard University Press).

Dawson, Hannah. 2007. *Locke, Language, and Early Modern Philosophy* (Cambridge: Cambridge University Press).

Daybell, James. 2012. *The Material Letter in Early Modern England: Manuscript Letters and the Culture and Practices of Letter-Writing, 1512–1635* (Basingstoke: Palgrave Macmillan).

de Vivo, Filippo. 2007. *Information and Communication in Venice: Rethinking Early Modern Politics* (Oxford: Oxford University Press).

della Dora, Veronica. 2008. "Mountains and Memory: Embodied Visions of Ancient Peaks in the Nineteenth-Century Aegean," *Transactions of the Institute of British Geographers, New Series*, **33**, 217–232.

della Dora, Veronica. 2009. "Performative Atlases: Memory, Materiality, and (Co-)-Authorship," *Cartographica: The International Journal for Geographic Information and Geovisualization* **44**, 240–255.

Denniss, John. 2008. "Learning Arithmetic: Textbooks and their Users in England 1500–1900," in Eleanor Robson and Jacqueline Stedall (eds.), *The Oxford Handbook of the History of Mathematics* (Oxford: Oxford University Press), 448–467.

Denniss, John. 2012. *Figuring It Out: Children's Arithmetical Manuscripts 1680–1880* (Oxford: Huxley Scientific Press).

Derrida, Jacques. 2003. *Paper Machine* (Stanford: Stanford University Press).

Desmond, Ray. 1994. *Dictionary of British and Irish Botanists and Horticulturists Including Plant Collectors, Flower Painters and Garden Designers* (London: Natural History Museum).

Didi-Huberman, Georges. 2005. *Confronting Images: Questioning the Ends of a Certain History of Art*, John Goodman (trans.) (University Park: Pennsylvania State University Press).

Dingwall, Helen. 1999. "The Power Behind the Merchant? Women and Economy in Late-Seventeenth Century Edinburgh," in Elizabeth Ewan and Maureen M. Meikle (eds.), *Women in Scotland c. 1100–c. 1750* (East Linton: Tuckwell Press), 152–162.

Dodds, Phil. 2020. "Geographies of the Book (Shop): Reading Women's Geographies in Enlightenment Edinburgh," *Transactions of the Institute of British Geographers*, **45**, 270–283.

Dondis, Donis A. 1973. *A Primer of Visual Literacy* (Cambridge, Mass.: MIT Press).

Donovan, Arthur. 1975. *Philosophical Chemistry in the Scottish Enlightenment: The Doctrines and Discoveries of William Cullen and Joseph Black* (Edinburgh: Edinburgh University Press).

Douglas, Aileen. 2001. "Making Their Mark: Eighteenth-Century Writing-Masters and Their Copy-Books," *Journal for Eighteenth-Century Studies*, **24**, 145–159.

Douglas, Aileen. 2017. *Work in Hand: Script, Print, and Writing, 1690–1840* (Oxford: Oxford University Press).

Downey, Greg. 2011. "Learning the 'Banana-Tree': Self Modification through Movement," in Tim Ingold (ed.), *Being Alive: Essays on Movement, Knowledge and Description* (London: Routledge), 76–90.

Draaisma, Douwe. 1995. *Metaphors of Memory: A History of Ideas about the Mind* (Cambridge: Cambridge University Press).

Duguid, Paul. 2015. "The Ageing of Information: From Particular to Particulate," *Journal of the History of Ideas*, **76**, 347–368.

Duncan, Alistair Matheson. 1996. *Laws and Order in Eighteenth-Century Chemistry* (Oxford: Clarendon Press).

Dunyach, Jean-François, and Ann Thomson. 2015. "Introduction," in Jean-François

Dunyach and Ann Thomson (eds.), *The Enlightenment in Scotland: National and International Perspectives* (Oxford: Voltaire Foundation), 1–19.

Eddy, Matthew Daniel. 2004. "The Rhetoric and Science of William Paley's Natural Theology," *Literature and Theology*, **18**, 1–22.

Eddy, Matthew Daniel. 2006. "The Medium of Signs: Nominalism, Language and Classification in the Early Thought of Dugald Stewart," *Studies in the History and Philosophy of Biological and Biomedical Sciences*, **37**, 373–393.

Eddy, Matthew Daniel. 2007. "The Aberdeen Agricola: Chemical Principles and Practice in James Anderson's Georgics and Geology," in Lawrence M. Principe (ed.), *New Narratives in Eighteenth-Century Chemistry* (Dordrecht: Springer), 139–156.

Eddy, Matthew Daniel. 2008. *The Language of Mineralogy: John Walker, Chemistry and the Edinburgh Medical School* (London: Routledge).

Eddy, Matthew Daniel. 2010a. "Tools for Reordering: Commonplacing and the Space of Words in Linnaeus' *Philosophia Botanica*," *Intellectual History Review*, **20**, 227–252.

Eddy, Matthew Daniel. 2010b. "The Alphabets of Nature: Children, Books and Natural History, 1750–1800," *Nuncius*, **25**, 1–22.

Eddy, Matthew Daniel. 2010c. "The Sparkling Nectar of Spas: The Medical and Commercial Relevance of Mineral Water," in Ursula Klein and Emma Spary (eds.), *Materials and Expertise in Early Modern Europe: Between Market and Laboratory* (Chicago: University of Chicago Press), 198–226.

Eddy, Matthew Daniel. 2011. "The Line of Reason: Hugh Blair, Spatiality, and the Progressive Structure of Language," *Notes and Records of the Royal Society*, **65**, 9–24.

Eddy, Matthew Daniel. 2013. "The Shape of Knowledge: Children and the Visual Culture of Literacy and Numeracy," *Science in Context*, **26**, 215–245.

Eddy, Matthew Daniel. 2014. "How to See a Diagram: A Visual Anthropology of Chemical Affinity," *Osiris*, **26**, 178–196.

Eddy, Matthew Daniel. 2015. "Useful Pictures: Joseph Black and the Graphic Culture of Experimentation," in Robert G. W. Anderson (ed.), *Cradle of Chemistry: The Early Years of Chemistry at the University of Edinburgh* (Edinburgh: John Donald), 99–118.

Eddy, Matthew Daniel. 2016a. "The Child Writer: Graphic Literacy and the Scottish Educational System, 1700–1820," *History of Education*, **45**, 695–718.

Eddy, Matthew Daniel. 2016b. "The Cognitive Unity of Calvinist Pedagogy in Enlightenment Scotland," in Ábrahám Kovács (ed.), *Reformed Churches Working Unity in Diversity: Global Historical, Theological and Ethical Perspectives* (Budapest: L'Harmattan), 46–60.

Eddy, Matthew Daniel. 2018. "The Nature of Notebooks: How Enlightenment Schoolchildren Transformed the *Tabula Rasa*," *Journal of British Studies*, **57**, 275–307.

Eddy, Matthew Daniel. 2019. "Family Notebooks, Mnemotechnics and the Rational Education of Margaret Monro," in Carla Bittel, Elaine Leong, and Christine von Oertzen (eds.), *Working with Paper: Gendered Practices in the History of Knowledge* (Pittsburgh: University of Pittsburgh Press), 160–176, 269–272.

Eddy, Matthew Daniel. 2022. "Society, Environment and the Chemistry and Daily Life during the Eighteenth Century," in Matthew Daniel Eddy and Ursula Klein (eds.), *A Cultural History of Chemistry in the Eighteenth Century* (London: Bloomsbury), 113–135.

Eddy, Matthew Daniel. 2021. "Diagrams," in Ann Blair, Paul Duguid, Anja-Silvia Goeing, and Anthony Grafton (eds.), *Information: A Historical Companion* (Princeton: Princeton University Press), 397–401.

Edney, Matthew H. 1994. "British Military Education, Mapmaking, and Military 'Map-Mindedness' in the Later Enlightenment," *Cartographic Journal*, **31**, 14–20.

Edinburgh University. 1867. *List of Graduates in Medicine in the University of Edinburgh* (Edinburgh: Neill).

Edwards, Elizabeth F., and Janice Hart (eds.). 2004. *Photographs Objects Histories: On the Materiality of Images* (London: Routledge).

Edwards, J. R. 2014. "'Different from What Has Hitherto Appeared on this Subject': John Clark, Writing Master and Accomptant, 1738," *Abacus*, **50**, 227–244.

Edwards, Nicholas. 1972. "Some Correspondence of Charles Webster (circa 1772–1844), Concerning the Royal Institution," *Annals of Science*, **28**, 43–60.

Eldesouky, Doaa Farouk Badawy. 2013. "Visual Hierarchy and Mind Motion in Advertising Design," *Journal of Arts & Humanities*, **2**, 148–162.

Elkins, James. 2000. *How to Use Your Eyes* (London: Routledge).

Ellerton, Nerida F., and M. A. Ken Clements. 2012. *Rewriting the History of School Mathematics in North America 1607–1861: The Central Role of Cyphering Books* (Berlin: Springer).

Ellerton, Nerida F., and M. A. Ken Clements. 2014. *Abraham Lincoln's Cyphering Book and Ten Other Extraordinary Cyphering Books* (Berlin: Springer).

Emerson, Roger L. 2008. *Academic Patronage in the Scottish Enlightenment: Glasgow, Edinburgh and St. Andrews Universities* (Edinburgh: Edinburgh University Press, 29 Apr).

Emerson, Roger L. 2016. *Essays on David Hume, Medical Men and the Scottish Enlightenment* (London: Routledge).

Engel, William E. 2016. *Chiastic Designs in English Literature from Sidney to Shakespeare* (London: Routledge).

Eskildsen, Kasper Risbjerg. 2012. "Exploring the Republic of Letters: German Travellers in the Dutch Underground, 1690–1720," in Kristian H. Nielsen, Michael Harbsmeier, and Christopher J. Ries (eds.), *Scientists and Scholars in the Field: Studies in the History of Fieldwork and Expeditions* (Aarhus: Aarhus University Press), 101–122.

Ezell, Margaret J. M. 1983. "John Locke's Images of Childhood: Early Eighteenth-Century

Response to Some Thoughts Concerning Education," *Eighteenth-Century Studies*, **17**, 139–155.

Fabian, Bernhard, and Karen Kloth. 1969. "The Manuscript Background of James Beattie's 'Elements of Moral Science,'" *Bibliotheck*, **5**, 181–189.

Faulkner, Charles H. 2005. *The Prehistoric Native American Art of Mud Glyph Cave* (Knoxville: University of Tennessee Press).

Fay, C. R. 1956. *Adam Smith and the Scotland of His Day* (Cambridge: Cambridge University Press).

Febvre, Lucien, and Henri-Jean Martin. 1976. *The Coming of the Book: The Impact of Printing 1450–1800*, David Gerard (trans.) (London: Verso).

Ferguson, Eugene S. 1992. *Engineering and the Mind's Eye* (Cambridge, Mass.: MIT Press).

Ferguson, Stephen. 1987. "Systems and Schema, Tabulae of the Fifteenth to Eighteenth Centuries," *Princeton University Library Chronicle* **69**, 9–30.

Fischel, Angela. 2015. "Drawing and the Contemplation of Nature—Natural History around 1600: The Case of Aldrovandi's Images," in Horst Bredekamp, Vera Dünkel, and Birgit Schneider (eds.), *The Technical Image: A History of Styles in Scientific Imagery* (Chicago: University of Chicago Press), 170–181.

Fleck, Ludwik. 1979. *Genesis and Development of a Scientific Fact*, Fred Bradley and Thaddeus J. Trenn (trans.) (Chicago: University of Chicago Press).

Fleet, Christopher. 2011. "Atlases, Map-makers and Map-engravers," in Stephen W. Brown and Warren McDougall (eds.), *Edinburgh History of the Book in Scotland*, vol. 2: *Enlightenment and Expansion 1707–1800* (Edinburgh: Edinburgh University Press), 91–102.

Fleet, Christopher, Margaret Wilkes, and Charles W. J. Withers. 2011. *Scotland: Mapping the Nation* (Edinburgh: Berlin).

Fleet, Christopher, Margaret Wilkes, and Charles W. J. Withers. 2016. *Scotland: Mapping the Islands* (Edinburgh: Birlinn).

Fleming, Juliet. 2015. "The Renaissance Collage: Signcutting and Signsewing," *Journal of Medieval and Early Modern Studies*, **45**, 443–456.

Fleming, Marjorie. 1934. *The Journals, Letters, and Verses of Marjory Fleming in Collotype Facsimile from the Original Manuscripts* (London: Sidgwick & Jackson).

Forceville, Charles. 2011. "Pictorial Runes in *Tintin and the Picaros*," *Journal of Pragmatics*, **43**, 875–890.

Forsyth, Sunderland. 2016. *Discover the Botanic Cottage*, Donna Cole (ed.) (Edinburgh: Royal Botanic Garden Edinburgh).

Fothergill, George A. 1908. "Notes on Scottish Samplers," *Proceedings of the Society of Antiquaries of Scotland*, **43**, 180–205.

Foucault, Michel. 1970. *The Order of Things: An Archaeology of the Human Sciences* (London: Tavistock).

Fowler, Caroline. 2019. *The Art of Paper: From the Holy Land to the Americas* (New Haven: Yale University Press).

Fox, Adam. 2020. *The Press and the People: Cheap Print and Society in Scotland, 1500–1785* (Oxford: Oxford University Press).

Franklin, James. 2017. "Early Modern Mathematical Principles and Symmetry Arguments," in Peter R. Anstey (ed.), *The Idea of Principles in Early Modern Thought: Interdisciplinary Perspectives* (London: Routledge), 16–44.

Frasca-Spada, Marina. 2002. *Space and the Self in Hume's Treatise* (Cambridge: Cambridge University Press).

Freedberg, David. 1989. *The Power of Images: Studies in the History and Theory of Response* (Chicago: University of Chicago Press).

Freedberg, David. 2003. *The Eye of the Lynx: Galileo, His Friends, and the Beginnings of Modern Natural History* (Chicago: University of Chicago Press).

Friday, Jonathan (ed.). 2004. *Art and Enlightenment: Scottish Aesthetics in the 18th Century* (Exeter: Imprint Academic).

Friedenberg, Jay. 2013. *Visual Attention and Consciousness* (London: Psychology Press).

Friedman, Michael. 2018. *A History of Folding in Mathematics: Mathematizing the Margins* (Berlin: Birkhäuser).

Friedman, Michael, and Wolfgang Schäffner. 2016. "On Folding: Introduction of a New Field of Interdisciplinary Research," in Michael Friedmann and Wolfgang Schäffner (eds.), *On Folding: Towards a New Field of Interdisciplinary Research* (Bielefeld: Transcript Verlag), 7–30.

Friedrich, Markus. 2018. *The Birth of the Archive: A History of Knowledge*, John-Noël Dillon (trans.) (Ann Arbor: University of Michigan Press).

Friend, Claire. 2016. "The Social Life of Paper in Edinburgh 1750–1820" (unpublished PhD thesis, Edinburgh University).

Friends of Duff House. 2011. "The Dunimarle Library at Duff House" (Banff: Duff House, Friends of Duff House).

Froide, Amy. 2015. "Learning to Invest: Women's Education in Arithmetic and Accounting in Early Modern England," *Early Modern Women* **10**, 3–26.

Funkhouser, H. Gray. 1937. "Historical Development of the Graphical Representation of Statistical Data," *Osiris*, **3**, 269–404.

Fyfe, Aileen. 2004. *Science and Salvation: Evangelical Popular Science Publishing in Victorian Britain* (Chicago: University of Chicago Press).

Fyfe, Aileen. 2012. *Steam-Powered Knowledge: William Chambers and the Business of Publishing, 1820–1860* (Chicago: University of Chicago Press).

Gallagher, Sam, and Rob Lindgren. 2015. "Enactive Metaphors: Learning through Full-Body Engagement," *Educational Psychology Review* **27**, 391–404.

Gardner, Howard. 1990. *Art Education and Human Development* (Los Angeles: The J. Paul Getty Museum).

Garrett, Aaron, and Colin Heydt. 2015. "Moral Philosophy: Practical and Speculative," in Aaron Garrett and James Anthony Harris (eds.), *Scottish Philosophy in the Eighteenth Century: Morals, Politics, Art, Religion*, vol. 1 (Oxford: Oxford University Press), 77–97.

Gaskell, Philip. 1995. *A New Introduction to Bibliography* (New Castle: Oak Knoll Press).

Genette, Gerard. 1997. *Paratexts: Thresholds of Interpretation* (Cambridge: Cambridge University Press).

Gianoutsos, Jamie. 2019. "A New Discovery of Charles Hoole: Method and Practice in Seventeenth-Century English Education," *History of Education*, **48**, 1–18.

Gibson, James Jerome. 1986. *The Ecological Approach to Visual Perception* (Hillsdale: Erlbaum).

Gibson, Jonathan. 1997. "Significant Space in Manuscript Letters," *Seventeenth Century*, **12**, 1–10.

Gibson, Jonathan. 2010. "Casting Off Blanks: Hidden Structures in Early Modern Paper Books," in James Daybell and Peter Hinds (eds.), *Material Readings of Early Modern Culture* (Basingstoke: Palgrave Macmillan), 208–228.

Gibson, Susannah. 2015. *Animal, Vegetable, Mineral? How Eighteenth-Century Science Disrupted the Natural Order* (Oxford: Oxford University Press).

Gibson-Wood, Carol. 2001–2003. "George Turnbull and Art History at Scottish Universities in the Eighteenth Century," *Canadian Art Review*, **28**, 7–18.

Gierl, Martin. 2012. *Geschichte als präzisierte Wissenschaft: Johann Christoph Gatterer und die Historiographie des 18. Jahrhunderts im ganzen Umfang* (Stuttgart: Rommann-Holzboog Verlag).

Gigerenzer, Gerd. 2008. *Rationality for Mortals: How People Cope with Uncertainty* (Oxford: Oxford University Press).

Gillispie, Charles Coulston. 1960. *The Edge of Objectivity: An Essay in the History of Scientific Ideas* (Princeton: Princeton University Press).

Ginzburg, Carlo. 1980. *The Cheese and the Worms: The Cosmos of a Sixteenth-Century Miller* (London: Routledge and Kegan Paul).

Gitelman, Lisa. 2014. *Paper Knowledge: Toward a Media History of Documents* (Durham: Duke University Press).

Glover, Katherine. 2011. *Elite Women and Polite Society in Eighteenth-Century Scotland* (Woodbridge: Boydell Press).

Goeing, Anja-Silvia. 2016. *Storing, Archiving, Organizing: The Changing Dynamics of Scholarly Information Management in Post-Reformation Zurich* (Leiden: Brill).

Goggin, Maureen Daly, and Beth Fowkes Tobin (eds). 2016. *Women and Things, 1750–1950: Gendered Material Strategies* (London: Routledge).

Gokcekus, Samin. 2019. "Elizabeth Hamilton's Scottish Associationism: Early Nineteenth-Century Philosophy of Mind," *Journal of the American Philosophical Association*, **2**, 1–19.

Golledge, Reginald G. (ed.). 1999. *Wayfinding Behavior: Cognitive Mapping and Other Spatial Processes* (Baltimore: Johns Hopkins University Press).

Gombrich, E. H. 1979. *The Sense of Order: A Study in the Psychology of Decorative Art* (Oxford: Phaidon).

Goody, Jack. 1977. *The Domestication of the Savage Mind* (Cambridge: Cambridge University Press).

Goody, Jack. 1986. *The Logic of Writing and the Organization of Society* (Cambridge: Cambridge University Press).

Goody, Jack. 1987. *The Interface Between the Written and the Oral* (Cambridge: Cambridge University Press).

Grafton, Anthony. 1999. *The Footnote: A Curious History* (Cambridge, Mass.: Harvard University Press).

Grafton, Anthony. 2001. *The Culture of Correction in Renaissance Europe* (London: British Library).

Grafton, Anthony. 2002. *Magic and Technology in Early Modern Europe* (Washington, D.C.: Smithsonian Institute Libraries).

Grafton, Anthony. 2012. "The Republic of Letters in the American Colonies: Francis Daniel Pastorius Makes a Notebook," *American Historical Review*, **117**, 1–39.

Grafton, Anthony, and Joanna Weinberg. 2016. "Johann Buxtorf Makes a Notebook," in Anthony Grafton and Glenn W. Most (eds.), *Canonical Texts and Editorial Practices: A Global Approach* (Cambridge: Cambridge University Press), 275–298.

Graham, Gordon. 2015. "Beauty, Taste, Rhetoric and Language," in Aaron Garrett and James Anthony Harris (eds.), *Scottish Philosophy in the Eighteenth Century: Morals, Politics, Art, Religion* (Oxford: Oxford University Press), 131–162.

Graham, Henry Grey. 1899. *The Social Life of Scotland in the Eighteenth Century*, vol. 2 (London: Adam and Charles Black).

Grant, I. F. 1924. *Every-Day Life on an Old Highland Farm, 1769–1782* (London: Longmans, Green and Co.).

Grant, James. 1876. *History of the Burgh and Parish Schools of Scotland* (London: William Collins).

Gregory, John. 1998. *John Gregory's Writings on Medical Ethics and Philosophy of Medicine*, Laurence B. McCullough (ed.) (Dordrecht: Kluwer).

Grenby, M. O. 2011. *The Child Reader 1700–1840* (Cambridge: Cambridge University Press).

Gubar, Marah. 2010. *Artful Dodgers: Reconceiving the Golden Age of Children's Literature* (Oxford: Oxford University Press).

Haakonssen, Knud. 2016. "The Lectures on Jurisprudence," in Ryan Patrick Hanley (ed.), *Adam Smith: His Life, Thought, and Legacy* (Princeton: Princeton University Press), 48–66.

Hacking, Ian. 1975. *Why Does Language Matter to Philosophy?* (Cambridge: Cambridge University Press).

Hacking, Ian. 1984. *The Emergence of Probability: A Philosophical Study of Early Ideas about Probability, Induction, and Statistical Inference* (Cambridge: Cambridge University Press).

Hacking, Ian. 1990. *The Taming of Chance* (Cambridge: Cambridge University Press).

Haldane, A. 1981. R. B. *The Great Fishmonger of the Tay: John Richardson of Perth and Pitfour* (Dundee: Abertay Historical Society).

Hallett, Michael, and Ulrich Majer. 2004. *David Hilbert's Lectures on the Foundations of Geometry 1891–1902* (Berlin: Springer).

Hamel, Christopher de. 2018. *Meetings with Remarkable Manuscripts* (London: Penguin).

Hammet, Iain Maxwell. 1985. "Lord Monboddo's *The Origin and Progress of Language*: Its Sources, Genesis, and Background, with Special Attention to the Advocates' Library" (unpublished PhD thesis, Edinburgh University).

Hans, Nicholas A. 1966. *New Trends in Education in the Eighteenth Century* (London: Routledge & Kegan Paul).

Harding, Diana. 1972. "Mathematics and Science Education in Eighteenth-Century Northamptonshire," *History of Education* **1**, 139–159.

Harding, Sandra. 1991. *Whose Science? Whose Knowledge?: Thinking from Women's Lives* (Ithaca: Cornell University Press).

Harris, James A. 2015. *Hume: An Intellectual Biography* (Cambridge: Cambridge University Press).

Hart, Rachel M. 1989. "Letters and Papers of Dr Robert Hamilton," *Northern Scotland*, **9**, 83–85.

Hartley, James. 1988. *Designing Instructional Text*, 2nd ed. (London: Kogan Page).

Hartley, James, and Joyce L. Harris. 2001. "Reading the Typography of Text," in J. L. Harris, A. G. Kamhi, and K. E. Pollack (eds.), *Literacy in African American Communities* (Mahwah: Erlbaum), 109–125.

Haskell, Francis. 1993. *History and Its Images: Art and the Interpretation of the Past* (New Haven: Yale University Press).

Hatch, Gary Layne. 1998. "Student Notes of Hugh Blair's Lectures on Rhetoric," in Lynée Lewis Gaillet (ed.), *Scottish Rhetoric and Its Influences* (Mahwah: Hermagoras), 79–94.

Haynes, Nick, and Clive B. Fenton. 2017. *Building Knowledge: An Architectural History of the University of Edinburgh* (Edinburgh: Historic Environment Scotland).

Hayward, Oliver S., and Constance E. Putnam. 1998. *Improve, Perfect, & Perpetuate: Dr.*

Nathan Smith and Early American Education (Hanover: University Press of New England).

Head, Randolph. 2019. *Making Archives in Early Modern Europe: Proof, Information, and Political Record-Keeping, 1400–1700* (Cambridge: Cambridge University Press).

Heal, Ambrose. 1931. *The English Writing-Masters and Their Copy-Books, 1570–1800* (Cambridge: Cambridge University Press).

Heesen, Anke Te. 2002. *The World in a Box: The Story of an Eighteenth-Century Picture Encyclopaedia* (Chicago: University of Chicago Press).

Heesen, Anke Te. 2005. "The Notebook: A Paper-Technology," in Bruno Latour and Peter Weibel (eds.), *Making Things Public. Atmospheres of Democracy* (Cambridge, Mass.: MIT Press), 582–589.

Heesen, Anke Te. 2014. *The Newspaper Clipping: A Modern Paper Object* (Manchester: Manchester University Press).

Hendriksen, Marieke M. A. 2015. *Elegant Anatomy: The Eighteenth-Century Leiden Anatomical Collections* (Leiden: Brill).

Henry, Lucy. 2011. *The Development of Working Memory in Children* (London: SAGE).

Hertel, Joshua T. 2016. "Investigating the Implemented Mathematics Curriculum of New England Navigation Cyphering Books," *For the Learning of Mathematics*, **36** (3), 4–10.

Hess, Volker, and J. Andrew Mendelsohn. 2010. "Case and Series: Medical Knowledge and Paper Technology, 1600–1900," *History of Science*, **48**, 287–314.

Higginbotham, J. J. 1874. *Men Whom India Has Known: Biographies of Eminent Indian Characters*, 2nd ed. (Madras: Higginbotham and Co.).

Higgins, Hannah B. 2009. *The Grid Book* (Cambridge, Mass.: MIT Press).

Higgs, Edward. 2004. *The Information State in England: The Central Collection of Information on Citizens since 1500* (New York: Palgrave Macmillan).

Hill, Jonathan E. 1999. "From Provisional to Permanent: Books in Boards 1790–1830," *Library*, **21**, 247–273.

Hill, William H. 1881. *History of the Hospital and School in Glasgow Founded by George and Thomas Hutcheson of Lambhill, A.D. 1639–41* (Glasgow: Royal Incorporation of Hutchesons' Hospital).

Hills, Richard. 1996. "James Watt and his Copying Machine," in Peter Bower (ed.), *The Oxford Papers: Proceedings of the British Association of Paper Historians Fourth Annual Conference* (Oxford: British Association of Paper Historians), 81–88.

Hirschman, Albert O. 1977. *The Passions and the Interests: Political Arguments for Capitalism before Its Triumph* (Princeton: Princeton University Press).

Hochberg, Julian. 2007. *In the Mind's Eye: Julian Hochberg on the Perception of Pictures, Films, and the World* (Oxford: Oxford University Press).

Hoffmann, Christoff. 2013. "Processes on Paper: Writing Procedures as Non-material Research Devices," *Science in Context*, **26**, 279–303.

Holmes, Frederic Lawrence, Jürgen Renn, and Hans-Jörg Rheinberger (eds.). 2003. *Reworking the Bench: Research Notebooks in the History of Science* (Berlin: Springer).

Holmes, Nigel. 2000–2001. "Pictograms: A View from the Drawing Board; or, What I have Learned from Otto Neurath and Gerd Arntz (and Jazz)," *Information Design Journal*, **10**, 133–134.

Hoolihan, Christopher. 1985. "Thomas Young, M.D. (1726?–1783) and Obstetrical Education at Edinburgh," *Journal of the History of Medicine and Allied Sciences*, **40**, 327–345.

Horner, Winifred Bryan. 1993. *Nineteenth-Century Scottish Rhetoric: The American Connection* (Carbondale: Southern Illinois University Press).

Hotson, Howard. 2007. *Commonplace Learning: Ramism and Its German Ramifications, 1543-1630* (Oxford: Oxford University Press).

Hottois, Gilbert. 2006. "Le Technoscience, de l'Origine du mot a son usage actuel," in Jean-Yves Goffi (ed.), *Regards sur les Technosciences* (Paris: Vrin), 21–38.

Houston, R. A. 1985. *Scottish Literacy and the Scottish Identity: Illiteracy and Society in Scotland and Northern England, 1600–1800* (Cambridge: Cambridge University Press).

Houston, R. A. 1993. "Literacy, Education, and the Culture of Print in Enlightenment Edinburgh," *History* **78**, 373–392.

Houston, R. A. 2002. *Literacy in Early Modern Europe: Culture and Education, 1500–1800*, 2nd ed. (New York: Routledge).

Howell, William Huntting. 2015. *Against Self-Reliance: The Arts of Dependence in the Early United States* (Philadelphia: University of Pennsylvania Press).

l'Huillier, H. 2003. "Practical Geometry in the Middle Ages and the Renaissance," in I. Grattan-Guinness (ed.), *Companion Encyclopedia of the History and Philosophy of the Mathematical Sciences*, vol. 1 (Baltimore: Johns Hopkins University Press), 185–191.

Hunter, Matthew C. 2013. *Wicked Intelligence: Visual Art and the Science of Experiment in Restoration London* (Chicago: University of Chicago Press).

Hutchins, Edwin. 2002. *Cognition in the Wild* (Cambridge, Mass.: MIT Press).

Ingold, Tim. 2000. *The Perception of the Environment* (London: Routledge).

Ingold, Tim. 2007. *Lines: A Brief History* (London: Routledge).

Ingold, Tim. 2009. "Stories against Classification: Transport, Wayfaring, and the Integration of Knowledge," in Sandra Bamford and James Leach (eds.), *Kinship and Beyond: The Genealogical Model Reconsidered* (Oxford: Berghan), 193–213.

Ingold, Tim. 2011. *Being Alive: Essays on Movement, Knowledge, and Description* (London: Routledge).

Ingold, Tim. 2015. *The Life of Lines* (London: Routledge).

Ingold, Tim. 2017. *Anthropology and/as Education* (London: Routledge).

Ingold, Tim, and Jo Lee Vergunst (eds.). 2008. *Ways of Walking: Ethnography and Practice on Foot* (Aldershot: Ashgate).

Irvine, James R. 1978. "Rhetoric and Moral Philosophy: A Selected Inventory of Lecture Notes and Dictates in Scottish Archives, 1700–1900," *Rhetoric Society Quarterly*, **8**, 159–164.

Irvine, James R. 1981. "Notes from Beattie's Lectures on Public Speaking: Marischal College," *Rhetoric Society Quarterly* **11**, 68–73.

J. O. 1853. "Replies," *Notes and Queries*, **7**, 68.

Jackson, H. J. 2001. *Marginalia: Readers Writing in Books* (New Haven: Yale University Press).

Jackson, H. J. 2008. *Romantic Readers: The Evidence of Marginalia* (New Haven: Yale University Press).

Jackson-Williams, Kelsey. 2020. *The First Scottish Enlightenment: Rebels, Priests, and History* (Oxford: Oxford University Press).

Jacobs, Edward, and Antonia Forster. 1995. "'Lost Books' and Publishing History: Two Annotated Lists of Imprints for the Fiction Titles Listed in the Circulating Library Catalogs of Thomas Lowndes (1766) and M. Heavisides (1790), of Which No Known Copies Survive," *Papers of the Bibliographical Society of America*, **89**, 260–297.

Jansen, Laura (ed.). 2014. *The Roman Paratext: Frame, Texts, Readers* (Cambridge: Cambridge University Press).

Jasanoff, Maya. 2006. *Edge of Empire: Lives, Culture, and Conquest in the East, 1750–1850* (New York: Vintage Books).

Jeffares, Neil. *Dictionary of Pastellists before 1800*, online edition at http://www .pastellists.com. Accessed 12 April 2020.

Johns, Adrian. 1998. *The Nature of the Book: Print and Knowledge in the Making* (Chicago: University of Chicago Press).

Johnson, Jeff. 2010. *Designing with the Mind in Mind* (Burlington: Morgan Kaufmann Publishers).

Jones, Ann Rosalind, and Peter Stallybrass. 2000. *Renaissance Clothing and the Materials of Memory* (Cambridge: Cambridge University Press).

Jordan, T. R., V. A. McGowan, and K. B. Paterson. 2012. "Reading with a Filtered Fovea: The Influence of Visual Quality at the Point of Fixation during Reading," *Psychonomic Bulletin & Review*, **19**, 1078–1084.

Jordanova, Ludmilla. 1993. *Sexual Visions: Images of Gender in Science and Medicine between the Eighteenth and Twentieth Centuries* (Madison: University of Wisconsin Press).

Jordanova, Ludmilla. 2012. *The Look of the Past: Visual and Material Evidence in Historical Practice* (Cambridge: Cambridge University Press).

Kaiser, David. 2009. *Drawing Theories Apart: The Dispersion of Feynman Diagrams in Postwar Physics* (Chicago: University of Chicago Press).

Kant, Immanuel. 1991. "On the First Ground of the Distinction of Regions in Space," in James van Cleve and Robert E. Frederick (eds.), *The Philosophy of Right and Left: Incongruent Counterparts in the Nature of Space* (Dordrecht: Kluwer), 27–33.

Kashatus, William C. 1994. "The Inner Light and Popular Enlightenment: Philadelphia Quakers and Charity Schooling, 1790–1820," *Pennsylvania Magazine of History and Biography*, **118**, 87–116.

Kassell, Lauren. 2016. "Paper Technologies, Digital Technologies: Working with Early Modern Medical Records," in Angela Woods and Anne Whitehead (eds.), *The Edinburgh Companion to the Critical Medical Humanities* (Edinburgh: Edinburgh University Press), 120–135.

Kaufman, M. 1996. H. "Monro Secundus and 18th Century Lymphangiography," *Proceedings of the Royal College of Physicians Edinburgh*, **26**, 75–90.

Kaufman, M. 1999. H. "Observations on Some of the Plates Used to Illustrate the Lymphatics Section of Andrew Fyfe's Compendium of the Anatomy of the Human Body, Published in 1800," *Clinical Anatomy*, **12**, 12–27.

Kemp, Martin. 2000. *Visualizations: The Nature Book of Art and Science* (Oxford: Oxford University Press).

Kennedy, Craig J., Tom Addyman, K. Robin Murdoch, and Maureen E. Young. 2018. "18th- and 19th-Century Scottish Laboratory Glass," *Journal of Glass Studies*, **60**, 253–268.

Kim, Mi Gyung. 2003. *Affinity, that Elusive Dream: A Genealogy of the Chemical Revolution* (Cambridge, Mass.: MIT Press).

Klee, Paul. 1961. *Notebooks*, vol. 1: *The Thinking Eye*, H. Norden (trans.), J. Spiller (ed.) (London: Lund Humphries).

Klee, Paul. 1973. *Notebooks*, vol. 1: *The Nature of Nature*, H. Norden (trans.), J. Spiller (ed.) (London: Lund Humphries).

Klein, Ursula. 1994. *Verbindung und Affinität: Die Grundlegung der neuzeitlichen Chemie an der Wende vom 17. zum 18. Jahrhundert* (Basel: Birkhäuser Basel).

Klein, Ursula (ed.). 2001. *Tools and Modes of Representation in the Laboratory Sciences* (Kluwer: Dordrecht).

Klein, Ursula. 2003. *Experiments, Models, Paper Tools: Cultures of Organic Chemistry in the Nineteenth Century* (Stanford: Stanford University Press).

Klein, Ursula. 2005a. "Technoscience avant la Lettre," *Perspectives on Science*, **13**, 226–266.

Klein, Ursula. 2005b. "Technoscientific Productivity," *Perspectives on Science*, **13**, 139–141.

Klein, Ursula. 2016. *Nützliches Wissen: die Erfindung der Technikwissenschaften* (Goettingen: Wallstein).

Klein, Ursula. 2019. "Paper Tools," in Jürgen Renn and Matthias Schemmel (eds.), *Culture and Cognition: Essays in Honor of Peter Damerow* (Berlin: Max Planck Institute for the History of Science), 155–159.

Klein, Ursula. 2020. *Technoscience in History: Prussia, 1750–1850* (Cambridge, Mass.: MIT Press).

Klein, Ursula, and Wolfgang Lefèvre. 2007. *Materials in Eighteenth-Century Science: A Historical Ontology* (Cambridge, Mass.: MIT Press).

Knappett, Carl. 2011. "Networks of Objects, Meshworks of Things," in Tim Ingold (ed.), *Redrawing Anthropology: Materials, Movements, Lines* (Farnham: Ashgate), 45–64.

Knappett, Carl, and Lambros Malafouris (eds.). 1998. *Material Agency: Towards a Nonanthropocentric Approach* (New York: Springer).

Knoles, Thomas, Rick Kennedy, and Lucia Zaucha Knoles (eds.). 2003. *Student Notebooks at Colonial Harvard: Manuscripts and Educational Practice, 1650–1740* (Worcester: American Antiquarian Society).

Krajewski, Markus. 2011. *Paper Machines: About Cards and Catalogues, 1548–1929* (Cambridge, Mass.: MIT Press).

Krajewski, Markus. 2015. *Thinking in Boxes: A BIT of Paper Machines* (Cambridge, Mass.: MIT Press).

Krauthausen, Karin, and Omar W. Nasim (eds.). 2010. *Notieren, Skizzieren: Schreiben und Zeichnen als Verfahren des Entwurfs* (Berlin: Diaphanes).

Kreider, Kristen. 2010. "'Scrap,' 'Flap,' 'Strip,' 'Stain,' 'Cut': The Material Poetics of Emily Dickinson's Later Manuscript Pages," *Emily Dickinson Journal*, **19**, 67–103.

Kurimay, Anita. 2020. *Queer Budapest, 1873–1961* (Chicago: University of Chicago Press).

Kusukawa, Sachiko. 2009. "Image, Text, and 'Observatio': The 'Codex Kentmanus,'" *Early Science and Medicine*, **14**, 445–475.

Kusukawa, Sachiko. 2012. *Picturing the Book of Nature: Image, Text, and Argument in Sixteenth-Century Human Anatomy and Medical Botany* (Chicago: University of Chicago Press).

Kwakkel, Erik, and Rodney Thomson. 2018. "Codicology," in Erik Kwakkel and Rodney Thomson (eds), *The European Book in the Twelfth Century* (Cambridge: Cambridge University Press), 9–24.

Laidlaw, Christine. 2010. *British in the Levant: Trade and Perceptions of the Ottoman Empire in the Eighteenth Century* (London: Tauris).

Laing, David. "On the Supposed 'Missing School of Design in the University of Edinburgh' 1784," *Proceedings of the Society of Antiquaries of Scotland Edinburgh*, **8** (1871), 36–40.

Lake, Crystal B. 2020. *Artifacts: How We Think and Write about Found Objects* (Baltimore: Johns Hopkins University Press).

LaMarre, Thomas. 2000. *Uncovering Heian Japan: An Archaeology of Sensation and Inscription* (Durham: Duke University Press).

Latour, Bruno. 1987. *Science in Action: How to Follow Scientists and Engineers through Society* (Cambridge, Mass.: Harvard University Press).

Latour, Bruno, and Steve Woolgar. 1979. *Laboratory Life: The Construction of Scientific Facts* (Beverly Hills: SAGE Publications).

Lauwereyns, Jan. 2012. *Brain and the Gaze: On the Active Boundaries of Vision* (Cambridge, Mass.: MIT Press).

Laver, A. Bryan. "Gregor Feinaigle, Mnemonist and Educator," *Journal of the History of the Behavioral Sciences*, **15** (1979), 18–28.

Law, Alexander. 1965. *Education in Edinburgh in the Eighteenth Century* (London: University of London Press).

Law, Alexander. 1988. "Latin in Scottish Schools in the 18th and 19th Centuries," *Bibliotheck*, **15**, 63–75.

Lawrence, Christopher. 1988. "Alexander Monro 'Primus' and the Edinburgh Manner of Anatomy," *Bulletin of the History of Medicine*, **62**, 193–214.

Lefèvre, Wolfgang. 2019. "Drawing Instruments," in Jürgen Renn and Matthias Schemmel (eds.), *Culture and Cognition: Essays in Honor of Peter Damerow* (Berlin: Max Planck Institute for the History of Science), 161–165.

Lehman, Christine. 2010. "Innovation in Chemistry Courses in France in the Mid-Eighteenth Century: Experiments and Affinities," *Ambix*, **57**, 3–26.

Lehmann, William C. 1960. *John Millar of Glasgow, 1735–1801: His Life and Thought and His Contributions to Sociological Analysis* (Cambridge: Cambridge University Press).

Lehmann, William C. 1970. "Some Observations on the Law Lectures of Professor Millar at the University of Glasgow (1761–1801)," *Juridical Review* **6**, 56–77.

Leong, Elaine. 2018a. *Recipes and Everyday Knowledge: Medicine, Science, and the Household in Early Modern England* (Chicago: University of Chicago Press).

Leong, Elaine. 2018b. "Read. Do. Observe. Take Note," *Centaurus* **60**.1–2, 87–103.

Lerer, Seth. 2012. "Devotion and Defacement: Reading Children's Marginalia," *Representations*, **118**, 126–153.

Lerner, Ralph. 2000. *Maimonides' Empire of Light: Popular Enlightenment in an Age of Belief* (Chicago: University of Chicago Press).

Leroi-Gourhan, André. 1993. *Gesture and Speech* (Cambridge, Mass.: MIT Press).

Levy-Eichel, Mordechai. 2015. "'Into the Mathematical Ocean': Navigation, Education, and the Expansion of Numeracy in Early Modern England and the Atlantic World" (unpublished PhD dissertation, Yale University).

Lidwell, William, Kritina Holden, and Jill Butler. 2003. *Universal Principles of Design* (Gloucester: Rockport Publishers).

Lilley, Keith D. 1998. "Taking Measures across the Medieval Landscape: Aspects of Urban Design before the Renaissance," *Urban Morphology*, **2**, 82–92.

Livingston, Eric. 2008. *Ethnographies of Reason* (London: Routledge).

Livingstone, David N., and Charles W. J. Withers. 1999. *Geography and Enlightenment* (Chicago: University of Chicago Press).

Livsey, Peter. n.d. *Napoleonic Encounters: The Waldies of Forth House, Newcastle* (Newcastle: Tynebridge Publishing and Newcastle Libraries).

Lochhead, Marion. 1948. *The Scots Household in the Eighteenth Century: A Century of Domestic and Social Life* (Edinburgh: Moray Press).

Long, Bridget. 2016. "'Regular Progressive Work Occupies My Mind Best': Needlework as a Source of Entertainment, Consolation, and Reflection," *Textile*, **14**, 176–187.

Lorimer, Hayden, and Katrin Lund. 2003. "Performing Facts: Finding a Way over Scotland's Mountains," *Sociological Review*, **51**, 130–144.

Lothian, John M. 1963. "Introduction," in Adam Smith, *Lectures on Rhetoric and Belles Lettres [1762–1763]*, John M. Lothian (ed.) (Edinburgh: Nelson).

Love, Harold. 1998. *The Culture and Commerce of Texts: Scribal Publication in Seventeenth-Century England* (Amherst: University of Massachusetts Press).

Lurie, David Barnett. 2011. *Realms of Literacy: Early Japan and the History of Writing* (Cambridge, Mass: Harvard University Asia Center).

MacGregor, Neil. 2011. *A History of the World in 100 Objects* (London: Penguin).

Macintyre, I. M. C. 2012. "A Sceptic and an Empiric in Medicine: George Young (1692–1757) and the Beginnings of the Scottish Medical Enlightenment," *Journal of the Royal College of Physicians of Edinburgh* **42**, 352–360.

Macintyre, I., and A. Munro. 2015. "The Monros: Three Medical Dynasties with a Common Origin," *Journal of the Royal College of Physicians of Edinburgh*, **45**, 67–75.

Macve, R., and K. Hoskin. 1994. "Writing, Examining, Disciplining: The Genesis of Accounting's Modern Power," in T. Hopwood and P. Miller (eds.), *Accounting as Social and Institutional Practice* (Cambridge: Cambridge University Press), 67–97.

Madrigal, Alixis C. 2020. "The Way We Write History Has Changed," *Atlantic*, 21 January.

Malafouris, Lambros. 2013. *How Things Shape the Mind: A Theory of Material Engagement* (Cambridge, Mass.: MIT Press).

Mangen, A., L. G. Anda, G. H. Oxborough, and K. Brønnick. 2015. "Handwriting versus Keyboard Writing: Effect on Word Recall," *Journal of Writing Research* **7**, 227–247.

Marglin, Stephen A. 1996. "Farmers, Seedsmen, and Scientists: Systems of Agriculture and Systems of Knowledge," in Frédérique Apffel-Marglin and Stephen A. Marglin (eds.), *Decolonizing Knowledge: From Development to Dialogue* (Oxford: Clarendon Press), 185–248.

Marschner, Joanna, David Bindman, and Lisa L. Ford (eds.). 2017. *Enlightened Princesses: Caroline, Augusta, Charlotte, and the Shaping of the Modern World* (New Haven: Yale University Press).

Martin, Alison E. 2011. "Society, Creativity, and Science: Mrs. Delany and the Art of Botany," *Eighteenth-Century Life*, **35** 102–107.

Martin, Henri-Jean. 1995. *The History and Power of Writing* (Chicago: University of Chicago Press).

Martin, Henri-Jean, and Bruno Delmas. 1988. *Histoire et pouvoirs de l'ecrit* (Paris: Perrin).

Masiyakurima, Patrick. 2020. *Copyright Protection of Unpublished Works in the Common Law World* (Oxford: Bloomsbury).

Mason, John. 1954. "Scottish Charity Schools in the Eighteenth Century," *Scottish Historical Review*, **33**, 1–13.

Matheson, Ann. 1979. "Theories of Rhetoric in the 18th-Century Scottish Sermon" (unpublished PhD thesis, Edinburgh University).

Mathew, W. M. 1966. "The Origins and Occupations of Glasgow Students, 1740–1839," *Past and Present* **33**, 72–94.

Matthews, William (ed.). 1984. *British Diaries: An Annotated Bibliography of British Diaries Written between 1442 and 1942* (Berkeley: University of California Press).

May, Steven W., and Heather Wolfe. 2010. "Manuscripts in Tudor England," in Ken Cartwright (ed.), *A Companion to Tudor Literature* (Chichester: Wiley), 125–139.

Mayhew, Robert J. 1998. "Geography in Eighteenth-Century British Education," *Paedagogica Historica*, **34**, 731–769.

Mayhew, Robert J. 2007. "Materialist Hermeneutics, Textuality, and the History of Geography: Print Spaces in British Geography, *c*. 1500–1900," *Journal of Historical Geography*, **33**, 466–488.

McCain, Charles Rodgers. 1949. "Preaching in Eighteenth Century Scotland: A Comparative Study of the Extant Sermons of Ralph Erskin (1685–1752), John Erskine (1721–1803), and Hugh Blair (1718–1800)" (unpublished thesis, Edinburgh University).

McCallum, Sandra. 2014. "A Case Study of the Moore Family, with Particular Reference to the Extended Tour Abroad of (Sir) John Moore with His Father Dr Moore (1772–1776)" (unpublished PhD thesis, University of Glasgow).

McCosh, James. 1875. *The Scottish Philosophy: Biographical, Expository, Critical, from Hutcheson to Hamilton* (London: Macmillan).

McKenzie, Donald F. 1985. *Bibliography and the Sociology of Texts* (London: British Library).

McKenzie, Donald F. 2002. *Making Meaning: "Printers of the Mind" and Other Essays*, Peter McDonald and Michael F. Saurez (eds.) (Amherst: University of Massachusetts Press).

McKinstry, Sam, and Marie Fletcher. 2002. "The Personal Account Books of Sir Walter Scott," *Accounting Historians Journal*, **29**, 59–89.

McKitterick, David. 2003. *Print, Manuscript, and the Search for Order, 1450–1830* (Cambridge: Cambridge University Press).

McLean, Ralph R. 2009. "Rhetoric and Literary Criticism in the Early Scottish Enlightenment" (unpublished PhD thesis, University of Glasgow).

McOmish, David. 2018. "The Scientific Revolution in Scotland Revisited: The New Sciences in Edinburgh," *History of Universities*, **31**, 153–172.

Meadows, Mark Stephen. 2003. *Pause & Effect: The Art of Interactive Narrative* (Indianapolis: New Riders).

Meek, Ronald L. 1977. "New Light on Adam Smith's Glasgow Lectures on Jurisprudence," in *Smith, Marx, & After: Ten Essays in the Development of Economic Thought* (Boston: Springer), 57–91.

Michael, Ian. 1987. *The Teaching of English: From the Sixteenth Century to 1870* (Cambridge: Cambridge University Press).

Middleton, Bernard C. 1963. *A History of English Craft Bookbinding Technique* (London: Holland Press).

Mijers, Esther. 2012. *"News from the Republick of Letters": Scottish Students, Charles Mackie, and the United Provinces, 1650–1750* (Leiden: Brill).

Miller, Nicholas B. 2017. *John Millar and the Scottish Enlightenment: Family Life and World History* (Oxford: Voltaire Foundation).

Mitchell, Samuel Latham. 1976. "Samuel L. Mitchell's Evaluation of the Lectures of Joseph Black," Herbert T. Pratt (ed.), *Journal of Chemical Education*, **53**, 745–746.

Mitchell, W. J. T. 1978. *Blake's Composite Art: A Study of Illuminated Poetry* (Princeton: Princeton University Press).

Mitra, Durba. 2020. *Indian Sex Life: Sexuality and the Colonial Origins of Modern Social Thought* (Princeton: Princeton University Press).

Moilliet, J. L. 1964. "Keir's 'Dialogues on Chemistry': An Unpublished Masterpiece," *Chemistry and Industry*, 2081–2083.

Monaco, James. 2009. *How to Read a Film* (Oxford: Oxford University Press).

Monaghan, E. Jennifer. 2005. *Learning to Read and Write in Colonial America* (Amherst: University of Massachusetts Press).

Monroe, Walter Scott. 1917. *Development of Arithmetic as a School Subject* (Washington, D.C.: Government Printing Office).

Moore, D. T., and M. A. Beasley. 1997. "The Botanical Manuscripts of Robert Brown," *Archives of Natural History*, **24**, 237–280.

Moore, Lindy. 2013. "The Value of Feminine Culture: Community Involvement in the Provision of Schooling for Girls in Eighteenth-Century Scotland," in Katie Barclay and Deborah Simonton (eds.), *Women in Eighteenth-Century Scotland* (London: Routledge), 97–114.

Moore, Lindy. 2015. "Urban Schooling in Seventeenth- and Eighteenth-Century Scotland," in R. Anderson, M. Freeman, and L. Freeman (eds.), *The Edinburgh History of Education in Scotland* (Edinburgh: Edinburgh History of Education in Scotland), 79–96.

Moore, Milcah Martha. 1997. *Milcah Martha Moore's Book: A Commonplace Book from Revolutionary America*, Catherine La Courreye Blecki and Karin A. Wulf (eds.) (University Park: Pennsylvania State University Press).

Morgan, Valerie. 1971. "Agricultural Wage Rates in Late Eighteenth-Century Scotland," *Economic History Review*, **24**, 181–201.

Morris, John. 1987. *Wheels and Herringbones: Some Scottish Bindings 1678–1773* (Dunblane: Bookbinder).

Mortera, Emanuele Levi. 2005. "Reid, Stewart, and the Association of Ideas," *Journal of Scottish Philosophy*, **3**, 157–170.

Morton, Charles. 1940. *Charles Morton's Compendium Logicae Secundum Principia*, Theodore Hornberger (ed.) (Boston: Colonial Society of Massachusetts).

Moss, Ann. 1996. *Printed Commonplace-Books and the Structuring of Renaissance Thought* (Oxford: Oxford University Press).

Mossner, Ernest Campbell. 1948a. "Beattie's 'The Castle of Skepticism': An Unpublished Allegory against Hume, Voltaire, and Hobbes," *Studies in English*, **27**, 108–145.

Mossner, Ernest Campbell. 1948b. "Hume's Early Memoranda, 1729–1740: The Complete Text," *Journal of the History of Ideas*, **9**, 492–518.

Mossner, Ernest Campbell. 1965. "Adam Smith: Lectures on Rhetoric and Belles Lettres," *Studies in Scottish Literature* **2**, 199–208.

Mueller, Pam A., and Daniel M. Oppenheimer. 2014. "The Pen Is Mightier Than the Keyboard: Advantages of Longhand over Laptop Note Taking," *Psychological Science*, **25**, 1159–1168.

Mulcahy, Ursula. 2019. "William Cullen, Consecrated Heretic: A Study in Eighteenth-Century Science and Patronage" (unpublished PhD thesis, Durham University).

Mulhern, James. 1938. "Manuscript School-Books," *Papers of the Bibliographical Society of America*, **32**, 17–37.

Müller-Wille, Staffan. 2017. "Names and Numbers: 'Data' in Classical Natural History, 1758–1859," *Osiris*, **32**, 109–128.

Müller-Wille, Staffan. 2018. "Linnaean Paper Tools," in H. A. Curry, N. Jardine, J. Secord, and E. C. Spary (eds.), *New Cultures of Natural History* (Cambridge: Cambridge University Press), 205–220.

Müller-Wille, Staffan, and Isabelle Charmantier. 2012a. "Natural History and Information

Overload: The Case of Linnaeus," *Studies in the History and Philosophy of the Biological and Biomedical Sciences*, **43**, 4–15.

Müller-Wille, Staffan, and Isabelle Charmantier. 2012b. "Lists as Research Technologies," *Isis*, **103**, 743–752.

Murray, David. 1913. *Robert and Andrew Foulis and the Glasgow Press with Some Account of the Glasgow Academy of Fine Arts* (Glasgow: Maclehose).

Myers, Robert Pooler. 1903. *Biographical Sketch of Levi Myers by his Grandson, Robert Pooler Myers, M. D.* (Princeton: Princeton University).

Nasim, Omar W. 2013. *Observing by Hand: Sketching the Nebulae in the Nineteenth Century* (Chicago: University of Chicago Press).

Neeley, Kathryn A. 2001. *Mary Somerville: Science, Illumination, and the Female Mind* (Cambridge: Cambridge University Press).

Nelles, Paul. 2007. "*Libros de papel, libri bianchi, libri papyracei*: Note-Taking Techniques and the Role of Student Notebooks in the Early Jesuit Colleges," *Archivum Historicum Societatis Iesu*, **76**, 75–112.

Nelles, Paul. 2010. "Seeing and Writing: The Art of Observation in the Early Jesuit Missions," *Intellectual History Review*, **20**, 317–333.

Nenadic, Stana. 1997. "Print Collecting and Popular Culture in Eighteenth-Century Scotland," *History*, **82**, 203–222.

Nenadic, Stana. 1998. "The Enlightenment in Scotland and the Popular Passion for Portraits," *Journal for Eighteenth-Century Studies*, **21**, 175–192.

New York Hospital. 1845. *A Catalogue of the Books Belonging to the Library of the New York Hospital* (New York: Craighead).

Nickelsen, Kärin. 2006. *Draughtsmen, Botanists, and Nature: The Construction of Eighteenth-Century Botanical Illustrations* (Dordrecht: Springer).

Noltie, Henry J. (ed.). 2011a. *John Hope (1725–1786): Alan G. Morton's Memoir of a Scottish Botanist*, rev. ed.(Edinburgh: Royal Botanic Garden Edinburgh).

Noltie, Henry J. 2011b. *John Hope, Enlightened Botanist: An Exhibition Guide* (Edinburgh: Royal Botanic Garden Edinburgh).

Noltie, Henry J. 2017. *Botanical Art from India: The Royal Botanic Garden Edinburgh Collection* (Edinburgh: Royal Botanic Garden Edinburgh).

Noltie, Henry J. 2019. "A Scottish Daughter of Flora: Lady Charlotte Murray and Her Herbarium Portabile," *Archives of Natural History*, **46**, 298–317.

Norman, Donald A. 2013. *The Design of Everyday Things*, rev. ed. (New York: Basic Books).

Norton, Arthur O. 1935. "Harvard Text-Books and Reference Books of the Seventeenth Century," *Publications of the Colonial Society of Massachusetts*, **28**, 361–438.

Noton, David, and Lawrence Stark. 1971. "Scanpaths in Saccadic Eye Movements While Viewing and Recognizing Patterns," *Vision Research*, **11**, 929–942.

Ogilvie, Brian W. 2003. "The Many Books of Nature: Renaissance Naturalists and Information Overload," *Journal of the History of Ideas*, **64**, 29–40.

Ogilvie, Skene. 1891. "The Consolations on the Death of Friends Arising from the Christians Hope of Immortality. A Sermon Occasioned by the Death of Robert Eden Scott, M.A., Professor of Logic and Moral Philosophy, in the University of King's College, Old Aberdeen," republished in Horace Edwin Hayden (ed.), *Virginia Genealogies* (Wilkes-Barre: Yordy), 640–642.

Oldroyd, David R. 1972. "Two Little Known Copies of Black's Lecture Notes," *Annals of Science*, **29**, 35–37.

O'Malley, Andrew. 2003. *The Making of the Modern Child: Children's Literature and Childhood in the Late Eighteenth Century* (London: Routledge).

Ong, Walter J. 1985. "Writing is a Technology That Structures Thought," in Gerd Baumann (ed.), *Written Word: Literacy in Translation* (Oxford: Clarendon Press), 23–50.

Ong, Walter J. 2004. *Ramus, Method, and the Decay of Dialogue: From the Art of Discourse to the Art of Reason* (Chicago: University of Chicago Press).

Otis, Jessica Marie. 2017. "'Set Them to the Cyphering Schoole': Reading, Writing, and Arithmetical Education, circa 1540–1700," *Journal of British Studies* **56**, 453–482.

Ovenden, Richard. 2020. *The Burning of Books: A History of Knowledge Under Attack* (London: Hachette).

Packham, Catherine. 2013. "Cicero's Ears, or Eloquence in the Age of Politeness: Oratory, Moderation, and the Sublime in Enlightenment Scotland," *Eighteenth-Century Studies*, **46**, 499–512.

Pallasmaa, Juhani. 2008. *The Eyes of the Skin: Architecture and the Senses* (Chichester: John Wiley and Sons).

Parkes, Joseph. 1828. *A History of the Court of Chancery* (London: Longman, Rees Orme, Brown and Green).

Pasanek, Brad. 2015. *Metaphors of Mind: An Eighteenth-Century Dictionary* (Baltimore: Johns Hopkins University Press).

Peer, Lindsay, and Gavin Reid (eds.). 2001. *Dyslexia: Successful Inclusion in the Secondary School* (London: David Fulton).

Perkins, David N. 1994. *The Intelligent Eye: Learning to Think by Looking at Art* (Santa Monica: The Getty Center for Education in the Arts).

Perl, Teri. 1979. "The Ladies' Diary or Woman's Almanack, 1704–1841," *Historia Mathematica*, **6**, 36–53.

Pickering, Samuel F. 1981. *John Locke and Children's Books in Eighteenth-Century England* (Knoxville: The University of Tennessee Press).

Pinna, Baingio and Gavin J. Brelstaff. 2000. "A New Visual Illusion of Relative Motion," *Vision Research*, **40**, 2091–2096.

Plant, Marjorie. 1952. *The Domestic Life of Scotland in the Eighteenth Century* (Edinburgh: Edinburgh University Press).

Poovey, Mary. 1998. *A History of the Modern Fact: Problems of Knowledge in the Sciences of Wealth and Society* (Chicago: University of Chicago Press).

Porter, Theodore M. 1986. *The Rise of Statistical Thinking, 1820–1900* (Princeton: Princeton University Press).

Porter, Theodore M. 1996. *Trust in Numbers: The Pursuit of Objectivity in Science and Public Life* (Princeton: Princeton University Press).

Porter, Theodore M. 2010. *Karl Pearson: The Scientific Life in a Statistical Age* (Princeton: Princeton University Press).

Potts, Alex. 1994. *Flesh and the Ideal: Winckelmann and the Origins of Art History* (New Haven: Yale University Press).

Powers, John. 2012. *Inventing Chemistry: Herman Boerhaave and the Reform of the Chemical Arts* (Chicago: University of Chicago Press).

Preston, John. 2010. "The Extended Mind, the Concept of Belief, and Epistemic Credit," in Richard Menary (ed.), *The Extended Mind* (Cambridge, Mass.: MIT Press), 355–368.

Price, Fiona. 2002. "Democratizing Taste: Scottish Common Sense Philosophy and Elizabeth Hamilton," *Romanticism* **8**, 179–196.

Prideaux, Sarah T. 1903. *Bookbinders and Their Craft* (New York: Scribner's Sons).

Priest, Greg. 2018. "Diagramming Evolution: The Case of Darwin's Trees," *Endeavour*, **42**, 157–171.

Prior, Katherine. 2004. "Hamilton [formerly Buchanan], Francis, of Buchanan (1762–1829)," *ODNB*.

Pryce-Jones, Janet E. (compiler), and R. H. Parker (annotator). 2014. *Accounting in Scotland: A Historical Biography* (London: Routledge).

Quaritch, Bernard. 1889. *A Collection of Facsimiles from Examples of Historic or Artistic Book-Binding* (London: Quaritch).

Raphael, Renée. 2017. *Reading Galileo: Scribal Technologies and the Two New Sciences* (Baltimore: Johns Hopkins University Press).

Raskin, Jef. 2000. *The Human Interface: New Directions for Designing Interactive Systems* (New York: Addison-Wesley).

Raven, James. 2014. *Publishing Business in Eighteenth-Century England* (Suffolk: Boydell and Bewer).

Reason, Matthew. 2006. *Documentation, Disappearance, and the Representation of Live Performance* (Basingstoke: Palgrave Macmillan).

Reid, John S. 1982. "The Castlehill Observatory, Aberdeen," *Journal for the History of Astronomy*, **13**, 84–96.

Reid, John S. 1983–1984. "Patrick Copland 1748–1822: Aspects of His Life and Times at Marischal College," *Aberdeen University Review*, **172**, 359–379.

Reid, John S. 1985. "Patrick Copland, 1748–1822: Connections Outside the College Courtyard," *Aberdeen University Review*, **51**, 226–250.

Reid, John S. 1990. "The Remarkable Professor Copland," *Bulletin of the Scientific Instrument Society*, **24**, 2–5.

Reid, Steven J. 2011. *Humanism and Calvinism: Andrew Melville and the Universities of Scotland, 1560–1625* (Farnham: Ashgate).

Reid, Thomas. 2004. *On Logic, Rhetoric and the Fine Arts*, Alexander Broadie (ed.) (Edinburgh: Edinburgh University Press).

Rendall, Jane. 1978. "Introduction: The Origins of the Scottish Enlightenment," in Jane Rendall (ed.), *Origins of the Scottish Enlightenment, 1707–76* (London: Macmillan). 1–27.

Rendall, Jane. 2005. "'Women that would plague me with rational conversation': Aspiring Women and Scottish Whigs, c. 1790–1830," in Sarah Knott and Barbara Taylor (eds.), *Women, Gender, and Enlightenment* (Basingstoke: Palgrave Macmillan), 326–347.

Rendall, Jane. 2012. "Adaptations: History, Gender, and Political Economy in the Work of Dugald Stewart," *History of European Ideas*, **38**, 143–161.

Renn, Jürgen. 2020. *The Evolution of Knowledge: Rethinking Science for the Anthropocene* (Princeton: Princeton University Press).

Renn, Jürgen, and Matthias Schemmel (eds.). 2019. *Culture and Cognition: Essays in Honor of Peter Damerow* (Berlin: Max Planck Research Library for the History and Development of Knowledge).

Rheinberger, Hans-Jörg. 1997. *Toward a History of Epistemic Things: Synthesizing Proteins in the Test Tube* (Stanford: Stanford University Press).

Rheinberger, Hans-Jörg. 2001. "History of Science and the Practices of Experiment," *History and Philosophy of the Life Sciences*, **23**, 51–63.

Rheinberger, Hans-Jörg. 2003. "Scrips and Scribbles," *MLN*, **118**, 622–636.

Rich, Anthony. 1860. *A Dictionary of Roman and Greek and Antiquities with Nearly 2000 Engravings on Wood* (London: Longman).

Richardson, Alan. 1994. *Literature, Education, and Romanticism: Reading as Social Practice, 1780–1832* (Cambridge: Cambridge University Press).

Richeson, A. W. 1939. "Unpublished Mathematical Manuscripts in American Libraries," *National Mathematical Magazine*, **13**, 183–188.

Richter, Antje. 2013. *Letters and Epistolary Culture in Early Medieval China* (Seattle: University of Washington Press).

Risse, Guenter B. 1986. *Hospital Life in Enlightenment Scotland: Care and Teaching at the Royal Infirmary of Edinburgh* (Cambridge: Cambridge University Press).

Roberts, Les. 2012. *Mapping Cultures: Place, Practice, Performance* (Basingstoke: Palgrave Macmillan).

Roberts, Lissa. 2005. "The Death of the Sensual Chemist: The New Chemistry and the Transformation of Sensuous Technology," in David Howes (ed.), *Empire of the Senses: The Sensual Cultural Reader* (Oxford: Berg), 106–127.

Robinson, Mairi (ed.). 1985. *The Concise Scots Dictionary* (Aberdeen: Aberdeen University Press).

Rock, Joe. 2000. "An Important Scottish Anatomical Publication Rediscovered," *Book Collector*, **49**, 27–60.

Rodger, Richard. 2017. "Capital Mapping: Geographies of Enlightenment Edinburgh" (unpublished PhD thesis, Edinburgh University).

Roos, Anna Marie. 2018. *Martin Lister and His Remarkable Daughters: The Art of Science in the Seventeenth Century* (Oxford: Bodleian Library).

Rosand, David. 2002. *Drawing Acts: Studies in Graphic Expression and Representation* (Cambridge: Cambridge University Press).

Rosenberg, Daniel. 2003. "Early Modern Information Overload," *Journal of the History of Ideas*, **64**, 1–9.

Rosenberg, Daniel, and Anthony Grafton. 2013. *Cartographies of Time: A History of the Timeline* (Princeton: Princeton Architectural Press).

Rosner, Lisa. 1991. *Medical Education in the Age of Improvement: Edinburgh Students and Apprentices 1760–1826* (Edinburgh: Edinburgh University Press).

Ross, Richard J. 1988. "The Memorial Culture of Early Modern English Lawyers: Memory as a Keyword, Shelter, and Identity," *Yale Journal of Law & the Humanities*, **10**, 229–326.

Rossi, Paolo. 1968. *Francis Bacon: From Magic to Science*, S. Rabinovitch (trans.) (Chicago: University of Chicago Press).

Rossi, Paolo. 2000. *Logic and the Art of Memory: The Quest for a Universal Language*, Stephen Clucas (trans.) (London: Athlone Press).

Rossignol, Marie-Jeanne. 2015. "Benjamin Franklin Bache's Childhood Diary: The 'Shaping' of a 'Self,'" *XVII–XVIII: Revue de la Société d'études anglo-américaines des XVIIe et XVIIIe siècles*, **72**, 197–212.

Rothschild, Emma. 2011. *The Inner Life of Empires: An Eighteenth-Century History* (Princeton: Princeton University Press).

Roy, William. 2007. *The Great Map: The Military Survey of Scotland 1747–1755* (Edinburgh: Birlinn).

Rubenstein, Richard. 1988. *Digital Typography: An Introduction to Type and Composition for Computer System Design* (New York: Addison-Wesley).

Rush, Benjamin. 1948. *The Autobiography of Benjamin Rush: His "Travels Through Life" Together with His Commonplace Book*, George W. Corner (ed.) (Princeton: Princeton University Press).

Sackman, Hal. 1968. "A Public Philosophy for Real Time Information Systems," *Pro-*

ceedings of the December 9–11, 1968, Fall Joint Computer Conference, Part II*, 1491–1498.

Saenger, Paul. 1997. *Space between Words: The Origins of Silent Reading* (Stanford: Stanford University Press).

Sanderson, Elizabeth C. 1996. *Women and Work in Eighteenth-Century Edinburgh* (London: Macmillan).

Sandford, Daniel Keyte, Thomas Thomson, and Allan Cunningham. 1836. *The Popular Encyclopedia: Being a General Dictionary of Arts, Sciences, Literature, Biography, History, and Political Economy* (Glasgow: Blackie & Son).

Sassoon, Joanna. 1998. "Photographic Meaning in the Age of Digital Reproduction," *Library Automated Systems Information Exchange*, **29**, 5–15.

Schaefer, Bradley E. 2001. "The Transit of Venus and the Notorious Black Drop Effect," *Journal for the History of Astronomy*, **32**, 325–336.

Schaeffer, Denise. 2014. *Rousseau on Education, Freedom, and Judgment* (University Park: Penn State Press).

Scharf, Sarah T. 2007. "Identification Keys and the Natural Method: The Development of Text-Based Information Management Tools in Botany in the Long 18th Century" (unpublished PhD thesis, University of Toronto).

Schimmelpenninck, Mary Anne. 1858. *Life of Mary Anne Schimmelpenninck*, vol. 1, *Autobiography*, Christiana C. Hankin (ed.) (London: Longman).

Schmidt-Biggemann, Wilhelm. 1983. *Topica universalis: Eine Modellgeschichte humanistischer und barocker Wissenschaft* (Hamburg: Meiner).

Schotte, Margaret E. 2013. "Regimented Lessons: The Evolution of the Nautical Logbook in France," *Annuaire de Droit Maritime et Océanique*, 91–115.

Schotte, Margaret E. 2019. *Sailing School: Navigating Science and Skill, 1550–1800* (Baltimore: Johns Hopkins University Press).

Schrantz, Kristen M. 2014. "The Tipton Chemical Works of Mr. James Keir: Networks of Conversants, Chemicals, Canals, and Coal Mines," *International Journal for the History of Engineering and Technology*, **84**, 248–273.

Schürer, Norbert. 2007. "Four Catalogues of the Lowndes Circulating Library, 1755–66," *Papers of the Bibliographical Society of America*, **101**, 329–357.

Scott, William Robert. 1900. *Francis Hutcheson: His Life, Teaching, and Position in the History of Philosophy* (Cambridge: Cambridge University Press).

Sebastiani, Silvia. 2013. *The Scottish Enlightenment: Race, Gender, and the Limits of Progress* (New York: Palgrave Macmillan).

Secord, James A. 1985. "Newton in the Nursery: Tom Telescope and the Philosophy of Tops and Balls, 1761–1838," *History of Science*, **23**, 127–151.

Secord, James A. 2003. *Victorian Sensation: The Extraordinary Publication, Reception, and Secret Authorship of "Vestiges of the Natural History of Creation"* (Chicago: University of Chicago Press).

Secord, James A. 2004. "Knowledge in Transit," *Isis*, **95**, 654–672.

Senchyne, Jonathan. 2019. *The Intimacy of Paper in Early and Nineteenth-Century American Literature* (Amherst: University of Massachusetts Press).

Sennett, Richard. 2008. *The Craftsman* (London: Allen Lane).

Serjeantson, Richard. 2013. "The Philosophy of Francis Bacon in Early Jacobean Oxford, with an Edition of an Unknown Manuscript of the *Valerius Terminus*," *Historical Journal*, **56**, 1087–1106.

Sexton, Alexander Humboldt. 1894. *The First Technical College: A Sketch of the History of "the Andersonian," and the Institutions Descended from It, 1796–1894* (London: Chapman and Hall).

Shapin, Steven, and Simon Schaffer. 1989. *Leviathan and the Air-pump: Hobbes, Boyle, and the Experimental Life* (Princeton: Princeton University Press).

Sheehan, Jonathon, and Dror Wahrman. 2015. *Invisible Hands: Self-Organization and the Eighteenth Century* (Chicago: University of Chicago Press).

Sher, Richard B. 2008. *The Enlightenment and the Book: Scottish Authors and Their Publishers in Eighteenth-Century Britain, Ireland, and America* (Chicago: University of Chicago Press).

Sher, Richard B. 2015. *Church and University in the Scottish Enlightenment: The Moderate Literati of Edinburgh* (Edinburgh: Edinburgh University Press).

Sherman, William H. 2009. *Used Books: Marking Readers in Renaissance England* (Philadelphia: University of Pennsylvania Press).

Silver, Sean. 2015. *The Mind Is a Collection: Case Studies in Eighteenth-Century Thought* (Philadelphia: University of Pennsylvania Press).

Simpson, Ian J. 1942. "Education in Aberdeenshire before 1872" (unpublished PhD thesis, Aberdeen University).

Simpson, Matthew. 2000. "'O man do not scribble on the book': Print and Counter-print in a Scottish Enlightenment University," *Oral Tradition*, **15**, 74–95.

Sinclair, William. 1995. "A Medical Student at Leiden and Paris: William Sinclair (1736–38)," in Kees van Strien (ed.), *Proceedings of the Royal College of Physicians of Edinburgh*, **25**, 487–494.

Slagle, Judith Bailey. 2002. *Joanna Baillie: A Literary Life* (London: Associated University Press).

Slauter, Will. 2012. "The Paragraph as Information Technology: How News Travelled in the Eighteenth-Century Atlantic World," *Annales. Histoire, Sciences Sociales*, **67**, 253–278.

Smeenk, Chris, and Eric Schliesser. 2013. "Newton's Principia," in Jed Z. Buchwald and

Robert Fox (eds.), *The Oxford Handbook of the History of Physics* (Oxford: Oxford University Press), 109–165.

Smets, Alexis. 2013. "Le concept de matière dans l'imagerie des chymistes aux XVIe et XVIIe siècles" (unpublished PhD dissertation, Radboud University Nijmegen).

Smith, Craig. 2018. *Adam Ferguson and the Idea of Civil Society: Moral Science in the Scottish Enlightenment* (Edinburgh: Edinburgh University Press).

Smith, Helen, and Louise Wilson (eds.). 2011. *Renaissance Paratexts* (Cambridge: Cambridge University Press).

Smith, J. V. 1983. "Manners, Morals and Mentalities: Reflections on the Popular Enlightenment of Early Nineteenth-Century Scotland," in H. M. Humes and H. M. Paterson (eds.), *Scottish Culture and Scottish Education, 1800–1980* (Edinburgh: John Donald), 25–54.

Smith-Gratto, Karen, and Mercedes M. Fisher. 1999. "Gestalt Theory: A Foundation for Instructional Screen design" *Journal of Educational Technology Systems*, **27**, 361–371.

Soll, Jacob. 2009. *The Information Master: Jean-Baptiste Colbert's Secret State Intelligence System* (Ann Arbor: University of Michigan Press).

Soll, Jacob. 2010. "From Note-Taking to Data Banks: Personal and Institutional Information Management in Early Modern Europe," *Intellectual History Review*, **20**, 355–375.

Soll, Jacob. 2014. *The Reckoning: Financial Accountability and the Rise and Fall of Nations* (New York: Basic Books).

Sommerlad, M. J. 1967. *Scottish "Wheel" and "Herring-Bone" Bindings in the Bodleian Library* (Oxford: Bodleian Library).

Sousanis, Nick. 2015. *Unflattening* (Cambridge, Mass.: Harvard University Press).

Spary, Emma C. 2004. "Scientific Symmetries," *History of Science*, **42**, 1–46.

Spence, Jonathan D. 1985. *The Memory Palace of Matteo Ricci* (New York: Penguin Books).

Spurr, Stephen H. 1951. "George Washington, Surveyor and Ecological Observer," *Ecology*, **32**, 544–549.

St. Clair, William. 2004. *The Reading Nation in the Romantic Period* (Cambridge: Cambridge University Press).

Stabile, Susan M. 2004. *Memory's Daughters: The Material Culture of Remembrance in Eighteenth-Century America* (Ithaca: Cornell University Press).

Stafford, Barbara Maria. 1996. *Artful Science: Enlightenment Entertainment and the Eclipse of Visual Education* (Cambridge, Mass.: MIT Press).

Stallybrass, Peter, Roger Chartier, J. Franklin Mowery, and Heather Wolfe. 2004. "Hamlet's Tables and the Technologies of Writing in Renaissance England," *Shakespeare Quarterly*, **55**, 379–419.

Starkey, Janet. 2018. *The Scottish Enlightenment Abroad: The Russells of Braidshaw in Aleppo and on the Coast of Coromandel* (Leiden: Brill).

Steinke, Hubert. 2005. *Irritating Experiments: Haller's Concept and the European Controversy on Irritability and Sensibility, 1750–90* (Amsterdam: Rodopi).

Stephens, H. M., and Roger T. Stearn. 2004. "Erskine, Sir William, Second Baronet (1770–1813), Army Officer," ODNB.

Steven, William. 1849. *History of the High School of Edinburgh* (Edinburgh: Maclachlan and Stewart).

Strasser, Susan. 2000. *Waste and Want: A Social History of Trash* (New York: Henry Holt).

Stray, Christopher. 2002. "A Pedagogic Palace: The Feinaiglian Institution and Its Textbooks," *Long Room*, **47**, 14–25.

Struthers, John. 1867. *Historical Sketch of the Edinburgh Anatomical School* (Edinburgh: MacLachlan and Stewart).

Suderman, Jeffrey M. 2001. *Orthodoxy and Enlightenment: George Campbell in the Eighteenth Century* (McGill-Queen's University Press).

Sutherland, D. J. S. 1938. "The History of Teaching Science in Scottish Schools" (unpublished PhD thesis, University of St. Andrews).

Suwa, Masaki, and Barbara Tversky. 2001. "Constructive Perception in Design," in John S. Gero and M. L. Maher (eds.), *Computational and Cognitive Models of Creative Design* (Sydney: University of Sydney Press), 227–239.

Sweeney, Douglas. 2020. "'The Most Important Thing in the World': Jonathon Edwards on Rebirth and Its Implications for Christian Life and Thought," in Chris Chun and Kyle C. Strobel (eds.), *Regeneration, Revival, and Creation: Religious Experience and the Purposes of God in the Thought of Jonathan Edwards* (Eugene: Pickwick Publications), 27–52.

Tave, Stuart M. 1952. "Some Essays by James Beattie in the London Magazine (1771)," *Notes and Queries*, **197**, 534–537.

Taylor, Douglas W. 1978. "The Manuscript Lecture Notes of Alexander Monro, Secundus (1733–1817)," *Medical History*, **22**, 174–186.

Taylor, Douglas W. 1979. *The Monro Collection in the Medical Library of the University of Otago* (Dunedin: University of Otago Press).

Taylor, Douglas W. 1986. "The Manuscript Lecture Notes of Alexander Monro Primus," *Medical History*, **30**, 444–467.

Taylor, Georgette Nicola Lewis. 2006. "Variations on a Theme: Patterns of Congruence and Divergence among 18th-Century Chemical Affinity Theories" (unpublished PhD thesis, University College London).

Taylor, Georgette. 2008. "Marking Out a Disciplinary Common Ground: The Role of Chemical Pedagogy in Establishing the Doctrine of Affinity at the Heart of British Chemistry," *Annals of Science*, **65**, 465–486.

Taylor, Georgette. 2014. "Pedagogical Progeniture or Tactical Translation? George Fordyce's Additions and Modifications to William Cullen's Philosophical Chemistry: Part II," *Ambix*, **61**, 257–278.

Taylor, Holly A., and Jeffrey M. Zacks (eds.). 2017. *Representations in Mind and World: Essays Inspired by Barbara Tversky* (London: Routledge).

Teevan, J., S. T. Dumais, D. J. Liebling, and R. Hughes. 2009. "Changing How People View Changes on the Web," *Proceedings of the 22nd Annual ACM Symposium on UIST*, 237–246.

Thaisen, Jacob. 2008. "The Trinity Gower D Scribe's Two Canterbury Tales Manuscripts Revisited," in Margaret Connolly and Linne R. Mooney (eds.), *Design and Distribution of Late Medieval Manuscripts in England* (York: York Medieval Press) 41–60.

Thomas, Keith. 1987. "Numeracy in Early Modern England: The Prothero Lecture," *Transactions of the Royal Historical Society*, **37**, 103–132.

Thomson, Duncan. 2004. "John Graham (1754–1817), Painter and Teacher of Art," ODNB.

Thomson, Thomas. 1853. *A Biographical Dictionary of Eminent Scotsmen, Division III. Dalrymple-Fordyce*, Robert Chambers (ed.) (Glasgow: Blackie and Son).

Thorne, R. G. 1986. *The History of Parliament: The House of Commons 1790–1820* (London: Secker and Warburg).

Thornton, Tamara Plakins. 1998. *Handwriting in America: A Cultural History* (New Haven: Yale University Press).

Timmermann, Anke. 2008. "Doctor's Order: An Early Modern Doctor's Alchemical Notebooks," *Early Science and Medicine*, **13**, 25–52.

Tomlinson, Louise M. 1997. "A Coding System for Notemaking in Literature: Preparation for Journal Writing, Class Participation, and Essay Tests," *Journal of Adolescent & Adult Literacy* **40**, 468–476.

Topham, Jonathon R. 1998. "Beyond the "Common Context": The Production and Reading of the Bridgewater Treatises," *Isis*, **89**, 233–262.

Towsey, Mark. 2008. "'Patron of Infidelity': Scottish Readers Respond to David Hume, c. 1750–c. 1820," *Book History*, **11**, 89–123.

Towsey, Mark. 2010. *Reading the Scottish Enlightenment: Books and Their Readers in Provincial Scotland, 1750–1820* (Leiden: Brill).

Tromans, Nicholas. 2007. *David Wilkie: The People's Painter* (Oxford: Oxford University Press).

Tsing, Anna Lowenhaupt. 2012. "On Nonscalability: The Living World Is Not Amenable to Precision-Nested Scales," *Common Knowledge*, **18**, 505–524.

Tufte, Edward R. 1983. *The Visual Display of Quantitative Information* (Cheshire: Graphics Press).

Turkle, Sherry. 2011. *Evocative Objects: Things We Think With* (Cambridge, Mass.: MIT Press).

Turnbull, George. 2014. *Education for Life: Correspondence and Writings on Religion and Practical Philosophy*, M. A. Stewart and Paul Wood (eds.) (Indianapolis: Liberty Fund).

Tversky, Barbara. 1981. "Distortions in Memory for Maps," *Cognitive Psychology*, **13**, 407–433.

Tversky, Barbara. 1993. "Cognitive Maps, Cognitive Collages, and Spatial Mental Modes," *Spatial Information Theory: A Theoretical Basis for GIS*, **716**, 14–24.

Tversky, Barbara. 2001. "Spatial Schemas in Depiction," in Meredith Gattis (ed.), *Spatial Schemas and Abstract Thought* (Cambridge: MIT Press), 79–111.

Tversky, Barbara. 2011. "Visualising Thought," *Topics in Cognitive Science*, **3**, 499–535.

Tversky, Barbara. 2017. "Diagrams: Cognitive Foundations for Design," in Alison Black, Paul Luna, Ole Lund, and Sue Walker (eds.), *Information Design: Research and Practice* (London: Routledge).

Tversky, Barbara. 2019. *Mind in Motion: How Action Shapes Thought* (New York: Basic Books).

Tversky, Barbara, and Diane J. Schiano. 1989. "Perceptual and Conceptual Factors in Distortions in Memory for Graphs and Maps," *Journal of Experimental Psychology*, **118**, 387–398.

Tversky, Barbara, Masaki Suwa, Maneesh Agrawala, Julie Heiser, Chris Stolte, Pat Hanrahan, Doantam Phan, et al. 2003. "Sketches for Design and Design of Sketches," *Human Behaviour in Design*, 79–86.

Tweed, Hannah C., and Diane G. Scott (eds.). 2018. *Medical Paratexts from Medieval to Modern: Dissecting the Page* (Cham: Palgrave Macmillan).

Tyner, Judith A. 2016. *Stitching the World: Embroidered Maps and Women's Geographical Education* (London: Routledge).

Underwood, E. Ashworth. 1977. *Boerhaave's Men at Leyden and After* (Edinburgh: Edinburgh University Press).

Uttal, David H., and Cynthia Chiong. 2004. "Seeing Space in More Than One Way: Children's Use of Higher Order Patterns in Spatial Memory and Cognition," in Gary L. Allen (ed.), *Human Spatial Memory: Remembering Where* (London: Lawrence Erlbaum Associates), 125–142.

Vance, Shona. 2001. "Schooling the People," in E. Patricia Dennison, David Ditchburn, and Michael Lynch (eds.) *Aberdeen before 1800* (East Linton: Tuckwell), 206–309.

Vickers, Brian. 1989. *In Defence of Rhetoric* (Oxford: Oxford University Press).

Vickery, Amanda. 1998. *The Gentleman's Daughter: Women's Lives in Georgian England* (New Haven: Yale University Press).

Vickery, Amanda. 2009. "The Theory and Practice of Female Accomplishment," in Mark Laird and Alicia Weisberg-Roberts (eds.), *Mrs. Delany and Her Circle* (New Haven: Yale Center for British Art), 94–109.

BIBLIOGRAPHY | 491

Vincent, David. 1989. *Literacy and Popular Culture: England 1750–1914* (Cambridge: Cambridge University Press).

Vine, Angus. 2017. "Search and Retrieval in Seventeenth-Century Manuscripts: The Case of Joseph Hall's Miscellany," *Huntington Library Quarterly*, **80**, 325–343.

Vine, Angus. 2018. *Miscellaneous Order: Manuscript Culture and the Early Modern Organization of Knowledge* (Oxford: Oxford University Press).

Voss, Julia. 2010. *Darwin's Pictures: Views of Evolutionary Theory, 1837–1874* (New Haven: Yale University Press).

Wallace, Mark C., and Jane Rendall (eds.). 2021. *Association and Enlightenment: Scottish Clubs and Societies, 1700–1830* (Lewisburg: Bucknell University Press).

Wardhaugh, Benjamin. 2012. *Poor Robin's Prophecies: A Curious Almanac, and the Everyday Mathematics of Georgian Britain* (Oxford: Oxford University Press).

Warwick, Andrew. 2003. *Masters of Theory: Cambridge and the Rise of Mathematical Physics* (Chicago: University of Chicago Press).

Waterston, Robert. 1949. "Further Notes on Early Paper Making Near Edinburgh," *Book of the Old Edinburgh Club*, **27**, 40–59.

Weeks, Dennis. 1991. "The Life and Mathematics of George Campbell, F.R.S.," *Historia Mathematica*, **18**, 328–343.

Werrett, Simon. 2019. *Thrifty Science: Making the Most of Materials in the History of Experiment* (Chicago: University of Chicago Press).

Wess, Jane. 2012. "Avoiding Arithmetic, or the Material Culture of Not Learning Mathematics," *Journal of the British Society for the History of Mathematics*, **27**, 82–106.

West, William N. 2006. *Theatres and Encyclopedias in Early Modern Europe* (Cambridge: Cambridge University Press).

Whitehouse, Tessa. 2015. *The Textual Culture of English Protestant Dissent 1720–1800* (Oxford: Oxford University Press).

Whiteside, Derek Thomas. 2014. "And John Napier Created Logarithms . . ." *Journal of the British Society for the History of Mathematics*, **29**, 154–166.

Whitmer, Kelly Joan. 2015. *The Halle Orphanage as a Scientific Community: Observation, Eclecticism, and Pietism in the Early Enlightenment* (Chicago: University of Chicago Press).

Williams, Drifi. 2009. "The Hamilton Gray Vase," in Judith Swaddling and Philip Perkins (eds.), *Etruscan by Definition: The Cultural, Regional and Personal identity of the Etruscans* (London: British Museum Research Publication), 12–20.

Williams, Tyler. 2018. "'If the Whole World Were Paper . . .': A History of Writing in the North Indian Vernacular," *History and Theory*, **57**, 81–101.

Williamson, Eila. 2002. "Schoolboy Scribbles on a Burgh Court Book," *Scottish Archives*, **8**, 96–112.

Wills, Hannah. 2019. "Charles Blagden's Diary: Information Management and British Science in the Eighteenth Century," *Notes and Records of the Royal Society*, **73**, 61–81.

Wilson, Duncan K. 1935. *The History of Mathematical Teaching in Scotland to the End of the Eighteenth Century* (London: University of London Press).

Winter, Sarah. 2011. *The Pleasures of Memory: Learning to Read with Charles Dickens* (New York: Fordham University Press).

Withers, Charles W. J. 1999. "Towards a History of Geography in the Public Sphere," *History of Science*, **37**, 45–78.

Withers, Charles W. J. 2000. "Toward a Historical Geography of Enlightenment Scotland," in Paul Wood (ed.), *The Scottish Enlightenment: Essays in Reinterpretation* (Rochester: University of Rochester), 63–97.

Withers, Charles W. J. 2001. *Geography, Science and National Identity: Scotland Since 1520* (Cambridge: Cambridge University Press).

Withers, Charles W. J. 2002. "The Social Nature of Map Making in the Scottish Enlightenment c. 1682–c. 1832," *Imago Mundi*, **54**, 46–66.

Withers, Charles W. J. 2006. "Eighteenth-Century Geography: Texts, Practices, Sites," *Progress in Human Geography*, **30**, 711–729.

Withers, Charles W. J. 2008. *Placing the Enlightenment: Thinking Geographically about the Age of Reason* (Chicago: University of Chicago Press).

Withers, Charles W. J. 2010. "Geography, Science, and the Scientific Revolution," in David N. Livingstone and Charles W. J. Withers (eds.), *Geography and Revolution* (Chicago: University of Chicago Press), 75–105.

Withers, Charles W. J., and Miles Ogborn. 2010. *Geographies of the Book* (Farnham: Ashgate).

Withers, Charles W. J., and Paul Wood (eds.). 2002. *Science and Medicine in the Scottish Enlightenment* (East Linton: Tuckwell Press).

Wittmann, Barbara. 2013. "Outlining Species: Drawing as a Research Technique in Contemporary Biology," *Science in Context*, **26**, 363–391.

Wolf, Jeffrey Charles. 2015. "'Our Master & Father at the Head of Physick': The Learned Medicine of William Cullen" (unpublished PhD thesis, Edinburgh University).

Wood, Paul B. 1993. *The Aberdeen Enlightenment: The Arts Curriculum in the Eighteenth Century* (Aberdeen: Aberdeen University Press).

Wood, Paul (ed.). 2000. *The Scottish Enlightenment: Essays in Reinterpretation* (Rochester: University of Rochester Press).

Wood, Paul. 2017. "Introduction," in Thomas Reid, *On Mathematics and Natural Philosophy*, Paul Wood (ed.) (Edinburgh: Edinburgh University Press), xvii–cxciv.

Wragge-Morely, Alexander. 2020. *Aesthetic Science: Representing Nature in the Royal Society of London, 1650–1720* (Chicago: University of Chicago Press).

Wright, Christopher, Catherine May Gordon, and Mary Peskett Smith (eds.). 2006. *British and Irish Paintings in Public Collections* (New Haven: Yale University Press).

Wright, Patricia. 1999. "The Psychology of Layout: Consequences of the Visual Structure of Documents," *American Association for Artificial Intelligence Technical Report FS-99-04*, 1–9.

Wright-St. Clair, R. E. 1964. *Doctors Monro: A Medical Saga* (London: Wellcome Medical Library).

Yale, Elizabeth. 2016. *Sociable Knowledge: Natural History and the Nation in Early Modern Britain* (Philadelphia: University of Pennsylvania Press).

Yates, Frances A. 1966. *The Art of Memory* (London: Routledge).

Yeo, Richard. 2001. *Encyclopaedic Visions: Scientific Dictionaries and Enlightenment Culture* (Cambridge: Cambridge University Press).

Yeo, Richard. 2003. *Defining Science: William Whewell, Natural Knowledge, and Public Debate in Early Victorian Britain* (Cambridge: Cambridge University Press).

Yeo, Richard. 2007. "Before Memex: Robert Hooke, John Locke, and Vannevar Bush on External Memory," *Science in Context*, **20**, 21–47.

Yeo, Richard. 2014. *Notebooks, English Virtuosi, and Early Modern Science* (Chicago: University of Chicago Press).

Yoshino, Yuki. 2013. "Desire for Perpetuation: Fairy Writing and Re-creation of National Identity in the Narratives of Walker Scott, John Black, James Hogg, and Andrew Lang" (unpublished PhD thesis, Edinburgh University).

Young, Ronnie. 2016. "Thomas Muir at Glasgow: John Millar and the University," in Gerard Carruthers and Don Martin (eds.), *Thomas Muir of Huntershills: Essay for the Twenty First Century* (Edinburgh: Humming Earth), 112–140.

Zachs, William. 2011. "Bindings," in Stephen W. Brown and Warren McDougall (eds.), *The Edinburgh History of the Book in Scotland*, vol. 2 (Edinburgh: Edinburgh University Press), 65–69.

Index

Page numbers in italics refer to figures.

Aberdeen University, 267, 285, 323, 374–75. *See also* King's College, Aberdeen; Marischal College, Aberdeen

abstraction, 171, 263, 268, 299, 304, 316, 336, 353; in realtime, 263

academies, 38, 40, 55–56, 67, 81, 83, 124–25, 138–39, 187, 206, 213–15, 231–32, 245, 258, 271, 278, 326, 328, 330, 338

accounting. *See* bookkeeping

aesthetics, 199, 211, 219–21. *See also* beauty; mnemoaesthetics; representation

agency, 7, 23–29, 35, 130, 417, 420–21; in editing, 130–36; through enactive artifacts, 44, 47; and enactive metaphors, 48, 70, 418

agriculture, 9, 58, 143–45, 203, 341

algebra, 103, 137, 188, 225, 278

alignment, 68–70, 164

Allan, David, 188, 190, 283

Alston, Charles, 385–86

altimetry, 201, 203, 207, 211; definition of, 201

amanuensis, 293, 395

anatomy, 37, 307, 343, 353, 375–81, 387; classrooms, 293–94, 333–35; diagrams, 326, 330–33, 335, 337–38; piracy in

print, 408–10; printing plates, 332; systems, 378

ancient history and languages, 3, 5, 61, 80, 111, 188, 227, 231, 250, 282, 289, 301, 324, 379; Greece, 227, 379; Greek, 54, 59, 111, 156, 289, 315; Greek notebooks, 289; Latin, 5, 37, 54–55, 58, 59, 82–83, 85, 88, 95, 103, 107, 124, 138, 156, 183, 243, 258, 282, 318; Mesopotamia, 250, 379; Rome, 227. *See also* Cicero; Euclid; humanism; Quintilian

Anderson, John, 289, 293; female students, 277–78

angularity, 5, 62, 76, 77, 85, 95, 132, 166, 174–77, 196–201, 204, 207, 212, 218, 222, 239–40, 242, 246, 250, 255, 266, 335–38, 340–41, 344, 350, 352, 373, 376, 380–81; equality of angles, 336

annotating, 14, 43, 114–46, 281, 314–16, 388, 416–17, 421

anthropology, 13, 19, 20, 75, 80, 226, 322. *See also* Ingold, Tim; Leroi-Gourhan, André

apprehension, 197–98

architecture, 16, 56, 195, 226

arithmetic, 56–57, 103, 137–38, 159, 236, 278, 334

Arrow, Jemima (schoolgirl), 107–8, 235–38, 249, 260, 328
art: and artifacts, 7–14, 416–18; and artificers, 7–8, 22–29, 420–23; as artificing, 7–8, 14–22, 418–20; of bookkeeping, 114, 138; of computation, 137; definition of, 4; of diagramming, 379; of memory, 15, 157; of omission, 363; of ranging, 73; systemizing, 4, 271; of writing, 45, 60, 63, 81. *See also* aesthetics; performance; representation
art history, 178, 220
art theorists, 212
association, 48, 67, 81, 139, 316, 318, 320, 325–26, 362, 368–69, 381; realtime aspects, 320
astronomy, 37, 199, 243, 270, 330, 341, 345, 385; diagrams, 199, 330. *See also* experimental instruments: telescopes; natural philosophy
attention, 4, 10, 18, 24, 30, 54, 61, 80–81, 83–84, 106, 111, 118, 124, 132, 163–64, 167, 170, 172, 176, 183, 233, 261, 263, 268, 275, 285, 289–90, 292, 296, 299–300, 316, 320, 326, 336, 353–55, 377, 398, 416, 418, 421, 423; with diagrams, 326; metacognitive aspects, 320; real-time aspects, 263, 275–76
Australia, 95, 277, 391
automaticity, 176, 178, 300; via transcribing, 322–23

Bacon, John (university student), 307
Baillie, Joanna (schoolgirl), 49–50
beauty, 59, 211, 219–20
Bell, Benjamin (university student), 408
Belsches, Williamina (schoolgirl), 161–62, 164, 180, 421
bifolia, 14, 16, 86, 92, 96–97, 99, 101–2, 104, 107, 208, 419; as mnemonic cues, 288
bifolium, 11–12, 20, 26; definition of, 96, 97, 101
Black, John (schoolboy), 186–87, 189, 328

Black, Joseph, 286–87, 296, 313, 328, 341–51, 355, 361–74, 379, 395, 397–99; diagrams, 326
blackboards. *See* chalkboards
Blagden, Charles (university student), 296–98, 304, 313, 346–48
Blair, Ann, 134–35, 280–81, 383
Blair, Hugh, 301, 377, 396–97; views on metacognition, 320
Blake, William, 222–23
Blane, Gilbert, 408–10
blankness, 31, 95, 123; blank box, 70; blank field, 149; blank grid, 122; blank index, 311–12; blank letter spaces, 50, 70–71, 281; blank mind, 80; blank page(s), 49, 68, 75, 113, 118, 122, 131, 198, 355–57, 364; blank paper, 95–97, 223; blank space and diagrams, 340; blank table, 70
bookkeeping, 36, 55, 57–58, 73, 76, 114–16, 137–46, 155, 195; double entry tables, 141–44; field books, 144–45
Borthwick, John (student), 389
botany, 37, 92, 186, 277, 330, 332, 351–60, 382, 385; diagrams, 330, 332–33, 335
boxes: of ephemera, 11, 426; of graphite, 21; inside indices, 311–12; as paratexts, 132; similarity to headings, 155–56; as storage metaphors, 4; for storing manuscripts, 92; inside tables, 70, 373; of watercolors, 211; of writing materials, 26. *See also* Napier's Box of Tables
boys, 34, 50–51, 54–57, 67, 70, 73, 76, 95, 114, 116–17, 159, 184, 189, 195, 197, 228, 240, 318, 388, 420, 422
Brougham, Henry (schoolboy), 85–86, 88
Bruce, George (university student), 353
Burns, Robert, 385

calculation, 86, 114, 118, 128, 136–37, 140–41, 201, 204, 211, 221, 238, 240, 243–44, 246–47, 249–52, 255–56, 260, 314, 318, 338, 344–45
calendarium, 58

Calvinism, 158–59, 271. *See also* Church of Scotland

camera obscura, 185, 204

capabilities, 24, 35, 77, 80, 82, 89, 130–36, 209, 226, 240, 268, 304–5, 318, 323, 418, 420; spatial, 209; mathematical, 209

Carr, Rosalind, 277

Carruthers, Mary, 172

casting of nines, 344–45

catechism, 49–50

categorizing, 30, 38, 68, 80–81, 132–34, 136, 144, 149, 151–80, 226, 270, 285, 313, 321–22

chalkboards (blackboards), 295, 331, 335, 374–75, 378–79, 381

charts, 50, 157, *179*, 348, 367–69

chemistry, 37, 55, 73, 199, 276, 286–87, 296, 307, 313, 353, 355; chemical affinity, 341–51, 361, 366–74; diagrams, 200, 326, 330, 341–51; substances, 341–51, 361–74; system, 361

China, 332

chirography, 128–29, 131, 162, 164–70, 172–74, 176; definition of, 60–61

chronograms, 231, 341, 366

Church of Scotland, 49–50, 137, 271, 279–80, 286, 289

Cicero, 36, 55, 83, 138, 258, 282

ciphering, 15, 136–46, 238, 245–49, 255

ciphers, 117, 140–41, 143–44, 146, 255, 256, 416; definition of, 137

circles, 109, 122, 159, 183, 193, 197–99, 208, 222, 230, 232, 325, 335, 337, 346–48, 352, 372, 379, 419

circlets, 346–48, 371–72

circulating (notebooks), 267, 281, 382–410

civil history, 388, 390

classics, 54–55

classification, 92, 157, 160, 221, 273, 322, 351, 421

classrooms, 79, 87, 227, 239, 243–46, 249, 256, 260, 273, 284–85, 289, 300–301, 312, 330, 336, 337, 340, 343, 349, 353–55, 359, 361, 369, 374–75, 377, 380; dis-

play of maps, 231–32; spatial aspects, 332–34

clavis, 313–14; of headings, 283–85

Cochrane, Thomas (university student), 29, 348–51, 361–66, 371–74, 421–22

codection, 15–16, 31, 40, 86, 226, 419–20

codex, 15, 43, 45–46, 57, 59, 73, 83, 86, 89–91, 96–99, 102, 106, 110–11, 226, 237, 271, 305, 418, 420–21; medieval, 70–71

codexing, 43, 85–113, 115, 146, 154, 238, 244, 246–48, 270–71, 418–19; definition of, 6

cognition, 3, 15, 18, 21, 31, 36, 43, 78, 80, 136, 154–55, 159, 175, 199, 225, 228, 230, 244, 263, 275, 285, 289, 292, 300, 302, 304, 316–18, 320–21, 323, 358, 377, 421; cognitive load, 243–44; metacognition, 317, 319, 381, 418

cognitive metaphors, 5, 34–35, 336, 418; bentness and thought, 318; chains of ideas, 318, 326; scrap paper metaphor, 159; straightness and thought, 317–18; train of ideas, 316–23; train of thought, 289, 317, 321. *See also* line of reason metaphor

colonialism, 11, 80, 125, 141, 152, 158, 196, 219, 374, 382–84, 388, 396, 404–5; East India Company, 289; maps, 224–25. *See also* Australia; India; New Zealand; North America

columns, 68–69, 71–78, 122–23, 141–43, 295, 366–74; and maps, 249; as thinking tools, 317

commonplace books, 58–59, 66–67, 113, 155–58, 321, 416; for students, 280

composition, 83, 137, 140, 116–17, 318

concentration, 282, 305, 348–49

conception, 111, 171

copies, 78–84

Copland, Patrick, 293–95, 333, 374–75, 379

copybooks, 50–51, 60–61, 79, 81, 108

copying, 46, 58, 70, 78–84, 137, 157–58, 171, 211–12, 249, 281, 285, 295, 305, 307,

copying (*continued*)
309, 319, 338, 369, 395, 420; images, 211, 330; maps, 249; metacognitive aspects, 319. *See also* transcribing; transcripts
copying machine, 395
corn books, 144–45
corrections, 54, 73, 79, 125–28, 137, 208, 314–15
Coventrie, Alexander (university student), 292, 306–7
Creech, William, 398
Cullen, William, 283, 288, 306, 341, 343–44, 353, 369–70, 394–96, 399–407; diagrams, 328
Cumming, Mary, 188, 210
Currie, James, 275–76, 300, 304–5
Currie, William (university student), 275–76, 300, 305
curvilinearity, 62, 94, 95, 104, 200, 209, 262, 328. *See also* circles

Darnton, Robert, 19–20
Darwin, Charles, and diagrams, 327
Daston, Lorraine, 76, 219
data bundle, 353–54
data visualization. *See* charts; diagrams; judgment: visual; layout; pictograms; posters; scanpaths; sightlines
decision-making, 45, 60, 131, 150, 158, 167, 170, 253, 299, 340, 359, 371, 400, 411, 417, 419–20, 422
de Hamel, Christopher, 12–13
demonstrations, 198, 232, 235, 273, 351, 363, 375
descriptions, 24, 181–223, 348–50, 353–54, 357–58, 363, 375–78, 385
desks, 64–65, 192, 233
developmental psychology, 29, 31, 35–36, 66, 71, 80–83, 111, 184, 197–99, 209, 219, 255, 323, 336, 353, 381, 420
diagrammatic knowledge, 324–27
diagramming, 15, 29, 267, 324–81, 418; Locke's view on, 199; young learners, 197–210

diagrams, 20, 71, 187, 226, 273, 290, 295, 324–81, 416–17, 421; cooking, 199; heliocentric, 199; historiography, 327–28; landscape, 211; and the mind, 325; pirated editions, 408–10. *See also* circlets; *clavis*; pictograms; transcribing: chiasmas; triangles
diaries, 125–28, 186, 301, 307, 393–94, 405
dichotomies, 156–57
dictates, 279
dimensionality, 154, 159–60, 197, 200, 250, 328, 352, 364, 418
directionality, 137, 316–23; in tables, 370–71
distraction, 293–95, 299, 316
divinity, 271, 283, 385, 397
doodles, 117, 123, 210, 282, 314
draftsmanship, 195, 205, 235
drawing, 3, 6, 12, 14–15, 19, 24, 39, 57, 67, 70–71, 91–93, 149, 181–223, 270–71, 276, 327–33, 351, 373, 378–79; connections to writing, 70, 132; as a developmental tool, 198; instructors, 188–90, 211–12, 245–46; as knowledge creation, 195–96; likeness of professors, 328–29; links to understanding, 206; in school notebooks, 181–223; technical drawings, 187; technical images, 206; as visual translation, 201, 253, 254–56, 358–60. *See also* draftsmanship; mapping; pictogramming; sketching
drawing instruments, 191–95, 242; calipers, 193; compasses, 105–7, 122, 193, 204, 208, 242, 418–19; drawing quills, 193; illustrations of, 194; pantographs (parallelograms), 193, 254; pen points, 192–93. *See also* measurement instruments
drawings. *See* chronograms; diagrams; figures; grids; landscapes; pictograms; polygons; schematic images; shapes; sketches
drawing schools: Academy of St Luke, 188–90; D'asti's Drawing Academy, 190;

Trustees Drawing Academy, 188–89, 210

Dunbar, James (schoolboy), 159, 385, 421

Duncan, Andrew, 38, 282–83, 287–88

Edinburgh High School, 51, 54, 56, 60, 85, 95, 100, 103, 107, 124

Edinburgh Royal Infirmary, 302–3

Edinburgh University, 29, 38, 45, 67, 190, 244, 267, 271, 275–77, 282–83, 286–92, 296, 301, 307, 313–14, 326, 331–35, 385

editing, 28, 124–36, 394–95

Elliot, Cornelius, 385

embodiment, 13, 18, 29–30, 48, 67–68, 260–63; and maps, 227–33

embroidery, 111, 189, 228

emotions, 59, 127, 161–62; anger, 77; happiness, 57, 162

encapsulation, 155–62

environment, 9, 13, 232, 239, 251. *See also* landscapes

epistemology, 87–88, 93, 226–27, 305, 384, 410; of images, 226; and maps, 227

epitomes, 151, 153, 299, 301, 304, 309

Erskine, John (of Torrie), 120–24

Erskine, Miss, 122–23

Erskine, William, 107, 110

Erskine of Torrie, children of, 118–24, 146

essays, 161–62, 183, 280, 393, 407

Euclid, 198, 243, 263, 324. *See also* geometry

exams, 59, 79, 280, 386

excerpts, 83, 306, 386

exemplars, 60, 80, 114, 116, 211–12, 233, 237, 245, 284, 313

exercise notebooks, 161–62

exercises, 79, 85, 103, 107, 143, 161, 201, 204, 213–14, 233, 249, 251, 255–56, 281, 300, 318

expenses: notebook transcripts, 392, 395, 397; notekeeping, 60, 92, 96–97, 306; syllabus, 282–83; tuition fees, 125, 187, 189–90, 238, 245, 277–78, 396

experimental instruments: air pumps, 204, *206*; flasks, 362, 364; Florentine flasks, 353; funnels, 353; furnaces, 328, 330, 361–62; glass bells, 364–65; glassware, 328, 330–31, 361, 364; lamps, 361; lenses, 218–20; magic lanterns, 204; microscopes, 204–6, 376, 385; prisms, 160, 204; retorts, 353; telescopes, 342, 385; thermometers, 204, 361, 366–67, 398; tubes, 353, 362, 364

experiments, 289, 296–97, 312, 349–50, 361–66, 376, 397–98; preparations, 332, 375, 379–80

extracts, 59, 82, 135, 143, 280, 304, 407

family books, 73–75

Ferguson, Adam, 286

figure and ground image, 163, 307–8

figures, 104–5, 150, 195–223, 226, 281, 295

Fleming, Marjory (child diarist), 125–28

flow of information, 9, 156, 171–80, 320, 322, 347, 370–71

focal points, 151, 153, 162–70, 172, 174, 419; labels, 114–16, 128–29, 153, 155–62, 200, 204, 208, 222, 237, 248, 262, 322, 364, 366, 418; maps, 225; scroll books, 295–96; serialized, 323; syllabi, 314; transcripts, 307–9, 311–12; verbal pictures, 67–78, 321–23; verticality, 170–80; visual coordinates, 172–74; visual cues, 153, 162–70, 256, 259–61; visual data, 225; visual navigation, 162–80. *See also* directionality

fold, 11, 14, 19–20, 85, 94, 107, 111, 251, 416

folding, 15–16, 20, 88, 90, 94, 96, 101, 106, 226, 251, 258, 419

foliation, 16, 31, 40, 416, 419–20

Forbes, James Ochoncar (student), 124–25

Fordyce, James, 318

Fowler, Caroline, 94

Fowler, James (schoolboy), 76–77, 102, 112, 181–83, 207–9

Fyfe, Andrew, 332, 357–60

games, 196, 332, 354; for children, 122–23
gardens, 239–40, 330; Royal Botanic
  Garden Edinburgh, 332, 355, 383
gauging, 56, 76, 137, 195, 201, 204, 207,
  210–11, 214, 226, 335
gender, 23–26, 28–31, 34–35, 46, 54, 79,
  111, 188, 190, 228–30, 240, 277, 278,
  420–21; and maps, 228–30, 235–38.
  See also boys; girls; women
geography, 55–57, 76, 82, 89, 187–88,
  196, 225–28, 230, 232–38, 249, 257–58;
  definition of, 227; globes, 208, 227,
  233, 258–59; instructors, 235–36;
  maps, 226, 228–30, 234–38, 249
geology, 374–75
geometry, 16, 24, 37, 55, 60, 71, 89–90,
  96, 105–6, 122, 124, 137, 158–59, 168,
  187–88, 196–97, 199–201, 203–5, 207,
  209–11, 214, 218, 221–22, 225, 242–44,
  246, 251–52, 255, 262–64, 278, 325, 329,
  335, 338, 347, 377–78; diagrams, 329;
  in realtime, 263–64. *See also* circles;
  curvilinearity; dimensionality; linearity;
  lines; polygons; rectilinearity; shapes;
  squares; triangles
geometry, practical. *See* altimetry; gauging; leveling; longimetry; planimetry;
  sailing; surveying
gestalt images, *72*, 15, 77, 101, 153, 162–
  70, 307–8, 369; definition of, 15; on
  transcript pages, 307–8; visual proximity, 368. *See also* angularity; curvilinearity; figure and ground image;
  kinesthetics; linearity; mnemonics;
  mnemotechnics; parallelity; preattentives; proportionality; rectilinearity;
  similarity; symmetry
girls, 2, 28, 34, 45–46, 50–51, 54, 56–57,
  66–67, 70, 73, 107, 114, 116–17, 130, 134,
  141, 184, 189, 195, 228–29, 235, 240, 270,
  328, 388, 420. *See also* Arrow, Jemima
  (schoolgirl); Baillie, Joanna (schoolgirl); Belsches, Williamina (schoolgirl);
  Erskine, Miss; Gray, Elizabeth Caroline
  Johnstone; Keir, Amelie (student);

Masson, Jeane (schoolgirl); Monro,
  Margaret (schoolgirl); Montgomery,
  Margaret (schoolgirl); Murray, Amelie
  (schoolgirl); Somerville, Mary (school-
  girl); Sym, Kath
Glasgow University, 190, 199, 267, 269–70,
  277–79, 283, 289–90, 324, 338, 341, 385
Goody, Jack (anthropologist), 75, 322
Grafton, Anthony, 88, 113, 154
Grant, Miss Grant of (university student),
  277
graphic design, 149, 171–74. *See also*
  layout
graphs, 170–71, 178–80
Gray, Elizabeth Caroline Johnstone,
  58–59
Greenfield, William (university student),
  381
Gregory, David, 243–45
Gregory, James, 385
Gregory, John, 307, 394, 396
Greig, John, 58, 118
Grew, Nehemiah, 359–60
Grey, Richard, 23–24, 83, 174
grids, 70–73, 77, 95, 122–23, 176, 186, 208,
  235, 309, 320, 351, 374
Gutenberg diagram, 171–74

Hamilton, Alexander, 287, 410
Hamilton, Elizabeth, 110–11, 136, 228–30
Hamilton, Francis (university student),
  382
Hamilton, James, 277
handouts, 212, 273, 295, 305, 312, 330,
  332
Haygarth, John (university student), 379,
  408
headings, 28, 38–39, 58, 65, 71–73, 77,
  128, 130–36, 151–80, 200, 204, 209,
  236–37, 255, 261, 270, 276, 278, 282–92,
  295–96, 301, 305–9, 311, 313–14, 319,
  321–23, 340–41, 350, 388, 398, 402–21;
  editing, 129–30, 132–34; as mnemonic
  labels, 155–62; pictographic, 364–
  66; as pictures, 321–23; as realtime

categories, 151–55; as scanpath cues, 170–80; as visual cues, 307–8; as word schemes, 158

Heesen, Anke te, 87

heuristics, 160, 172, 341, 363

Hill, John, 314–15

Hill, Peter, 385, 392

Hogarth, William, 139–41

Holland, 236, 275

Home, Henry (Lord Kames), 80, 82–83, 139, 198

home learning, 24, 38, 40, 49–50, 54, 60–61, 227–31, 239, 271, 328

Hope, John, 332–33, 353–55, 357; female students, 277

Hope, Thomas Charles (university student), 355–61

household books (for accounts), 73–75

household management, 134–35, 137; data storage, 139

human development, 6, 35–36; and drawing, 195–210

humanism, 23, 54–55, 82–83, 135, 138–39, 155–56, 184, 255, 257–58, 282, 416–17

human nature, 26

Hume, David (law professor), 295, 309, 313–14, 391

Hume, David (philosopher), 151–53, 161, 171, 179, 385

Hutcheson, Frances, 80

ideas, 8, 31, 35, 44, 48, 66–67, 80–83, 101, 135, 158, 183, 199, 210–11, 222, 251, 258, 301–2, 305, 316–23, 326–27, 335–36, 341, 377, 381, 394, 418, 420; and diagrams, 326, 335; and language, 377; on paper, 101; words as ideas, 48. *See also* cognitive metaphors

identity, 26, 28, 389, 394, 405, 421. *See also* agency

imagination, 319

inattention, 94, 316

index, 39, 87, 134–36, 296, 298, 311–12, 416, 421–22

indexing, 134–36, 311–12

India, 26, 95, 193, 197, 212, 289, 382; Tipu Sultan, 383

information, 3–4, 9, 11, 21, 38, 44, 46, 49, 59, 70–71, 73, 75, 79, 82, 90, 92, 95–96, 98, 100, 102, 109, 113, 116, 123, 140, 143–45, 153–57, 159, 162–63, 167, 170–74, 176, 178, 203, 206–7, 209, 218, 220–21, 231, 237–38, 244, 247, 249, 256–60, 271–72, 275–76, 280, 285–86, 288, 295, 300, 303–4, 306, 309, 312, 314–16, 322, 331, 340–41, 347, 350–51, 354, 357–58, 360–61, 363, 366–67, 381–82, 384, 404, 417, 419, 421; definition of, 10. *See also* flow of information; grids; Gutenberg diagram; index; realtime interface; tables

information management, 11, 23, 39, 44, 71, 73, 75, 96, 109, 115, 123, 136, 145, 155–56, 280, 306, 340, 361

information networks, 382–410

Ingold, Tim, 13–14, 80, 325

intellectual property, 408

interactivity, 4, 8, 18, 19, 31, 34, 48–49, 80, 86–87, 106, 111, 114, 130, 131, 159, 162, 177, 179, 183, 209, 218–19, 223, 225, 227, 230, 244, 300, 305, 323, 357, 359, 407

interface. *See* realtime interface

interleaving, 29, 281, 364, 385, 419

Jackson, Robert (schoolboy), 158–59, 164, 168–69, 197–98

Johnstone, David (university student), 309, 315–16

Jordanova, Ludmilla, 20, 253–54

journalizing, definition of, 140–41

judgment, 43, 48, 80, 82, 132, 144, 150, 158, 196, 209, 275, 320, 355; and copying, 82–83; realtime, 275–76, 299, 320; visual, 351–66

Kames, Lord. *See* Home, Henry

Keir, Amelie (student), 73, 199–200, 276, 421

Keith, Isabella (governess), 125–28

keywords, 72, 77, 83, 132, 157, 171, 174, 312, 421

kinesthetics, 13, 18, 38, 43, 45, 48, 77, 117, 151–52, 154–55, 170, 179, 221, 230, 233, 246, 268, 275, 305–6, 313, 347, 361, 373, 418–20; and tables, 366–74
King's College, Aberdeen, 385
knowledge communities, 87, 417, 421
knowledge economy, 155, 267, 392, 405, 410, 421
knowledge formation, 276, 353–54
knowledge system, 8–9, 37, 43, 87, 136, 146, 180, 224, 226, 267–68, 270, 272–74, 282, 286, 288–89, 292, 313, 321–23, 329, 331, 341, 351, 355, 361, 374, 385, 419–21; metacognitive aspects, 321–22; schematized, 313–14, 355; scrolling, 292; visual data, 331. *See also* system; systemizing
Krajewski, Markus, 6, 87

labor, 36, 99, 306, 394, 403, 410, 415, 419
landscapes, 16, 185–86, 190, 201, 207, 210–18, 221–22, 226, 244, 252, 264, 325; instructions, 213; mountainscapes, 210, 212; riverscapes, 210; seascapes, 210, 218
law, 58, 151, 156, 186, 269, 270–71, 290, 295, 300, 309, 312–13, 315, 320, 383–84, 390–91, 393, 397, 399–404, 407
layout, 46, 57, 66–67, 68–78, 130–34, 137, 144–45, 149–50, 170–72, 176, 178, 249, 419; for maps, 225; for scroll books, 295–96; serialized, 323; for a syllabus, 314; for transcripts, 307–9, 311–12. *See also* columns; focal points; foveal vision; gestalt images; grids; Gutenberg diagram; headings; margins; matrix; *mise-en-page*; modules; paragraphs; preattentives; ranging; rows; scanning; scanpaths; schematic images; sightlines; significant space; standard module; tableaux; tables
learning disability, 23, 94, 231
lecture notebooks, 274–323; definition of, 280; diagrams in, 324–81; outlines, 276, 282–83, 289, 306; stages of creation, 280–81
ledgers, 46, 58, 68, 73–74, 114, 137, 139–46, 303–4, 416; as databases, 143–44
Lee, John (university student), 314–15
Leechman, William, 385, 406–7
Leroi-Gourhan, André, 13, 19
letters, 125, 144; majuscules, 168–70, 296; spacing, 60–61, 70. *See also* chirography; orthography; typography
leveling, 137, 195, 204, 210–11, 335; diagrams, 201–3
liberal education, 184, 188, 232
libraries, 10–11, 20, 87–88, 117, 146, 190, 281, 283, 291, 295, 301, 344, 383–92, 397, 401, 407; children as users, 117–24
linearity, 101, 143, 149, 170–80, 181–223, 336; descriptions, 375–77; madness, *32*; maps, 233–35; surveying, 260–61. *See also* angularity; curvilinearity; rectilinearity; tessellations
line of reason metaphor, 317–23. *See also* cognitive metaphors
lines, 48, 50, 61–62, 65, 70–71, 77, 91, 95, 109, 112, 122, 132, 141, 143, 149, 171, 176, 178, 183–84, 191–97, 200–201, 206, 209, 212–15, 222–23, 228, 230–35, 237–38, 245, 250, 252, 255, 260–63, 276, 296, 312, 316–23, 326, 328, 336, 338, 340–41, 348, 350–52, 355, 373–81, 421; heuristics, 350–51; links to mental operations, 316–23; as material artifacts, 13–14, 317; as memory tools, 317; as metonym for homework, 54–55, 59, 82, 318; as signs, 325. *See also* gestalt images; linearity; line of reason metaphor; memory: memorial lines; mnemotechnics: technical lines; sightlines
Linnaeus, Carolus, 156, 282, 284, 351
listening, 290, 300–301, 377. *See also* orality
listing, 46, 135, 311

lists, 57, 81, 92, 120, 122, 134–36, 151, 158, 171, 236, 249, 277, 282, 288, 303, 305, 311–13, 319, 321–23, 368, 370–71, 390, 392

literacy, 4, 14, 45–84, 136, 140–41, 153, 210, 418, 421

Locke, John, 30–31, 36, 48, 64, 66, 70–71, 80–82, 86, 111, 154, 198, 219, 255, 336, 353, 379, 381, 415, 418, 420; cognitive model of, 316–17; commonplace index, 311; on diagrams, 199, 325–26, 336, 379; on drawing, 184, 210–11; theory of mind, 325, 377. See also *tabula rasa*

logic, 30, 158–59, 269, 293; children, 159; topical, 102–3, 156–57

London, 89, 121, 192, 211, 242, 346, 375, 386, 390, 399, 402, 408

longimetry, 203, 211; definition of, 201

Lowndes, Thomas, 400–404

Maclaurin, Colin, 243

manuscript book covers, 106–10, 382

manuscript book tooling, 111–13

manuscript textbooks, 279–80

map-mindedness, 227–33

mapping, 149, 224–64

maps, 102, 104, 187, 208, 224–64; on cloth, 228–29; desk maps, 223–38, 249; field maps, 233, 239, 244–56, 260; jigsaw puzzles, 230; as memory theaters, 256–64; university lectures, 295. *See also* geography; surveying

marginalia, 11, 125, 128, 136, 213, 275, 314–16, 328–29, 344, 385, 389; children, 117–24

margins, 61, 66, 73, 95, 119, 130–34, 144, 162, 164, 210, 282, 307–8, 316, 328

Marischal College, Aberdeen, 158, 293, 318, 407

Masson, Jeane (schoolgirl), 117–18

materiality, 19, 31. *See also* notebook components

materia medica, 283, 394, 400–406, *400, 401, 406*

mathematics, 24, 37, 50, 75–76, 102, 120–22, 128–30, 140, 144, 159, 181, 197, 201–5, 207, 209, 212–14, 221–22, 225, 235, 240, 242–44, 256, 278, 324–26, 335, 337–38, 385; in academies, 55–57. *See also* algebra; arithmetic; bookkeeping; calculation; casting of nines; geometry; geometry, practical; trigonometry

matrix, 28, 75–77, 121, 141, 196–97, 237, 311, 366, 416. *See also* tables

maxims, 81, 117

measurement instruments, 204; definition of, 242; measurement chains, 242, 260–62; protractors, 193; quadrants, 90–91, 204, 207; rulers, 43, 90, 105, 192, 204–5, 242, 348; sectors, 192; sundials, 90, 181–83, 204, 207–9; theodolites, 242, 260–62

mechanical philosophy, 55–56, 169, 204, 278, 290, 338, 399; diagrams, 330. *See also* astronomy

media. *See* notebook components; notekeeping instruments; notekeeping media

medicine, 9–10, 29, 38, 45, 271, 276–77, 282–83, 287–88, 292, 295, 298–304, 306–7, 320, 329, 331, 341, 348, 351, 367, 379, 383–88, 390, 392, 394–95, 398; medical piracy, 399–410. *See also* anatomy; botany; *materia medica*; midwifery; natural history; physiology; surgery

memoranda books, 94, 96–97, 116, 137, 141, 153, 171

memory, 18, 45, 66, 83, 113, 219–23, 237, 256, 275–76, 299–300, 305–6, 332, 335, 349–50, 352, 357, 362, 364, 369, 377, 379, 381; act of, 257–58; *aide-mémoire*, 304; art of, 157–58; of diagrams, 374–81; limitations, 37; memorial lines, 83; memory theaters, 238, 256–64; technical memory, 258; transcription of, 305–16; working memory, 122, 176,

memory (*continued*)
203–4, 243–44, 256. *See also* kinesthetics; mnemonics; mnemotechnics;
recollection
mensuration, 137, 196, 203–5, 213, 225
mental operations, 30, 48, 81–82, 267–68,
299, 316–23, 336, 353, 355, 377, 418;
and maps, 263. *See also* abstraction;
apprehension; association; attention;
conception; imagination; inattention;
judgment; misperception; perception;
recollection; sensation; volition
metacognition, 316–23, 381, 418. *See also*
mental operations; mindfulness
metaphysics, 30
metrology, 89, 128–30, 137, 141–43, 242,
244, 246, 250–52, 255–56, 260–61; definition of, 201–5, 209; and simultaneity,
263–64
midwifery, 277, 287, 387, 410; female students, 277
Mill, James, 161–62
Millar, James, 296, 338–41
Millar, John, 269–70, 290
Miller, Hugh, 108–9, 388
mimesis, 219–20, 409
mind, 4–5, 12, 15, 19, 21, 23, 29–36, 34–35,
39, 44–46, 48, 59, 66–67, 70, 77–78,
80–82, 86, 111, 118, 127, 130, 135, 155,
158–59, 161, 163, 183–84, 187, 198–99,
209, 220, 223, 226, 228–29, 239, 257,
262–64, 267, 275–76, 288, 292, 299,
305, 316–21, 325–27, 336, 348, 353, 355,
364, 366, 374–75, 377, 379–81, 415, 418,
420; children's, 48. *See also* apprehension; automaticity; cognition; cognitive
metaphors; concentration; developmental psychology; gestalt images;
ideas; kinesthetics; memory; mental
operations; metacognition; mnemonics; philosophy of mind; reason;
science of mind; *tabula rasa*
mindfulness, 59, 81–82, 130, 144, 154–55,
158, 161–62, 176, 213, 318
mineralogy, 273, 341

mise-en-page, 75
misperception, 316
Mitchell, Samuel Latham (university student), 296–97, 361, 392, 397
mnemoaesthetics, 219–21
mnemonics, 15, 72, 152–53, 155–62, 172,
176, 232, 237, 246, 256–64, 286, 288–
89, 300–301, 309, 343, 350–51, 362,
366, 374, 380–81, 421; architectural
mnemonic, 172, 256–57, 321–23
mnemotechnics, 15, 22, 34, 84, 123, 149,
209, 220–21, 238, 256–64, 273, 283,
313–14; layout, 151–80; locational
scheme, 162–70; multivalent meanings, 179; technical lines, 174, 421;
technics, 219–21
modules, 71–72, 75–77, 162, 164, 167, 196,
261, 307–8, 416. *See also* standard
module
Monro, Isabella, 45, 73–75
Monro, Margaret (schoolgirl), 28–29,
45–46, 57, 71, 73–75, 130–36, 145, 158,
270, 388, 421
Monro Primus, Alexander, 57, 130–34
Monro Secondus, Alexander, 45–46, 288–
89, 293, 307, 326, 331–38, 353, 375–81,
383, 387–88, 391–92, 394–97, 407–10;
anatomy diagrams, 326
Montgomery, Margaret (schoolgirl),
141n57, 228–30
Moore, Mary (schoolgirl), 57, 125
morality, 49, 57, 59, 81–82, 125, 139, 158,
161–62, 318; links to copying, 81–82;
relationship to numbers, 139
moral philosophy, 36, 66–67, 161–62,
253, 269–70, 286, 289–90, 317–18, 320,
322, 326, 385–86, 407; metacognitive
aspects, 320
morphology, 332, 357–59
multiplication tables, 120–22
multisensory, 18, 43, 226, 231–33, 244,
362, 364
Murray, Amelie (schoolgirl), 57
Murray, Charlotte, 92–93
museums, 10, 19, 330

music, 121–22

Myers, Levi (university student), 298–99

Napier's Box of Tables, 89, 121–22

Nasmyth, Alexander, 212

Native American art, 22

natural history, 120, 136, 186, 273, 295, 306, 309, 351, 385; diagrams, 330

natural philosophy, 55, 144, 152, 205–6, 225, 277–78, 290, 292, 337, 374, 384–85; classrooms, 293–95; diagrams, 336–37; measurement, of force, 345–46

nature, 92, 199, 213, 318

navigation, 199, 208, 232; nautical, 224–25; visual, 244, 305

needlework, 228–29; needlepoint, 111. *See also* embroidery; sewing; stitching

Neville, Sylas (university student), 301, 307, 393, 405

Newton, Isaac, 326

New Zealand, 277, 391–92

North America, 22, 118, 158, 277, 279, 282, 296, 298, 391; Florida, 382; New York, 390n28; Pennsylvania, 298, 390; South Carolina, 299n91; Virginia, 405. *See also* Mitchell, Samuel Latham (university student); Native American art; Rush, Benjamin; universities: Harvard University; universities: University of Pennsylvania

notebook components. *See* bifolia; bifolium; drawings; fold; headings; index; keywords; layout; manuscript book covers; manuscript book tooling; maps; modules; pages; paratexts; quires; sidenotes; spines; standard module; tables; title pages; watercoloring; writing (noun)

notebook genres. *See* bookkeeping: field books; codex; commonplace books; copybooks; diaries; dictates; exercise notebooks; family books; household books (for accounts); lecture notebooks; ledgers; manuscript textbooks;

memoranda books; paper books; pocket books; recipe books; scroll books; sketchbooks; surveying: field books; transcripts

notebooks as artifacts, 8–14

notekeepers as artificers, 22–29

notekeeping: affordances, 6, 9, 87, 104, 106, 178, 235, 271, 415; as artifice, 14–22; definition of, 16–18

notekeeping communities, 24, 268, 277; collective notekeeping, 389–91; consortia, 307; manuscript communities, 323; *Schreibechor*, 274–75

notekeeping instruments. *See* drawing instruments; experimental instruments; measurement instruments; painting instruments; writing instruments

notekeeping media: adhesives, 211; bindings, 12, 14, 16, 88, 101–2, 104, 106–7, 109–13, 203, 213, 228, 230, 251, 382, 386, 416–19; cardboard, 90, 95–96, 110; cards, 87–88, 92; carmine, 193; chalk, 14, 95, 181, 190, 193, 208, 210–11, 232–33, 295, 324, 331–32, 335, 375; clay, 379; cloth, 68, 228–30, 237; crayons, 190, 193, 210–11; drawing supplies, 193–94; dyes, 211; embossing, 111–12; glue, 89, 106, 113; graphite, 21, 43, 61, 70, 72, 77, 102, 116, 119, 122, 143, 177, 181, 190, 192–93, 197, 208, 213, 222, 226, 237, 245, 251–52, 262, 296, 309, 326, 332, 340, 348, 364, 373, 391, 418–19; gum (erasers), 61; ink, 14–16, 43, 58, 61–64, 68, 81, 102, 129, 131–32, 143, 187, 193, 197, 199, 208, 212–14, 221–22, 226, 237, 244, 251–52, 293, 296, 332, 340, 348, 373, 375, 391, 395; leather, 11, 20, 46, 86, 88–89, 102–4, 106, 111–13, 251, 363; linen, 71, 94–95, 189, 195; paint, 15, 19, 188, 210–11, 213–14, 221–22, 237; parchment, 61, 93, 107; paste, 15, 315–16; powder, 61, 81, 418; silk, 193, 228; slate boards, 50, 95–98, 141, 233; stitches, 104, 106–7, 109, 111, 228, 230, 416; string, 20, 90–91, 106–7, 109, 229, 237, notekeeping media (*continued*) 380, 418; tackets, 107, 237; thread, 68, 104, 109, 111; watercolors, 14, 16, 102, 119, 187, 193, 208, 211–14, 220–22, 238, 245, 250–52, 328, 332, 352. *See also* chalkboards (blackboards); paper

notekeeping skills. *See* annotating; bookkeeping; categorizing; ciphering; circulating (notebooks); codexing; copying; diagramming; drawing; editing; folding; indexing; interleaving; journalizing; listening; listing; mapping; observing; pictogramming; quiring; ranging; reading; scaling; scrolling; sewing; shading; sketching; stitching; systemizing; tracing; transcribing; watercoloring; writing (verb)

numeracy, 14, 49, 136–46, 418

observation, 4, 8, 9–10, 12–13, 15, 24, 31, 36–37, 45, 59, 64, 66, 70–71, 78, 104–5, 119, 127, 151, 155, 196–97, 199, 210, 218–23, 225, 245, 251, 261, 263, 268, 273, 280, 282, 285, 287, 290, 294, 299–302, 304–7, 314, 320, 325, 334, 357, 361, 374–75, 381, 387–88, 393, 403, 416–17

observing, 66, 128, 191, 218–23, 296, 330, 348, 350, 366, 417, 421–22; collective forms, 307; as diagramming, 374, 381; as drawing, 210–23; as mapping, 251–52; realtime, 209, 263, 374–81; as surveying, 225, 261; in university notes, 300–301; utility of, 218–23; as writing, 59–68, 320

omission, 363, 419; as agency, 132–34

Ong, Walter, 102, 155–56

optics, 169, 218–19, 385. *See also* experimental instruments: prisms

orality, 59, 289–90, 301, 303–4; dictation, 279–80, 303–4. *See also* listening

orthography, 50–57, 59–61, 67, 76, 235, 249, 348–49, 393; definition of, 59–62

pages: as maps, 155, 167; as objects, 88; as paper machines, 149; as pictures, 66–67, 321–23. *See also* bifolia; layout; *mise-en-page*; modules

painting, 12, 14, 19–21, 185, 190, 192, 201, 211–13, 220–22, 418; anatomical, 332

painting instruments: brushes, 212, 214, 221, 250, 418; easels, 192

Panton, Paul (university student), 328, 341, 370–71

paper, as conceptual tool: directionality of, 235; epistemology of, 11–12; logic of, 159; mapping connections, 237, 239; metacognitive aspects, 316–23; "paperness," 88; paperwork, 9; slips as mnemonic cues, 286–88; white space, 163. *See also* paper tools

paper, as material, 43–44, 47, 419; bonded, 107; brown, 108, 110; for drawing, 193–95; formats, 92; grey, 108; gum flower, 193; index cards, 87–88; linen-based, 71, 94–95; loose-leaf, 50, 107; marbled, 107, 112; as a medium, 89–96; scrap, 85–87, 95–97, 118, 295, 314; size, 96, 130–32, 139; size for chemical tables, 370–71, 373; size for field maps, 251–53; size of posters, 359; size of scroll books, 296; size of transcripts, 305; slips, 87, 137, 141, 273; stacked, 88, 419; stationery, 95–96, 102; thickness, 104, 107, 110, 132

paper-based instruments, 90–93

paper books, 96–97, 100, 102, 108

paper cubes, 258

paper machines, 4, 6–8, 16, 31, 36, 39, 43, 84–85, 87–89, 92–93, 98, 104, 106, 113, 159, 267, 272–73, 319, 419–20; components of, 85–113; comprised of slips, 286; definition of, 6; metacognitive aspects, 319; multivolume, 273–74, 323; relationship to maps, 226

paper systems, 143, 289, 315, 327

paper technologies, 87

paper tools, 48–49, 357, 368

paragraphs, 14, 71–72, 77, 132, 144

parallelity, 70, 77, 95, 101, 160, 196n38, 223, 230, 376, 377, 380 paratexts, 124–36, 281; in notebook transcripts, 309–16

paths, 13–14. *See also* scanpaths

Patricks, William (university student), 313–14

pedagogy, 29, 38, 49, 54, 64–65, 80, 83, 110–11, 139, 144, 156–58, 174, 197–99, 210, 221, 228, 316, 319, 369, 379

penmanship. *See* orthography

perception, 18, 48, 59, 81, 268, 299, 316, 336; diagrams, 331; realtime, 320; utility of, 218–23

performance, 64–67, 81, 111, 154, 305, 317, 320, 379

perpendicularity, 77, 196, 381

Perth Academy, 11, 14, 16, 20, 51, 55–56, 108, 113, 123, 143–44, 169–70, 176, 187, 204, 206, 213–14, 221, 245, 251–52, 256, 270

philosophy, 37, 66, 151, 263, 299, 314. *See also* epistemology; logic; mechanical philosophy; metaphysics; moral philosophy; natural philosophy; philosophy of mind; propositions; science of mind

philosophy of mind, 37, 161–62, 326. *See also* mind; science of mind

physiology, 330, 332, 335, 357; diagrams, 330, 335

pictogramming, 15, 351–66, 418

pictograms, 176, 335, 351–66, 421; as diagrams, 351–66; as schema, 351–53; in realtime, 354–55, 357, 363, 366

piracy, 396–410; pirated editions, 394–95, 397–98

plagiarism, 407–10

planimetry, 211; definition of, 201

plates, 353, 359–60

Playfair, John, 336–37

Playfair, William, 170–71, 180

pocket books, 9, 70, 275

political economy, 178, 270

Pollock, David (university student), 309–10

polygons, 119, 200, 246, 344, 418

posters, 60, 273, 284, 330, 332–33, 335, 351–54, 357–60

powder, 61, 81, 418

preattentives, 163–70

prehistory, 18–19

Priestley, Joseph, 157, 174

print culture: ABC books, 49–50; advertising, 59, 138, 398–99, 402, 408; Bibles, 227, 271; bibliographies, 273; book market, 382–410; bookplates, 389–90; books as manuscripts, 106–13; booksellers, 96, 102; bookshops, 71, 89, 312; copyright, 399–404, 407; dislocation of figures in books, 203, 243–44; printing industry, 188–89, 225, 309, 392–99; prints, 111, 139–40, 193, 196, 211–12, 227–28, 332, 358–60; stationers, 108. *See also* Creech, William; Elliot, Cornelius; libraries; Lowndes, Thomas

proportionality, 104–5, 159, 193, 196, 223, 231, 246, 262, 336, 373

propositions, 270, 325–26, 375

Purdie, Peter (schoolboy), 110, 128–30

Quintilian, 36, 83, 258, 282. *See also* humanism

quires, 102–4, 107, 125, 237, 372–73, 417, 419; definition of, 96–101; flattened, 104–6

quiring, 16, 96–106, 276, 417; definition of, 96

Quynn, William (university student), 306

Ramsay, Allan, 190–91, 210, 221

Ramus, Petrus, 283–84

ranging, 15, 68–78, 80, 119, 141–42, 145, 176, 249, 261, 373, 417, 419–20; as an art, 73; metacognitive aspects, 320, 323–24

ratiocination, 30, 31, 48, 262, 336

rationality, 145, 158; relationship to linearity, 317

reading, 139, 153, 174, 176, 179; advanced skills, 139; aloud, 320; autodidactic,

reading (*continued*)
151; banned in lectures, 286; correspondence letters, 45; difficulties, 77; headings, 153, 156, 157, 162, 179; horizontal, 76; lessons, 67; library books, 301; in mapping, 90; pre-reading, 281; professors reading student notebooks, 334–35, 389, 394, 404, 408; reading public, 92; relation to ciphering, 141; rereading, 218, 311; right-angle reading, 76–77; scanpaths, 174, 176; as supervision, 171; tables, 369–73; written corrections, 125, 134. *See also* scanning

realtime interface, 6, 20, 34, 40, 49, 83, 104, 196, 267, 273, 275–76, 354, 357, 366, 375, 378, 381, 419–21; composting, 308; connections to reason, 30; definition of, 154, 156; drawing, 187; in editing, 131; interactive layouts, 151–80; metacognition, 317–18; scanpaths, 170–80; scroll books, 300; *tabula rasa*, 150; training, 81; transcribing, 319; word pictures, 323

realtime learning, 292–305

reason, 8, 21, 29–31, 34–36, 44–46, 82, 111, 262, 267, 278, 289, 292–93, 317, 325, 336, 338, 379, 415, 418, 421–22; as codexing, 111; definition of, 29–36; as interface, 316–23; with maps, 262–64; as scrolling, 293; as a skill, 30, 415. *See also* cognition; line of reason metaphor; mental operations; metacognition; rationality

recipe books, 58, 135

recollection, 34, 48, 81, 83, 151, 252, 268, 275–76, 299–300, 304, 316, 320, 326, 336, 353–54, 355, 379, 380–81, 418; with diagrams, 326; metacognitive aspects, 320

rectilinearity, 50, 68–71, 77, 95, 101, 167–68, 209, 223, 309, 311, 375–76, 378–79, 419; surfaces, *98. See also* geometry; squares; triangles

recycled materials, 93, 109–10, 123. *See also* repurposed media

Reid, Thomas, 317–20; diagrams, 336

religion, 118, 227, 271. *See also* Calvinism; Church of Scotland; divinity; theology

Renaissance, 23, 83, 89, 154–56, 167, 171, 199, 258, 282, 306, 330

Renn, Jürgen, 273

Rennie, J. (student), 407

representation, 14, 18, 24, 37, 48, 67, 71, 141, 172, 178–79, 183, 197, 200, 222–23, 225, 231, 235, 248, 268, 283, 326, 332, 341, 350–51, 366, 370, 381, 419; links to madness, 32. *See also* aesthetics; art; beauty; kinesthetics; mnemonics; mnemotechnics

repurposed media; 109–10; shapes, 300; visualizations, 335, 358–60

revisibilia, 114–46; definition of, 116–17

revising, 137, 140. *See also* editing

Rheinberger, Hans-Jörg, 9

rhetoric, 30, 37, 301, 320, 377, 396

Richardson, Robert (academy student), 115–16, 143–44

Robertson, Hannah, 185–86, 211, 213

Robison, John, 290, 337, 392; on the mind, 292–93

rote learning, 79–80, 149, 415

rough notes, 98, 271. *See also* scroll books; scrolling

Rousseau, Jean-Jacques, 35–36

rows, 68–69, 75–78, 122–23, 143–44, 366–74; as thinking tools, 317

Royal Society of London, 350, 408

Rush, Benjamin, 298–99, 304

Rush, James (university student), 298

Rutherford, Daniel (university student), 380

sailing, 210–11, 214; diagrams, 201–2

Sandby, Paul, 238, 260

Sandy, George (legal apprentice), 186–87, 189, 328, 393

scaling, 351–52, 359, 374–75

scanning, 170–80

scanpaths, 153, 170–80; examples of, *175, 177, 179*

schematic images, 196–97, 220–23, 328–29, 341–51, 357, 364–66, 381

scholarly culture, 23, 305, 351. *See also* humanism

schools, 187, 271, 326, 328; Academy of St. Luke Drawing School, 188, 190; burgh, 40, 50–54, 231–32, 258, 330; charity, 50; Dalkeith Grammar School, 232–33; day, 38; Edinburgh Merchant Maiden Hospital, 50, 66, 70; finishing, for girls, 56–57; Foulis Academy of Art, 188–91, 210; girls', 38, 125; grammar, 38, 54–56, 60, 83, 124, 232, 258, 278, 318; hospital, 40, 50–51, 66; Nicholson's Commercial Academy, 235; parish, 40, 50–54, 231–32; Royal Military Academy Chelsea, 124–25; sabbath, 50, 98; sewing, 228; Trongate Academy, 187–88; village, 50; Warrington Academy, 157; workhouses, 50. *See also* Edinburgh High School; Perth Academy

science, 8–9, 34, 36–39, 87, 154, 159, 273, 341, 351, 401; definition of, 36–37; as a scheme, 158–59; as a system, 8–9, 37–39; technoscience, 55–56, 86, 195–96, 199, 204, 237. *See also* algebra; altimetry; anatomy; botany; chemistry; cognition; developmental psychology; experimental instruments; experiments; gauging; geography; geometry; leveling; longimetry; *materia medica*; mathematics; mechanical philosophy; medicine; mensuration; metacognition; metrology; midwifery; mineralogy; natural history; natural philosophy; navigation; optics; planimetry; sailing; surgery; surveying; system; trigonometry

science of mind, 35, 262–63, 317, 336

Scott, Robert Eden, 265–86, 299–300, 302

Scott, Sir Walter, 60, 107

Scottish Enlightenment, 34–36, 267

screens, digital, 155

scribbles, 108, 114, 338, 421

scribes, 70–71

scroll books, 82, 94, 98, 212, 273, 280, 292–307, 328, 333, 338, 340–41, 346–47, 357, 373, 379; connection to handouts, 313; definition of, 79; layout, 295–96

scrolling, 58, 79, 97, 128, 271, 273, 281, 290, 292–96, 298–305, 340, 352; definition of, 79; diagrams, 338–41; metacognitive effects of, 317–18; pictograms, 352; realtime aspects, 302; strategies, 301–2

self: self-directed, 120; self-help, 190; self-knowledge, 161–62; self-made artifacts, 35, 422; self-observation, 64–66; self-organized, 14, 18, 33, 35, 121, 317; self-reference, 80; self-reflection, 125–27; self-training, 355

sensation, 299, 361; senses, 362–63. *See also* multisensory

sentences as thinking tools, 317–20

sewing, 12, 16, 89–90, 102, 104, 195; cognitive aspects, 228; maps, 227–30; needles, 228–30; schools, 228; as thinking, 111. *See also* embroidery; needlework; stitching

shading, 197, 212–18, 364, 418

shapes, 197–200, 246, 252, 325–26, 335; as diagrams, 335–51

shorthand, 77, 387; disadvantages of, 300–301

short notes, 296–97. *See also* scroll books

short passages, 299–300, 302. *See also* scroll books

sidenotes, 314–15

sightlines, 153, 170–80, 209, 260, 261–62, 421–22; definition of, 173–80

significant space, 144, 209

similarity, 66, 149, 163–64, 178, 196, 201, 209, 212, 214, 246, 251, 325, 343, 351, 366, 374, 379; equal linear lengths, 196, 223, 336

simplicity, 90, 197, 221–22, 286, 305, 335, 346, 351, 381

Simson, Robert, 324–25, 327

simultaneity, 6, 30, 31, 33, 154–55, 184, 254, 263–64, 267–68, 275, 316–23, 332, 340, 354; cognition, 31, 81, 320; metacognitive aspects, 319–20

sketchbooks, 190–91

sketches, 186, 191, 197, 208, 209, 222, 265, 293, 328, 333, 335, 346, 364

sketching, 15–16, 184, 191, 201, 208–9, 211, 213–14, 221–22, 233, 234, 238, 249, 293, 328, 333, 335, 340, 346–47, 352, 364–66, 418. See also draftsmanship

skill, 15–22; definition of, 15; superskills, 39, 418; transferable skills, 71. See also notekeeping skills

Sligo, George (university student), 295

Smith, Adam, 269–71, 312; view of systems, 272–73

Smyth, James Carmichael (university student), 338

social reproduction, 30, 34, 415; of knowledge systems, 323

sociomateriality, 43, 415

Somerville, Mary (schoolgirl), 67, 212, 221, 224–25

spatiality, 18, 20

specimens, 332, 334, 351

spines, 107, 113, 416–17

squares, 5, 68, 85, 183, 197–99, 201, 208, 222–23, 239, 255, 325, 335, 347–48, 352, 376–78

stable technologies, 71–72, 75

standard module, 71–73, 162, 164, 307–8; visualization of, 72, 308

St. Andrews University, 267, 385; notekeeping, 275

Steuart, Robert, 384–85

Stewart, Dugald (moral philosophy professor), 37, 66–67, 75, 262–64, 289–92, 317, 320–23; metacognition, 320; view on diagrams, 326, 336

stitching, 106–7, 109, 228, 416, 419. See also notekeeping media: string; sewing

surgery, 277, 293, 353–54, 383, 410; surgical instruments, 353

surveillance, 124–25

surveying, 16, 55, 137, 169, 195, 204, 211, 226, 232, 238–56, 335; embodiment aspects, 260–63; field books, 244, 246; field-mindedness, 238–50; instructors, 260; memory aspects, 260–63. See also maps

survey maps, 226, 244–56, 260

syllabus, 280, 403–4; diagrams, 327; images, 286, 329; layout, 309; lecture headings, 311; in libraries, 283; as machine, 269; maps, 329; metacognition, 321–22; as outlines, 276; as a scheme, 158; scrolling, 294, 301; student alterations, 314–15; synonyms of, 282; as a system, 269–74; as a technology, 282–92; transcribing, 305–6; typography of, 308–9. See also lecture notebooks

Sym, Kath, 58–59

symmetry, 77, 94, 101, 105, 163–65, 196, 199, 239–40, 313–14, 373

system, 4, 9, 18, 36–38, 43, 80, 82, 87, 94, 103, 110, 124, 136, 143, 146, 154–55, 159, 162, 172, 180, 224, 226, 267–68, 269–74, 276, 282, 286, 288, 305–6, 314, 321–23, 327, 329, 335, 341, 351, 355, 360, 363, 374, 376, 385, 389, 403, 420–21; based on paper, 273; of chemistry, 373–74; expandability, 306; of information, 154–55; as machines, 272–73; maps, 224; mathematical, 250; on paper, 143, 226, 315; realtime interface, 154–55; schemes, 159; science as a system, 36–39; scientific, 273. See also knowledge system

systemizing, 40, 267, 269–323; and transcribing, 305

tableaux, 14, 416

table of contents, 134–36, 296; for transcripts, 309–10

tables, 28, 58, 72, 75–78, 89, 120–23, 145, 157–59, 242, 273, 295, 335, 366–74, 417; of chemical affinity, 366–74; microtables, 371–74; multiplication tables, 120–22; thermometric tables, 366–67

tabula rasa, 3–7, 11, 31, 43–44, 48, 70, 80, 86, 94–95, 149–50, 154, 183–84, 223, 225, 267, 316–17, 326, 336, 381, 415, 418, 420; and diagrams, 326–27; *tabula figura*, 183–84, 198, 208, 336; *tabula folia*, 86, 94–96; *tabula memoria*, 262; *tabula verba*, 59, 95, 183, 326–27, 381; and transcribing, 316–23
tacit knowledge, 235, 249
tactility, 289; with maps, 230
tessellations, 246–56, 260; as memory aid, 262
theology, 37, 288, 291; systematic, 271
thinking tools, 88, 187, 209, 288, 325, 335, 338, 351, 355, 358
Thorburn, John, Jr. (university student), 334–35, 376, 387–99
title pages, 26, 296, 309
tonality, 165, 213–18. *See also* shading
tracing, 70, 184, 193, 208–9, 213, 249, 254, 340, 373–81; maps, 227, 230–31, 237
trade secrets, 62, 285
transcribers (professional), 311–12, 393–94
transcribing, 134–37, 292, 305–16; as abridgement, 305; chiasmas, 348–49; cognitive aspects, 318–23; diagrams, 338; filling in notes, 305–7, 375
transcription, 106, 116–17, 271
transcripts, 273, 281, 302, 305–16, 338, 354–55, 357, 366, 375, 379, 382–410, 421; definition of, 305–16; with headings, 305–8
translations, 43, 54–55, 57, 83, 85, 95, 102
transmission, 7, 14, 44, 87, 209, 327, 383, 423; via diagrams, 327–28
travel, 210–11, 213, 225, 227, 245, 258, 306, 382, 385–86, 397–98
triangles, 7, 197–99, 201, 222, 239, 246, 255, 325, 335–37, 340, 352, 379–81
trigonometry, 176, 225, 328–41
Turnbull, George, 195–96, 212, 220
Tversky, Barbara, 18–19, 209
typography, 60, 162–70, 308–9, 311

Tytler, Alexander Fraser, 301, 303, 306, 366

universities, 38, 40, 55, 269–323; Cambridge University, 275; female students; 277–78; German, 274–75; Harvard University, 275; Leiden University, 284, 386, 398; St. Andrews University, 267, 385, 275; University of Jena, 283–84; University of Paris, 275, 373n94; University of Pennsylvania, 298–99; Uppsala University, 351. *See also* Aberdeen University; Edinburgh University; Glasgow University
user-centered interface, 4, 7, 75, 157, 391, 421–22; user-training, 86; user-friendly, 6, 16, 44, 204, 256, 415; user-powered technologies, 88; user-testing, 75
utility, 34, 77, 82, 101, 125, 171, 176, 189, 210, 218–19, 221, 243, 285, 288, 290, 292, 302, 327, 344

volition, 81

Waldie, John (university student), 389–90
Walker, John, 273, 309–10, 385
watercoloring, 191, 208, 210, 212–18, 220–23, 250, 418
Watt, James (inventor), 395
Watts, Isaac, 157
wayfaring, 13, 14, 325
Webster, Thomas (university student), 374–75
women, 24, 56, 221, 228–29, 277
writing (noun). *See* annotating; ciphers; copies; corrections; descriptions; doodles; epitomes; essays; exercises; excerpts; extracts; letters; marginalia; maxims; scribbles; sentences
writing (verb), 15, 19, 43, 45–84, 417–18, 420; as an art, 60; instructors, 57, 60, 66, 70, 249; on paper surfaces, 94–95; posture, 64–68; as a superskill, 45–84; writtenness, 45–46

writing instruments: erasers (rubbers), 3, 26, 193, 197, 212; ink pots, 293; knives, 106; lead pens (tracers), 43, 61, 70–71; magazine cases, 194; nibs, 62–64, 132, 212, 293, 418–19; pencil cases, 193; pencils, 26, 43, 61, 190, 193, 211, 242, 250, 418; penknives, 26, 62–63, 81, 418; pens, 20, 43, 61–64, 66, 71, 95, 187, 197, 210, 250, 293, 320, 348, 418; pins, 107, 353; pocket cases, 193; quills, 26, 62–64, 66, 81, 92, 94, 132, 293, 418; razors, 61, 81; scissors, 106–7, 353; steel pens, 192

Young, Thomas, 277, 387